SUN AND PLANETARY SYSTEM

ASTROPHYSICS AND
SPACE SCIENCE LIBRARY

A SERIES OF BOOKS ON THE RECENT DEVELOPMENTS
OF SPACE SCIENCE AND OF GENERAL GEOPHYSICS AND ASTROPHYSICS
PUBLISHED IN CONNECTION WITH THE JOURNAL
SPACE SCIENCE REVIEWS

VOLUME 96
PROCEEDINGS

SUN AND
PLANETARY SYSTEM

PROCEEDINGS OF
THE SIXTH EUROPEAN REGIONAL MEETING IN ASTRONOMY,
HELD IN DUBROVNIK, YUGOSLAVIA, 19–23 OCTOBER 1981

Edited by

W. FRICKE

Astronomisches Rechen-Institut, Heidelberg, F.R.G.

and

G. TELEKI

Astronomical Observatory, Belgrade, Yugoslavia

D. REIDEL PUBLISHING COMPANY

DORDRECHT : HOLLAND / BOSTON : U.S.A.

LONDON : ENGLAND

Library of Congress Cataloging in Publication Data

European Regional Meeting in Astronomy (6th : 1981 : Dubrovnik,
    Croatia)
    Sun and planetary system.

    (Astrophysics and space science library ; 96)
    Includes index.
    1.   Astronomy–Congresses. I.   Fricke, W. (Walter Ernst), 1915–
1915–   . II. Teleki, G. III. Title. IV. Series.
QB1.E86   1981   523.2   82–7578
ISBN-13:978-94-009-7848-5   e-ISBN-13:978-94-009-7846-1
DOI: 10.1007/978-94-009-7846-1

Published by D. Reidel Publishing Company,
P.O. Box 17, 3300 AA Dordrecht, Holland.

Sold and distributed in the U.S.A. and Canada
by Kluwer Boston Inc.,
190 Old Derby Street, Hingham, MA 02043, U.S.A.

In all other countries, sold and distributed
by Kluwer Academic Publishers Group,
P.O. Box 322, 3300 AH Dordrecht, Holland.

D. Reidel Publishing Company is a member of the Kluwer Group.

TABLE OF CONTENTS

Preface   xiii

Address by G. Teleki   1

Opening Address by R. Kontič   3

Address by R. M. West   7

W. FRICKE / New Impetus to the Exploration of the Solar
    System (Inaugural Address)   9

Z. KOPAL / The Solar System - Known Facts and Unsolved
    Problems (Opening Invited Lecture)   13

SECTION I - SUN FROM THE ASTRONOMICAL AND PHYSICAL
            POINTS OF VIEW

J.-C. PECKER / The Star "Sun" (Invited Review Paper)   25
F. CHOLLET / Resultats d'Observations du Soleil a
    l'Astrolabe du CERGA   35
C. SMITH, D. MESSINA / The Effect of Personal Equation on
    Measurements of the Solar Semi-diameter by Transit
    Circle Observers   39
H. TÜG, TH. SCHMIDT-KALER / A Direct UBV Color Measurement
    of the Sun   45
G. E. KOCHAROV / Nuclear Astrophysics of the Sun
    (Invited Paper)   47
G. MARX / The Solar Neutrino Puzzle (Invited Paper)   53
N. KIZILOĞLU, D. EZER / Solar Models and Neutrino Problem   59
M. STIX / The Rotation-Magnetism-Convection Coupling in
    the Sun (Invited Review Paper)   63
R. A. ZAPPALÀ, F. ZUCCARELLO / Relationship between Photo-
    spheric and Chromospheric $\Omega(\vartheta,t)$ deduced by Tempo-
    rarily and Spatially Correlated Tracers   71
A. KUBIČELA, M. KARABIN / Line-cf-sight Velocity Field of
    Synodic Solar Rotation   73
J. TUOMINEN, I. TUOMINEN, J. KYRÖLÄINEN / Solar Rotation
    and Meridional Motions Derived from Sunspot Groups   77

M. TERNULLO / Subphotospheric Velocity Fields Inferred
by Sunspots Motions   79
V. A. KRAT / Granulation, Supergranulation and Atmos-
pheric Waves (Invited Review Paper)   81
J. PERDANG / A Noisy Sun ?   89
V. E. MERKULENKO, V. I. POLYAKOV, V. S. LOSKUTNIKOV /
Spectral Analysis of Wave Motions in the Region of
Temperature Minimum of the Sun's Atmosphere   93
A. R. ABBASOV, L. F. GOLUBEVA, L. B. TZIRULNIK / On the
Detection of the Periodic Oscillations of the Incli-
nation of the Frequency Spectrum of S-Component of
Solar Radio Emission   97
E. A. GURTOVENKO / Some Photospheric Characteristics of
the Sun as a Star   99
M. S. DIMITRIJEVIĆ / Stark Broadening of Heavy Ion
Solar Lines   101
N. G. STCHUKINA / The Analysis and Observation of the
Neutral Spectrum of Potassium in the Solar Atmosphere   103
R. I. KOSTIK / Mesoturbulence in the Solar Atmosphere   105
C. J. MACRIS, B. CH. PETROPOULOS / Voigt Profile for the
CO Emission Lines in the Solar Sunspot Spectrum   107
G. B. GELFREIKH, V. M. BOGOD, A. N. KORZHAVIN,
E. V. KONONOVICH, O. B. SMIRNOVA, S. V. STARTSEV,
V. V. PIOTROVICH / Solar Radio Granulation at
Microwaves and its Optical Identification   109
M. MARIK / A Theoretical Model of the Solar Chromo-
sphere over a sunspot   113
F. CHIUDERI DRAGO / Space-born and Ground-based Ob-
servations of a Solar Active Region and a Flare   115
V. M. GRIGORYEV, V. S. PESHCHEROV, M. L. DEMIDOV / The
Beginning of Observations of Large-scale Solar
Magnetic Fields at the Sayan Observatory:
Instrument, Plans, Preliminary Results   119
G. Ya. SMOLKOV / The Solar Complex of SibIZMIR:
Observatories, Instruments and the Main Results of
the Investigations   123
S. I. GOPASYUK / Motions in Sunspots Like Torsional
Oscillations   125
Z. B. KOROBOVA / Sunspots Proper Motions and Positions
of H-Alpha Filaments in the Active Region of
August 1979   127
I. SATTAROV / On the Umbral Dots   129
E. JENSEN, O. ENGVOLD / Turbulent Velocity Fields in
Quiescent Prominences   131
M. SH. GIGOLASHVILI / Determination of the Electron
Density in a Quiescent Prominence   133
TS. S. KHETSURIANI / A Polarization Investigation of
Prominences   135
C. CHIUDERI / The Formation of Solar Prominences   137
V. M. CADEŽ / Radiation from Surface Wave Packet
and Type II Solar Radiation Emission   139

G. EINAUDI / Magnetohydrodynamics and Thermodynamics of
Coronal Loops   141
E. RIBES / Temperature and Steady Flows in Slender
Magnetic Tubes   143
E. WOYK (CHVOJKOVÁ) / Gravito-magnetic Explanation of the
Temperature of Stellar Coronae and of Planetary
van Allen Belts   145

SECTION II - ASTRONOMICAL, GEOPHYSICAL AND GEODETIC
            PROBLEMS RELATED TO THE EARTH

J. KOVALEVSKY, Ya. S. YATSKIV / Astronomical and
Geophysical Problems of the Earth's Rotation
(Invited Review Paper)   149
G. A. WILKINS / A Note on the Initial Results and Future
Plans of Project MERIT (Invited Review Paper)   163
M. FEISSEL / Steps towards the Determination of Earth
Rotation Parameters not Dependent on the Observation
Technique (Invited Paper)   165
L. V. MORRISON, F. R. STEPHENSON / Secular and Decade
Fluctuations in the Earth's Rotation :
700 BC - AD 1978 (Invited Paper)   173
P. BROSCHE / Oceanic Tides and the Rotation of the
Earth (Invited Paper)   179
N. CAPITAINE / Effects of the Non-Rigidity of the Earth
Derived from Astronomical Observations   185
D. DJUROVIĆ / Correlation between Solar Activity
and Universal Time Variations   189
B. KOLACZEK, G. TELEKI / On Investigations of Mean
Latitude Variations   191
M. DJOKIĆ / Closing Errors of the Belgrade Latitude
Observations and Temperature Influences   195
P. FARINELLA, A. MILANI, A. M. NOBILI, F. SACERDOTE /
Reference Systems Linkage from Space   197
P. FARINELLA, A. MILANI, A. M. NOBILI / High Precision
Tracking of Geosynchronous Satellites:
Oceanographic Applications   199
S. CATALANO, P. FARINELLA, A. MILANI, A. M. NOBILI /
Optical Tracking of Synchronous Satellites for
Geophysical Purposes   201
D. ARABELOS, L. N. MAVRIDIS, I. TZIAVOS / Gravimetric
Geoid Determination for the Area of Greece   203
G. BARTA / Coincidence of Some Magnetic and Gravity
Field Characteristics   205
V. OUMLENSKY, V. SHKODROV / Precession and Nutation Influence
on the Harmonic Coefficients of a Planetary Potential   209
G. G. MATESHVILI / Aerosol Stratification of the
Stratosphere and Mesosphere   213
B. P. TRITAKIS / Interplanetary Magnetic Field and
Solar Cycle Modulation of the Tropospheric
Circulation over West Mediterranean   215

A. S. SOCHILINA / On the Evolution of Nearly Circular
Orbits of Satellites with Critical Inclination   217

SECTION III - PHYSICS OF PLANETS, MINOR PLANETS,
            SATELLITES AND INTERPLANETARY MEDIUM

G. ARRHENIUS, M. J. CORRIGAN, R. W. FITZGERALD,
C. SCHIMMEL / Excitation in the Early Solar Nebula -
New Experimental Findings (Invited Paper)   221
H. J. VÖLK / Physical Processes of Relevance to the
Formation of the Planetary System (Invited
Review Paper)   233
N. F. NESS / Magnetospheres of Jupiter and Saturn
(Invited Paper)   243
V. ČELEBONOVIĆ / A Physical Detail Relevant to the
Savić-Kašanin Theory of Behaviour of Materials
under High Pressure   249
A. CHENG, L. J. LANZEROTTI, V. PIRRONELLO / On the
Lifetime of E-ring Grains and their Nature   251
V. PIRRONELLO, G. STRAZZULIA, G. FOTI / H_2 Enrichment
of Interplanetary Medium   253
R. ROBLEY, A. BÜCHER, S. KOUTCHMY / Study of the Inter-
planetary Dust at High Ecliptic Latitudes:
Doppler-Fizeau Shifts   255
A. MAMPASO, C. SÁNCHEZ MAGRO, J. BUITRAGO / Near In-
frared Emission from the Solar Corona   257
V. VUJNOVIĆ / Associative Ionization and Sodium in
the Atmospheres of Planetary System Bodies   259
I. V. GAVRILOV / On Physical Interpretation of
Hypsometric Characteristics of the Moon and Planets   261
V. V. SHEVCHENKO / The Lunar Photometric Constant in
the System of True Full Moon   263
H. J. SCHOBER / Surface Properties of Asteroids
(Invited Paper)   265
E. F. HELIN / Earth-Crossing Asteroids: New Discoveries
(Invited Paper)   269
F. SCALTRITI, V. ZAPPALÀ, E. ANDERLUCCI / Synthetic
Lightcurves of Asteroidal Binary Systems   277
T. GEHRELS, R. S. MCMILLAN / CCD Scanning for
Asteroids and Comets (Invited Paper)   279
H. J. SCHOBER, A. SCHROLL / Color Variations of
Asteroids During Rotation   285
G. E. TAYLOR / Results from Occultations by Minor
Planets   287
C.-I. LAGERKVIST, H. RICKMAN / On the Rotation of M
Asteroids   289
P. PAOLICCHI, P. FARINELLA, V. ZAPPALÀ / Asteroid
Rotation Rates: Comparison Between Theory and
Observations   291
P. PAOLICCHI, P. FARINELLA, V. ZAPPALÀ / Asteroid
Collisional Evolution: Outcomes of Catastrophic Impacts   295

Z. KNEŽEVIĆ, V. ZAPPALÀ / An Improved Representation
of the Average Opposition Magnitudes of Asteroids   299
V. ZAPPALÀ, F. SCALTRITI / A Coordinate Program for
Pole Determination of Asteroids   303
H. J. SCHOBER, L. K. KRISTENSEN / The Worldwide
Photoelectric Campaign for the Asteroid 51 Nemausa   305
J. RAHE, J. C. BRANDT, L. D. FRIEDMAN, R. L. NEWBURN Jr. /
Halley's Comets and Plans for its Observation
During its Return in 1986 (Invited Paper)   307
J. SVOREŇ / Perihelion Asymmetry in the Photometric
Parameters of Long-period Comets at Large Helio-
centric Distances   321
J. RAHE / Ultraviolet Spectroscopy of Comets   323
H. J. FAHR, H. W. RIPKEN, G. LAY / Plasma-Dust Inter-
actions in the Solar and Cometary Environment   331
A. HAJDUK / The Total Mass and Structure of the
Meteor Stream Associated with Comet Halley   335
S. GASKA, P. GRONKOWSKI / Outbursts of Comets   337
J.-P. J. LAFON / On the Electric Charge of Inter-
planetary Grains   339
A. S. ASAAD / Observations of Zodiacal Light at Abu-
Simbel, Egypt, During the Period 1975-1979   343
J. BUITRAGO, R. GÓMEZ, F. SÁNCHEZ / The Brightness
Integral Equation. A Different Approach to the
Study of the Zodiacal Light   345
V. MILIČEVIĆ, M. MUMINOVIĆ / Sarajevo Observatory
Fireball Patrol   349

SECTION IV - MOTIONS IN THE PLANETARY SYSTEM

P. J. MESSAGE / Review of the Dynamics of Satellites
and Planetary Rings (Invited Review Paper)   353
L. KRESÁK / Dynamical Evolution and Disintegration of
Comets (Invited Review Paper)   361
J. A. FERNÁNDEZ / A Dynamical Study of Possible
Birthplaces of Comets   371
W. M. NAPIER / An Interstellar Origin for Comets   375
A. CARUSI, G. B. VALSECCHI / Strong Perturbations at
Close Encounters with Jupiter (Invited Paper)   379
A. CARUSI, G. B. VALSECCHI / Statistics of Close
Encounters of Minor Bodies with the Outer Planets   385
H. RICKMAN, J. KARM / Keplerian Estimates of
Pre-discovery Encounters with Jupiter for Short-
Period Comets   389
S. VAGHI, H. RICKMAN / The Motions of Comets near the
2/1 Resonance with Jupiter   391
H. RICKMAN, A. M. MALMORT / Temporary Satellite
Captures by Jupiter for Orbits Resembling the
One of Comet P/Gehrels 3   395
D. BENEST, R. BIEN, H. RICKMAN / On the Long-term Orbital
Evolution of Comet P/Boethin   397

P. B. BABADZHANOV, YU. V. OBRUBOV / On the Displacement of Meteor Showers Activity   401

M. KUZMANOSKI / Some Characteristics of the Asteroid
Belt Structure   403

H. SCHOLL / Resonances in the Motions of Minor Planets
and their Use for the Determination of Masses
(Invited Paper)   405

T. KIANG / Stability of Real Hecuba and Hilda Asteroids   411

W. HÖPPNER / First-order Theory of Ceres, Pallas, Juno
and Vesta   413

YU. V. BATRAKOV / Methods of Computation of the
Perturbed Motion of Small Bodies in the Solar
System (Invited Paper)   415

V. G. SHKODROV / A New Method for Expression of the
Perturbation Function   421

V. A. BRUMBERG, T. V. IVANOVA / Relativistic
Dynamics of the Earth-Moon System   423

V. G. IVANOVA / On the Secular Effects in the Motion
of a Planetary Satellite   429

B. POPOVIČ / L'Invariabilité Séculaire des Grands
Demiaxes des Orbites Planétaires   431

R. K. SHARMA / On Linear Stability of Triangular
Libration Points of the Photogravitational
Restricted Three-Body Problem when the More
Massive Primary is an Oblate Spheroid   435

P. BRETAGNON, J. CHAPRONT, M. CHAPRONT-TOUZÉ /
Nouvelles Theories des Planetes et de la Lune
dans les Ephemerides Francaises   437

R. DVORAK / New Planetary Ephemerides back to 4000 B.C.   441

S. DÉBARBAT, M. SÁNCHEZ, M. STANDISH / Ephemerides et
Observations de Mars a l'Astrolabe   443

S. SADŽAKOV, M. DAČIĆ, D. ŠALETIĆ, B. ŠEVARLIĆ /
Observations of the Sun and Inner Planets with the
Large Meridian Circle in Belgrade   445

V. BENISHEK-PROTITCH / Some Results of the Mercury
Transit Observations in 1970 and 1973 at Belgrade   447

L. S. KOROLEVA, V. I. ORELSKAYA / Contribution of the
Pulkovo Observatory to the Improvement of
Orientation of the FK4 System Using Observations
of Selected Minor Planets   449

SECTION V - THREE-DIMENSIONAL REFRACTION

G. TELEKI, J. SAASTAMOINEN / Problems of Three-
dimensional Refraction in Astrometry
(Invited Paper)   455

J. A. HUGHES, S. DELATEUR / The Use of Lidar to Obtain
Three-dimensional Refraction Data (Invited Paper)   463

C. SUGAWA, I. NAITO / Final Refraction Problems in
Time and Latitude Observations through Classical
Techniques (Invited Paper)   471

R. FUKAYA, H. YASUDA / Astronomical Refraction
    Calculated from Aerological Data in Japan   475
A. POMA, E. PROVERBIO, S. MANCUSO / Experimental
    Model for Diurnal Astronomical Refraction   477
G. TELEKI / Astrometric Site Selection (Invited
    Review Paper)   483
T. A. TH. SPOELSTRA / The Influence of Ionospheric
    Refraction on Radio Astronomy Interferometry   493
E. WOYK (CHVOJKOVÁ) / Disregard of Some Ray
    Tracing Principles in Optical and Radio
    Spectral Bands   497
L. HRADILEK / Refraction Effects on Geodetic
    Measurements in Three-dimensional Terrestrial
    Nets (Invited Paper)   499
V. S. MILOVANOVIĆ / Experimental Investigation of
    Refraction above Water Crossings   503
F. K. BRUNNER / Atmospheric Turbulence and its
    Effects on Direction Measurements (Invited
    Review Paper)   505
E. TENGSTRÖM / Local Geodynamics with Two-colour
    Instruments (Invited Paper)   511
L. N. MAVRIDIS / Terrestrial Refraction and Vertical
    Temperature Gradient (Invited Paper)   519
A. I. GOUNARIS, L. N. MAVRIDIS, A. L. PAPADIMITRIOU /
    Terrestrial Refraction and Vertical Temperature
    Gradient in the Area of Thessaloniki   523
P. SAVAIDIS / A Study of the Refractivity N of the
    Air in the Area of Athens   529

LIST OF PARTICIPANTS   531

SUBJECT INDEX   534

PREFACE

The Sixth European Regional Meeting in Astronomy (VI ERMA) on
"Sun and Planetary System" was held in Dubrovnik (Yugoslavia),
19-23 October 1981, sponsored by the Federal Executive Council of
Yugoslavia, the IAU, IUGG, and the European Physical Society. The
Scientific Organizing Committee included W. Fricke (Chairman),
H. Alfvén, H. Haupt, H. Kautzleben, Z. Kopal, J. Kovalevsky,
V.A. Krat, L. Kresák, G. Marx, P. Melchior, P.J. Message, J.-C. Pecker,
G. Sitarski, B. Sevarlić, G. Teleki, and R.M. West. The Local
Organizing Committee representing the Union of Societies of
Mathematicians, Physicists and Astronomers, Yugoslavia, included
G. Teleki (Chairman), I. Pakvor (Vice-Chairman), Z. Knezević
(Secretary), B. Jovanović, L. Randić, P. Ranzinger, V. Ruzdjak,
B. Popović, and B. Sevarlić. The Meeting was attended by 216
participants from 27 countries (5 continents).

The scientific programme was divided into the following five
sections: (1) Sun from the astronomical and physical points of
view (main organizer: J.-C. Pecker); (2) Astronomical, geophysical
and geodetic problems related to the Earth (main organizer:
J. Kovalevsky); (3) Physics of planets, minor planets, satellites
and interplanetary medium (main organizer: Z. Kopal); (4) Motions
in the planetary system (main organizer: L. Kresák); (5) Three-
dimensional refraction (main organizer: G. Teleki).

Ten invited review papers were presented in 5 plenary sessions;
20 invited papers and about 100 contributed papers were presented in
parallel sessions.

The editors are indebted to the main organizers of the various
sections for their great efforts in establishing a presentation of
present-day research. They also wish to express warmest thanks to
the authors of papers and to referees who have made a timely
completion of the proceedings possible.

January 1982   Walter Fricke
                                                    George Teleki
                                                    Editors

W Fricke and G. Teleki (eds.), Sun and Planetary System, xiii.
Copyright © 1982 by D. Reidel Publishing Company.

ADDRESS BY

G. Teleki
Chairman of the Local Organizing Committee

Ladies and Gentlemen,
Dear Colleagues,

Welcome to Dubrovnik!

It is a great honour for me to greet all of you in the name of the Union
of Societies of Mathematicians, Physicists and Astronomers of Yugoslavia,
of the Yugoslav National Committee of Astronomy, and of the Local
Organizing Committee of this Meeting.

I especially welcome:

Mr Radoje Kontić, the representative of the Federal Executive Council of
                  Yugoslavia, the domestic sponsor of the Meeting,
Dr Richard West, the Assistant General Secretary of the International
                  Astronomical Union, the main sponsor of the regional
                  meetings in astronomy,
Mr Josip Bazeli, the president of the Executive Council of Dubrovnik
                  Assembly.

Another pleasant duty has fallen to me. In the name of all us I would
like to greet Professor Walter Fricke, the chairman of the Scientific
Organizing Committee of the Meeting, who has helped greatly in the
shaping of the profile and the organization of scientific part of the
Meeting. He helped us from the beginning to the end with his worthy
advice, and therefore we are indebted to Professor Fricke. We also owe a
debt of gratitude to Professor Pecker, Dr. Kovalevsky, Professor Kopal and
Professor Kresák, the main organizers of the Meeting sections, as well as
to other members of the Scientific Organizing Committee for their contribution.

The Federal Executive Council of Yugoslavia, the Union of Republican and
Provincial Self-Managed Communities of Interests for Scientific Activi-
ties in Yugoslavia, the Belgrade Astronomical Observatory and the
Yugoslav Airlines have penuciarily helped the Local Organizing Committee

1

W. Fricke and G. Teleki (eds.), Sun and Planetary System, 1–2.
Copyright © 1982 by D. Reidel Publishing Company.

to be able to carry out the Meeting. The subsistance aids have been given by the International Astronomical Union, the Belgrade Astronomical Observatory, the Hvar Observatory, the Department of Astronomy of Belgrade University, the Department of Geodesy of Belgrade University and the Local Organizing Committee of this Meeting. I thank all of them for their important contributions.

Last but not least, let me to express my personal thanks to my co-workers. The activity and laborious work of my two colleagues were of dominant importance. I have Mr Ivan Pakvor, the vice-chairman, and Mr Zoran Kneževíc, the secretary, in mind. I have to express, from the depth of my heart, my gratitude to both of them. I should also like to thank all the other members of the Local Organizing Committee as well as the technical staff for their contribution. It is my pleasant duty to offer thanks to Mrs Zdenka Talajić and Mrs Marina Bender, the representatives of the Atlas Travel Agency, for their permanent care and good co-operation.

After these expressions of gratitude, let me mention that when we decided to organize this Meeting we hade two goals in mind: firstly, to guarantee good conditions for a scientific meeting, and, secondly, to promote friendly relationships. For this reason we selected Dubrovnik and the Hotel Libertas. We should be most pleased if our goals would be realized. Therefore, it is my great pleasure to welcome you once again and wish you, apart from success in your work, a very pleasant stay in Dubrovnik.

Thank you!

OPENING ADDRESS BY

R. Kontić
Member of the Federal Executive
Council of Yugoslavia

Ladies and Gentlemen,
Comrades,
Dear Guests,

It is a pleasant duty and an honour for me, to greet the participants
of the VI European Regional Meeting in Astronomy on behalf of the
Federal Executive Council and I would like to express our satisfaction
because this important gathering of scientists and researchers is being
held in Yugoslavia this year.

I am glad that such a large number of scientists and researchers, particularly younger ones, from nearly all the countries of Europa and other
parts of the world have accepted the invitation to participate in the
work of this meeting, as this shows that scientific thought in the field
of astronomy is constantly expanding and developing, confirming at the
same time the readiness of the participants to publicize the results of
research carried out and knowledge gained to date, and to exchange
experiences and arrive at a critical scientific position through comprehensive discussions.

The fact that the Federal Executive Council is sposoring this gathering
attests to the importance that our country, as a developing country,
attaches to the development of scientific thought, primarily that
intended to be in the service of the progress of peaceful mankind.

This year's meeting shall be the venue for summing up scientific and
professional achievements on the subject "The Sun and the Planetary
System". During the next several days, employing various methods of work,
you will be able to view the solar system, not only from the standpoint
of astronomy, but also physics, geophysics, geodesy and other sciences.

Bearing in mind how extensive the scientific programme of this meeting
of yours is comprising 12 lectures at plenary sessions, 29 lectures in
groups and 117 papers, the importance of this gathering is self-evident,
as are the possibilities of its contributing to the further development

3

W. Fricke and G. Teleki (eds.), Sun and Planetary System, 3–5.
Copyright © 1982 by D. Reidel Publishing Company.

of astronomy and other directly related scientific disciplines. It is to
be expected that your research and its results shall be instrumental for
advancing scientific thought in general, and, in particular, research
into the Sun and use of solar energy to the benefit of mankind, which
is today, more than ever before, faced with the problem of providing
energy.

Astronomy is one of the oldest sciences which has long been studying
celestial bodies. However, at the beginning of the space age this
interest in studying cosmic phenomena has extended to other sciences
as well, so that today not only astronomy, but many other sciences also
engage in space research and the application of research results for
peaceful purposes.

Our country is exerting efforts to join in such research, as are the
countries from which you have come, as it has become evident that the
results of such research are today of multiple use, and that they are in
the service of the progress of mankind and its happier future.

The energy of the estimated stocks of liquid fuels in the world being
equivalent to the energy our planet receives from the Sun in only 12
hours, this surely warrants the efforts exerted in, and the importance
attached to research on the Sun, its energy potential and the possibilities of its use benefiting mankind.

This, like other types of research, calls for ample finances. Nevertheless, I have to state that, regrettably, instead of investing
resources in such and similar research, which would be of use to and in
the service of man, the international community spends over US $ 500
billion annually on manufacturing and purchasing arms in quantities
sufficient to destroy contemporary civilization ten times over.

On this occasion I would like to point out that the quality of astronomic research carried out in our country improves from year to year
and in certain research fields meets international standards, and to
this effect has certainly gained from the expansion of the network of
astronomic centres in our country and their better staffing. Since this
is the first major gathering of astronomers held in our country, it will
lend a strong impetus to the further development of astronomy in Yugoslavia and contribute towards our inclusion in various forms of international cooperation, on bilateral and multilateral bases alike.

I would like to reiterate my conviction that this gathering shall make
possible a fruitful exchange of scientific knowledge on the topic: The
Sun and the Planetary System, and beyond, while concurrently enhancing
better knowledge, closer ties, and friendship between you and your
countries.

I wish to extend my gratitude to the organizers, particularly to the
members of the Scientific Organizing Committee headed by Prof. Walter
Fricke, and to the members of the Local Organizing Committee headed by

Dr. G. Teleki, for the efforts they have invested and shall continue to invest these days, to ensure the best possible conditions for your work, as well as to the representatives of Dubrovnik, the host city which has for decades been extending its hospitality to guests from all over the world.

I will be exceptionally pleased if while staying here you get a chance to make use of your spare time to learn something about the realities of our socialist self-management Yugoslavia, its working people and their achievements.

With the best wishes for your successful and fruitful work, for the enrichment of your knowledge, for opening up new vistas, for the further development of astronomy - all in the interest and for the wellbeing of a man. and peace, I declare the VI European Regional Meeting in Astronomy open.

ADDRESS BY

R.M. West
Assistant General Secretary of the
International Astronomical Union

Ladies and Gentlemen,
Friends and Colleagues,

On behalf of the IAU Executive Committee and its President, Professor
M.K.V. Bappu, it is my pleasant privilege to extend to all of you the
very best wishes for a successful meeting.

As you probably know, there are several specific aims of IAU Regional
Astronomy Meetings. One of these is to bring together a large number of
astronomers in a geographic region, in particular younger astronomers,
for the exchange of scientific ideas and results. At the same time,
these meetings are also very useful for getting acquainted. And we should
not forget that they may help to further astronomy in the host countries.

Accepting the sponsorship of the Sixth European Regional Astronomy
Meeting, the IAU Executive Committee is convinced that the chosen time
and place exceptionally well fullfil these aims. Here in Dubrovnik,
astronomers have come from East and West, as well as from North and
South and I believe that many of those present are about to begin their
career in our science.

Astronomy in this country is obviously in a very active state. This is
evident from the increasing number of Yugoslavian astronomers who part-
participate in meetings abroad and also the many papers published in
national and international journals. It is indeed very gratifying to
feel the broad support of astronomy in Yugoslavia, just now so amply
expressed in the substantial cortributions by many agencies towards the
organization of this meeting.

Three different factors are necessary to make a meeting successful.
First of all, there is what I would call the intrinsic one; that is the
choice of subjects by the Scientific Organizing Committee, the response
by the participants and the resulting scientific programme. Thanks to
the great efforts of all involved, this factor is certainly well in hand.

W. Fricke and G. Teleki (eds.), Sun and Planetary System, 7–8.
Copyright © 1982 by D. Reidel Publishing Company.

Secondly, we should not forget the <u>extrinsic</u> one, defined by the Local
Organizing Committee, and visible in the surroundings and the social
programme. Anybody who would care to look out the window may assure
himself that Dubrovnik is a most magnificent town - and also a very
sunny one - to meet in. The third factor is what I would term <u>luck</u>.
Although we can not take care of this to the same degree as the two
others, the impressively thorough preparations by so many people
certainly tilt the scales in our favour.

Let me close with a very recent, personal experience, made less than two
days ago when I entered Yugoslavia, at the airport of Zagreb. Here, a
customs officer courteously asked to see my passport. When he become
aware of my profession, his face lit up with a very bright smile and he
said: "I see, an astronomer. That is a wounderful science. You are really
a lucky man." Hearing such words from the first person I met in your
country, I can only agree that Yugoslavia is indeed the right place for
a large astronomical meeting!

NEW IMPETUS TO THE EXPLORATION OF THE SOLAR SYSTEM
(INAUGURAL ADDRESS)

Walter Fricke
Astronomisches Rechen-Institut
Heidelberg, Federal Republic of Germany

On behalf of the Scientific Organizing Committee it is an honour
and great pleasure for me to extend to all participants and guests my
warmest welcome to this Meeting. We owe much gratitude to our Yugoslav
colleagues for the initiative they have taken in organizing the
Meeting in their beautiful country and in particular for having chosen
as our meeting place the old cultural centre Dubrovnik. We owe thanks
to our Yugoslav friends for their great efforts made in the local
organization. We also thank the Federal Executive Council of Yugoslavia
and the International Astronomical Union for their support given to
this Meeting. Finally, in speaking as a participant, I am happy to
meet here in Dubrovnik experts and scholars from many European
countries, all eager to learn more in the field of research related
to the "Sun and Planetary System".

Regional meetings in astronomy were initiated by the International
Astronomical Union about ten years ago. The leading idea was the
promotion of contacts between astronomers of neighbouring countries
for the purpose of encouraging joint discussions across national
borders. Such aims cannot be achieved by General Assemblies of the
Union which take place only once every three years. The series of
European regional meetings began in 1972 with a meeting in Athens and
has successfully been continued during all the years with the
exception of those, in which General Assemblies of the IAU were held.

In devoting our present meeting to the "Sun and Planetary System"
we have chosen a broad area of research in which unprecedented
progress has been made within the past few decades. At the time when
I was a student of astronomy at the universities in Berlin and
Göttingen - this was about 45 years ago - the senior astronomers told
their students that the planetary system does no longer offer a
subject for astronomical research which would lead to significant new
findings. The planetary system was considered to be well-known such
that further exploration could at best lead to minor refinements in
the description of some well-established facts.

W. Fricke and G. Teleki (eds.), Sun and Planetary System, 9–11.
Copyright © 1982 by D. Reidel Publishing Company.

This misleading view can now-a-days be understood quite well, because powerful telescopes had come into operation after 1900 and had contributed magnificently to the exploration of the stars and extra-galactic objects. One believed to be near the point where the structure of the universe as a whole can be explained completely on the basis of observations and of relativistic theories of gravitation.

New impetus to the exploration of the solar system was soon given by observations of the surface phenomena of the Sun which have led to the discovery of a general magnetic field and of local magnetic field phenomena in the outer parts of the solar atmosphere and the corona. At about the same time Hannes Alfvén developed a theory of the evolution of the Solar System in which electromagnetic forces are of decisive importance, because they may exceed by far the gravitational force of the Sun on the matter in regions of the solar system. Alfvén has concentrated on the interaction of ionized matter with electro-magnetic fields and thereby opened a new field of research, plasma physics. He has encouraged work on the physics of the early solar nebula, the magnetospheres of planets and other phenomena.

The "golden age" of solar system exploration began with the missions to the Moon, the study of lunar material on the Moon and in laboratories on the Earth giving impetus to research in many fields, such as physics, geophysics, chemistry, mineralogy, etc. Within the past decade the space missions to our nearest neighbour planets Mars and Venus have not only yielded hundredths of photographs, the most attractive ones appeared in all news-papers of the world, but have also provided results of measurements which by far not yet are fully analysed and exploited. The whole story of the Voyager visits to Jupiter and its satellites and the latest Voyager encounter with Saturn is not yet written except in headline news. Never before has the whole human race had an opportunity to receive such impressive information on the marvellous world about us.

In our admiration of the technological wonders of this so-called space age we should not forget that the impetus to technical develop-ments came from individuals. Scientists of many fields have contributed to the foundation of the so-called space sciences: theoreticians, experts in laboratories, engineers and technicians, and last not least, observers with optical and radio telescopes, or laser equipments.

I feel obliged to draw your attention to the important contribution to recent progress which came from common astronomers who have made high precision observations of positions of the Sun, Moon, planets and satellites since a long time. Another contribution of equal importance has been the construction of theories of motion which allow one - after fitting with the observations - to compute ephemerides. In parallel, one may now apply numerical integrations as a powerful tool and may compute orbits almost for the whole life time of certain objects.

In the exploration of the solar system the Earth is included for more than technical reasons. Most of the astronomical observations made with ground-based instruments not only yield information on the observed objects but also to some extent on the Earth's body and its atmosphere. Inherent in astrometric measurements are the effects of the non-rigid Earth, of movements in the Earth's crust, of the secular and irregular variation of the Earth's rotation, and the effects of latitude variations. Inspite of the recent progress made in improving measuring techniques, one may safely conclude that all these effects and their sources will remain to be subject of research in the future, and geophysics and geodesy, now already rapidly on the move, will increasingly attract the attention of scientists. As far as the Earth's atmosphere is concerned, astronomers must consider the study of three dimensional refraction as one of their lasting tasks.

In coming to the end I honestly should say that my own contribution to the exciting research field related to the Sun and planetary system is only a marginal one. It is even one outside of the terms of reference of this meeting. I have employed more than 100 000 observations of the Sun, Moon and planets made since about 1900 for the determination of the new international reference coordinate system, the FK5, to be introduced in 1984. This system shall approximate the inertial system as defined by the laws of dynamics of the solar system. As you know there is the hypothesis involved that our planetary system forms an isolated dynamical system. This may not be entirely correct for the whole life time of our solar system, but as I hope for the next 25 years which is so far the mean life time of a conventional celestial reference coordinate system.

After having spoken so much on new impetus to the exploration of the solar system, I sincerely wish that this Meeting may give impetus to a fruitful cooperation of European astronomers in this field and beyond its borderlines in all fields of sciences.

THE SOLAR SYSTEM - KNOWN FACTS AND UNSOLVED PROBLEMS

Zdeněk Kopal
Department of Astronomy,
University of Manchester, England.

I greatly appreciated the invitation of the Organizing Committee of the VIth ERMA meetings to deliver an introductory address to this conference - mainly as an opportunity to raise certain questions, fundamental to problems of solar-system research, without moral obligation to provide their answers (and in the hope that papers presented at subsequent sessions of our congress may contribute towards their clarification). The actual number of questions concerning unsolved problems which one is tempted to voice is indeed much larger than one could bring out in the time available for this introductory session. I shall, therefore, limit myself only to certain fundamental aspects of our subject - in the hope that they may be of interest also to non-specialists in the respective problems.

The observational evidence at the disposal of students of the solar system at the present time - augmented greatly by spacecraft of the past 20 years - is very extensive, and provides a much more complete picture of our celestial home than that known to our ancestors, for whom ground-based telescopes were the only source of information. In an attempt to discern the initial design and subsequent evolution of the solar system, let us first survey certain basic facts, as well as "invariants" of this system which are unlikely to have undergone appreciable changes in the course of its long astronomical past.

First, the age of the system (i.e., the time elapsed since the condensation of the primordial "solar nebula" from pre-existing material) is known today (from radiometric dating of chondritic meteorites as well as of the lunar soil) to be close to 4.6 billion years; and the uncertainty of this age does not exceed 1 per cent of its absolute value. This, therefore, sets the time-span available for the evolution of the system from its pristine stage up to the present. It is not a short time; being equal to almost one-half of the age of our Galaxy; and since its formation the Sun with all its attendants has revolved around the galactic centre not less than 25 times. In contrast, the duration of the actual collapse of the primordial solar nebula into the system we know today appears to have been relatively short: the abundances, in meteoritic material, of certain short-lived nuclides of volatile elements indicate that the principal formative epoch of our system did not last

13

W. Fricke and G. Teleki (eds.), Sun and Planetary System, 13–21.

more than 1% of its total age; and was possibly as short as 10^7 years.

Second, from extensive dynamical studies (in particular, by Hagihara) it transpired that, if the principal planetary constituents of the solar system are regarded as a mechanical "clockwork", the present scale of the system could not have changed as a result of mutual gravitational perturbations in 10^9 years by more than a few per cent. Therefore, the absolute sizes of planetary orbits around the Sun (as well as their proportions) could scarcely have undergone any appreciable changes since the time of their formation.

Third - and last but not least - it is sufficiently well-known now (for a summary of the relevant evidence cf., e.g., Newkirk, 1980) that our Sun as a zero-age Main Sequence star emitted only about 60% of its present energy output, and its effective temperature was close to 5000K, in contrast with 5700K that it is today. As a result, the archaic planetary surfaces received only 60% of the heat they receive from the Sun today; and their temperatures should have been 10% lower. This should, of course, be true only of surfaces unprotected by any atmosphere - such as those of Mercury or of our Moon; while for the Earth (or, to a lesser extent, Mars) this need not have been literally the case, because of the "greenhouse effects" of their evolving atmospheres.

These, are, in brief, the principal constraints which should be kept in mind in any attempt to reconstruct the past of the solar system from its observed present. Efforts to do so - from the days of Kant and Laplace to our own time - fell so far short of their goal; and their main contribution to our knowledge has, in fact, been in the demonstration of how the solar system could *not* have originated; and to lead us gradually to our goal by narrowing down the range of admissible possibilities. A part of the difficulties obstructing the work of our ancestors had, in fact, been their preoccupation with accounting for the properties of the solar system as we see it today - without allowing sufficiently for the fact that, in their maturity, some of them may be quite different from those obtaining at the time of the origin. In other words, our distinguished predecessors of the 18th and 19th centuries may not have sufficiently realized that the ageing adult of today may, in many respects, be quite different from what it was in its early childhood!

It is true that some of its properties - such as the distances of individual planets from the Sun and from each other - may have weathered the times better than others, and remained largely invariant in the course of the past 4.6 billion years. As we shall point out later, however, this need not have been the case of planetary masses or dimensions; and their chemical composition - once common to both - could likewise have undergone a secular change But before we list reasons why this could have been the case, let us introduce the principal "dramatis personae" of our narrative, to make them to testify on their origin and relation with others. In doing so, we shall note that the general population of the solar system splits up in front of our eyes into several distinct groups, or "sub-systems", characterized by common physical (kinematic) or chemical properties. A closer examination of these sub-systems may disclose to an inquisitive mind their past relations, as well as the extent of the symbiosis in which they live with others today.

SOLAR SUB-SYSTEMS.

The total amount of information characterizing the present solar system
- much of which came in our possession very recently thanks to deep-space
planetary probes - makes it clear that the picture of this system as a single
unit would be grossly oversimplified today. In the light of our more recent
knowledge, this system naturally splits up into several distinct sub-units
which can be categorized on different grounds.

To begin with, let us consider first the *dynamical characteristics* of
different members of the solar system, because these are in general better
conserved in the course of time and, therefore, provide a more reliable link
with its distant past. As is well known, the motions of most constituents of
the solar system around its central star observe a certain set of rules which
can be summarized as follows:

1)   their orbits deviate but little from circles;

2)   the plans of such orbits cluster closely around the invariable
       plane of the system (though their inclinations to the solar equ-
       ator is considerable - amounting, in fact, to several degrees!);

3)   the celestial bodies revolving in such orbits do so in the same
       direction.

The larger the mass of the respective body, we may add, the more
closely these rules are obeyed. Major planets follow them more closely than
the asteroids - let alone comets (which disregard them all). It is also easy
to see what could have led to the establishment of the foregoing set of "traf-
fic rules" among more permanent members of the system after a sufficiently
long time: namely, the need of self-preservation. Bodies which disobeyed
them could perish more readily in collisions (the origin of metallic asteroids?),
or be ejected from the system by perturbations - as is still happening to
the comets. The fact that this latter class of bodies disobeys all rules
testifies to their impermanence; and to the need of continuous replacement
if any are to be seen in the sky at all! The occasional occurrence of relatively high eccentricities and inclinations among small asteroids (among the
Earth-crossers, for instance, for Tantalus we find that $e = 0.30$ and $i =
64°$) can be accounted for if these are extinct cometary cores - an explan-
ation which would, however, be surely inapplicable to Pallas ($e = 0.26$ and
$i = 35°$).

When we turn next to the *chemical composition* of different members of
the solar system, a completely different situation is encountered. As is
well known, 99.87% of the mass of the solar system is stored in the Sun;
the atmosphere of which (reflecting probably the original composition of
the primordial "solar nebula"), consists of 71% hydrogen, 26.5% helium, and
only 2.5% of all heavier elements lumped together (cf. Allen, 1973). The

second largest reservoir of mass in the solar system is stored in the major
planets - from Jupiter to Neptune - representing together only 0.13% of the
mass of the Sun (but 224 times the mass of the rest of the system). The
composition of this mass continues to be akin to that of the Sun; its main
constituents being hydrogen and helium, supplemented (especially in Uranus
and Neptune) by increasing fractions of heavier volatiles (such as nitrogen,
oxygen and carbon).

Recent measurements performed by the Voyagers of 1979-1981 *in situ*
established that the outer layers of the Jovian atmosphere consist (by mass)
of 80% hydrogen and 19% helium; while for Saturn these abundances turned
out to be 88% for hydrogen and 11% for helium; all heavier elements lumped
together account for the balance. Therefore, the hydrogen contents of the
Jovian and especially Saturnian atmosphere proved to be higher than for
the Sun! These figures need not, of course, pertain literally to the interi-
ors of all these bodies (which can so far be inferred by only indirect methods); but an overwhelming proportion of hydrogen and helium is required
for all to reconcile their theoretical models with observational constraints.
Indeed, in more senses than one all four major planets of the solar system
can be regarded as "mini-stars" rather than planets; for they emit 2-3 times
more energy than they receive from the Sun by irradiation; whether the
source of excess energy is mechanical (terminal stage of Kelvin contraction)
or chemical (through exothermic reactions in the interior) remains as yet
conjectural - both are distinct possibilities!

But the exploration of Jupiter and Saturn by the Voyagers in 1979-
1981 held further cosmochemical surprises for us in store: namely, by their
discovery of the fact that the permanent satellites of these planets, Jupiter
I-IV and the first Saturnian satellites from Mimas to Japetus possess (with
the possible exception of Io) mean densities between 1.8 - 3.3 g/cm³ in the
Jovian family, and 1.0 - 1.9 g/cm³ in the Saturnian family - i.e., values
so low that astronomical bodies of small mass (comparable with, or smaller
than those of our Moon) can scarcely be anything else but "icy snowballs",
consisting largely of frozen compounds of hydrogen with heavier volatiles
of the C-N-O group. It is true that the absolute values of these densities
are comparable with those of the planets around which these satellites re-
volve (1.34 g/cm³ for Jupiter, and 0.69 g/cm³ for Saturn); but these satellites are 10^{-5} to 10^{-6} times less massive than their central planets; and
their ability to retain so large a fraction of volatile elements within their
mass was all the greater surprise.

Indeed, the permanent satellites of Jupiter and Saturn (to which we
shall no doubt have to add those of Uranus and Neptune, as well as Pluto)
represent celestial bodies of structure unknown to us before the historic
exploits of the Voyagers in 1979-1981. An exception to their near-monopoly
in exploration of the outer parts of the solar system was the 1978 discovery
by Christy and Harrison (using the methods of ground-based photographic
astrometry) that Pluto (one hesitates to call it a planet any more) is a binary system consisting of two bodies of comparable mass (cf. Harrington,
1979) - a fact which led to a first reliable determination of its mass (equal
to 1.6×10^{25} g or only one-quarter of that of our Moon); while ground-

based speckle interferometry specified Pluto's diameter to be close to 4000 km (i.e., 1.15 that of our Moon); leading to a bulk density of only 0.5 g/cm^3 (cf. Bonneau and Foy, 1980). More exact data on the Uranian and Neptunian satellites (in particular, Triton) will have to await the (hopefully successful) fly-bys of Voyager 2 in 1986 and 1989 in their proximity. However, there is but little room for doubt that several of them (in particular, Neptune's Triton) will turn out to possess the structure akin to that of Jupiter's Ganymede or Saturn's Titan; and their material to consist largely of "dirty ices" of frozen hydrocarbons - i.e., bodies not previously encountered anywhere within the inner precincts of the solar system.

But major planets and their satellites are not the only celestial bodies known to consist of "dirty ices". Another group belonging by their chemistry to the same category are, of course, the comets - small of mass, but innumerable in bulk - and those which venture into the deeper precincts of the solar system bear ample evidence of their volatile composition at the expense of their longevity. Whether or not cometary nuclei are floating icebergs of frozen hydrocarbons - travelling through space at velocities which may enable some to escape the solar system altogether - or ice-encrusted asteroids is a question which we can only ask, but not yet answer; but their large contents of frozen hydrocarbons evaporating under the effect of solar heat has been spectroscopically established beyond any doubt.

Summarizing the present state of our knowledge we conclude that bodies consisting primarily of light volatiles account for an overwhelming part of the material which constitutes the solar system; and their individual masses range by 17 orders of magnitude - from 2×10^{33} g for the Sun to 10^{16} g for a typical comet. If the mass is sufficiently large (as it is for the Sun or major planets), its bulk will consist of hydrogen (in plasma, atomic or molecular form); while in small masses hydrogen can be retained only in compounds of liquid or solid state.

In the next section of this paper we shall attempt to identify the reasons why this may be the case. In what follows we only wish to characterize another group of planets of the solar system of very different composition from those which we described so far - namely, those mainly of silicon and iron - which are generally referred to as the "terrestrial planets".

This name stems, of course, from the fact that our Earth happens (though only just) to be the largest, densest and most massive of this class of bodies, whose mean distance from the Sun ranges between 1.52 A.U. for mars to 0.39 A.U. for Mercury - almost within a belt of 1 ± 0.5 A.U. for them all. Their mean density ranges from 3.34 g/cm^3 for the Moon to 5.53 g/cm^3 for our Earth - increasing (though not without exception) with increasing mass of the respective configuration. These densities are considerably higher than those of the major planets located at greater distances from the Sun; and the reason is an almost complete absence of hydrogen and helium in their composition (the terrestrial hydrosphere amounts to only about 10^{-4} times of the mass of the entire planet).

The internal structure and chemical composition is well known for the

Earth, and only somewhat less so for our Moon - two extreme types of planetary bodies of this class: we know the bulk of their masses consist of silicon and iron, largely in oxidized form. For the Earth, iron accounts for almost 39% of its total mass; oxygen, 27%; silicon, 14%; magnesium, 11%; and heavier elements (led by sulphur and nickel) proportionally less. For our Moon, the respecrive estimates are broadly similar; its lower bulk density being due (apart from lessened effects of self-compression) to a diminished proportion of iron, and increased proportion of silicon and magnesium.

An additional reservoir of silicate-iron material in the solar system can be found in the asteroidal belt, located mainly (but not exclusively) between the orbits of Mars and Jupiter. The total mass of all asteroids within this belt has recently been estimated (cf. Kresak, 1977) to 3.0×10^{24} grams - i.e., 2.5×10^{-4} times the combined mass of the terrestrial planets - to which the three largest asteroids (Ceres, Pallas and Vesta) contribute more than a half of the total. But, judging by their reflectivities in different parts of the spectrum, only about 20% of all asteroids (including, however, Vesta) happen to be of silicatious type; the remaining 80% appear to be carbonaceous - of very light composition (with densities between 1.5 and 2.5 g/cm³) - well known to us from meteoritic samples occasionally intercepted by our Earth.

The Martian satellites Phobos and Deimos - probably asteroids intercepted by that planet in the earliest state of its evolution - belong to this latter type; for their very low bulk densities (less than 2 g/cm³) together with characteristic reflectances investigated by the Viking Orbiters in 1976 is characteristic of the chondrites. These chondritic bodies (generally of very small calibre) may, in fact, represent a third chemical category of the material constituting our solar system - side by side with hydrogen-helium and silicate-iron bodies - light solids (though largely devoid of volatiles) which probably represent the oldest type of material that solidified at the earliest stage of the collapse of the primordial solar nebula; and parts of them (chondrules in carbonaceous chondrites?) may hail back to its presolar past.

It will not be our aim to follow up these last remarks in greater detail at this time. Instead, we wish to return to the basic dichotomy between the "major" and "terrestrial" planets of the solar system - so similar in their kinematic characteristics, and yet so different in their chemical composition. The bulk of the planetary mass in the solar system is, we repeat, similar in composition to that of the Sun; while the terrestrial planets show a marked deficiency of the volatiles. It is also found that the hydrogen-rich bodies (of large or small masses) are met only at large relative distances from the Sun - in excess of 5 A.U. - while the bulk of silicate-rich bodies are encountered in the inner precincts of the solar system - at distances generally less than 1.5 A.U. from our central luminary - enjoying warmer climate. To what extent can this latter fact alone account for their different chemical composition?

WERE THE PRESENT TERRESTRIAL PLANETS ONCE MAJOR PLANETS?

The clue to a possible answer to this question may rest in the cosmic climate prevalent at different distances from the Sun in the course of its long astronomical past. At the time of their origin 4.6 billion years ago, all planets condensed from matter constituting the primordial solar nebula, of composition approximated by that of the present solar atmosphere - un-adulterated by anything that has happened since - for several reasons. The large mass of the Sun has been more than sufficient to prevent any selective escape of different elements from its gravitational field at temperatures prevalent in its atmosphere - temperatures too low to give rise to nucleosynthesis by equilibrium processes on any appreciable scale. No efficient mixing of atmospheric gases with those below the limit of the outer convective zone of the Sun occurs today; and this situation is unlikely to have been different in the course of its past.

The chemical composition of the solar atmosphere is known to us in considerable detail (cf., e.g., Aller, 1961) from quantitative spectroscopic analysis; and on the basis of the results furnished by it let us pose to ourselves the following question: how large an amount of gas of solar atmospheric composition would be needed to produce, on condensation, a planet of the terrestrial mass and composition? Allowing for appropriate escape of lighter constituents in a weaker gravitational field, we find that the masses so determined show some dispersion if different elements are used for its determination. That deduced from the relative abundance of oxygen is probably too low because of high volatility of this gas; while for iron it comes out much too high because of an apparent overabundance of this element in the terrestrial interior.

But if we disregard these extremes, the average deduced from the rest for a hypothetical mass of the proto-Earth comes out to about 300 present terrestrial masses ($\oplus$), and no less than 100 $\oplus$ for the initial masses of other terrestrial proto-planets. It is that much of solar atmospheric gases that we need to condense if their residue is to equal our Earth in mass. It may be of interest to note that a mass of 300 $\oplus$ happens to come close to the present mass of Jupiter; and 100 $\oplus$ is not far from that of Saturn. May we, therefore, conjecture that our Earth - and other terrestrial planets - began their cosmic careers as configurations of masses comparable with those of Jupiter and Saturn; and that their present disparity in mass arose subsequently - due to the fact that while the Earth and other terrestrial planets have gradually lost most of their original endowment of light volatile elements, Jupiter and other major planets have managed to retain theirs? This question was raised by your present speaker already several years ago (cf. Kopal, 1972a, b); and nothing that transpired since has lessened a possibility that the answer may indeed be in the affirmative; the problem certainly deserves further study.

What could have brought about so large a difference in the loss of volatile elements between these two groups of planets? At least a part of the answer may go back to the location of these groups in space: for while the terrestrial planets are confined to a region within 1.5 A.U. from the

Sun, the distances of the major planets range between 5 and 30 A.U.; and
we no longer have valid reasons to suspect any appreciable change in their
distribution in the past 10^9 years. This, in turn, would have exposed
these bodies to a very different climate: while the surfaces of the terrestrial planets are maintained by sunlight at mean temperatures ranging
from 200 to 700K, those of the major planets range from 50 to 100K. In
particular, the mean temperature of Jupiter is lower by some 160K than that
of the Earth today. Lower temperatures are bound to inhibit escape of the
volatiles through the exospheres of the respective configurations; and the
fact that our Earth lost most part of their initial endowments while Jupiter
retained its own may be due at least partly to this cause.

There may, of course, also have been others. For instance, even if
the original masses of Jupiter and of the Earth were originally comparable,
the rate of escape of the volatiles at equal temperatures would have been
different for initial condensation of different size. Suppose, for the sake
of argument, that the condensation of gas (and solid particles) which gave
rise to the Earth was larger in size, or contracted more slowly, than proto-
Jupiter: this could have deprived our Earth of its volatiles (because of
diminished gravity) just as effectively as warmer cosmic climate. In other
words, Jupiter and other major planets may have retained their original
(i.e., early solar) composition more faithfully not only because their material was cooler, but also if they may have been formed in a more compact form;
or again because they collapsed more rapidly to their present size due to
internal reasons.

This whole argument presupposes that the material of the primordial
solar nebula was well mixed, and homogeneous in the main features of its
composition at any distance from the Sun. But that this was so seems probable. The isotopic ratios of common elements (such as oxygen or nitrogen),
ascertained in recent years by spacecraft exhibit only marginal differences
between Venus to Saturn; and the same was probably true of the initial
distribution of the elements themselves as well. Is it also not reasonable
to assume that the initial condensations which gave rise to different proto-
planets may have been largest in the proximity of the Sun - the largest
condensation of them all?

A problem of especial interest would arise, in this connection, with
the twin system Earth-Moon. If the masses of both these partners were
originally much larger than they are today, many difficulties connected with
the origin of the system - such as the capture at a later stage of their
developments - would completely disappear. If the picture proposed for
consideration is on the right track, the Earth and the Moon could have
originated as a really close binary system, and made to appear well-separ-
ated later as a result of progressing contraction caused by escape of the
volatiles. The same may, to a lesser extent, have been the history of the
Neptune-Triton system (and possibly of the triple system Neptune-Triton-
Pluto if the latter is a former satellite of Neptune).

But whatever may have been the case, arguments advanced so far
lend some weight to a possibility that all planets of the solar system origin-

ated as bodies of comparable masses; and that their presently manifest disparity in mass or chemical composition may have developed later in the course of time - as a result of protracted exposure to very different cosmic climate prevalent at different distance from the Sun.

REFERENCES

Allen, C. W.: 1973, *Astrophysical Quantities* (3rd ed.), Athlone Press, London, p.164.

Aller, L. H.: 1961, *The Abundances of the Elements*, Interscience Publ. Inc., New York; Table 5-5.

Bonneau, D. and Foy, R.: 1980, Astron. Astrophys., *92*, L1.

Harrington, R. S. and B. J.: 1979, Mercury, *8*, 1.

Kopal, Z.: 1972a, Annals New York Acad. Sci., *187*, 108.

Kopal, Z.: 1972b, *The Solar System*, Oxford Univ. Press; pp.120-122.

Kresák, L.: 1977, Bull. Astr. Inst. Czechoslovakia, *28*, 65.

Newkirk, G.: 1980, in *The Ancient Sun* (R. O. Pepin, J. E. Eddy and R. B. Merrill, eds.), Pergamon Press, pp.293-320.

SECTION I

SUN FROM THE ASTRONOMICAL AND PHYSICAL POINTS OF VIEW

THE STAR "SUN"   L'ETOILE "SOLEIL"

Jean-Claude PECKER
Collège de France, Paris.
Institut d'Astrophysique du CNRS.

Résumé : Dans cette présentation bibliographique, l'auteur envisage le
Soleil du point de vue de Sirius, en quelque sorte...: 1. Le rayon du
Soleil. Déterminations astrométriques. Le Soleil est-il aplati? Recherches et perspectives nouvelles. 2. La masse du Soleil, selon les
déterminations les plus récentes. Perte de masse séculaire du Soleil.
3. La luminosité du Soleil, et la "constante solaire". Ses variations
rapides et séculaires. Température effective. 4. Composition chimique
du Soleil. Rôle de la diffusion; les inhomogénéités de composition.
Une conclusion, aussi philosophique qu'astronomique, tentera de resti-
tuer les recherches solaires dans l'éventail des disciplines de l'astronomie, comme l'un des centres majeurs de l'interdisciplinarité.

Note : Some general references are referred to by letters and appear
in the first part of bibliography, other appear as usual and refer to
the second part of bibliography.

As seen from Sirius, the Sun would appear as a very ordinary
star. This idea has been overused, for years (see E, F, G); it has at
least accustomed the stellar physicist to the idea that any star is
more complex than one can describe it through the two-dimensional physical approach of its structure; in the recent years, the space research has given us many additional observational arguments; the stellar world displays both complexity and diversity. The Sun is only one
amongst many stars; we know it slightly better than others, and this
remains our main reason to study it; another reason is that it directly influences the physical conditions on the Earth.
But we know it as seen from the Earth, namely from a point located
in its equatorial plan; in other terms, cos i is very near unity.
Moreover, being in the equatorial plane may mean we have biased ideas
about some solar phenomena strongly depending upon the heliographic
latitude, such as solar wind, brightness distribution, or photospheric
magnetic field... It is thus quite possible that we may learn many
things from stellar studies that would enable us indeed to understand
better the physical processes at work in the Sun.
A star is generally characterized by its mass $\mathfrak{M}_{\odot}$, its radius $\mathcal{R}$,

25

W. Fricke and G. Teleki (eds.), Sun and Planetary System, 25–34.
Copyright © 1982 by D. Reidel Publishing Company.

its luminosity $\mathcal{L}$, its chemical composition (X : Y : Z). The Vogt-Russel theorem claims this even overdetermines the star; this might indeed well not be sufficient: we know at the contrary that the theorem applies only to gaseous spheres <u>in equilibrium</u>, and chemically <u>homogeneous</u>; but at birth, not only mass and composition determine <u>what will become</u> of the star: one has also to consider its momentum of rotation, its global magnetic field, some environmental conditions,... which should allow the evolving star to diversify considerably. Hence, speaking of the Sun as a star imposes a short discussion of the classical numbers $\mathcal{M}_\odot$, $\mathcal{R}_\odot$, $\mathcal{L}_\odot$, and X, and Y or Z, but also of other global properties, such as the possible time-variations of these quantities, or the average rotation and magnetism of the star Sun, or again the out-of-equilibrium outer layers, and their properties, or again the neutrino flux.

1. <u>The radius</u> $\mathcal{R}_\theta$ of the Sun depends upon the angular diameter θ of its disk, upon the solar parallax ϖ, upon the Earth's radius $\mathcal{R}_\oplus$. The first correct determinations of solar diurnal parallax were done during the XVIIIth century; nothing has been essentially changed till 1940; the admitted value, that of Sir Harold Spencer Jones, was then 8"790, with 3 correct significant figures at the most. The solar parallax is derived from direct measurements of the apparent position of a planet, such as Venus, or an asteroid with an appropriate orbit, passing close to the Earth (Eros, Amor, Adonis...) and the application of the third Kepler's law, written as a simple relation between orbital great axes and orbiting period of both planet and the Earth.

At that time, the solar angular diameter was measured with a reasonable accuracy, and the numbers then published were accurate at about one second of arc. Typical is the 1891 value of Auwers, equal to 1919".3, when the Earth is at the mean distance from the Sun.

The Earth's radius $\mathcal{R}_\oplus$ was known, through geodetical determinations. During the last thirty years, the radius of the Earth and the solar parallax have been improved very much. The astronomical unit of distance (UA) is, strictly speaking, the semi major axis of a planet of mass zero, unperturbed, and of the same orbital period as the Earth. The relation between 1 UA and the meter was first determined by the angular measurements leading to the value of the solar parallax. Since the sixties, better results have been achieved by direct measurements of planetary distances, using the powerful techniques of radarastronomie. Table I gives the main constants used in this now coherent system of masses and distances.

The determination of the solar radius has however natural limitations as soon as one reaches a sufficient accuracy:(i) The first obvious problem is the physical ambiguity in the very <u>definition of the solar limb</u>; it is of the order of only a few kilometers, even less if the wavelength is strictly defined. (ii) More serious is the assumed <u>oblateness</u> of the Sun; it imposes to define the equatorial radius, and the polar radius. (iii) The periodical or irregular <u>variation of the solar radius</u>, within small times may strongly affect the value of the determination; some "average" radius should be defined from the measures.

1a. The problem of <u>solar oblateness</u> has been discussed in the recent years.

We should not forget than this important type of measurements may be a clue to the study of internal rotational structure of the Sun, and from there, to cosmological theories.

As a matter of history, I would like to recall of the early findings of Maunder, from sunspots studies (1919), and those, more recent, of Roberts and myself (1954), using the M-regions, or of Trellis and Fréon (1958, 1960), using again the coronal streamers and the cosmic rays: it exists regions of a given Carrington longitude, systematically avoided for several successive years, and at all latitudes, by birth of sunspots groups, or by M-regions; this led to think that the deep interior could move as a solid body with a period of about 27,3 days, in the regions where sunspots originate. More recently, the X-ray study of coronal holes have fully confirmed the "solid rotation" of the deep layers, as suggested by the latitude behaviour of coronal holes (see Hundhausen, 1979).

A similar solid rotation, but much faster, was implied by other authors, from a completely different line of thought. Jordan, Thiry, Bergmann, but mostly Brans and Dicke, advocated the introduction, in the gravitation theory, of a coupling between Mercury's revolution and solar rotation. The coupling is function of the flattening of the Sun, and linked with the rotation of its deep interior. Without entering into the details, we may say that on the observational side, a flattening of $5.10^{-5} \pm 0.7$ was found by Dicke (1967) and coauthors, -and denied by Hill and coauthors (1975-1979). As the technique of measurements is strongly influenced by secondary effects, such as atmospheric refraction or combination of the effects of oblateness and of differential limb darkening, these results have still to be considered mostly as stimulating for further research. Rösch and Yerle (G. p 292) at the Pic du Midi, are now putting in operation a new method of measuring oblateness, which would be free from these criticisms. The oblateness, found by Dicke, would correspond to an additional 4"/century advance in the motion of Mercury perihelion; it would lead to think of a fast rotating central core, 20 times faster than superficial layers. Without entering in the discussion, we want to note the contradiction between the fast rotating core implied by Brans and Dicke, and the slower solid rotation speed implied by coronal holes observations.

1b. The problem of short-time radius variations is closely linked with that of <u>solar oscillations.</u> Such oscillations were first observed in 1960 by Leighton, and soon after, with a different type of technique by Evans, Michard (1962) and others. The five-minutes oscillations, since that time, have been extensively studied, observationally and theoretically. Solar sismology is progressing fast...

In the particular case of the solar 5-minutes oscillations (see in F the reviews by Deubner and by Leibacher and Stein), one has been able to show they originate in the convective zone (Deubner 1975); they are standing, being trapped, roughly speaking, between the bottom of the convection zone, and the chromospheric layers; and they are typically acoustic (p-modes). Although one observes the fundamental mode, it contains little energy. Most of the energy is in the highest values of the azimutal quantum number ℓ (which characterizes the number of modes on the solar surface, or so to say, the non-radial character of the oscillations) and in the low values of the principal quantum number n (which characterizes the number of modes along the solar radius). The harmonic analysis on sufficiently long sequence allows to distinguish the various ℓ-components within the n components. One has thus been able (Fossat, Grec) to determine the harmonic structure of the global 5-minutes oscillations with a great accuracy; the existence of non-radial modes, in itself, puts a limit to the knowledge of solar radius. The velocity of the matter is of the order of 0.3 to 1.6 m s^{-1}, which implies amplitudes of the order of several hundred meters, i.e. of the order of milliseconds of arc (see in this book, the astrometric determination by Chollet). The 160 minutes-oscillations, determined from indirect angular diameter data, by various observers (Kosovichev and Severny, in this book) (see also Deubner in F) are still of a smaller amplitude, of 0.3 to 0.4 m s^{-1}; they are quite comparable to the effects of differential refraction, and possibly affected by atmospheric effects, although the phase relations between different observers seem to plead for their true astrophysical reality.

The determination of some average solar radius can thus be done only at the condition of defining the averaging process. In other terms, we face in the case of the Sun a problem similar to the one we face in the case of the Earth; but whenever the "geoïd", and the very many terms of its potential, can be considered as rather constant with time, during at least many years, the potential of the "helioïd" has many terms indeed, continuously changing, and for which we must know not only the amplitude, but the periodicity, which is of the order of minutes, possibly less. This is the price to pay for the Sun being made of gases!

1c. But a question remains: is the solar radius submitted to some secular variation? The hypothesis was revisited recently, after an analysis of Greenwich solar radius measurements, by Eddy and Boornazian; and it has initiated a very important program of systematic measurements of the solar diameter at the University of Arizona and at the HAO. It seems, according to Shapiro's analysis of transit of Mercury measurements, that the radius of the Sun is reasonably constant, or has been so, for the last 250 years. Data by Dunham seem to indicate, from eclipse data, a small shrinkage of 0".7±0".4 in 250 years for the diameter, or 0".0015/year for the radius. The debate is still open but its very existence is, in itself, a sign of the new era, during which one tries to make a full use of the historical data, now more than two centuries old, and some times still older.

2. The mass of the Sun is not determined in an completely independant way more than its diameter, in the sense that both determinations rest upon the measurement of the solar parallax. Around the early forties, the mass of the Sun, as derived from the observations of asteroïd Eros, was known with only 3 significant figures: 1.99 10^{30}kg.

Again, the recent years have seen some improvement in this determination, due to the use of the precisely defined motions of closer
asteroids, of artificial satellites and of solar system planetary probes. The present value of the solar mass is put in table I; the larger
limitation rests in the estimation of the gravitational constant G,
the value of the ratio of the solar mass to the earth's mass being
known with 6 or 7 significant figures.

Again one might ask: is the solar mass constant? And the reply is,
of course, no! First of all, the energy production of the Sun is to be
accounted for by the transformation, every second, of $6 \ 10^{38}$ atoms of
hydrogen into helium. In other terms, the Sun is at present loosing
$6 \ 10^{-14}$ of its mass per year, in form of radiation. The solar wind,
directly measured at Earth's distance, shows us that a mass loss of
$2 \ 10^{-14}$ has to be taken into account, in form of electrons, protons
and various ions, atoms, molecules, or dust grains. But we have no
clear idea about its real value, as this is inferred from measurements
made near the solar equatorial plane, and do not necessarily represent
a correct value for the whole solar surface.

3. The luminosity of the Sun, in its turn, depends also of our knowledge of its distance. But it primarily depends upon the measurement
of the so-called "solar constant". Abbott and his associates reached,
in the early forties, a value of 1.90 cal cm^{-2} mn^{-1}. But corrections
for the IR and UV were needed; in the fifties, the admitted value was
of 1.97. The space era has allowed to slightly improve this situation;
on one side, one can measure the energy in the invisible parts of the
spectrum, non observable from the ground, IR or XUV...; on the other
side, the transparency in the observable parts is not any more affected by climatological conditions. The results are accordingly much
better, and the accuracy of 1% can be reached; strangely enough, this
is still considered as a good determination (see table I).

The limitation rests now in the spottedness of solar surface. The
real relative accuracy of determination reaches really a value of 10^{-4}
(see Wilson et al. 1980). But when a spot appears, or disappears at
the limb of the disk, its size being of the order of $20 \ 10^{-6}$ of the
solar surface, a small number of spots on the solar disk (R of the order of 10 during minimum) are sufficient in the best observing conditions to put a limit of the order of $2 \ 10^{-4}$ to the accuracy. One could
think (as earlier thought) that the total energy radiated by the visible part of the Sun is however not much affected, as one might think
that the energy blocked by the spot is going in the form of a bright
ring around the spot. However actual measurements (Wilson, 1981) do
not reflect this. It might be that the blocked energy goes out in the
polar region, and the polar brightening might escape global radiation
measurements because of the foreshortening. Hence the spottedness is a
real limitation, as long as one does not perform measurements of all
points of the solar surface with an equal weight, using necessarily
out-of-ecliptic vehicles.

3a. In spite of this limitation, variations of a few percents can
be considered as affecting the solar constant, during times comparable

to the duration of the solar cycle, as shown by Fröhlich (in C). It is
quite possible that this does not concern the Sun as a whole, but only
the Sun as seen from the Earth, precisely because of the preceeding
considerations. But this has still to be determined; in one extreme
case, the distribution of brightness from equatorial zones to polar
one varies during one cycle just as it should to explain the observed
changes; in the opposite case, the variation in the total luminosity
of the Sun is the only origin of the observed changes; we are now inclined to exclude the latter, for reasons already said. One has other
reasons to exclude it: if it were true, then the variations observed
in various wavelengths should be in phase; it appears that such is not
the case, as shown by Müller and associates through measurements of T_b
(brightness temperature) in the IR; these last data would favor a
change in spectral distribution with cycle, hence a change in effective temperature or total luminosity. This last point however is contro-
versial. Considering thus that our situation of being equatorial observers of the Sun is of some disadvantage, and allows only measurements of the variable solar constant, or of the variable spectrum, as
seen from the Earth, we have to consider that the effective temperature deduced from these data may be badly known. The luminosity itself
can be considered still as badly known, as long as the question marks
I have put are not removed.

3b. Strictly speaking, the production of energy which defines the
solar luminosity cannot be described only by the photonic flux. One
has also to take into account, in principle, the neutrino flux. One
can say that it depends upon the production of energy at Sun's center
as it is now, whenever the photon flux depends upon what was this
energy production several 10^8 years ago. Any secular change in luminosity would make the use of such data quite meaningless. One can also
add that, energetically, the neutrino flux is carrying a little fraction of the energy produced in the central regions. A complete cycle
of pp reactions is producing, on the average, a total of radiative
average energy of 26.10 MeV, for which the precise counting of photons
of various energies has no meaning, in view of the very fast process
of interaction leading to thermal equilibrium (the photon output goes
out from the Sun, as radiant energy, after some 10^8 years). The same
cycle, taking into account the probability of the 3 forms (P_1, P_2, P_3)
of the pp-cycle produces also, on the average, 2 neutrinos of total
average energy of 0.53 MeV –i.e. a fraction of 2.03% of the radiant
energy. The true luminosity should be increased by about 2%. This result is questionable; the respective probabilities of cycles P_1, P_2,
P_3 are function of chemical composition in the central parts of the
Sun.

4. 4a. The line spectrum of the Sun allows us to assign it a spectral
type and a luminosity class: it is a star G 2 V, apparently rather
normal (see Strömgren, in A).

4b. The solar spectrum allows us to determine various quantities;
and the fact that we observe it from center to limb, and during eclip-

ses, only outside the disk, allows a great accuracy. Actually, the chemical composition of the atmosphere, as determined by various authors, taking into account, of course, the important departures from ETL, is slightly different from the central composition, as deduced from global structure conditions. The simplest ideas to explain this in a qualitative way is to say that nuclear reactions, at work in the solar central regions, modify locally the chemical composition, whenever the composition of the outer layers reflects that of the medium from which the Sun has been condensing. This is however not sufficient. Other processes are at work; convection is favoring a mixing, whenever diffusion may have some parting tendancy. We must note that the chemical composition at the center of the Sun can be determined from the total radiation flux, from the total neutrino flux, and from the ratio of helium isotopes in the cosmic rays. It seems that the turbulent diffusion has to increase the hydrogen content in the central region to produce the observable flux of neutrinos, and the observed ratio ^{3}He/^{4}He (Schatzman, Maeder, 1980).

4c. Strictly speaking, the continuous spectrum of the Sun should enable us to know the distribution of temperatures and densities in the solar atmosphere; but we must confess that this diagnosis is still quite provisional, and that, in order to allow us to determine from it the atmospheric "effective temperature" and "effective gravity", we would need a much more refined theory of out-of-equilibrium stellar atmospheres than at present available. Models are rather good in the deep layers, already bad in upper photosphere and quite inappropriate for chromosphere and corona; moreover they do not allow abondance determinations of sufficiently good accuracy. Still much has to be done, along the line of a physics out-of-equilibrium; but we still ignore some basic physic phenomena such as the true cause of the heating of the corona, as the origin of the wind, or as the mechanisms of keeping up the inhomogeneities in density or temperature (see F).

4d. Has the Sun a planetary system? This question may quite reasonably be asked by an astronomer from Sirius...

4e. The same anthropocentrism may apply to such questions as "Is the Sun a variable star?" (read: "would an Earth astronomer consider the Sun as a variable star, if seen from Sirius?). Or: "Is the Sun a magnetic star?".
We shall not go very far in the reply to such questions. Solar planets would not be detected; but one knows more important planetary systems around close-by stars. One knows the recent detection of solar-like activity cycles, in several stars, by Wilson and associates. No doubt that such a type of variability would be detected. But relative changes of luminosity of the order of 10^{-3} would probably go undetected. One is now thinking to observe stellar oscillations; but it is still a very daring type of measurements (Fossat, in E).
The magnetism of the Sun, and its rotation, are weak; the broadening of spectral lines are definitely commanded by collisional damping; only very fine structures studies have allowed the determination of a

strong magnetic field; it is quite unlikely that, from far aside, our star would be known as a rotator or as a magnet.

However the careful study of locations of spots has allowed Trellis (1963) to detect a very minute effect showing the Sun as an oblique rotator, the magnetic axis precessing around the rotation axis by a period of about 50 years.

We have turned, from rather far, around the star Sun. One would like to conclude. What is the situation of solar research nowadays? Frankly, I do not know! Obviously, it is still, and will be for long, a very fascinating subject; and many questions of physical importance can be studied only through the solar studies. But it seems to me that a more profound conclusion may be drawn. The Sun is not an isolated object of study. Sun is a star; as any star ..., it is variable, magnetic, rotating, affected by turbulences, winds, oscillations, radial or non-radial, ...; it is neither stationnary nor in physical equilibrium; as for any star, the approximations of radiative equilibrium, hydrostatic equilibrium, local thermodynamical equilibrium, ... are valid only in very narrow regions. All this is well documented in the solar case; the fast progress of stellar studies often came from the solar research; and many coming improvements in the knowledge of solar physics might well come from the stellar spectroscopy. The Sun is a star, like many others, or unlike many others; the likeness is indeed without importance; what counts is that we get more and more, through its study, the very intense feeling of the connections between the various parts of a whole system. First, the solar interior and its atmosphere are deeply coupled; but how could we separate the physics of planets from that of the Sun? And now that we see many new studies developing, -the flattening, the neutrino flux, the relativistic effects around the Sun-, many appear as more or less linked with our views of the Universe as a whole. One one side, we come to the picture of a Sun which extends for beyond the usual understanding of this word. The solar-terrestrial, or solar-planetary interactions are typical of a coupling internal to the solar system; the role of the main planets in triggering the activity cycle may possibly have to be considered. The whole thing, internal layers, photosphere, chromosphere, gas corona, dust corona, wind, planets, comet ring... constitute one single object, one single star. The degree of degenerescence varies regularily from inside to outside, from quasi-equilibrium conditions to "very-far-from-equilibrium" conditions. One cannot easily treat the various layers of this object without considering the others. This complex object, this all-connected, all-coupled body, it is our Sun. But, on the other side, the Sun is a highly condensed region imbedded in the diluted interstellar medium; through this interaction, its takes part through the evolution of the Galaxy as a whole. And so forth... The solar studies indeed open us to the universal solidarity which gives a sense to the word "Universe", and to its physics.

TABLE I. Constants of the star "Sun".

1. Solar radius $\mathcal{R}_\odot$ = 9.960 10^8 m   ± 0.001 10^8 m

   Earth radius $\mathcal{R}_\oplus$ = 6378140 m   ± 5 m
   Angular solar diameter θ = 1919.28"   ± 0.1"
   Solar parallax (annual) ϖ = 8.794148 "   ± 0.000 007 "
   Astronomical Unit   1 UA = 1.49597870 10^{11} m ± 0.000 000 04 m

Note that the astronomical unit is not equal to the semi major axis of
the earth's orbit. The value of θ and ϖ correspond to 1 UA.

2. Solar mass $m_\odot$ = 1.9891 10^{30} kg   ± 0.0012 kg

   Gravitational Newton constant G= 6.672 10^{11} $m^3kg^{-1}s^{-2}$ ± 0.004
   Earth's mass $m_\oplus$ = 5.9742 10^{24} kg   ± 0.0036
   Ratio $m_\odot/(m_\oplus+m_{\mathbb{C}})$ = 328900.5   ± 0.3
   Solar atmospheric gravity   g = 2.7396 10^2 m s^{-1} + 0.001

Note that the accuracy on the product $Gm_\odot$ is of the order of 10^{-6}.

3. Solar luminosity $\mathcal{L}_\odot$ = 3.861 10^{26} W   ± 0.030 10^{26} W

   Solar constant   S = 1.97 cal cm^{-2} min^{-1} ± 0.03
                           or   S = 1373 W m^{-2} ± 20 W m^{-2}
   Absolute magnitude (visible) M_V= 4.71
   Apparent magnitude (visible) m_V= – 26.86
   Effective temperature T_{eff}= 5783 K ± 20 K

Units are expressed in the system SI (m,kg,s). The astronomical units
are : 1 UA; 1 solar mass; 1 day of 86400 s. The astronomical unit
corresponds, by definition, to the value of the gaussian gravitational
constant defined as k=0.017 202 098 95 $m^{3/2}kg^{-1/2}s^{-1}$.

4. Chemical composition.

Outer layers: H :$\log A_H$=12 (convention)
D :A_D/A_H=5 10^{-6} B :$\log A_B$ = 2.80   N :$\log A_N$=8.19±0.07
Li:$\log A_{Li}$=0.30±0.30   He:$\log A_{He}$=10.92   O :$\log A_O$=8.90±0.15
Be:$\log A_{Be}$=1.15   C :$\log A_C$=8.61±0.12   Fe:$\log A_{Fe}$=7.50±0.30

or X:Y:Z = 0.738 : 0.244 : 0.018 (according Grevesse).

BIBLIOGRAPHY.

1. Classical and general recent reference books on the Sun.

A. Kuiper,G.P., 1952, The Sun, Chicago University Press, U.S.A.
B. Clemence,G.M., 1967, Dynamics of the Solar System, in Condon and
   Odishaw, Handbook of Physics, Mc Graw Hill, 2nd ed. A.S.A.
C. White,O.R., 1977, The Solar Output and its variation, University
   Press, Boulder, Colorado.
D. Rösch,J., 1978, Pleins feux sur la physique solaire, CNRS, Paris.
E. Bonnet,R.M., Dupree,A.K., 1981, Solar phenomena in stars and
   stellar systems, Reidel, Dordrecht, The Netherlands .
F. Jordan,S., 1981, The Sun as a star, NASA-CNRS monograph, NASA
   press, Washington, U.S.A.
G. Hénoux,J.-C., Moriyama,F., 1981, Proc. Japan-France Seminar on
   solar Physics.; and the present volume.
H. Jordan,C., 1981, Oxford Solar conference of EPS.

2. Other references quoted in review.

Ando,H., Osaki,Y., 1975, Publ. Astron. Soc. Japan, 27, 581.
Boornazian,A., Eddy,J., 1979, Bull. Am. Astron. Soc., 11, 437.
Brueckner,G.E., 1981, Space Research XXI, in press.
Brueckner,G.E.,Cook,J.W., Vanhoosier,M.E., 1981, Astron. Astrophys. in
   press.
Caudell,T.P., Hill,H.A., 1979, M.N.R.A.S., 186, 327.
Deubner,F.L., 1975, Astron. Astrophys., 44, 371.
Dicke,R.H., Goldenberg,H.M., 1967, Phys. Rev. Lett. 18, 313.
Duncan,C.H., Geist,J., Willson,R.C., 1980, Science, 207, 177.
Dunham,D.W., Fiala,A.D., Herald,D., Muller,P.M., Sofia,S., 1980,
   Science, 210, 1243.
Endal,A.S., Lesh,J.R., O'Keefe,J., Sofia,S., 1979, Science, 204, 1306.
Evans,J.W., Michard,R., 1962, Astrophys. J., 136, 493.
Fréon,A., Legrand,J.-P., Trellis,M., 1960, Cr. Acad. Sc. Paris, 250,
   2550.
Hill,H.A., Oleson,J.L., Stebbins,R.T., 1975, Astrophys. J., 200, 472
   and 484.
Hudson,H.S., Willson,R.C., 1981, Astrophys. J., 244, L85.
Hundhausen,A.J., 1979, Rev. Geophys. Sp. Phys., 17, 2034.
Kneubühl,F.K., Müller,E.A., Rast,J., Stettler,P., 1980, Astron.
   Astrophys., 87, L3.
Leighton,R., Noyes,R.W., Simon,G.W., 1960, Astrophys. J., 135, 474.
Maeder,A., Schatzman,E., 1980, Cr. Acad. Sc. Paris, 291, Sér.B, 81;
   1981, Astron. Astrophys.
Maunder,E.W., 1919, M.N.R.A.S., 79, 457.
Pecker,J.-C., Roberts,W.R., 1954, J. Geosphys. Res., 60, 33.
Shapiro,I.I., 1980, Science, 208, 51.
Trellis,M., 1958, Cr. Acad. Sc. Paris, 247, 1964.
Trellis,M., 1963, Cr. Acad. Sc. Paris, 256, 2300.
Wilson,O., 1978, Astrophys. J., 226, 379.

RESULTATS D'OBSERVATIONS DU SOLEIL A L'ASTROLABE DU CERGA

F. Chollet
Observatoire de Paris
75014 Paris – France

ABSTRACT.   The Solar Astrolabe is described, and the results of observations made in 1978 and 1979 are presented. Obtained were corrections to the orbital elements of the Sun and a correction to the equinox of the FK4.

Après une description de l'astrolabe solaire, les résultats des observations effectuées durant les campagnes de 1978 et 1979 sont présentés. Une série de corrections aux éléments de l'orbite solaire, ainsi qu'à l'équinoxe du FK4 sont obtenues.

C'est en 1956 qu'entrait en fonction, à Paris, le premier astrolabe conçu essentiellement, par A. Danjon, pour la détermination des paramètres de la rotation de la Terre. Les qualités intrinsèques de l'instrument permettaient une extension du champ de ses activités et, très rapidement, les premiers catalogues d'étoiles étaient publiés (4) ainsi que les résultats de la première campagne d'observations de Mars (1). Les seuls objets brillants du système solaire qui restaient inaccessibles à l'astrolabe étaient le Soleil et les planètes inférieures.

Alors que des méthodes modernes, plus précises, surpasseront bientôt les instruments terrestres d'observation dans bien des domaines, seules les mesures de position du Soleil, malgré leur caractère fondamental, restent encore hors de portée des nouvelles techniques. Le nombre d'instruments dévolus à ce type de mesures n'est d'ailleurs plus réduit qu'à quelques unités et, pour toutes ces raisons, il nous a semblé important et possible d'entreprendre une expérience d'observation du Soleil à l'astrolabe.

L'ASTROLABE SOLAIRE

Il fallait évidemment adapter l'instrument à ce nouveau genre d'observations. On sait que l'astrolabe moderne est constitué d'une lunette horizontale munie d'un prisme diviseur placé en avant de l'objectif. Un

35

bain de mercure qui définit l'horizon, complète le système optique. Ce
montage ne permet que l'observation des astres dont la direction fait,
avec la surface du mercure, un angle fixé par le prisme diviseur. Un
micromètre biréfringent, dernier perfectionnement apporté par A. Danjon,
relié à un chronographe électronique permet la mesure de l'instant de
passage de l'objet observé tout en compensant certains défauts optiques
dû à la séparation des pupilles d'entrée. Enfin, la possibilité d'observer suivant différents azimuts autorise les observations extraméridiennes
et, par conséquent, un même objet peut être observé à ses passages Est
et Ouest, lorsque sa trajectoire coupe le cercle de hauteur.

L'adaptation de l'astrolabe aux observations du Soleil exigeait que
l'on résolve plusieurs problèmes :
- Assurer sa protection par un filtre obligatoirement placé en avant de
l'instrument. Le micromètre au foyer ne résisterait pas au flux concen-
tré d'énergie solaire.
- Préserver au mieux les qualités instrumentales et, essentiellement, la
stabilité de la distance zénithale d'observation.
- Augmenter la durée annuelle des campagnes limitée à trois mois au
CERGA pour la distance zénithale classique de 30°.

L'adoption d'un filtre constitué d'une lame de silice dont la face
extérieure a été recouverte d'une couche de chrome-nickel, a permis de
satisfaire aux deux premières conditions. Le prisme classique, en verre
et travaillant en transmission, a été recouvert d'un traitement semi-
réfléchissant afin de permettre la mesure de l'angle de ses faces d'entrée dont dépend directement la distance zénithale d'observation. L'extension de la durée des campagnes a été obtenue en effectuant des observations à 45° et 60° de distance zénithale. Deux prismes diviseurs sup-
plémentaires, en céramique micro-cristalline donc très stables, travaillant en réflexion permettraient ces nouvelles observations (3 bis). Le
même filtre protecteur, par une simple rotation, pouvait être employé
quelle que soit le prisme diviseur utilisé. Il est ainsi possible chaque
année, d'observer durant 9 mois à 60° de distance zénithale, 6 mois à
45° et 3 mois à 30°.

LES RESULTATS

L'analyse des résidus d'observations de chaque bord du Soleil a été effectuée en deux temps. Un premier calcul, possible presque immédiatement
après la séance complète de mesures quotidienne pour une distance zénithale donnée, fournit une valeur de la correction à l'ascension droite,
au demi-diamètre du Soleil ainsi qu'une combinaison des corrections à la
distance zénithale et à la déclinaison. A la fin de chaque campagne, ces
résultats sont analysés afin d'en tirer les corrections aux paramètres
de l'orbite solaire ainsi que les erreurs systématiques du catalogue de
référence (0). Ce sont ces résultats, tirés des observations effectuées
en 1978 et 1979 qui figurent sur le tableau ci-après.

On remarque immédiatement la disparité de certains résultats, entre
les deux campagnes d'observations. Cela est dû, à la fois, à quelques

SOLEIL CERGA			
EPHEMERIDES A.E. REFERENCE FK4			
CAMPAGNE	1978	1979	AUTRES RESULTATS
Nbre données	55	52	
E	$+ 0^s043 \pm 0^s039$	$+ 0^s087 \pm 0^s020$	$\begin{cases} 0^s052 & (2) \\ 0^s069 & (5) \\ 0^s060 & (3) \end{cases}$
ΔL	$.0^s025 \pm 0^s038$	$0^s050 \pm 0^s015$	0^s043
$\Delta \varpi$	$- 10''5 \pm 4''4$	$- 4''9 \pm 7''4$	$- 9''64$
Δe	$- 0''04 \pm 0''07$	$- 0''29 \pm 0''12$	$- 0''14$
$\Delta \varepsilon$	$- 1''45 \pm 0''65$	$- 0''24 \pm 0''89$	$\begin{cases} 0''01 & (5) \\ 0''04 & (7) \end{cases}$
Δz_{30}	$- 0''71 \pm 1''32$	$0''84 \pm 1''56$	
$\Delta z_{45} - \Delta z_{30}$	$- 2''34 \pm 0''57$	$- 2''38 \pm 0''27$	
D	$- 0''40 \pm 1''11$	$- 0''36 \pm 1''25$	$0''01 \quad (7)$
$\sigma (\Delta \alpha)$ $\sigma (\Delta \delta)$	0^s040 $0''52$	0^s037 $0''37$	

Note: the column "AUTRES RESULTATS" also carries a brace marked (6) grouping the ΔL, $\Delta \varpi$, Δe values (0^s043, $-9''64$, $-0''14$).

problèmes instrumentaux et à la forme des équations de conditions. L'astrolabe solaire a été construit à partir d'un instrument en assez mauvais état. La qualité des images, particulièrement lors des mesures de constantes, était mauvaise. C'est pourquoi nous avons dû, à la fin de 1978, démonter entièrement l'instrument et le remettre en état. La différence des dispersions en α et δ entre 1978 et 1979 confirme, a posteriori, la nécessité de cette révision. Lors du remontage de l'instrument, le filtre protecteur n'a pas été replacé dans sa position initiale, oubli qui n'avait pas une grande importance, mais qui nous a fourni un test indirect de la qualité de l'instrument. La différence des corrections

aux distances zénithales d'observations de 45° et de 30° est restée, à
0"04 près, la même en 1978 et 1979, alors que les corrections elles-
mêmes subissaient une translation de l'ordre de 1"5. C'est exactement ce
que l'on devait attendre d'un instrument à constantes stables.

Pour ce qui concerne le demi-diamètre solaire, les résultats sont
affectés d'une erreur systématique. L'instrument mesure, en fait, la
différence de distance zénithale entre les deux bords observés successi-
vement. Cette différence n'est égale au diamètre solaire que si les points
observés sont effectivement ceux pour lesquels la tangente au bord solaire est horizontale. Cette erreur peut être évaluée et nous espérons,
peut-être dès 1980, fournir chaque année une ou deux valeurs du demi-
diamètre solaire à quelques 0"01 près.

La dernière colonne du tableau présente les mêmes résultats obtenus
par d'autres méthodes. Il faut, compte tenu de ce qui précède, les com-
parer aux seuls résultats de 1979. L'accord est, en général satisfaisant
malgré le très petit nombres de mesures dont nous disposons. Nos résultats semblent affranchis, à la précision obtenue, d'erreurs systématiques
ou tout au moins affectés des mêmes erreurs que les autres déterminations... On notera enfin la valeur assez faible et tout à fait satisfai-
sante de la dispersion.

CONCLUSIONS

Cet instrument prototype a répondu aux conditions indispensables
pour sa mise en observation de routine. On remarquera enfin, que des résultats d'observation du Soleil sont obtenus seulement 2 ans après les
observations. Le seul retard important est dû aux mesures de temps et de
latitude qui ne sont disponibles qu'en fin d'année. Après une dernière
campagne, en 1981, le prototype sera remplacé par l'instrument défini-
tif dans lequel, en particulier, le prisme classique diparaîtra au pro-
fit d'un prisme travaillant en réflexion, plus stable.

REFERENCES

(0) - Chollet F. Thèse. Université de Paris VI. 1981
(1) - Débarbat S., Kovalevsky J., Bull. Astro., 1963, vol.24 n°1,p.69-73.
(2) - Fricke W., 1979, in "Colloquium on European Satellite Astrometry",
      C. Barbieri, P.L. Bernacc Eds., Padoue, Italie.
(3) - Fricke W., 1981, in "Reference Coordinate Systems for Earth Dyna-
      mics", E.M. Gaposchkin, B. Kolaczek Ed. Reidel.
(3 bis) - Laclare F., C.R. Acad. Sc. Paris. 1980, 291 série B, p. 189.
(4) - Guinot B. Thèse. Fac. Sci. Univ. Paris 1958.
(5) - Morrison L.V. Monthly Not. R. Astr. Soc. 1979, 187, p. 41.
(6) - Stumpff P. Astron. Astrophys. 1981, 101, p. 52-71.
(7) - Van Flandern . Astron. J. 1971, 76 n° 1, p. 81.

THE EFFECT OF PERSONAL EQUATION ON MEASUREMENTS OF THE SOLAR SEMI-DIAMETER BY TRANSIT CIRCLE OBSERVERS

Clayton Smith and Daniel Messina
U.S. Naval Observatory

Systematic trends in the semi-diameter of the Sun from measurements made with the Cape Observatory transit circle in 1834 and from 1861 to 1871 and 1884 to 1892 (the only years for which raw observations of the solar limbs were available to us) should probably be attributed to large variations in the personal equation of the individual observers rather than to any real variation of the solar diameter.
An analysis of Washington six-inch observations from 1899 to 1903, 1911 to 1918, 1956 to 1971 and 1975 to 1981 does not support any significant secular change in the solar diameter.

INTRODUCTION

In a recent discussion of Greenwich transit circle observations, Eddy and Boornazian (1979) seem to have found evidence for a significant secular decrease in the solar semi-diameter of the order of magnitude of one to two seconds of arc per century. In an attempt to broaden the scope of the discussion, we surveyed the transit circle literature with the hope of finding other series of observations which would be useful from the point of view of examining solar semi-diameter variations and found Cape and Paris series which had not previously been discussed. A preliminary analysis of the Paris series did not yield a conclusion because of changes of instrumentation and attendant systematic changes of the measured solar semi-diameter. The Cape series was useful for studying the effects of personal equation, so we completely discussed all observations available to us.
In order to confirm a "Naval Observatory" result given by Eddy and Boornazian (1979) and quoted by Lubkin (1979) wherein a secular decrease in the solar semi-diameter exceeding two seconds of arc per century in both the horizontal (SD_H) and vertical (SD_V) coordinates, we opened a new discussion of Naval Observatory material in which we restricted our attention to observations made only with the six-inch transit circle. We did not include observations made with the nine-inch transit circle because the series was flawed by numerous changes in the instrumentation rendering it unsuitable for a study of the solar semi-diameter.

W. Fricke and G. Teleki (eds.), Sun and Planetary System, 39–43.

C. SMITH AND D. MESSINA

TABLE I-ANNUAL MEAN SEMI-DIAMETER BY OBSERVER FROM OBSERVATIONS AT CAPE

Year	Observer	SD_H	ME	N	SD_V	ME	N
1861	GM	961″.05	±0″.14	55	962″.35	±0″.17	37
1862	GM	960.75	0.37	33	962.26	0.25	27
1863	GM	961.19	0.46	42	962.43	0.13	33
1864	GM	961.56	0.16	53	962.23	0.15	52
1865	GM	961.53	0.15	66	962.14	0.15	64
1866	CF	962.61	0.67	7	963.26	0.29	7
	GM	961.87	0.18	56	962.27	0.14	59
1867	CF	962.85	0.60	6	962.67	0.69	6
	GM	961.67	0.15	50	962.24	0.12	54
1868	GM	961.95	0.15	61	962.46	0.12	66
1869	JS	961.35	0.40	10	962.55	0.52	3
	GM	961.70	0.13	55	962.01	0.14	60
1870	JS	961.65	0.14	12	962.26	0.35	9
	GM	961.84	0.12	43	962.32	0.16	36
1884	RP	960.38	0.22	27	961.97	0.15	28
	GM	960.00	0.36	11	961.37	0.20	14
	WC	960.80	0.27	26	962.80	0.17	29
1885	GM	960.30	0.58	9	961.46	0.40	9
	WC	961.80	0.54	6	962.74	0.47	6
1886	RP	961.23	0.21	27	962.11	0.15	32
	GM	960.45	0.16	43	962.37	0.14	47
	WC	962.25	0.35	53	962.58	0.13	53
	WF	961.13	0.29	18	961.43	0.24	18
1887	RP	960.75	0.24	13	961.48	0.19	14
	GM	960.00	0.14	70	962.25	0.12	72
	WC	962.18	0.21	67	962.33	0.09	68
	WF	961.28	0.21	25	961.30	0.19	25
1888	WC	962.85	0.16	45	961.97	0.10	50
	RP	961.05	0.23	28	961.61	0.19	30
	GM	960.45	0.16	42	961.91	0.13	44
1889	RP	961.50	0.26	17	961.62	0.17	18
	WC	962.40	0.17	40	961.83	0.12	43
	GM	960.75	0.19	30	962.09	0.13	32
1890	WC	961.80	0.14	19	961.64	0.14	21
	GM	961.05	0.46	14	962.38	0.26	14
	RP	960.75	0.33	11	961.10	0.26	11
	JP	961.80	0.45	10	962.43	0.27	11
	AC	962.85	0.92	7	962.35	0.21	13
1891	WC	961.65	0.12	16	961.75	0.18	18
	RP	960.90	0.12	15	961.51	0.18	15
	GM	961.65	0.16	23	962.55	0.15	25
	AC	961.50	0.25	14	962.21	0.22	15
	RW	961.80	0.28	11	962.61	0.29	12
	JP	961.20	0.27	9	961.47	0.44	9
1892	RW	961.65	0.26	13	962.54	0.36	13
	WC	961.65	0.14	20	961.65	0.17	21
	GM	962.40	0.31	7	963.64	0.36	7
	JP	961.20	0.25	14	961.70	0.20	15
	RP	960.90	0.17	14	960.86	0.22	15

CAPE RESULTS

Cape observations of the solar limbs were culled from an incomplete sub-set of the annual volumes of Cape meridian observations. Available to us were the results from 1834 (SD_H only), 1861 to 1871 and 1884 to 1892. As described in a progress report by Smith and Messina (1981), observations were corrected for the following effects: 1. Changes in the Earth-Sun separation, 2. Motion of the Sun in right ascension during the time of observation between meridian passage of the east and west limbs, and 3. A term in $\cos\delta$ in right ascension to refer observations at different declinations to the equator. The resulting annual mean SD_H and SD_V values are given for each observer in Table I, except for a value of SD_H for 1834 which was equal to $961\rlap{.}''50$ with 132 observations (123 by one observer, nine by two others).

A linear regression with time, T, in units of a century as the independent variable performed on the values in Table I yields the following results:
$$SD_H(Cape)=(961\rlap{.}''45\pm0\rlap{.}''08)-(0.15\pm0.72)(T-18.762)$$
$$SD_V(Cape)=(962\rlap{.}''16\pm0\rlap{.}''03)-(1.08\pm0.29)(T-18.762)$$
The discordance between the horizontal and vertical results is seen to originate in the personal equation of 'GM' (who was the principal observer during the interval 1861 to 1870) relative to the observers 'JP', 'RP' and 'WC', whose contributions became increasingly important relative to the observer 'GM' in the later years 1884 to 1892.

WASHINGTON SIX-INCH RESULTS

Observations of the solar semi-diameter made in the years 1899 to 1918 were available to us in published volumes. No observational limb material was available to us in any form from 1919 to 1955. From 1956 on, detailed results in computer readable form were again available.

Observations were corrected as described earlier for the Cape observations, and annual means of SD_H and SD_V were formed and are given in Table II with the number of observations and the number of observers contributing to the mean. The mean error of the mean is about $\pm0\rlap{.}''15$ in each year for both SD_H and SD_V and hence is omitted from Table II.

The values of SD_H in the years 1899 to 1903 given in Table II clearly differ in a systematic way from the later results. The reason for this is instrumental, in that those results were obtained with a fixed wire micrometer. All subsequent results were obtained with a moving wire micrometer. Values of SD_V from 1899 to 1903 do not differ significantly from the later values as one would expect, since no change was made in the method of measuring declinations.

From the values given in Table II, but excluding SD_H and SD_V from 1899 to 1903, solutions for linear rates with time were made and gave the results:
$$SD_H(6")=(961\rlap{.}''34\pm0\rlap{.}''04)+(0.14\pm0.15)(T-19.541)$$
$$SD_V(6")=(961\rlap{.}''29\pm0\rlap{.}''03)-(0.10\pm0.12)(T-19.527)$$

TABLE II – ANNUAL MEAN SEMI-DIAMETER FROM WASHINGTON SIX-INCH TRANSIT
CIRCLE OBSERVATIONS

YEAR	SD_H	N_H	SD_V	N_V	OBSERVERS
1899	962."49	86	961."81	86	9
1900	961.96	7	960.96	7	4
1901	962.54	25	962.07	25	4
1902	961.91	129	961.36	127	5
1903	961.65	89	961.07	90	3
1911	961.11	105	961.28	110	4
1912	961.12	139	961.27	142	3
1913	961.17	118	961.40	117	4
1914	961.40	177	961.45	172	5
1915	961.31	171	961.32	170	4
1916	961.52	89	961.36	96	3
1917	961.40	92	961.27	97	3
1918	961.30	64	961.25	67	3
1956	961.41	56	961.41	60	7
1957	961.34	112	961.61	114	5
1958	961.18	104	961.31	108	5
1959	960.97	143	960.85	142	6
1960	961.10	135	961.09	142	7
1961	961.17	114	961.18	133	8
1962	961.28	82	961.43	77	9
1963	961.56	59	961.52	55	7
1964	961.45	93	961.18	83	7
1965	961.29	103	961.30	104	9
1966	961.36	83	961.03	65	9
1967	961.48	76	961.15	78	9
1968	961.44	105	961.28	100	9
1969	961.56	115	961.46	120	11
1970	961.48	117	961.29	109	12
1971	961.72	117	961.45	118	10
1975	961.47	91	–	–	8
1976	961.56	170	–	–	12
1977	961.76	109	961.35	37	10
1978	961.44	154	961.30	133	11
1979	961.05	153	961.30	148	12
1980	961.20	200	961.25	191	12
1981	960.88	22	961.06	21	10

We also see from Table II that six-inch observations never involved fewer than three observers. It is also a fact that since 1911, no individual observer has ever dominated the final system. It should also be mentioned that even in our most recent results, persistent personal equation in the measurement of the solar semi-diameter amounting in some cases to more than 1.5 is still to be found. However, the effect of any one observer on the final result is strongly diminished by the contributions of many observers.

CONCLUSIONS

The Cape results are inconclusive because of the influence of personal equation, and the small numbers of participating observers.

The Washington six-inch transit circle results clearly do not support any secular change whatsoever in the solar semi-diameter. In our opinion, the "Naval Observatory" results given by Eddy and Boornazian have come about by presuming that fixed wire and moving wire SD_H results are systematically similar, and by mixing six-inch and nine-inch transit circle results.

It is appropriate to mention here the conclusion by J.H. Parkinson, et al. (1980) that the Greenwich meridian circle observations are not suitable for the investigation of changes in the solar semi-diameter and the conclusion by A. Wittman (1980) who finds from the agreement between the mean of Tobias Mayer's observations of the Sun, 1756-1760, and photoelectric results obtained in the 1970's no support for a secular change in the solar semi-diameter.

REFERENCES

Eddy, J.A. and Boornazian, A.A. 1979: Secular Decrease in the Solar Diameter. Bull. of the Amer. Ast. Soc., vol. 11, no. 2, p. 437.
Lubkin, G.B. 1979: Analyses of Historical Data Suggest Sun is Shrinking. Physics Today, vol. 32, no. 9, Sept., pp. 17-19.
Parkinson, J.H., Morrison, L.V. and Stephensen, F.R. 1980: The Constancy of the Solar Diameter Over the Past 250 Years. Nature, vol. 288, no. 5791, December 11, pp. 548-551.
Smith, C.A. and Messina, D. 1981: The Horizontal and Vertical Semi-diameters of the Sun Observed at the Cape of Good Hope (1834-1887) and Paris (1837-1906): A Report on Work in Progress. NASA Conference Proceedings CP-2191 on "Variations of the Solar Constant, S. Sophia, ed.
Wittman, A. 1980: Tobias Mayer's Observations of the Sun: Evidence Against a Secular Decrease of the Solar Diameter. Solar Physics, vol. 66, pp. 223-231.

ACKNOWLEDGEMENT

This work was supported in part by a grant from the National Aeronautics and Space Administration, contract no. PCN 961-72273(1C).

A DIRECT UBV COLOR MEASUREMENT OF THE SUN

H. Tüg, Th. Schmidt-Kaler
Astronomisches Institut der
Ruhr-Universität Bochum, FRG

For many astronomical applications the Sun has to be placed on a stellar scale. The main problem of a direct comparison between Sun and stars is to cover the range of more than 10^{12} (30 mag) in brightness. Additional complications arise from the time delay between day and night and the fact, that the Sun is an extended light source while the stars appear as point sources. Ideally the intervening optical elements and the path of light should be the same for solar and stellar observations.

Our approach to measure the solar color indices B-V and U-B is based on a technique shown in the figure. The instrument mainly consists of a 5 m long tube with a movable achromatic lens and a small photometer mounted at the backend. While the stars are observed through an entrance aperture of about 5 cm, for the Sun this aperture is reduced either to 1 mm or 10 μm. The illuminated areas of lens, filters, and photocathode are the same for all observations.

By slight modifications of the instrument the large brightness range was covered in three almost independent modes:

   (a) by a pinhole (1 mm $\emptyset$) and two additional reflecting
       filters ($\tau = 10^{-2}, \tau = 10^{-3}$)
   (b) using the moon as a natural link between Sun and stars
   (c) in a diffraction mode with a 10 μm entrance aperture

The Sun was compared with 25 bright stars including 3 primary and 4 secondary Johnson-Morgan standards. If the three observing modes are given equal weight the final results are $+0^{m}.686 \pm 0^{m}.011$ for B-V and $+0^{m}.183 \pm 0^{m}.020$ for U-B. A detailed error analysis and description of the procedure is given by Tüg and Schmidt-Kaler (1981). A comparison of our results with the only direct measurements of Gallouët (1964) shows good agreement. Unfortunately Gallouët's work was not taken seriously, because it did not fit to the well established conception of spectroscopists, that the Sun is a G2V star with a color index

45

W. Fricke and G. Teleki (eds.), Sun and Planetary System, 45–46.
Copyright © 1982 by D. Reidel Publishing Company.

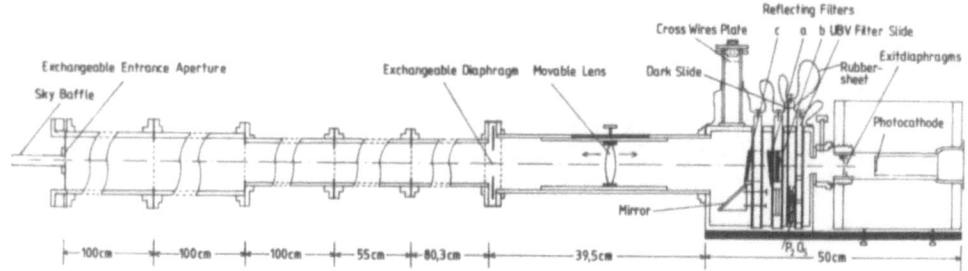

Sun reducer and photometer device

B-V = $0.^m62$. But this cannot be maintained any more.

The diffraction mode (c) might be of more common interest, because it allows to measure and compare the Sun directly with stars in the same optical path of light and with the same optical elements. It is based on the reduction of the solar brightness by a pinhole of typically 5 - 20 μm in size. Only the innermost part of the solar diffraction pattern is taken out by the lens. The strong wavelength dependence of the pattern maximum requires a correction which for a small round aperture is very well understood. A detailed description of this technique is given by Tüg (1981).

In connection with a small Cassegrain telescope the method is promising for many astronomical applications like the comparison of solar and stellar energy distributions, albedo measurements of planetary objects, photometry of the Sun in any filter system, and variations of the solar irradiance. A small version of the Sun reducer might also be useful for various observations from space.

REFERENCES

Gallouët, L., 1964, Ann. D'Astrophys. <u>27</u>, 423
Tüg, H., Schmidt-Kaler, Th., 1981, in press
Tüg, H., 1981, in press

NUCLEAR ASTROPHYSICS OF THE SUN

G.E. Kocharov
Physico-Technical Institute of the Academy
of Sciences, Leningrad, 194021, USSR

ABSTRACT.   An analysis is given of the state of the art and
prospects of investigations of high energy processes in
solar matter.

SOLAR NEUTRINOS

According to the available experimental data, the rate of the reaction
$^{37}Cl(\nu, e^-)^{37}Ar$ is at least three times lower than predicted by the
recent solar models. To remove the obvious discrepancy between the
standard theory and observations, several possibilities were proposed.
We will consider only the possibilities connected with the topics of
this meeting (for details see Kocharov, 1980).

The observed solar neutrino count rate can be explained if the
core of the Sun is colder than it is usually assumed. It was shown
(Kocharov and Starbunov, 1970) that this possibility is real if the
abundance of helium-3 in the solar interior is higher than given by
the standard models. For the weight abundance $X_3 = 1$, 0.1 and 0.01 %
the rates of the reaction $^{37}Cl(\nu, e^-)^{37}Ar$ are   0.034, 0.048 and
0.85 SNU accordingly. To explain the value (2.2 $\pm$ 0.4) SNU which may
be considered as the solar neutrino effect X_3 should be 0.005 %.
Therefore, if the abundance of helium-3 is of the same order as in the
solar wind the discrepancy between the experiment and theory will be
removed. Is it possible? If helium-3 burning $^3He(^3He, 2P)$ ^4He provided
the solar luminosity during all the history of the Sun the ^3He
abundance in the primary Sun would have to be a few percent which is
not ruled out now but it appears unlikely. If ^3He is the primary
nuclear fuel over part of the Sun's life, the hydrogen-burning age is
reduced and would decrease the solar neutrino flux. Even if the
primordial X_3 was negligible the role of helium-3 may be essential
because the hydrogen burning inevitably leads to production of
helium-3. In the frame of the current solar models the weight
abundance of helium-3 increases with increasing the distance from the
centre of the Sun reaching ~ 1 % at half the solar mass. If a

47

W. Fricke and G. Teleki (eds.), Sun and Planetary System, 47–52.

continuous or periodical flux of helium-3 to the central region took
place the central temperature of the present Sun would be lower and
accordingly the fluxes of high energy neutrinos would be lower too.

The main factor determining the opacity is the heavy-element
concentration (Z) in the solar interior. It is usually assumed that
the primary Sun was homogeneous throughout its volume and that the
composition of the primary Sun was the same as that of the surface at
the present time. The solar neutrino puzzle has forced us to sacrifice
this generally accepted assumption; solar models have been constructed
in which the heavy element content in the interior have been taken
15 times less than the value for the photosphere, which reduced the
predicted neutrino count rate to 1.4 SNU (Bahcall and Ulrich, 1971).

It is important to implement an experiment for the detection of
low energy neutrinos using the reaction $^{71}Ga(\nu, e^-)$ ^{71}Ge. In the
frame of the current solar models PP neutrinos give the main
contribution (65 %) to the rate of this reaction.

These two possibilities of removing the discrepancy between the
theory and experiment should manifest themselves in different ways in
experiments with a low threshold detector. Therefore the availability
of results in even two detectors should enable theoretical progress
to be made.

Experiments based on the reactions $^7Li(\nu, e^-)$ ^7Be, $^{71}Ga(\nu,e^-)$ ^{71}Ge,
$^{115}In(\nu, e^-)$ ^{115}Sn have been developed now. Taken together, the
results of the four experiments (^7Li, ^{38}Cl, ^{71}Ga, ^{115}In) should allow
us to measure the physical parameters of the solar interior
(temperature, density, composition).

SOLAR FLARES ENRICHED IN HELIUM-3

The extra high ratio of the ^3He fluxes should be considered as
remarkable in the corresponding solar flares. Comparing with the
solar wind abundance the enrichment of ^3He attains even 10^4. How
and where the isotope separation occurs are the most important
questions.

A possible mechanism of helium isotope separation in the solar
atmosphere due to plasma effects has been proposed and developed
during the recent 4 years (see Kocharov, L. and Kocharov, G., 1981 a,b).
As a result, a concrete model of the corresponding physical processes
in the solar plasma has been proposed. Firstly, the electrons are
accelerated up to $\geq$ 10 keV. If a sufficiently large number of these
electrons penetrates the cold plasma of the upper chromosphere, the
electron shock will be formed. An ion-acoustic turbulence ahead of the
wave exists. High-energy electrons ($\geq$ 20 keV) will shoot through the
ion-acoustic front, generating before the front a dense beam, which is
unstable. As a result a strong Langmuir turbulence is generated. It
heats quickly the thermal electrons and forms a layer of non-

isothermal plasma just before the ion-acoustic front, which propagates through this layer. Preferential ^{3}He preheating will be produced in this front by nonlinear scattering of ion-sound turbulence.

The results of the comparison of the theory with the experiments are as follows:
1. Large enrichment of helium-3 relative to helium-4 and the absence of detectable fluxes of deuterium and tritium are explained in the frame of the developed model.
2. According to experimental data the highest helium-3 enrichment occurs in events with small absolute fluxes of particles: $K_{3,4} \sim I(P)^{-0.4}$ and $K_{3,4} \sim I(^3He)^{-0.8}$. The plasma theory predicts these dependences not only qualitatively but also quantitatively.
3. In the frame of the considered model the flares enriched in helium-3 should be accompanied by emission of X and radio-waves. It is shown that there should be a correlation between the value $Y = [I(^3He)/I(^4He)]^{16/7} I(^4He)$ and the fluxes of X-ray and microwave radiation. An analysis of the available data shows that such a correlation does indeed exist. This means that in a ^{3}He-rich solar flare there is a connection between the acceleration of nuclei and electrons.
4. A plasma mechanism in principle can explain the simultaneous enrichment in helium-3 and heavy elements. It is important to measure the degree of ionization and the energy spectra of heavy nuclei in the low energy region.
5. After the quick heating of the upper chromosphere the excitation of ^{4}He occurs simultaneously with ionization by electrons. As a result the heating of the chromosphere should be accompanied by emission of UV radiation in He II line. The intensity of this line should be connected linearly with Y. Thus observation of the He II line give us directly the ^{3}He preheating region.

In order to check and define more precisely the proposed helium-3 rich flare model it is desirable to observe hard and soft X-ray bursts (especially with a high temporal and spatial resolution), microwave continuum, radio-bursts in different wave-length regions, UV radiation in He II line and traditional experiments on isotopic and charge composition and the energy spectrum of accelerated particles.

SOLAR γ-RAYS

Solar gamma rays contain information about the timing of acceleration of protons and nuclei in flares, in situ energy spectrum and angular distribution of solar cosmic rays, density, temperature and composition of solar matter.

Solar gamma lines were first observed by Chupp et al. (1973) during the flare of August 4, 1972. Very important results were obtained recently during the operation of the Solar Maximum Mission spacecraft (February-December, 1980). Here we will consider mainly the results of the X- and γ-radiation from the solar flare, June 7, 1980 (Evenson et al., 1981; Forrest et al., 1981; Pesses et al., 1981; von Rosenvinge et al., 1981; Simnett, 1981).

An analysis of the available data leads to the following conclusions:
1. In all hard X-ray energy intervals 26-52 keV, 52-118 keV, 118-228 keV, 228-386 keV there are series of multiple impulsive spikes at around 03.12, 03.14 and 03.16 UT. The width of the individual spikes is qualitatively similar in all sets. The first series lasted about 70 s at the event onset. The number of subpulses in the first series is 7. The modulation depth is typically a factor of 4 during the first series of spikes. Forrest et al. (1980) noted, that the energy interval 4.1-6.5 MeV (covering the prompt lines 4.43 MeV and 6.15 MeV) also exhibited a deep modulation, like X-rays at least for the first 5 bursts.

It is clear that both the high energy electrons and ions interacted with the solar atmosphere simultaneously (within 1 s) with some subpulses of ~ 4 sec duration each and with a quasi-period between subpulses of 10 seconds.
2. Taking into account the resolution time (16.38 s) in the frame of the theoretical model of neutron generation, their energy spectrum and the character of thermalization with using the observed time history of the 2.223 MeV line we can say that the generation of neutrons began at 03.12 UT and lasted 45 sec. Therefore, the onset of neutron generation coincides with the onset of the first burst and the duration includes the first 4 spikes. The maximum flux is $(7.1 + 1.2)$ 10^{-2} cm^{-2} s^{-1}. Total duration is 500 sec. Production of 4.10^{23} neutrons is required if they were produced isotropically (Chupp et al. 1981).
3. The time history of the prompt lines 4.43 MeV (^{12}C), 6.13 MeV(^{16}O) and unresolved lines from Mg,Si and Fe(1-2 MeV) and CNO(4-7.1 MeV) in 45 sec shows the presence of four quasiperiodic spikes each of a few second duration. The fluxes of 4.43 and 6.13 MeV are $(2.7 \pm 1.3)10^{-2}$ and (2.7 ± 0.8) 10^{-2} cm^{-2} sec^{-1} respectively.
4. The flare spectrum consists of a power law continuum with the index -3.2 together with a collection of resolved and unresolved gamma ray lines extending up to 7.1 MeV.
5. The hard X-ray continuum accounts for ~100 % of the observed counting rate near 300 KeV, ~50 % near 1 MeV and ~10 % in the 4.1-6.4 MeV range. Hence, Chupp et al. (1981) concluded that the counting rate up to ~300 keV will map the interaction rate profile of energetic electrons while the counting rate in the 4.1-6.4 MeV range will map the interaction rate of energetic ions >30 MeV.
6. The total energy of all protons estimated from the observed flux of 4.43 MeV line equals $3 \cdot 10^{28}$ erg (Rosenvinge et al., 1981). This value is 100 times more than the value obtained using the data of direct measurement of protons by means of ISEE-3 and HELIOS-1 space-craft. Therefore, only a small part of all accelerated particles

escaped into interplanetary medium.

7. Very interesting results were obtained by Evenson et al. (1981)
in experiments on the detection of electrons (5-100 MeV) and protons
(25-145 MeV) during SMM. An analysis of the data for 17 flares shows
that when flares are accompanied by γ-rays, the ratio of electrons
and protons is too high, ~ 0.2 (usually $Ne/Np \simeq 0.01$).

The similarity of fine temporal structures of X- and γ-radiations
and the enhanced electron to proton ratio in the wide energy interval
allow us to suggest that in the present case a model similar to that
for a helium-3 rich solar flare may operate. Firstly electrons are
accelerated, which penetrate the dense layer of solar atmosphere and
create the necessary conditions for the acceleration of protons and
various nuclei.

The most unexpected result is the simultaneous acceleration of
electrons and protons. This unexpectedness is connected with the
opinion that the delay (a few minutes) between the prompt gamma-rays
and X-radiation was experimentally proved for the solar flare of
August 4, 1972. An analysis, however, shows that the onset times of
the X-rays 29-41 keV, gamma-rays 0.35-8 MeV and microwave fluxes
37 GHz were the same. The difference was in the rise time of the
radiations. A slow increase of γ-ray intensity may be a result of
proton and heavy nuclei propagation effect. It is shown by Kocharov
et al. (1981) that the experimental data on the solar flare of
August 4, 1972 may be explained in the frame of assumption that
electrons and heavy particles were accelerated simultaneously and
generation of gamma-rays took place in rare medium with 10^{10} particles
cm^{-3}. With increase of density the rise time of γ-rays decreases
approaching the rise time of X-rays (the case of the solar flare of
June 7, 1980). It is important that the main part of accelerated
particles escaped into interplanetary medium in the flare of
August 4, 1972, and in the flare of June 7, 1980 only a very small
part of particles escaped from the Sun. These data support the above
possibility because for the former flare γ-ray generation took place
in rare medium, and in the latter flare, in dense medium.

CONCLUSION

Out of a number of interesting problems concerning the nuclear
astrophysics of the Sun we have considered only three. It is not
accidental. Namely in the discussed cases unexpected results were
obtained which immediately lead to a necessity of a critical re-
consideration of conventional notions about the physical processes
occuring in different regions of the Sun. As a result, both the
theory and experiment were developed considerably. In all cases it is
clear that integrated experiments with a detailed investigation of
all the accompanying phenomena are necessary. For the neutrino
astrophysics of the Sun and solar flares enriched in helium-3 such
kind of experiments are being done. As to unexpected results on the
fine structure of solar flare X- and γ-radiation, undoubtedly they

will provide a new impetus for theory and experiments.

REFERENCES

Bahcall, J.N., Ulrich, R.K.: 1971, Astrophys.J., 170, 593.
Chupp, E.L. et al.: 1973, Nature, 241, 333.
Chupp, E.L. et al.: 1981, Astrophys.J. Lett. 244, L171.
Evenson, P., Meyer, P., Yanagita, S.: 1981, Proc. XVII ICRC, 3, 32.
Forrest, D.J. et al.: 1980, Bull.AAS, 12, 890.
Forrest, D.J. et al.: 1981, Proc. XVII ICRC, SH 1.2-2.
Kocharov, G.E.: 1980, Nuclear Astrophysics of the Sun. Verlag
Karl Thiemig, München.
Kocharov, G.E., Starbunov, Yu.N.: 1970, Acta Phys. Hung., 29, 353.
Kocharov, L.G., Kocharov, G.E.: 1981 a, Preprint FTI-722; b. Proc.
IUPAP IAU, Symp. No. 94, Origin of Cosmic Rays, p. 393.
Kocharov, G.E., Kovaltsov, G.A., Kocharov, L.G.: 1981, in press.
Pesses, M.E. et al.: 1981, Proc. XII ICRC, 3, 36.
von Rosenvinge, T.T. et al.: 1981, Proc. XVII ICRC, 3, 28.
Simnett, G.M.: 1981, Invited Lecture at the XVII ICRC.

THE SOLAR NEUTRINO PUZZLE

George Marx
Department of Atomic Physics,
Eötvös University, Budapest

ABSTRACT: The thermonuclear fusion H→He is generally accepted as main source of solar luminosity. The neutrinos created
in the fusion chain contain direct information about the composition and temperature in the core of the Sun. Experiments
performed by Raymond Davis indicate a significant solar neutrino flux, but the detection rate is smaller than predicted
by the standard solar model. The talk discusses the possible
reasons of this discrepancy and the line of future actions.

After a century long controversy Bethe has called the attention to the fact, that nuclear matter of the universe is
far from equilibrium composition, so thermonuclear reactions
can explain the energy generation of stars. In the case of
Sun and other main sequence stars the fusion of hydrogen into
heavier elements (like helium) is the source of energy. This
has led to the construction of the standard solar model. A
gas sphere of the mass of the Sun, with the chemical composition of the present solar photosphere is supposed to be in
thermal and mechanical equilibrium. The weight is balanced
by the gas pressure, this pressure presupposes a high temperature, at such a high temperature thermonuclear reactions become possible. Thermonuclear reactions change gradually the
chemical composition of the central region, which results a
stellar evolution. When the slowly increasing luminosity
reaches the present value of 3.86×10^{26} watt, the integration
stops. One considers the present state of the Sun to have
been explained. The original chemical composition was

$$\text{H: } 73\% \qquad \text{He: } 25\% \qquad \text{else: } 2\%$$

The theoretical integration stops at 4.7 billion years, when
the central temperature of the Sun amounts 15 million degrees,
the dominating energy generating reaction is the H→He fusion,
and during the 4.7 billion years of solar history the composition of the central regions changed to

$$\text{H: } 41\% \qquad \text{He: } 57\% \qquad \text{else: } 2\%$$

The primordial hydrogen is made of protons. Helium and

W. Fricke and G. Teleki (eds.), Sun and Planetary System, 53–58.

other composite nuclei consist 50% protons and 50% neutrons. So the thermonuclear fusion is connected with the transmutation of half of the protons to neutrons. In the transmutation positrons are produced, but due to the conservation of leptonic charge each positron is accompanied with a neutrino:

$$4H \rightarrow He^{++} + 2e^+ + 2\nu.$$

The formation of a helium nucleus liberates 23.28 MeV. In order to provide the present luminosity, 10^{37} new helium nuclei must be formed inside the Sun each second, which are associated with the emission of $2 \cdot 10^{37}$ neutrinos per second. The Sun is practically transparent for neutrinos, so at the Earth one expects a flux of 10^{11} neutrinos per cm^2 per sec. Their observation offers a test of the standard solar model, especially a very direct test of the assumption that the solar energy is supplied by H→He fusion.

THE SOLAR NEUTRINO DETECTOR

Neutrinos can be and have been detected by reaction recommended by Bruno Pontecorvo (1946) and Louis W. Alvarez (1949):

$$\nu + {}^{37}Cl \rightarrow {}^{37}A + e^-.$$

These neutrino induced reactions can be detected by counting the radioargon decays:

$$^{37}A \rightarrow {}^{37}Cl + e^+ + \nu.$$

The Cl nucleus is stable against spontaneous beta decay. The captured neutrino needs a threshold energy of 0.81 MeV to induce the beta transition (1). Unluckily enough, the main fusion sequence does not give neutrinos of such a high energy:

$$^1H + {}^1H \rightarrow {}^2H + e^+ + \nu \qquad \text{(pp neutrinos, } E \leq 0.42 \text{ MeV)}$$
$$^2H + {}^1H \rightarrow {}^3He + \gamma$$
$$^3He + {}^3He \rightarrow {}^4He + {}^1H + {}^1H.$$

But 0.25% of the ^2H formation goes via triple collision:

$$^1H + {}^1H + e^- \rightarrow {}^2H + \nu \qquad \text{(pep neutrinos, } E = 1.44 \text{MeV)}$$

and these neutrinos would produce reaction (1) on Earth with a probability 0.3×10^{-36} radioargon per ^{37}Cl atom per sec. Let us call 10^{-36} radioargon per ^{37}Cl per sec = 1 SNU, this unit will be used later on. To visualize it: it means the production of a single radioargon nucleus per day in a swimming-pool quantity of chlorine compound. Too bad!

William Fowler discovered that if the central solar temperature is 15 million degrees indeed, and if 50% of the central materials is ^4He, there is a chance also for the

$$^3He + {}^4He \rightarrow {}^7Be + \gamma$$

reaction. ^7Be is not stable at this temperature, it captures either an electron:

$$^7Be + e^- \rightarrow {}^7Li + \nu \qquad \text{(Be neutrinos, } E = 0.861 \text{ MeV in 90\%)}$$
$$^7Li + {}^1H \rightarrow {}^8Be \rightarrow {}^4He + {}^4He;$$

or it captures a proton:

$$^7Be + {}^1H \rightarrow {}^8B + \gamma,$$

$$^8B \rightarrow \,^8Be + e^+ + \nu \qquad\qquad \text{(B neutrinos, } E \lesssim 14.06 \text{ MeV)}$$
$$^8Be \rightarrow \,^4He + \,^4He.$$

The energetic boron neutrinos have a capture cross section
by four orders of magnitude higher than the pep neutrinos,
so if each He fusion went via the boron branch, the chlorine
capture rate would be 4000 SNU (Marx-Menyhárd 1960). Let us
take e.g. $1m^3$ of C_2Cl_4 (an inexpensive fluid used in "dry
cleaning"): one would expect the production of two radioargon atoms per day in this volume! The first trial experiment was performed by John Galvin and Raymond Davis by $4 m^3$
C_2Cl_4 in a limestone mine 700 m deep and they found 3 ± 5 radio-
argons per day (1960).

The "boron branch" involves proton capture by Be nucleus
(charge number 4), this can happen only via Gamow tunnelling
through the Coulomb barrier, what is very temperature sensitive and very much suppressed. According to the standard solar model only 14% of the fusion goes via Be-Li and only
0.02% via Be-B. This has the consequence that the expected
detection rate is (according to John Bahcall, 1981)

> 0.0 SNU from pp neutrinos
> 0.3 SNU from pep neutrinos
> 1.7 SNU from Be neutrinos
> <u>5.0 SNU from B neutrinos</u>
> 7 ± 2 SNU from all the solar neutrinos.

The large scale experiment started in 1970. Raymond Davis
of Brookhaven built a $380 m^3$ tank, he filled it with purified
C_2Cl_4 and waited to reach the saturation value of radioargon.
(The half life of ^{37}A is 35 days.) After 100 days he collected
the argon from the tank by He bubbling and filled a miniature
portional counter with it, so he became able to observe each
radioargon decay with an efficiency better than 90%. The expected signal is a few ^{37}A atoms/day. In order to suppress the
cosmic radiation background (due to muons and cosmic ray produced neutrinos), the whole neutrino observatory was installed
in the Home Stake Gold Mine of South Dakota, 1500 m deep,
where the cosmic radiation background amounts about 1/10
^{37}A/day. The experimental results are shown on the Figure.
In the past decade the solar neutrino detector had its ups
and downs. The signal was below noise level for years. Now
a signal value 1.0 ± 0.2 SNU is accepted, which is five times
the statistical error. But it is only a fraction of the expected value! This is the solar neutrino puzzle.

POSSIBLE EXPLANATIONS

Different specialists tried to find error at different
places. As a rule: everyone was pointing to the field of
someone else. One possible candidate was argon chemistry.
(The argon atom is indifferent chemically, but it is produced

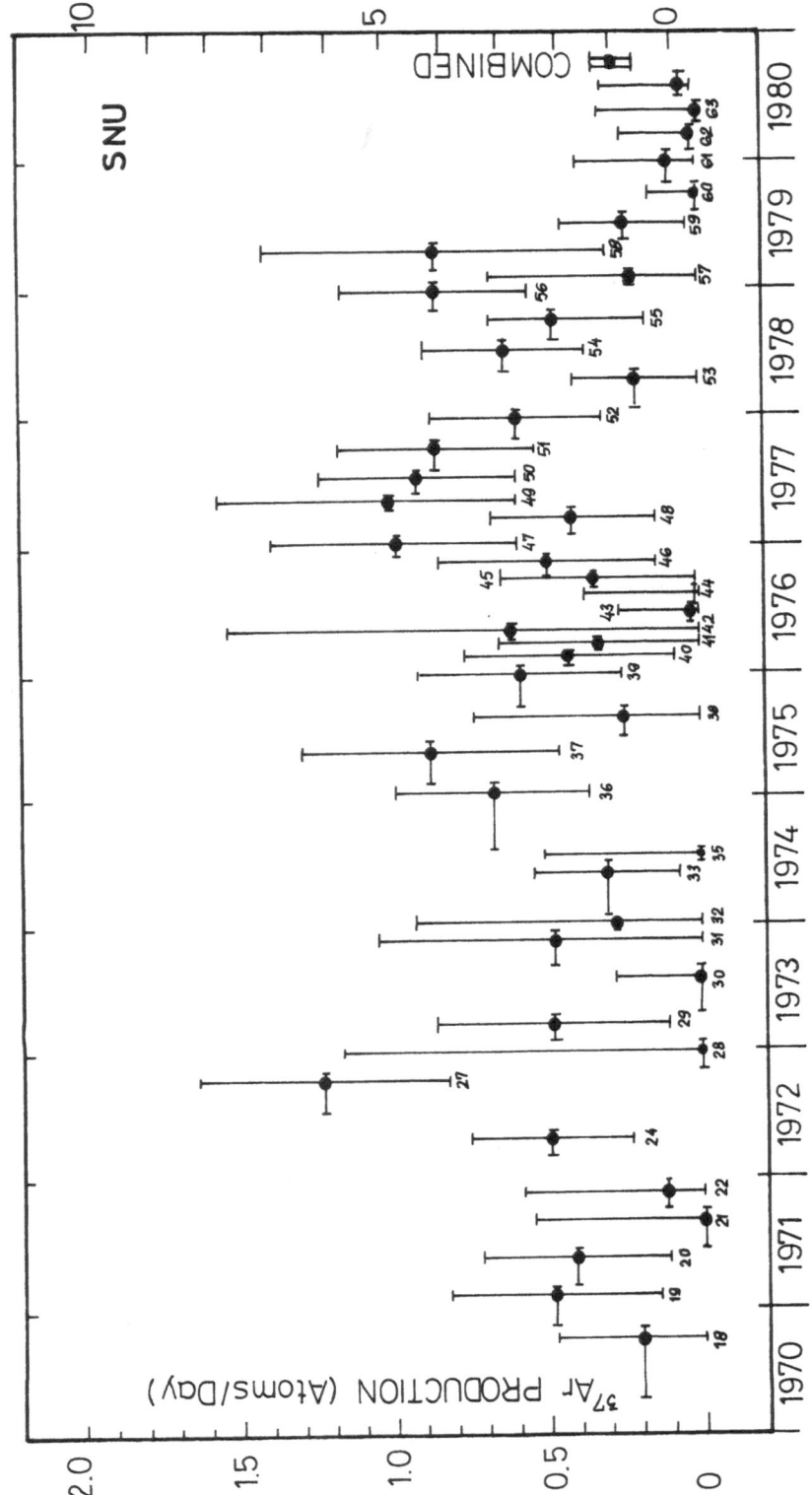

as argon ion! The ion may break up the double bonds of C_2Cl_4 and the produced polymer may encage the argon ion. But control experiments with A^+ ions eliminated this explanation.) An other suggested place of error was nuclear physics. (E.g. if the cross section of $^3He+^3He$ is larger at very law energies, than expected, e.g. due to the formation of ^6Be resonance, the $^3He+^3He$ branch may be enhanced with respect to the $^3He+^4He$ channel. But in other reactions no indication was found for the existence of a ^6Be resonance.) A third accused partner was the neutrino, and it is still sitting in the prisoner's box. (Up to now the fate of neutrino was studied in lab only on a few meters of pathway. The solar neutrinos have to run 8 minutes with the speed of light, to reach Earth. If neutrinos were unstable, they would never reach us. But to become unstable, the neutrinos have to have a finite rest mass, a new hyperweak interaction to make them decay or oscillate, and new light particles as their decay products. Three assumptions seem to be high price for the explanation of a single observational evidence! Since the last year, however, there are several indications for a neutrino mass, - see G. Marx-G. Szalay 1977 and Trètyakov et al. 1980 - consequently this escape route is not completely blocked. One still needs a firm proof that neutrinos are longlived enough to serve as astronomical information carriers.)

If one accepts the chlorine solar neutrino experiment, its message says that less high energy neutrinos are produced in the Sun, than predicted by the standard solar model. The simplest explanation is that the very temperature sensitive B branch is supressed. The neutrino thermometer shows the central part of the Sun cooler than expected.(Allowing only 12 million degrees instead of 15 million degrees.) But at lower temperature the observed solar luminosity has to be explained on an other way, e.g. by importing more H fuel to the central region. An other possibility is, eliminating He somehow from the central region, to block the $^3He+^4He \rightarrow ^7Be$ pathway. Both could be realized by a mixed-up Sun. But these modifications would mean giving up the consistency of the standard solar model.

For theoretical astrophysicists it is certainly bad news that the first consistency check of the standard solar model has gone wrong. How about other consistency checks?

A widely discussed possibility is the solar seismology. The eigenfrequences of the Sun depend on its density distribution, obtainable from the solar model. But the unique identification of the observed frequencies with definite oscillations modes is not yet available up to now.

An other interesting testing field is paleoclimatology. The standard solar model predicts a 5% increase of the solar luminosity per billion year, connected with the increase of He aboundance. This might result observable consequences in the climatic history of planets. (E.g. a smaller solar constant might give an ocean temperature below freezing point. The large albedo of snow and ice would prevent the absorbtion of heat enough to melt the oceans. (G. Marx 1978 and 1980)

The most dramatic idea, suggested by the discrepancies of the standard solar model, is that the Sun is not stable at all. As Dilke, Gough (1972) guessed that a sudden mixing and homogenizing the solar material might produce considerable changes in the neutrino luminosity and modest changes in the optical luminosity.

NEW EXPERIMENTS

Davis is firm that ^{37}A atoms are created in his tank in the number given above. But he cannot be sure that they are produced by solar neutrinos, proving the stability of these messengers. Davis is unable to run a controll experiment with the Sun turned off. The detector is not sensitive enough to observe the small oscillation of neutrino intensity, due to the excentric orbit of Earth. So there is evidently a need to make other solar neutrino experiments.

Option one: To repeat the chlorine experiment with a tank ten times larger, to increase its sensitivity. But the sensitive detector must be moved to the deepest mine of the Earth, to suppress the cosmic ray background.

Option two: using gallium instead chlorine as target.

$$\nu + {}^{71}Ga \to {}^{71}Ge + e^- \qquad {}^{71}Ge \to {}^{71}Ga + e^+ + \nu \qquad \text{(threshold: 0,233 MeV),}$$

Due to the very low threshold this experiment can catch all types of neutrinos. A Brookhaven-Irvine-Pennsylvania-Heidelberg-Revohot collaboration plans to perform this experiment with 50 tons of gallium. (Gallium is a precious metall, used in light emitting diodes. E.g. Hungary produces it as by product of aluminium as a level 1 ton per year, for a price 500 Dollars/kg.) An other gallium experiment is in preparation in the USSR.

Option three: investigation of geologic deposits of thallium, to search for an evidence of

$$\nu + {}^{206}Tl \to {}^{206}Pb + e$$

transition, which can give an average value for the solar neutrino luminosity in the past few million years (Mel Freedman 1978), which is a valuable information concerning the variable Sun idea.

REFERENCES

Marx G.-Szalay A.S.1977:J.of Astronomy a.Astrophys.49.p.437.
Marx G.-Menyhárd N.: 1977, Science 131, p 299
Marx G. 1978: Acta Astronautica 6.p.221
Marx G.-Miskolci F. 1981: Adv. in Space Research 1.p.5

SOLAR MODELS AND NEUTRINO PROBLEM

N. KIZILOĞLU and D. EZER
Department of Physics, Middle East Technical University
Ankara, Turkey

Abstract: The discrepancy between observed and predicted neutrino capture rate has led us to consider another possibility about the solar
interior. The evolutionary sequence of standard models and contaminated
solar models are constructed. Both evolutionary study begins with the
threshold of stability. For the contaminated models, the surface is
assumed to accrete material from the interstellar medium during the
evolution at a rate to increase the heavy element abundance to its present observed value. The predicted total solar neutrino fluxes are
3 .48 SNU and 4.68 SNU for accreted and standard solar models, respectively, which are still larger than the observed in David's Cl-37
experiment.

The discrepancy between predicted and observed neutrino capture
rate from the sun is still a problem. The origin of this diwagreement
is unknown; whether the fault is in modelling of the sun or in the physical theory for the propagation of neutrinos. Many variations of solar
models have been tested in order to reduce the total predicted neutrino
flux to the observed value(See review articles by e.g. Bahcall and
Davis, 1976; Kuchowitz, 1976). Some of them are the sudden mixing mechanism suggested by Ezer and Cameron (1972), rotating core model(Ezer
and Cameron,1968), the effect of radiative opacities(Carson, Ezer and
Stothers,1974), the magnetic field effect(Chitre, Ezer and Stothers,
1973) and the effect of heavy element abundance in the solar composition
(Bahcall and Ulrich,1971). All the above suggestions have yielded a low
value of neutrino flux for the sun.

In the construction of solar models, it is always assumed that
the primordial solar abundances are identical with the present photospheric abundances. It has been suggested that the sun has accreted
materials from the interstellar medium during its evolution(Hoyle and
Lyttleton,1939; Aumann and McCrea,1976; Newmann and Talbot,1976).
Therefore, surface abundances might not reflect the abundances deep in
the interior. Based on this point of view, solar models with low initial heavy element abundances are constructed assuming that the surface
is contaminated by accreting interstellar material during the evolution

59

W. Fricke and G. Teleki (eds.), Sun and Planetary System, 59–62.

at a rate to increase the heavy element abundance to its present observed
value throughout the convection zone(Christensen-Dalsgaard et al.,1979).
In this study, the evolutionary calculation starts from the zero-age
main sequence and the p-p chain and the CNO-bicycle reactions are assumed to be always in nuclear equilibrium throughout the sun. In the
present work, the evolutionary sequence of standard solar models and the
so called contaminated solar models are constructed by calculating
in detail 1) the pre-main sequence stages of evolution starting from
the threshold of energy stability, and 2) the approach to the equilibrium
of nuclear reactions in various branches of the p-p chain and CNO-bicycle
throughout the interior of the sun. Hence, the energy sources are: gravitational contraction, the p-p chain, and the CNO-bicycle reactions.
But energy generation from different branches of p-p chain and CNO-
bicycle, are considered seperately until He-3 reaches its steady-state
value in the p-p chain and until C-12 is depleted toward its equilibrium
value with N-14, respectively, at each mass shell of the solar model.
When He-3 comes into equilibrium with other nuclei in the p-p chain and
equilibrium abundance is reached between C-12 and N-14 nuclei, then the
energy generation rates are governed by their nuclear equilibrium values.
Throughout the evolutionary study the formation and destruction of H,
He-4, C-12, N-14, O-16, and He-3 which is important for neutrino fluxes
are followed explicitly. The nuclear parameters are obtained from
Fowler et al. (1975) and the pep reaction is also included. The electron
screening factors for the thermonuclear reactions are calculated in the
way described in Ezer and Cameron (1967). The rest of the input physics
is the same as in Ezer and Cameron (1971). The ratio of mixing length
to pressure scale height,α: l/h is varied in order to get right standard
model for the present sun. The same value of α is used for both standard
and contaminated solar models.

       The evolutionary study was carried out by the Henyey method
which was applied from the center all the way to the surface as described
by Ezer and Cameron (1971). In this way the accretion of matter onto the
surface of the sun can be handled in a better way. Two different sets
of opacity tables are used. The first set for three hydrogen-helium
mixture in which the hydrogen-helium ratios are 4, 1, and 0; is required
for the standard solar model. These opacities were made available by
running the 1970 version of the Cox opacity code at the Institute for
Space Studies. The second set of opacities is for the accreted solar
models and obtained from the tables of Cox and Stewart (1970) for
different hydrogen-helium and hydrogen-heavy element ratios.

       Initial solar composition for the standard model is taken from
Cameron's table of abundances (1973). Using this composition a sequence
of solar models was evolved up to its present age 4.5×10^9 years at
which point the model should produce the observed luminosity and radius
of the sun. The present characteristics of the standard model reported
in this work is well represented by the initial composition,
       X: 0.760, Y: 0.221 and Z: 0.019 and for α: 1.3,
where X , Y, and Z are the fractional contents (by mass) of hydrogen
helium and heavy elements, respectively. The heavy element amount is

Age 4.5x10^{9}yr.	X:0.760 Z:0.019	X:0.780 Z:0.005	X:0.800 Z:0.005	X:0.800 Z:0.005	X:0.800 Z:0.005
Total acc. mass(grs)	___	3x10^{30}	1x10^{30}	5x10^{30}	6x10^{30}
Cent. den. (gr/cm^3)	142	158	140	139	138
Cent.temp. (K°)	1.49x10^7	1.55x10^7	1.47x10^7	1.47x10^7	1.47x10^7
Cent. hyd. amount	0.418	0.392	0.455	0.462	0.461
Mass out. conv.z.(M$_\odot$)	3.8x10^{-3}	1.4x10^{-3}	2.5x10^{-3}	3.4x10^{-3}	3.6x10^{-3}
Dept.out. conv.z.(R$_\odot$)	0.185	0.172	0.168	0.192	0.198
Eff.temp. (K°)	5791	6000	5910	5891	5883
Surf. Z value	0.019	0.018	0.007	0.015	0.018
L / L$_\odot$	1.00	1.23	1.04	1.04	1.04
R / R$_\odot$	1.00	1.03	0.98	0.99	0.99

Table-I: Some physical charecteristics of solar models.

taken as Z:0.005 by mass, for initial composition of the contaminated
model. This amount is given as a lower limit for the heavy element abundance in the sun at birth, by Aumann and McCrea (1976). The program is
so adjusted that during the evolution, the material is added at every
1.25x10^{8}years, to the outer convection zone of the sun at a rate to
increase Z to the present photospheric value taken as 0.019. The physical charecteristics of some solar models at the age of 4.5x10^{9}years,
are summarized in Table-I. The second column gives the physical properties defined under column one, for the standard solar model. The third,
fourth, fifth, and sixth columns indicate the same quantities for the
accreted solar models calculated for different composition and various
mass accretion rates. As seen in the third column, although the surface
Z value was reached, the correct solar luminosity was not obtained, for
the composition under consideration. The almost correct luminosity and
radius values are obtained when we choose the initial composition as,
X:0.800 and Z:0.005 by mass; but the surface Z value is reached when
the total accreted material was 6x10^{30}grams. It seems from the table
that the inner boundary of convection zone moves inward as the total
amount of accreted mass is increased. The amount of material accreted
onto the sun should be greater than 3x10^{-2}M$_\odot$ to obtain the observed
photospheric value of Z, which is higher than the one predicted by
Aumann and McCrea (1976). It should be noted that they also do not
exclude the possibility of higher values.

The predicted neutrino flux for standard and accreted models are 4.68 and 3.48 SNU, respectively, where 1 SNU is 10^{-36} solar neutrino captures per second per target atom. Neutrino fluxes were calculated using the Cl-37 capture cross sections given by Bahcall (1978). Ezer and Cameron (1971) have found the total neutrino flux as 5.2 SNU, for standard model. In this study there is a reduction in the total neutrino flux, due to the difference in nuclear reaction rates; but it is still larger than the present observed value of 2.2 ± 0.4 SNU (Bahcall et al., 1980). A lower value of total neutrino flux could be obtained for contaminated model, if the luminosity would be adjusted to the right value; but even that value of the total neutrino flux would be higher than the obtained one by Christensen-Dalsgaard et al. (1979). More investigations are required to remove the discrepancy between predicted and observed rate.

Support from Turkish Research Council (TUBITAK) is acknowledged.

References

Aumann, J.R., and McCrea, W.H.:1976, *Nature* 262,pp.560-61.
Bahcall, J.N.:1978, *Rev. of Mod. Phys.* 50,pp.881-903.
Bahcall, J.N., and Davis,R.:1976, *Science* 191,pp.264-67.
Bahcall, J.N., and Ulrich, R.K.:1971, *Astrophys.J.* 170,pp.593-603.
Bahcall, J.N., Lubow, S.H., Huebner, W.F., Magee, N.H., Merts, A.L.,
        Argo, M.F., Parker, P.D., Rozsnyai, B., and Ulrich, R.K.:1980,
        Phys. Rev. Lett. 45,pp.945-48.
Cameron, A.G.W.:1973, *Space Sci. Rev.* 15,pp.121-46.
Carson, T.R., Ezer, D., and Stothers, R.:1974, *Astrophys.J.* 194,pp.743-44.
Chitre, S.M., Ezer, D., and Stothers, R.:1973, *Astrophys. Lett.* 14,
        pp.37-40.
Christensen-Dalsgaard, J., Gough, D.O., and Morgan, J.G.:1979, *Astron.*
        Astrophys. 73,pp.121-28.
Cox, A.N., and Stewart, J.N.:1970, *Astrophys.J. Suppl.* 19,pp.243-59.
Ezer, D., and Cameron, A.G.W.:1967, *Can.J. of Phys.* 45,pp.3461-477.
Ezer, D., and Cameron, A.G.W.:1968, *Astrophys.Lett.* 1,pp.177-79.
Ezer, D., and Cameron, A.G.W.:1971, *Astrophys. Spa. Sci.* 10,pp.52-70.
Ezer, D., and Cameron, A.G.W.:1972, *Nature Phys. Sci.* 240,pp.180-82.
Fowler, W.A., Caughlan, G.R., and Zimmerman, B.A.:1975, *Ann. Rev. Astron.*
        Astrophys. 13,pp.69-112.
Hoyle, F. and Lytleton, R.A.:1939, *Proc. Camb. Phil. Soc. Math. Phys.*
        Sci. 35,pp.405-15.
Kuchowitz, B.:1976, *Rep. Prog. Phys.* 39,pp.291-343.
Newmann, M.J., and Talbot, R.J.:1976, *Nature* 262,pp.559-60.

THE ROTATION-MAGNETISM-CONVECTION COUPLING IN THE SUN

M. Stix
Kiepenheuer-Institut für Sonnenphysik
Schöneckstr.6, D-78oo Freiburg, FRG

ABSTRACT

The rotational and magnetic evolution of a star depends on its initial mass and angular momentum, but not directly on its initial magnetic field if the star has an outer convection zone. The present field is generated by a self-excited dynamo. Rotation influences convection and so generates the two main ingredients of the dynamo process, non-uniform rotation and helicity. Several competing descriptions have been proposed: A two-level picture where first single convective cells are distorted by rotation and then the net effect on the mean flow and field is considered, and a direct numerical approach, including rotational effects, where small-scale motions enter only as isotropic diffusivities of heat, momentum, and field. Difficulties arise in particular from the fact that the magnetic field is concentrated into fibrils of large field strength. Predictions of a quantitative relationship between a star's rate of rotation and its magnetic field amplitude are also difficult, although observations of late-type stars suggest such a relationship. A transition between two modes of dynamo operation may occur as the star's rotation slows down by magnetic braking.

THE INITIAL STATE

The rotation-magnetism-convection coupling plays an important rôle during a very early phase of stellar formation. These are the problems: First, an interstellar cloud of, say, $4 \cdot 1o^{37}$ g initial mass and $1.73 \cdot 1o^{-23}$ gcm^{-3} initial density, rotating with the galactic rate $\omega = 1o^{-15} s^{-1}$ (Dorfi, 1981), has a total angular momentum of $\sim 1o^{62} gcm^2/s$; on the other hand the solar system, comprising $5 \cdot 1o^{-5}$ of the cloud mass, has an angular momentum of $3 \cdot 1o^{5o} gcm^2/s$, i.e. only $3 \cdot 1o^{-12}$ of what the initial cloud had. How and where is all this angular momentum lost? Secondly, assuming conservation of magnetic flux, the same cloud would compress an initial field of, say, 3 μG to a field strength much in excess of anything observed in the solar system. Thus, how does the cloud,

W. Fricke and G. Teleki (eds.), Sun and Planetary System, 63–70.

and later the protostar, loose nearly all of its magnetic flux?

Dorfi has partially answered these questions through his numerical modelling of the initial cloud collaps in a magnetic field. In the case where the axis of rotation is perpendicular to the initial field, magnetic braking is particularly effective: 99% of the cloud's angular momentum is transferred to the ambient medium after only one tenth of the free fall time, and the loss of angular momentum, achieved by Alfvén waves, continues.

The explanation of magnetic flux loss from the cloud after the initial contraction is less straightforward. The field, originally badly needed to remove the angular momentum, presumably gives rise to instabilities in the cloud and thus helps to create turbulence which in turn very effectively removes the flux. This latter process, often described by a "turbulent" electromagnetic diffusivity of order $\beta \sim v^2 \tau$, where v and τ are typical values of velocity and time scale in the turbulence, is however poorly understood. Dorfi finds values of v and τ such that the associated time scale L^2/β, where L is the extent of the turbulent region, is comparable to the free-fall time.

Dorfi's calculations end after ca. 2 free-fall times with a cloud of still 4oo solar masses. Fragmentation, further transfer of angular momentum (probably by turbulence) and the formation of the planetary system will occur between this state and the initial Sun. The initial values of solar mass and total angular momentum are determined during this period. As the Sun's mass loss is insignificant, and its time scale of braking is comparable to its age, the present values also depend on the initial state. In contrast to this, the Sun's magnetic field has long forgotten its initial form and strength. Its time scale is now observed to be 11 years, and the time scale of turbulent transport, L^2/β, where L is now the solar radius (or the depth of the convection zone) and β is estimated using convection zone values of v and τ, is of similar, or even smaller, order, in any case very short compared to the age of the Sun.

We have, then, a star with a given mass and angular momentum and ask what kind of magnetic field this star will generate. The generation proceeds via the dynamo process according to the well-known induction equation

$$\frac{\partial B}{\partial t} = curl\left(v \times B\right) - curl\frac{1}{\mu\sigma}curl\,B \qquad (1)$$

where μ and σ are the permeability and conductivity of the stellar gas, and v is the velocity field. The form of v determines whether a field, B, will be generated. Two important contributions to v are (non-uniform) rotation and convection. We first consider the latter.

Table 1. Parameters for convection zones of main sequence stars

Spectral Type	R (m)	d/R	κ_0 (m^2/s)	d^2/κ_0 (years)
F5	$8.4 \ 10^8$	0.10	$12.0 \ 10^9$	0.02
Go	7.3	0.20	9.0	0.07
G5	6.5	0.36	6.9	0.25
Ko	5.9	0.42	2.7	0.72
K5	5.2	0.47	1.4	1.35
Mo	4.4	0.50	0.9	1.7c

CONVECTION

According to the theory of stellar structure main sequence stars later than FO possess outer convection zones, mainly because partial ionization of hydrogen causes a superadiabatic temperature gradient. Using Prandtl's mixing length theory such convection zones have been calculated e.g. by Baker and Temesváry, and Table 1 contains some of the parameters relevant to the present subject, as would be obtained for a ratio $\alpha = 1.5$ of mixing length to scale height. R is the stellar radius, d is the depth of the convection zone, and κ_0 is a heat transport coefficient obtained in recent models by Belvedere, Paternò and myself under the assumption that the transport is entirely by convection. κ_0 is of the order $v_c l$, where v_c is the convective velocity, determined by the buoyancy of the rising gas, and l is the mixing length. The last column of Table 1 contains the time scale of turbulent mixing. It increases towards later stars. Since mean magnetic fields also are subject to turbulent transport, we may speculate that cycles of later stars may have longer periods.

In its simplest form the transport coefficient enters the equations of energy, momentum, and magnetic field as a scalar diffusivity. In the case of homogeneous isotropic turbulence this would be the sole effect. It turns however out that the most interesting effects in the present context occur when the turbulent flow is inhomogeneous and (or) anisotropic, which certainly is the case in a stellar convection zone where gravity and rotation provide preferred directions. One of these effects is differential rotation, which we consider next.

DIFFERENTIAL ROTATION

In contrast to the mean rate, ω , of rotation, which still depends on the conditions during the formation of our star, the particular form of differential rotation does no longer reflect the initial state (except, of course, through its dependence on ω itself). It is governed by processes occurring on the short, or mixing, time scale

because this is the time required to restore solid body rotation if no
other process is at work. One possibility is the generation of non-uniform rotation together with (or even via) an axisymmetric meridional
circulation by anisotropic turbulent viscosity, or by a latitude-
dependent heat transport coefficient. Both mechanisms have been used
in many model calculations, most recently by Schmidt (1981). In the
case of latitude dependent heat transport he finds that the meridional
circulation and the concomitant pole-equator difference in surface
temperature exceed the upper limits set by solar observations unless
the Prandtl number is smaller than, say, o.2. This means that turbulent
transport of mean momentum should be less efficient than transport of
heat. For anisotropic turbulent viscosity the meridional circulation
appears to be of secondary importance. Its speed at the solar surface
is moderate or large ($\gtrsim$ 1o m/s) if the condition of vanishing pressure
perturbation, p', is imposed at the lower boundary of the model convection zone, while it can be very slow ($\sim$ 1 m/s) otherwise, even if
the Prandtl number is unity. Although it would appear that in the
latter case a singularity in the volume force would occur at the lower
boundary and that, therefore, a boundary with p'$\neq$o would be unphysical,
Spruit (1981 , private communication) has suggested that this is not
so: a small geometrical adjustment, by $\delta r = p'(dp/dr)^{-1}$, where p is the
undisturbed pressure, would remove the singularity. In Schmidt's model
δr could be as small as 1o^{-6} of a scale height. - For each of the two
driving mechanisms there seems to be no difficulty to generate the observed equatorial acceleration of the Sun.

Why - and how - are the transport coefficients for heat and momentum
variable in space and anisotropic? Of course, because turbulent convection in the Sun is neither homogeneous nor isotropic. A recent treatment of this problem is by Durney and Spruit (1979) who describe convection cells by their dimensions l_r, l_θ , and l_φ in the three space
directions. Both the onset of convection and the form of the convective flow depend on rotation: the former since the temperature gradient
must exceed the adiabatic gradient by at least $4(\omega.k)^2 T g^{-1}(k_\theta^2 + k_\varphi^2)^{-1}$
where ω is the vector of angular velocity, T the temperature, g the
acceleration of gravity, and k the wave vector, $k_r = 2\pi/l_r$ etc.;
the latter, the convective flow, with components v_i, is obtained
from a linearized equation of motion. As in conventional mixing length
theory l_r is set equal to the scale height. The ratios l_θ/l_r and
l_φ/l_r should depend on ω ; Durney and Spruit in their first discussion
treat these ratios as free parameters, depending on latitude. They
finally derive the thermal and kinematic diffusivities as tensors of
the form $\tau \langle v_i v_j \rangle$, where τ is the lifetime of the convection cell
and the average is over the cell volume. According to this picture the
two transport coefficients become equal, i.e. the Prandtl number would
be 1. The above-mentioned results of Schmidt must be seen in this context.

It has also been pointed out by Durney and Spruit and by Gough
(1978) that, besides the term linear in the mean flow,$\langle v \rangle$, which yields
the kinematic turbulent viscosity, there appears a term in the viscous

stress tensor which does not depend on $\langle \mathbf{v} \rangle$, the "turbulence pressure". Because of its tensorial form it cannot in general be balanced by a static pressure and thus also helps to drive $\langle \mathbf{v} \rangle$. The recent model of Durney (1981) takes this into account, but so far functions only at rotation rates smaller than the Sun's. Rotation would presumably do less damage to this model if the convection cells were elongated parallel to the axis of rotation so that $\mathbf{k} \cdot \boldsymbol{\omega}$ would be small. Durney and his colleagues at Sacramento Peak Observatory presently try to find indications at the solar surface for such a rotational influence upon convection.

An alternative approach to the rotation-convection coupling is direct numerical simulation. Instead of the two steps considered in the above scheme, namely first to treat small-scale motions and to calculate the rotationally influenced transport coefficients, and then to use these in order to drive a large-scale axisymmstric flow, in this approach everything is calculated in one numerical model. The point here is of course that the largest cells are accessible to numerical modelling; because of their long life time these are just the ones most strongly influenced by rotation. The calculations cannot be carried out with solar values of all the hydrodynamic parameters. The Rayleigh number on the Sun would be of order 10^{10}! However, the hope is that the expected hydrodynamic processes occur already at moderate values of these parameters. Also, isotropic small-scale motions can be taken into account as scalar diffusion coefficients; this equally leads to tractable parameter values.

Gilman has made the most extensive calculations of such models. His main result is that an equatorial acceleration such as seen on the Sun can easily be produced. The angular momentum is transported towards the equator by means of Reynolds stresses, i.e. by a correlation $\langle v_\theta \, v_\varphi \rangle$ of the latitudinal and longitudinal velocity components in the "global" (but non-axisymmetric) pattern of convection.

It is not clear presently whether the solar convection zone is closely modelled in these numerical calculations. Until recently only incompressible (Boussinesq) convection has been treated, but now Glatzmaier and Gilman (1981) investigate compressible models as well. In most models the velocity components v_θ, v_φ of the non-axisymmetric global convection are of the same magnitude as the differential rotation they produce ($\sim$ 1oo m/s), while on the Sun an upper limit of order 1o m/s has been set by LaBonte et al.(1981). Screening of the global convection by a strongly turbulent layer near the surface is probably not sufficient to explain the difference (Stix, 1981a), although differences in heat flux are very effectively screened by the turbulence (Spruit, 1977; Rüdiger, 1981).

MAGNETISM

Explicit numerical modelling as described at the end of the preceding section can be extended to the magnetic case. The Lorentz force
then must be added to the equation of motion, and Eq.(1) must be used
to calculate the evolution of the magnetic field. Like the Rayleigh
number, the magnetic Reynolds number in the convection zone is so large
that the turbulent electrical conductivity must be used for a numerically tractable model. Still, the effects due to inhomogeneity and
anisotropy are not parametrized; in particular, no α-effect (see below)
is used. In their model Gilman and Miller (1981) indeed find self-excited fields. Regular cyclic behaviour as on the Sun, and the strong
toroidal field which on the Sun manifests itself in bipolar spot groups,
is however not yet obtained.

More traditional in modelling the solar cycle is the mean-field
approach. The generation of the toroidal field by means of non-uniform
rotation (" ω ") is well understood. More difficult is the complementary process of regeneration of the (reversed) poloidal field from the
toroidal, mainly because small-scale motions, $\mathbf{v}'$, and fields, $\mathbf{B}'$, are
involved: The Coriolis force renders the convective flow helical; using
the picture of frozen-in magnetic lines of force we can envisage twisted
loops of fieldlines being formed out of an originally toroidal field,
with a net contribution to the new poloidal field. Formally this process has been described as a mean e.m.f. α$\mathbf{B}$ parallel to the (predominantly toroidal) mean field, and has been termed the "α-effect". Parker,
Steenbeek and Krause, and many others have used such "αω-dynamos" to
model the solar cycle, and I have reviewed these models recently (Stix,
1981b). The problem lies in the quantitative determination of α. Most
estimates in the past have used the concept of "first order smoothing",
neglecting the terms $\mathbf{v}' \times \mathbf{B}'$ - $\langle \mathbf{v}' \times \mathbf{B}' \rangle$ in the equation for the fluctuating part of the magnetic field. This means that the case where local
fields are strongly concentrated into flux tubes and so many evade (or
resist) much of the twisting effect of the helical flow, is not properly treated. An overestimate of α is the probable consequence for the
mean field models. It seems clear that any approach to the solar dynamo
problem must incorporate the intermittent nature of the magnetic field
in the convection zone. Parker (1981) recently undertook steps into
this direction: he found, for example, that the magnetic stress of an
ensemble of "fibrils" is larger by the factor m = B'/$\langle$B$\rangle$ compared to
the stress of an evenly distributed field. Equally important are studies
of magnetohydrodynamic turbulence up to moderately large magnetic Reynolds numbers (~1oo), employing either closure techniques (Léorat et al.,
1981) or direct numerical calculation (Meneguzzi et al., 1981).

It is presently unclear how an oscillatory mean stellar magnetic
field determines its amplitude. One guess which I have used was that,
locally, an equilibrium between the Coriolis and Lorentz forces is
reached, and that the α-effect is then quenched. Another attempt
(Durney and Robinson, 1981) is to equate the time of rise of a buoyant
magnetic flux tube in the convection zone with the amplification time

of the dynamo process. Both ideas imply a field amplitude growing with
the rate of the stellar rotation; a detailed prediction does however
depend on the knowledge of how differential rotation and α-effect themselves depend on that rate.

Observations clearly indicate that rotation is the dominating influence. Figure 4 of Vaughan et al.(1981) shows that for stars of any
one spectral type there is a relationship between the flux, S, in the
emission cores of the H and K lines of CaII and the period of rotation,
as determined by the rotational modulation of S. If, as on the Sun
(Skumanich et al., 1975), S varies linearly with the magnetic flux, then
we have the desired relation between field and rotation.

There seem to exist two distinct groups of late type stars with Calcium emission: Rapid rotators, with rather irregular emission of large
intensity, and slow rotators, with less intense and rather smooth emission (Vaughan and Preston, 1980). Durney et al.(1981) suggest that these
differences indicate a transition from a regime where several inter-
fering magnetic modes exist, to a regime where only the fundamental mode
is excited. Figure 1, composed from results obtained by Möller (1974),
shows that such behaviour also occurs in a non-linear theory (the "cut-
off α-effect" model mentioned above). Even in one and the same star,
like our Sun, the magnetic field might have changed from one mode to
another, without any change in mean rotation. The Maunder minimum of the
17th century is the most prominent example. "Strange attractors", i.e.

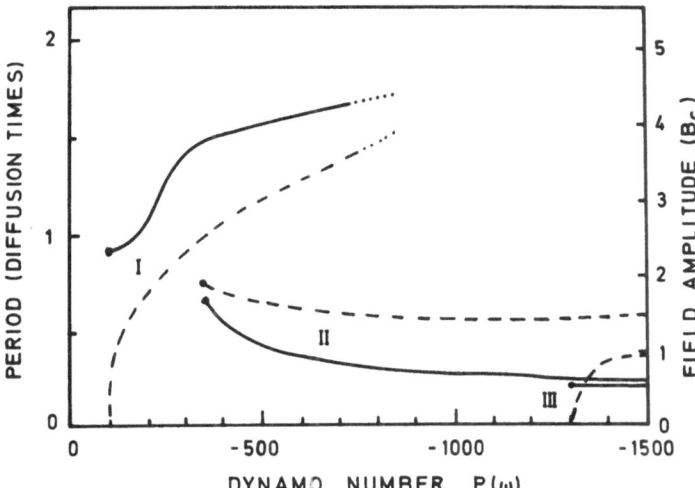

Figure 1. Periods (solid) and amplitudes (broken) of the first 3 anti-
symmetric (with resp. to the equator) modes as functions of the dynamo
number P(ω), a dimensionless measure of $\alpha \cdot \partial \omega / \partial r$. The field unit B_c
is the critical field strength above which the α-effect is quenched.
Both P and B_c are expected to increase with ω. Composed after Möller
(1974).

magnetohydrodynamic systems which display a stochastic branching into
different types of solutions have been proposed as an explanation to
this phenomenon.

REFERENCES

The list below contains only papers which have not been quoted explicitly in Stix (1981). Further review articles are by Monin (198o),
Cowling (1981), and Weiss (1981). A new monograph by Vainshtein et al.
(198o) has been published recently. Almost all authors listed helped
with suggestions and unpublished material; I wish to express my thanks
to them.

Cowling, T.G.:1981,Ann.Rev.Astron.Astrophys.$\underline{19}$,115.
Dorfi,E.A.:1981,PhD Thesis,Univ.Wien.
Durney,B.R.:1981,Astrophys.J.$\underline{244}$,678.
Durney,B.R.,and Spruit,H.C.:1979,Astrophys.J.$\underline{234}$,1067.
Durney,B.R.,and Robinson,R.D.:1981,preprint.
Durney,B.R.,Mihalas,D.,and Robinson,R.D.:1981,Publ.Astron.Soc.Pacific,
     in press.
Gilman,P.A.,and Miller,J.:1981,Astrophys.J.Suppl.$\underline{46}$,211.
Glatzmaier,G.A.,and Gilman,P.A.:1981,Astrophys.J.Suppl.$\underline{45}$,351.
Gough,D.O.:1978,in G. Belvedere and L. Paternò(eds.),"Workshop on Solar
     Rotation",Catania,p.337.
LaBonte,B.J.,Howard,R.,and Gilman,P.A.:1981,preprint
Léorat,J.,Pouquet,A.,and Frisch,U.:1981,J.Fluid Mech.$\underline{1o4}$,419.
Meneguzzi,M.,Frisch,U.,and Pouquet,A.:1981,Phys.Rev.Lett.$\underline{47}$,1o6o.
Möller,W.:1974,Diplomarbeit,Univ.Göttingen.
Monin,A.S.:198o, Usp.Fiz.Nauk $\underline{132}$,123 (Engl.$\underline{23}$,594).
Parker,E.N.:1981,preprint.
Rüdiger,G.:1981,preprint.
Schmidt,W.:1981,PhD Thesis, Univ.Freiburg.
Skumanich,A.,Smythe,C.,and Frazier,E.N.:1975,Astrophys.J.$\underline{2oo}$,747.
Spruit,H.C.:1977,Astron.Astrophys.$\underline{55}$,151.
Stix,M.:1981a,in R.B.Dunn(ed.)"Solar Instrumentation:What's next?"
     Sunspot,p.25.
Stix.M.:1981b,Solar Phys. $\underline{74}$, 79.
Vainshtein,S.I.,Zel'dovich,Ya.B.,and Ruzmaikin,A.A.:198o, "Turbulent
     Dynamo in Astrophysics", Moscow (in Russian).
Vaughan,A.H.,and Preston,G.W.:198o,Publ.Astron.Soc.Pacific $\underline{92}$,385.
Vaughan,A.H.,and six others:1981,Astrophys.J., in press.
Weiss,N.O.:1981,Geophys.Astrophys.Fluid Dyn., in press.

Mitteilungen aus dem Kiepenheuer-Institut Nr. 212

RELATIONSHIP BETWEEN PHOTOSPHERIC AND CHROMOSPHERIC $\Omega(\vartheta,t)$ DEDUCED BY TEMPORARILY AND SPATIALLY CORRELATED TRACERS

R. A. Zappalà and F. Zuccarello
Istituto di Astronomia - Osservatorio Astrofisico
Viale Andrea Doria - Città Universitaria - 95125 CATANIA

1. INTRODUCTION

In previous works we pointed out that solar mean angular velocity measured by tracers (sunspot-groups and K faculae) depends not only on
latitude but also on the age of the tracer (Belvedere et al. 1977;Ter-
nullo et al. 1981).
On the basis of those results it seemed interesting to analyze the motion of "spatially" and "temporarily" correlated photospheric and chromospheric features, whose ages ranged from 2 - 9 days; thus we could
obtain information on the kinematic behaviour of the same young active
region in two different layers of the solar atmosphere.

2. DATA ANALYSIS

The data analysis of non-recurring photospheric tracers (sunspot-groups)
is based on daily drawing and the data concerning non-recurring chromospheric tracers (faculae) is based on daily K spectroheliograms; all
data were obtained during the Catania solar patrol from January 1,1972
to November 27, 1977 (Godoli 1969).
In order to recognize the first appearence of the tracer as a birth, we
restricted our analysis to the part of the solar disk inside 70° from
the central meridian of the Sun.
Moreover, in order to study just those phenomena which were temporarily
and spatially correlated, we selected photospheric and chromospheric
phenomena having the same age and whose geometrical "center of mass"
were overlapping within ± 2°.
From the differences in Carrington longitudes we measured separately and
day by day the velocity of the sunspot-groups and of the K faculae.
The average values have been calculated over 5° latitude strips and the
velocity have been divided into 8 classes according to the age.
From the comparison of the curves of rotation obtained for correlated
sunspot-groups and K faculae, there emerges a similar behaviour pattern,
within the experimental errors, for the two levels; in addition the young

W. Fricke and G. Teleki (eds.), Sun and Planetary System, 71–72.

active regions show variations in velocity which depend not only on latitude but also on the age.

3. DISCUSSION

It is well known that the tracers are indicators of strong and localized magnetic fields (Parker 1977), so the similar behaviour of photospheric and chromospheric $\Omega(\mathcal{V}, t)$ deduced by spatially and temporarily
correlated tracers give us a confirmation of the magnetic linkage existing between the two layers in the same active region.
Our results also contribute to the reliability of the hipothesis (Stix
1976) of the existence of a sub-photospheric convective region diffe-
rentially rotating at a speed higher than that observed in the photospheric and chromospheric layers.
Moreover, as in a given moment there are present in the Sun active regions of different ages, on the basis of our results, it can be seen
that the relative velocities of these active regions are different.
This means that, in a fixed latitude, active regions of different ages,
have different velocities too.
An analysis of two solar cycles is in progress at the Catania Astrophysical Observatory, in order to confirm quantitatively the result we have obtained. In fact, the existence of the magnetic linkage between different active regions has been shown by Sky-lab X-ray picture, and this,
together with our result, could give information about solar transients.
We think that in order to obtain some information on the triggering mechanism of the solar transient, it is very interesting to evaluate the
effects of twisting and shearing due to the motions of sunspots belonging to different active regions but which are magnetically linked.

REFERENCES

Belvedere,G.,Godoli,G.,Motta,S.,Paternò,L.,and Zappalà,R.A.:1977,
 Astrophys. J.214,L91.
Godoli,G.:1969, Solar Phys.9,246.
Parker,E.N.:1977, Ann.Rev.Astron.Astrophys.15,45
Stix,M.:1976, Astron.Asstrophys.47,243
Ternullo,M.,Zappalà,R.A.,and Zuccarello,F.:1981, Solar Phys.74,111.

LINE-OF-SIGHT VELOCITY FIELD OF SYNODIC SOLAR ROTATION

A.Kubičela
Astronomical Observatory, Belgrade, Yugoslavia
M.Karabin
Institute of Astronomy, Faculty of Sciences,
Belgrade, Yugoslavia

ABSTRACT. Angular velocity of synodic solar rotation have been treated as vectors. A colinear and a perpendicular components with respect to the solar rotation axis have been found. Consequently, factor $\cos 7°25$ has been introduced into the apparent effect of the Earth's orbital motion. An alternating term has been also added in the expression of the synodic line-of-sight velocity of an arbitrary photospheric point.

1. INTRODUCTION

The treatment of the synodic solar rotation in literature is often incomplete in the sense that the non-orthogonality of the solar rotation axis with respect to the ecliptic plane has been neglected. On the other hand, Beckers (1978) expressed some doubts in the existence of the apparent yearly precession effect reported by Kubičela and Karabin (1977). Therefore, in this paper a more general consideration of the synodic solar rotation as well as another explaination of the apparent yearly precession have been attempted.

2. COLINEAR COMPONENT OF THE SYNODIC SOLAR ROTATION

The line-of-sight component of solar rotation is usually expressed as:

$$V_1 = R\omega_1 \cos\phi_m \sin\lambda_m \cos B_o, \tag{1}$$

where ω_1 is the angular velocity of sidereal solar rotation, R is radius of the Sun, ϕ_m and λ_m are heliographic latitude and longitude of the observed photospheric point, M, and B_o is heliographic latitude of the centre of the solar disk.

The Earth's revolution introduces an apparent angular velocity of the opposite direction to the solar rotation. Its line-of-sight component

73

expressed in the ecliptic coordinate system is analogous to (1):

$$V_a = R\omega_2 \cos b_m \sin(l_m - L_e),$$

where ω_2 represents the sidereal orbital angular velocity of the Earth, b_m and l_m are heliocentric ecliptic latitude and longitude of any given point at the Sun, and L_e is heliocentric longitude of the Earth. Angular velocity ω_2 is a periodic function of time according to the second Kepler's law. Such a variability is to be assumed in all angular velocities derived from it later in this paper.

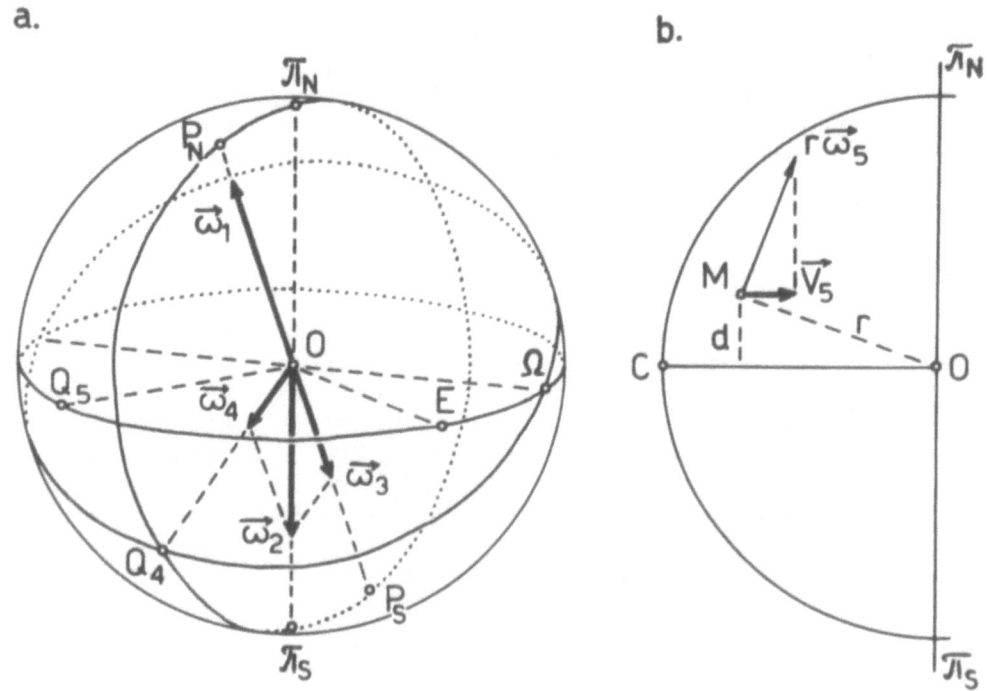

Figure 1.   a) The solar rotation angular velocity $\vec{\omega}_1$, the apparent
            angular velocity effect of the Earth's revolution $\vec{\omega}_2$
            and its two components.
         b) The line-of-sight component of $\vec{\omega}_5$.

In order to clarify the influence of $\vec{\omega}_2$ on $\vec{\omega}_1$ it is necessary to project $\vec{\omega}_2$ onto the rotation axis of the Sun. Let vector $\vec{\omega}_1$ (Figure 1a) be the angular velocity of sidereal solar rotation. The apparent rotation velocity introduced by the Earth's revolution, $\vec{\omega}_2$, lies in the direction toward the south ecliptic pole π_S. This velocity can be represented by its two components: $\vec{\omega}_3 = \vec{\omega}_2 \cos 7°.25$ colinear with $\vec{\omega}_1$, and $\vec{\omega}_4 = \vec{\omega}_2 \sin 7°.25$ perpendicular to the solar rotation axis. It is now obvious that ω_3 can be readily subtracted from ω_1 in (1), resulting in:

$$V_s = R(\omega_1 - \omega_2 \cos 7°25) \cos\phi_m \sin\lambda_m B_o. \tag{2}$$

Taking into account the numerical value of ω_2, one finds the mean angular velocity $\bar\omega_3 = 1.975072 \times 10^{-7}$ rad/s, or as a peripheral velocity at the solar equator $V_3 = 137.5$ m/s. This value is only about 1 m/s smaller than the one usually applied, namely 138.6 m/s.

3. The $\vec{\bar\omega}_4$-ROTATION AND ITS LINE-OF-SIGHT ASPECT

However, so far the component $\vec{\bar\omega}_4$ has not been taken into account. It represents an apparent rotation of the solar globe around a fixed axis defined by the direction OQ_4 in Figure 1a, pointing somewhere in Aquarius. In the plane of solar nodal meridian this apparent rotation has the direction from the ascending node toward solar rotation north pole. Its mean angular velocity is:

$$\omega_4 = 2.512609 \times 10^{-8} \text{ rad/s}, \tag{3}$$

and the corresponding peripheral photospheric velocity along the solar nodal meridian $V_4 = 17.5$ m/s. If visible $\vec{\bar\omega}_4$-rotation would complete one full apparent turn in 7.92 years. However, due to the fact that during the revolution of the Earth we see the Sun from different longitudes, various aspects of $\vec{\bar\omega}_4$-rotation will be seen from the Earth (Kubičela, 1982).

In order to obtain photospheric line-of-sight components of the apparent $\vec{\bar\omega}_4$-rotation, vector $\vec{\bar\omega}_4$ has to be projected into the ecliptic plane, onto direction OQ_5 in Figure 1a, perpendicular to the line-of-sight, CE. This procedure gives:

$$\omega_5 = \omega_4 \cos 7°25 \cos(L_e - L_\Omega), \tag{4}$$

where L_Ω is longitude of the ascending node of the solar equator. Actually, the apparent rotation $\vec{\bar\omega}_4$ never completes a single period of rotation. Angular velocity $\vec{\bar\omega}_4$ exists due to the revolution of the Earth, but it is seen from our planet as ω_5-changing its sign with $\cos(L_e - L_\Omega)$ namely every six months (with $\vec{\bar\omega}_5$ pointing westward from begining of March till begining of September and pointing eastward for the rest of the year). Such a phenomenon can be interpreted as an angular oscillation of the solar globe around its E-W ecliptic diameter (Kubičela and Karabin, 1977, Figure 1b).

Now the line-of-sight component of the oscillation $\vec{\bar\omega}_5$ as a function of position within the solar disk can be found. The necessary relations are easily seen in Figure 1b - the projection of visible solar hemisphere into the plane containing centre of the Sun, 0, centre of the solar disk, C, and the direction toward north, π_N, and south, π_S, ecliptic poles. If $\vec{\bar\omega}_5$ points, for example, eastward, the peripheral velocity of any photospheric point M in Figure 1b will be $r\omega_5$, where r represents the distance of M from the ecliptic E-W diameter of the Sun. Its line-of-sight component, V_5, can be found from $V_5 = d\omega_5$. On the oder side $d = R \sin b_m$ and V_5 becomes:

$$V = R\omega_5 \sin b_m , \tag{5}$$

corresponding to Figure 3b in Kubičela and Karabin (1977).

Finally, the full expression for synodic line-of-sight velocity V_m of an arbitrary photospheric point M is obtained by adding (5) to (2):

$$V_m = R(\omega_1 - \omega_2 \cos 7^\circ.25)\cos\phi_m \sin\lambda_m \cos B_o + R\omega_5 \sin b_m . \tag{6}$$

Line-of-sight velocites obtained by (6) differ from previously used ones by the factor $\cos 7^\circ 25$ in the first term and by the existence of the second term at right-hand side. For everyday use it is convenient to express b_m as a function of ϕ_m and λ_m.

REFERENCES:

Beckers,J.M.: 1978, Osservatorio Astrofis. Catania, Publ.162, pp.166-177.
Kubičela,A.: 1982, Hvar Obs. Bull. Suppl., 1. in press.
Kubičela,A. and Karabin,M.: 1977, Solar Phys. 54, pp.505-509.

SOLAR ROTATION AND MERIDIONAL MOTIONS DERIVED FROM SUNSPOT GROUPS

Jaakko Tuominen, Ilkka Tuominen and Juhani Kyröläinen
Observatory and Astrophysics Laboratory, University of
Helsinki, Tähtitorninmäki, 00130 Helsinki 13, Finland

Latitudinal and longitudinal motions of sunspot groups have been studied using the positions of recurrent sunspot groups of 103 years published by Greenwich observatory. In order to avoid any limb effects, only positions close to the central meridian have been used.

To begin with, the data were divided into two parts: those belonging to the years around sunspot maxima and those belonging to the years around sunspot minima. Using several different criteria it was ascertained that sunspot groups show meridional motions and that their drift curves as a function of latitude are different around maxima and around minima. In addition, also the angular velocity, as a function of latitude, was found to be different around maxima and minima. In these investigations medians of the values are always used. It is shown that the results do not change essentially even when several of the extreme individual velocities in one or the other direction are excluded, as may be the case when average values are used. Further, the drift curves are symmetrical with respect to the equator, which also is a proof for their correctness.

After having obtained this result, it was natural to expect that the drifts of sunspots in latitude and longitude change continuously during the solar cycle. This is to be expected also on the basis of a diagram published by Howard and LaBonte (1980). This diagram is based on spectroscopic observations of solar rotation. Howard and LaBonte find that there are zones on the sun rotating with greater or smaller angular velocity than the average velocity at the latitude in question, and that these zones displace themselves towards the equator with the period of solar activity. A similar phenomenon has been studied by Stenflo (1977).

To find out whether the same phenomenon could be traced in latitudinal and longitudinal motions of sunspot groups, the sunspot cycle from each minimum to the next minimum was divided into four parts, each of which was three years long (the longest cycles being 12 years). The drifts in latitude and longitude in latitudinal zones 4° broad

W. Fricke and G. Teleki (eds.), Sun and Planetary System, 77–78.

were then derived in these four phases for the whole material of 103
years. The result was that there really are zones with motions towards
the poles and towards the equator, and rotating with smaller and
greater angular velocities, respectively, than the average velocity,
and that these zones displace themselves towards the equator similarly
as in the diagram given by Howard and LaBonte (1980). The zones are
symmetrical with respect to the equator and the velocities in the two
coordinates are of the same order of magnitude, i.e. a few meters per
second.

The complete papers describing these investigations will be
published elsewhere.

REFERENCES

Howard,R. and LaBonte,B.J.: 1980, Astrophys. J. 239, L33-L36
Stenflo,J.O.: 1977, Astron. Astrophys. 61, 797

SUBPHOTOSPHERIC VELOCITY FIELDS INFERRED BY SUNSPOTS MOTIONS

Maurizio Ternullo
Osservatorio Astrofisico di Catania

The angular velocity of solar rotation tracers, has been observed to decrease systematically with their age (Godoli et al., 1978); such a decrease is different at different latitudes and results there= fore in variations of both faculae (Belvedere et al., 1977) and sun= spots (Ternullo et al., 1981) velocity profiles.
A research is in progress at the Catania Astrophysical Observatory in order to gain a deeper insight into this complex phenomenon.
It appears now that spots velocity seems to depend, for each latitude, on the solar cycle phase and, in a less simple way than till now ob= served, on their age.
During the declining phase of the 20th cycle (1971-1975), I observed that the angular velocity of the sunspot groups lying in the two bands contiguous to the equatorial belt (latitude between 5° N and 10° N, and between 10° S and 15° S), was almost constant with respect to the spots age, while it decreased at all of the other latitudes.
When objects of the XXIst cycle began to be observed (1976-1979), I noticed that at some latitudes (around 17° both N and S, and around 32° N), spots velocity either had a local maximum or seemed to increa= se with age, while decreased at all of the other latitudes.
It seems then that some bands can be recognized in the Sun, at least at the latitudes where spots are formed. In these bands, which appear to be parallel to each other and symmetrically lying with respect to the equator, spots velocity either increases or decreases with the age of the objects, while in a few cases remains statio= nary on values which are local maxima or minima. The bands seem to move latitudinally while the solar cycle goes on. The equator is a zone of conspicuous slowing down (Ternullo et al., 1981).
Spots behaviour suggests the hypothesis that the magnetical flux tubes suffer, before emerging into the photosphere, the impact with a mate= rial flux, able to alter their velocity; such a velocity variation is

W. Fricke and G. Teleki (eds.), Sun and Planetary System, 79–80.
Copyright © 1982 by D. Reidel Publishing Company.

different at   different latitudes, which can be explained assuming that
there are different velocity zones - fast or slow - with which the
flux tubes interfere. After the top of a flux tube has emerged, and
its intersection with the photosphere has become visible as a spot
group (Babcock, 1960), the submerged velocity fields make the spot
group either to slow down or to accelerate. A slowing down is easier
to be observed, because it is superimposed on the general slowing
down, which makes the tracers to "loose memory" slowly of the higher
velocity of the layers they come up from, and to assume, when old,
the same velocity of the surface plasma (Foukal, 1972; Golub et al.,
1981). Positive acceleration is not completely reliable, because is
very close to the statistical uncertainty level; anyway,it seems to be
necessary to postulate an accelerating agent which acts in competi=
tion with the mentioned mechanism of general slowing down, in order
to explain why, in certain latitude strips, velocity does not decrease.
Such a model explains how the relative velocities between spots, lying$_1$
at latitudes close to each other, reach also very high values (100ms^{-1})
provided that such a comparison is made between spots which are at
least seven/eigth days old.
This model recalls strongly the results, obtained spectroscopically by
Howard and LaBonte (1980); they observed in the photosphere alterna=
ting zones of fast and slow rotation, originating at polar latitudes
and drifting to the equator in roughly 22 years; the equator itself
is almost always centered in a slow zone; the average amplitude of the
motions is about 3 ms^{-1}.
The last of the mentioned properties, makes difficult to explain
the observed behaviour of the spots, as due to the impact between
flux tubes and the Howard's velocity fields.
We can reconciliate our observations with the Howard's ones, by assu=
ming, as it is done in this paper, that velocity fields do exist at
depth inaccessible to the spectroscopic detection, which are similar
to those observed by Howard, but affected by velocities one or two or=
der of magnitude higher. Howard's velocity fields would therefore be
the damped, surface display, of a stronger subphotospheric phenomenon.

References

Babcock, H. W.: 1961, Ap. J. 133, p.572.
Belvedere, G., Godoli, G., Motta, S., Paternò, L., Zappalà, R. A.: 1977,
    Ap. J. 214, p.L91.
Foukal, P.: 1972, Ap. J. 173, p.439.
Godoli, G., Mazzucconi,F.: 1978, Proceedings of the Workshop on Solar
    Rotation, Osservatorio Astrofisico di Catania No. 162, p. 135.
Golub,L., Rosner,R., Vaiana,G., Weiss, N.:1981, Ap. J. 243, p. 309.
Howard, R., LaBonte, B. J.: 1980, Ap. J. 239, p.L33.
Ternullo, M., Zappalà, R.,Zuccarello,F.:1981.Sol. Phys. 74, 111.

GRANULATION, SUPERGRANULATION AND ATMOSPHERIC WAVES

V.A.Krat
Pulkovo Observatory,196140, Leningrad, USSR

ABSTRACT. A short review of the last years investigations of the fine and large scale structure of the solar
atmosphere and connected with this structure atmospheric
waves is given. The intergranular network and the supergranulation are considered as general strong magnetic
field and the measured faint magnetic field as a statistical mean of quasiopen fields of granulation. There is a
tendency of systematical formation of large scale structures beginning from the mesogranulation to the configurations of 10^5 km scale size.

The solution of problems of solar dynamics and solar
activity depends on our knowledge of fine structure of the
solar atmosphere. Even in the case of large scale configurations their borders seem to be very narrow as comparable with the breadth of the widest parts of the intergranular network. The interaction between the structure of
active and "undisturbed" regions can be either of regular
or statistic character. The same can be said about the
dynamical processes including waves containing enormous
energy exceeding the energy of active processes. Moreover,
at some phases of solar activity waves can be a basic reservoir of the activity energy.

Our task consists not only in reviewing the present
state of physics of the "undisturbed" solar atmosphere but
also of a preliminary discussion of some problems associated with the further progress of solar studies in general.
Magnetic fields and motions determine the solar atmospheric structure, since there seem to be no other phenomena
responsible for the formation of inhomogeneities. We shall
begin our review with the problems of solar granulation.

W. Fricke and G. Teleki (eds.), Sun and Planetary System, 81–87.

I. GRANULATION

The wide spred concept of granulation presents it as a picture of convection. But it is based on few facts derived from spectroscopic observations of motions above the granular field (granules and intergranular network) and theoretical considerations of the problem of the general subphotospheric convection identified with the structure of granulation. This picture contained a great deal of uncertainty due to a large scatter in numerical values of the used fundamental atmospheric parameters, though the existence of the subphotospherical convection zone now is unquestionable. It was doubtful only whether the upper boundary of this zone is screened by the rapidly increasing opacity with depth or it is situated in the visible part of the photosphere. The recent model of the photosphere - HSRA (Gingerich et al,1971) gives the rise of the optical depth (τ) in the continuum from I to IO in the geometric depth interval only 56 km. Thus, in reality the position of the convection zone in its upper part with respect to τ remained to a great extent indefinite.Nevertheless many theoreticians have considered the granulation as a direct picture of convection.

Our statements will be founded mainly on the most accurate observational data. A study of the granulation which is the finest structure on the Sun can be done successfully only when the effective optical resolution is about 0.″2 - 0.″3. With a poorer resolution the contrasts of granules and intergranular network are considerably and sometimes irregularly (due to the inhomogeneities of seeing) blurred. Theoretical reductions using a transfermodulation function usually give no positive effect containing some arbitrary elements, though the methods of speckle interferometry in some cases can be successfully applied. It may be mentioned also that series of photographic frames must present a homogeneous time sequence. Presently these conditions can be fulfilled only in the series of strato-spheric photographs. Even in this case the spatial resolution on spectrograms is so far not better than 0.″5. Simultaneous direct photography and spectrography of the same region on the solar disc gives an improvement of photometric measurements on spectrograms.

The most important results of stratospheric solar investigations can be summarized in the following statements: I. the structure of granulation field is bimodal near the disc center showing up in two different structures - granular and intergranular network. The latter gradually disappears when approaching the limb of the disc. There is a global dark network configuration locating about 200 km

lower bright granules (Pravdjuk et al.,1974).
2. the mean contrast of central parts of granules to the
"bottom" parts of intergranular network filaments gives
the temperature difference about 500 K for the disc center
five times smaller at the limb. On the extreme limb we see
granules at the boundary of scarce dark granulelike features (Pravdjuk et al.,1974). The horizontal temperature
gradient at the network border is approximately equal to
the vertical temperature gradient (Karpinsky,1981).Using
the best stratospheric spectrograms Sobolev (1974) found
the temperature difference derived from the equivalent
widths of spectral lines also to be about 500 K.
3. the features of granulation (granules and network) have
their prolongation in the upper photosphere and chromosphere and usually are inclined to the horizontal plane
showing evidence of a complicated structure of magnetic
fields comparable (or greater) in strength with the field
of gravity. The kinetic energy of separate currents must
be also taken into account (Krat,1973). In chromospheric
regions the structural elements are at least twice as wide
(Krat, 1972).
4. dark objects in the network - porules being in size
comparable with typical granules (800 km) - change their
contrast and position in the network during the "life-time
about 7 min (Kitai,Kawaguchi,1979). As it is known from
ground observations inside porules or large knots of the
network there are very small bright objects (filigree)
which are also well seen on the stratospheric photographs.
A study of these objects being at the limit of optical resolution is very hard (Mehltretter, 1974).
5. the investigation of radial velocities in the field of
granulation shows that the motions are well correlated
with brightness only at higher photospheric levels which
probably suggest the existence of forced convection above
the temperature inhomogeneities of granulation (Krat,
Shpitalnaya,1974). Basing on the new data on the fine photospheric structure we have proposed a new hypothetical
model of this phenomenon (Krat, 1977) correlating with
ideas of Stenflo (1973,1976). Large temperature gradients
at the border between granules and the network which are
underestimated due to some probable influence on their determination of a small discrepancy between the observed
and real optical picture of granulation contradict to the
concept of the visible connection zone. As it was shown
by Altrock and Musman (1976) even twice smaller temperature differences in the granulation field do not agree with
theoretically calculated possible differences at the interface between the convective zone and the photosphere.
The same discrepancy between theory and observation we see
when dealing with limb darkening and the structure of limb
granulation. But it was explained using spectroscopic data.

We explain the origin of the dark network in the same way
as it is usually done in the case of sunspots and penum-
brae by the influence of strong magnetic fields on convective transfer of energy in the upper parts of the convection zone, nevertheless the theory of this effect confronts
many theoretical difficulties.But we cannot doubt the reality of this effect being the only possible cause of the
temperature decrease in sunspots. The dark network can be
explained on the assumption that it is a globaly spread
configuration of filamentary strong and as a rule horizontal magnetic field of the intensity of IO^3 g to some extent
lowering in the upper part of the convection zone. Fainter
magnetic fields cannot damp the convective motion. This
network configuration includes also large scale structures
as supergranulation and greater features. The existence
of magnetic knots with the magnetic field intensity of
about $2.IO^3$ g (Sheely,I969, Tarbell and Title,I979) can be
considered as evidence in favour of our hypothesis. It may
be noted that the borders of these large structures do not
differ in brightness from the general intergranular network,though some regular arrangements of fine structure
(lanes, circles et al) can be detected.
     Presently we cannot measure accurately objects of the
size $0.2 - 0.4$. We know that when using "mean" values of
the field intensity averaged over ten and more square seconds of arc the depolarization of the Zeeman line components considerably reduces the resulting "mean" field intensity due to the different directions of elementary magnetic features. Horizontal fields in this case do not contribute to the measured magnetic field intensity. Of course, the above said characterized only the dark network.
As to granules we have no reason to ascribe to them magnetic fields greater than IO g. It is also probable that in
granules we can meet plasma overshots directly from the
convective zone.

2. LARGE SCALE CONFIGURATIONS

     The present concept of the chromosphere as a transient
(intermediate) region between the photosphere and corona
is similar to the first models of the inhomogeneous chromosphere outlined in Krat and Sobolev (I957), Krat (I968).
Inhomogeneity rapidly increases with height forming a
large system of upward and downward currents associated
with small scale magnetic structures. The typical elements
-fibrils are presumably oriented horizontally and hence
the measurements of the magnetic field intensities there
are of poor accuracy. Since the dark photospheric network
lies deeper than the granules we cannot expect to find

strong magnetic fields in the "undisturbed" chromosphere.
Thus the intensity of the chromospheric field probably is
of the order of several gauss and nowhere exceeds ten gauss. Only in chromospheric facular network in H_α, H and
K (CaII) the field can reach IO g which can be considered
as an evidence for inhomogeneity of the magnetic field
having its origin in the granular field which consists in
the chromosphere of many arcs and loops. Leighton was the
first to discover large scale objects - supergranulation
in the chromosphere. This fact gave rise to many new pro-
blems.For instance, whether the cells of supergranulation
are bound to the subphotospheric convection or they arise
as a result of large scale gasodynamic processes and especially of long waves similar to tsunami (Kassinsky and
Krat,I973). We suppose that the overall existing horizontal magnetic field can be lifted up by some impulses going from the convection zone. At these places vertical
components of strong photospheric field occur. Strictly
speaking this means that supergranulation arises as a consequence of perturbations of the granular network. From
this point of view the occurence of spicules at the borders of supergranulation cells seem to be quite natural.
This hypothesis is confirmed by the fact that the pattern
of granulation does not sensibly vary with respect to supergranulation. Moreover, we can say that the magnetic
field does not penetrate deeply in the convection zone
otherwise we should have observed at the borders of supergranulation cells, pores or spots, which is not the case.
    Naturally, a discovery of supergranulation and there-
after of 5 min oscillations was followed by many papers on
the large scale solar configurations and waves. Recent extensive review papers were published by Canfield and
Beckers (I976) and Dubov (I978). Therefore we shall not
discuss all papers mentioned in the latter review.
    The first problem concerns the existence of other
than supergranulation large scale objects. Thus, it seems
reasonable to consider first the distribution of brightness and radial velocities fluctuations on the solar disc.
The precise photoelectric photometry of the "undisturbed"
photosphere enabled to discover large scale modes (Yudina,
I976) in white light and at two effective wavelengths λ_1 =
=4200 A and λ_2 =5300 A with a large scatter from $7 \cdot 10^3$ km
to $I8 \cdot 10^3$ km in size, the mean value being $I3.2.I0$ km. The
influence of supergranulation is only slightly sensible
(size $3.2 \cdot 10^4$ km). The amplitudes of fluctuations for λ_1
and λ_2 are I.9% and I.6% respectively. It is of interest
that on stratospheric photographs obtained at the altitudes $3.I0$ km and $3.3 \cdot 10$ km (Herse, I979) we can see neither
granulation nor large scale features. It can be explained
if we account for an effective shift of the continuum upwards in the linear scale by I60-I70 km at $\lambda \approx 2000$ A which

leads to the disappearance of the photospheric dark network lying deep. It may be mentioned that Kubicela (I97I)
applied a very sensitive method for a study of the supergranulation. Further Kubicela and Karabin found considerably greater large-scale features. Recent paper by Krat,
Makarov and Tavastsherna (I980) containing a statistical
analysis of the spectrographic data obtained by use of
the large coronograph of the Pulkovo High Altitude Solar
Station (near Kislovodsk) leads to the conclusion that
the size of large scale inhomogeneities of the solar atmosphere are the same in the range of heights characterized
by the lines H_α and Fe-lines (approximately about $2.5 \cdot 10$
km). Measurements were made in the central parts of the
lines H_α , H_β , D (Na),   VI 4379.2 A, Fe 4383.6 A,
CH I4377.2 and 4387.3 A. The maxima of the distribution
curve have been obtained at $I.2 \cdot 10^4$ km and $3.3 \cdot 10^4$ km. The
former two maxima are more pronounced. The existence of
features of the size $I \cdot 10^4$ km was recently confirmed by
November et al.(I98I). These objects were named mesogranules. It is interesting that the size of characteristic
fields of 5 min wave propagation varying from $4 \cdot 10^3$ km to
$8 \cdot 10^3$ km are qualitatively similar to the size of mesogranules. The rms of the radial velocity in mesogranules
is 60 m/sec but the value is uncertain due to the effect
of superimposed flows of supergranulation. The life time
of mesogranules is not less than 2 hours being considerably shorter than that of supergranules (20 hours).
    The second maximum coincides with the chromospheric
network in the H and K lines (CaII) (Singh and Bappu,
I98I) and slightly varies during the solar cycle. Thus
all three modes of large-scale objects distribution can
be considered as certain.
    Our ideas at first glance seem to disagree with the
pattern of weak general solar magnetic field which has
its prolongation in the interplanetary space and for instance the sectorial field (Svalgaard at al.,I975).
The variation of this field during the cycle of Solar activity questions its existence as an object independent
of activity processes and it can be considered as a statistical active formation. The discrepancy between the
"old" (weak global field) and "new" field (strong global
field) can be solved with the statement that generally
the fields of granular penetrating in the outer layers of
the Sun are actually weak. There remains a dilemma. Is
this field regulated by the motions in the chromosphere
and corona or is it only a statistical mean of elementary
 weak photospheric fields? It can also be a residual field of active processes and in this case its cyclic variations would be natural.
    In conclusion I should like to note that this is a
short rewiew paper far from being a detailed one but it

seemed to me reasonable to discuss the problem of the solar atmosphere structure from one point of view. Of course,
further observations will introduce many corrections to
the proposed picture.

REFERENCES

Altrock R.C.,Musman S.,1976, Astrophys.J.203, p.533.
Canfield R.,Beckers J.M.1976. Coll.Int.CNRS N 250,
        p.291.
Dubov E.E. 1978. Itogi nauki i tekhniki. Ser.Astron.14,
p.148.
Gingerich O.,Noyes R.W.,Kalkofen W. 1971,Solar Phys.,
18,p.347;
Herse M. 1979, Solar Phys. 63, p.35.
Karpinsky V.N. 1981 Solnechnije Dannije N 1,p.88.
Kassinsky V.V., Krat V.A. 1973, Solar Phys. 31,p.219.
Kitai R.,Kawaguchi I. 1979,Solar Phys. 64,p.312.
Krat V.A.,Sobolev V.M.1957, Izv.Pulk.Obs.160,p.116.
Krat V.A.1968, Solar Phys.I,p.71.
Krat V.A. 1972, Solar Phys. 27, p.312.
Krat V.A. 1973, Solar Phys. 32, p.37.
Krat V.A. 1977, Radiophysika 20, p.1303.
Krat V.A.,Makarov V.I.,Tavastsherna K.S. 1980, Solar
Phys. 68, p.237.
Krat V.A.,Shpitalnaja A.A. 1974, Solnechnije Dannije
N° 2, p.63.
Kubicela A. 1972, Bull.Obs.Astron.Beograd v.28,p.187.
Kubicela A.,Karabin M. 1977, Solar Phys. 52, p.199.
Mehltretter I.P. 1974,Solar Phys.,38, p.43.
November L.J.,Toomre J.,Gebbie K.B.,Simon G.W. 1981
Astrophys. Lett.45, p.2.
Pravdjuk L.M.,Karpinsky V.N.,Andrejko A.V. 1974,
Solnechnije Dannije N 2, p.70.
Sheeley N.R. Jr. 1969 Solar Phys.,9, p.347.
Singh I.,Bappu M.K. 1981, Solar Phys.,71,p.161.
Sobolev V.M. 1974, Solnechnije Dannije N 2, p.55.
Stenflo I.O. 1973, Solar Phys.,32, p.97.
Stenflo I.O. 1976, IAU Symp.N 71,p,75.
Svalgaard L.,Wilcox I.M.,Scherer P.H.,Howard R. 1975,
Rep. N 639.
Tarbell T.D.,Title A.M. 1977,Solar Phys. 52, p.13.
Yudina I.V. 1976,Solnechnije Dannije N° 9, p.94,N° 10,
p.67.

A NOISY SUN ?

J. Perdang
Institut d'Astrophysique
B-42oo Cointe-Ougrée, Belgium

Abstract : We present theoretical and observational arguments
suggesting that the solar oscillations contain a 'chaotic'
component. On the basis of experiments on nonlinear mode
coupling we conjecture that the structure of the low frequency
end of the observed power spectra is indicative of such chaotic motions.

1. INTRODUCTION

The most generous estimate of the relative radial surface
displacement of the reported solar oscillations does not exceed 10^{-4} (Brown et al 1978). For solar theorists such a
small figure is seemingly regarded a sufficient condition
for strict validity of linear oscillation theory. Manifestly
however such a theory cannot account for the shape of the
observed power spectra. It is currently argued that the observed peaks correspond to vibrationally unstable linear
modes; in the context of this idea the spectra are featured
by the dissipative (positive and negative) necessarily non-
linear processes occurring in the Sun. Since at present even
linear dissipation in the outer solar layers is only poorly
understood the prospects of developing a satisfying theory
of the form of the solar power spectrum leaning on the idea
of nonlinear dissipation look meager.
A more promising avenue is traced out by the following analogy. The particles of a gas in a box at moderate densities
are essentially interaction-free; however mutual dissipation-
less coupling is instrumental in securing ergodicity i.e.
validity of thermodynamics. Sinai (cf Arnold 1976) has proved
that the presence of the slight collisional coupling in a 2-
particle 'gas' is sufficient to generate a class of 'chaotic'
motions of the particles accessible to a statistical description. This analogy has led us to study more closely the role
of small nonlinear mode coupling on the global motions with
dissipation being disregarded.

89

W. Fricke and G. Teleki (eds.), Sun and Planetary System, 89–92.

2. THE MECHANICAL BACKGROUND

A pioneering note by Woltjer (1935) established that the equations of stellar nonlinear adiabatic oscillations can be recast into a Hamiltonian form, with generalised coordinates q_j, $j=1,\ldots,F$, defined as the expansion coefficients of the finite relative displacement in a series of the linear eigenfunctions $\xi_j(m)$:

$$\delta r/r = \sum_j q_j(t)\, \xi_j(m) \qquad . \qquad (1)$$

On the other hand it is now known for some 2o years that the majority of nonlinear Hamiltonian oscillators of $F>1$ degrees of freedom can display besides multiperiodic or 'regular' oscillations a second type of bounded motions of a partially noisy or random character; the latter, not representable by multiple Fourier series, are referred to as 'Kolmogorov unstable'(Izraelev and Chirikov 1966) or 'chaotic' oscillations. The genealogy of the latter is captured by the Kolmogorov-Arnold-Moser (KAM) theorem which in essence teaches us that chaos is the outcome of internal resonances. If H_0 is the Hamiltonian of a system of F degrees of freedom whose motions are all known to be multiperiodic of frequencies Ω_j, $j=1,\ldots,F$, then almost all motions of a system described by a perturbed Hamiltonian $H_0 + \varepsilon h$ remain multiperiodic provided that the frequencies Ω_j are nonresonant, i.e. there exist no integers k_j such that

$$\sum_{j=1}^{F} k_j\, \Omega_j \sim o \qquad , \qquad (2)$$

in the limit of small ε. As the nonresonance requirement is violated in general a sizeable fraction of the totality of bounded motions is found to be chaotic, experimental instances being known where chaos prevails at arbitrarily small values of ε. It is also observed that with increasing ε the width of the approximate resonance bands for which chaos occurs is increasing (cf Perdang 1981b for a brief review).

In applying these ideas to stellar oscillations one notices that low order approximate resonances among the linear oscillation frequencies violate KAM so that they can generate chaotic motions (Perdang 1978) in the actual nonlinear Hamiltonian oscillations. But such approximate resonances are shown to appear generically in the linear spectrum of all stellar models (a) among adjacent radial asymptotic modes, (b) among the $(2\ell+1)$ x degenerate nonradial modes, (c) among nonradial modes (k,ℓ) and $(k-1,\ell+2)$ in the asymptotic k range. A closer analysis then reveals that any small amount of energy - provided it is injected into sufficiently high asymptotic resonant linear solar modes - should lead to chaos (Perdang 1981b).

3. ASSESSING CHAOS

A bounded motion $f(t)$ - the solar surface displacement or velocity curve - observationally known over a sufficiently long time period T can in principle be identified as regular or chaotic by investigating its power spectrum $P(\omega;T)$. (a) For a regular motion this spectrum is made up of sharp lines located

at F basic frequencies Ω_j and at the combination frequencies of the latter $\Omega_{k_1 \ldots k_F} = |k_1 \Omega_1 + \ldots + k_F \Omega_F|$ of order $k = \sum^F |k_j|$; the peak heights of regular oscillations decreasing exponentially with order k, the spectrum will reveal just a few combination frequencies. For the <u>chaotic</u> motion however a highly involved broad band structure appears in the spectrum the centre of each band lying close to an Ω_j or $\Omega_{k_1 \ldots k_F}$ of low order (Noid et al 1977, Blacher and Perdang 1981a); moreover each band possesses a hierarchical substructure revealed at improved resolutions. Hence mere inspection of a power spectrum can already allow us to distinguish regular motions and chaos. (b)A less subjective procedure consists in resorting to a structure index of the spectrum capable of quantifying its complexity; we proposed the <u>fractal dimension</u> d of the renormalised power spectrum (in the limit $T \rightarrow \infty$, with the highest peak normalised to unity; Perdang 1981a; Blacher and Perdang 1981a) as an adequate measure. In fact for a regular motion d is unity, so that a spectrum with a $d > 1$ discloses an underlying chaos. Obviously a convincing empirical determination of d requires a high resolution power spectrum. In the solar context a minimum resolution $2\pi/T$ of $\sim 1\,\mu$Hz has been estimated (Perdang 1982); the present South Pole program shows that such a resolution is now within reach (Grec et al 1980).
A chaotic time behaviour f(t) is not just a purely random time series. (a)Numerical experiments disclose that chaos due to 1-1 resonances is well represented by a linear superposition of a deterministic oscillation with however a probabilistic frequency, and a noisy correction responsible for a random amplitude fluctuation (Perdang 1981b):

$$f(t) = a \left\{ \cos\omega(1+x)t + \eta \cos\omega y t \right\} \quad ; \quad (3)$$

(a, ω and η a fixed amplitude, resonance frequency and 'noise' effect respectively; x and y are dimensionless random variables of average 0). The power spectrum of (3) then contains a main band centered at frequency ω and a secondary band at the origin. Power spectra of numerically computed Hamiltonian chaotic motions are found to conform to this structure (Blacher and Perdang 1981ab,1982). (b)Eq(3) also reveals that for η and an x-dispersion not too large ($\lesssim .1$ say) the chaotic time behaviour still remains essentially phase-coherent; this property has been demonstrated numerically for Hénon-Heiles chaos and for chaotic oscillations in polytropes (Blacher and Perdang 1981b, 1982). So far phase coherence has been regarded as a distinguishing attribute of normal mode oscillations (Caudell and Hill 1980).

4. THE SHAPE OF THE SOLAR SPECTRUM

On the assumption that a mechanism is operating in the Sun that energises a broad band of asymptotic linear modes the following structure of the low frequency end of the power spectrum is to emerge : Due to the 1-1 resonances of neighbouring pairs of modes a chaotic motion sets in which is basically represented by a superposition of a huge number of components of form (3).

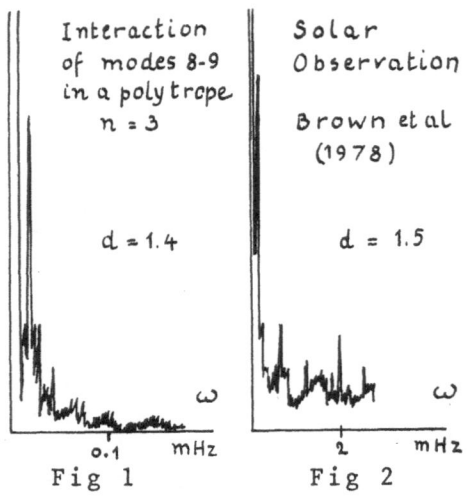

Fig 1   Fig 2

For each pair the resonance frequency differs while the 'intrinsic noise' contribution remains centered at the same frequency namely at the origin. Fig 1 provides an illustration of the shape of such a chaotic contribution due to the resonance among two adjacent radial modes of a polytropic model (Blacher and Perdang 1982). Consequently at the origin a very high structured band is due to build up; the chaotic character of the latter implies a fractal dimension d > 1.

The SCLERA power spectra (cf Fig 2) all display the expected hierarchical structure near the frequency origin $\omega \lesssim$ 1 mHz; moreover estimates of the fractal dimension of the latter yield d~1.5. A similar 'noisy' band at the low frequency end has been observed in the high resolution power spectra recorded at the South Pole (Grec et al 1980).

Therefore we feel entitled to conclude that present observations are not incompatible with a noisy ringing of the Sun, the intrinsic noise component being the result of resonant interactions of the linear modes. Since the latter secure an efficient energy exchange among the modes one expects that equipartition of energy should hold in the asymptotic linear spectrum. The knowledge of the density of modes in the asymptotic range then enables us to derive the asymptotic part of the power spectrum analytically (Perdang 1982).

REFERENCES

Arnold VI 1976 'Méthodes mathématiques de la mécanique' Mir, Moscow

Blacher S,Perdang J 1981a Physica 3D 512; 1981b MNRAS 196 1o9; 1982 (in prep.)

Brown TM, Stebbins RT, Hill HA 1978 Ap J 223 324

Caudell TP, Hill HA 1980 MNRAS 193 381

Grec G, Fossat E, Pomerantz M 1980 Nature 288 541

Izraelev FM, Chirikov BV 1966 Sov Phys Dokl 11 3o

Noid DW, Koszykowski ML, Marcus RA 1977 J Chem Phys 67 4o4

Perdang J 1978'Stellar Oscillations: The Asymptotic Approach' FNRS,Brussels; 1981a Astrophys Space Sci 74 149; 1981b Comm IAU Coll 66,Crimea; 1982 Astrophys Space Sci 83, 311.

Woltjer J 1935 MNRAS 95 26o

SPECTRAL ANALYSIS OF WAVE MOTIONS IN THE REGION OF TEMPERATURE MINIMUM OF THE SUN'S ATMOSPHERE

V.E.Merkulenko, V.I.Polyakov, and V.S.Loskutnikov
SibIZMIR, Irkutsk-33, P.O.Box 4, 664033, USSR

ABSTRACT-We present some results of an analysis of the spatial spectrum and two-dimensional distribution of horizontally running waves of five-minute oscillations.

1. INTRODUCTION

High-resolution observations of the photospheric lines made by Deubner et al. (1979) and by other researchers revealed a fine structure of the spectrum of five-minute oscillations, in agreement with the ρ-mode distribution on the (K, ω) diagram (Ulrich and Rhodes, 1977).

In the first part of the present work an analysis of filtergrams in the Ba II 4554 + 0.05 Å arising in the temperature minimum region has been made, comparing the spatial spectrum of oscillations to theory. Using the same line, the second part studies the two-dimensional distribution of directions of horizontally running waves.

2. SPATIAL SPECTRUM OF OSCILLATIONS

A filtergram of 1" arc resolution has been used, taken at disk center of a quiet chromosphere. On a 256"x 200" area, 50 photometric scans have been carried out with the photometer slit 0.5"x1". The calculation of the spectrum from the scans has made use of the correloperiodogram-analysis method, which has consisted in calculating the correlation coefficient between the initial realization and a set of harmonic functions. Spatial spectra for photometric scans have been plotted in Fig. 1. The abscissa gives the wavelength, the ordinate indicates the distance in seconds of arc normal to photometric scans. The degree of shading corresponds to two consecutive values of the correlation coefficient in excess of a 90% confidence level. A periodogram averaged over 50 scans is shown in Fig. 2 by a solid line. On the ordinate the mean value of the product of the correlation coefficient has been plotted

93

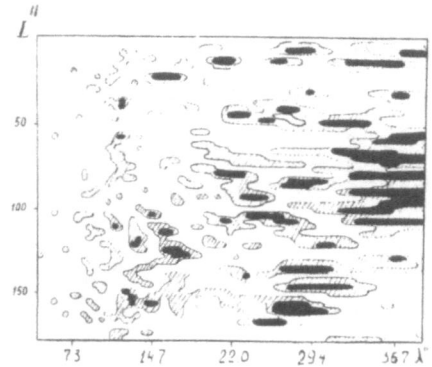

Figure 1. Diagram of spatial spectra of consecutive photometric scans.

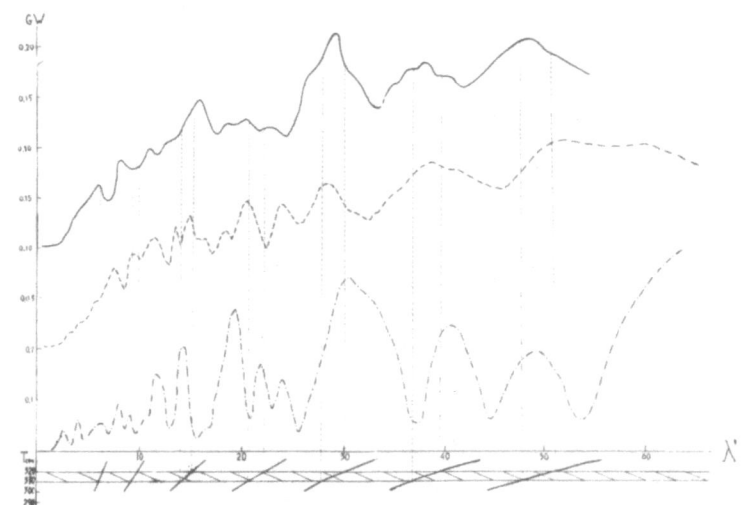

Figure 2. Averaged spatial spectra:
        over 50 scans - solid line;
        over 25 scans - dotted line;
        over 1   scan - dash-dot line.

versus the related value of probability. The dotted line shows a periodogram, averaged over 25 scans on a filtergram taken 3 hours later. The dash-dot line corresponds to the periodogram for one scan. The best coincidence of the main peaks of the spectrum in space with the theoretically calculated spectrum of general oscillations of the Sun takes place in 3.22 to 3.12 MHz frequency band. A fragment of the diagram of Ulrich and Rhodes (1977) for this band is shown on the lower part of Fig. 2.

3. THE TWO-DIMENSIONAL DISTRIBUTION OF DIRECTIONS OF HORIZONTALLY RUNNING WAVES

The analysis has been made of a series of filter-grams taken in the course of 32 minutes at disk center on a 128"x128" scale. The interval between the frames is 30 sec. The microphotometer slit is 2"x2". For each line and each column of a three-dimensional matrix, a two-dimensional spectrum ($\omega, \pm K_{x,y}$) has been calculated with a version of FFT program. The degree of asymmetry of the spectrum on the coordinates with respect to the   axis has been determined by the relation

$$F_{x,y} = \frac{\sum P_{+\kappa} - \sum P_{-\kappa}}{\sum P_{+\kappa} + \sum P_{-\kappa}}$$

where $P_{\pm\kappa}$ is the product of the level of spectral distribution by the relevant area for the left and right parts of the diagram. The arrows in Fig. 3 show the value of $F_{x,y}$

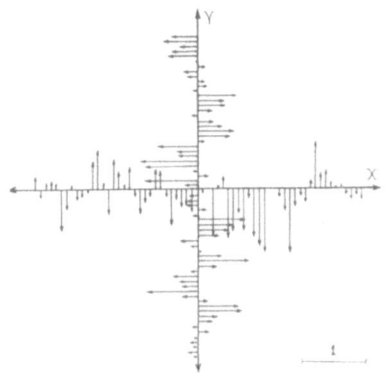

Figure 3. Variation of the direction of horizontally running waves on the x and y axes.

as calculated for each scan along the x and y axes. The length of the arrow equals unity, if the waves are running along the scan in the same direction but it tends to zero in the case of standing wave. Fig. 4 shows the variation of the value F, obtained from the vectorial addition of the values F_x and F_y .

4. DISCUSSION

The fundamental harmonics of the spatial spectrum of oscillations in the temperature minimum region for small values of the wave number k are confidently identified with the p -modes of nonradial oscillations of the entire Sun. The distribution of horizontally running waves shows an anisotropy which can be determined from the spectra taken at short intervals of observation. As for long intervals, such anisotropy must disappear as a consequence of the averaging, in which case only the drift velocity

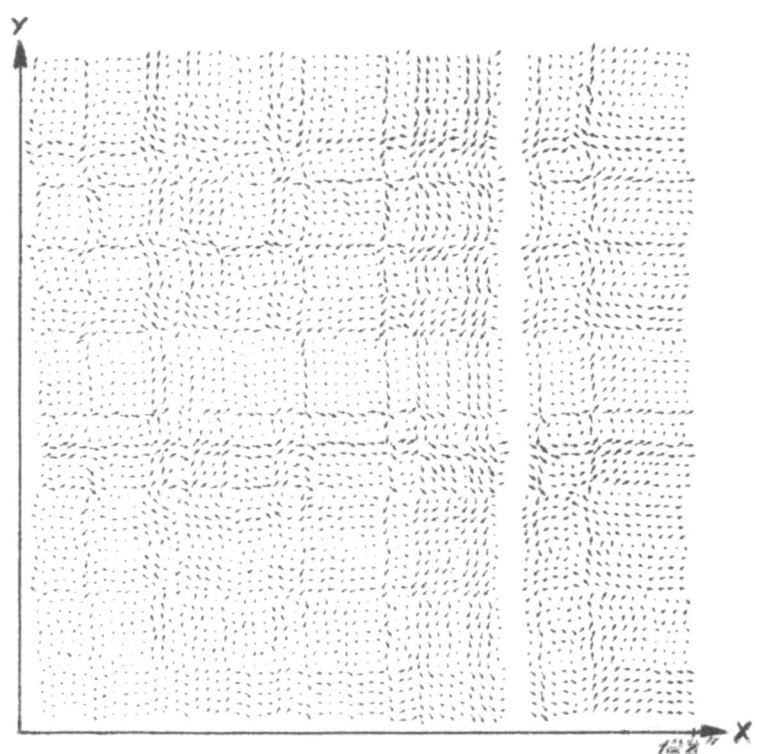

**Figure 4. Two-dimensional picture of variation of
the direction of horizontally running waves.**

of oscillations is measured due to solar rotation (Rhodes
et al., 1979). The direction of propagation of these waves
reverses periodically both in the Θ coordinate and in the
γ coordinate, with a characteristic scale of variation of
about 40".

Acknowledgement

We gratefully acknowledge beneficial comments
of Prof. G.Marx on the style of this paper.

REFERENCES

Deubner, F., Ulrich, R.K., and Rhodes, E.J.: 1979, Astron.
    and Astrophys., 72, p. 177.
Rhodes, E.J., Deubner, F., and Ulrich, R.K.: 1979, Astro-
    phys. J., 227, p.629.
Ulrich, R.K., and Rhodes, E.J.: 1977, Astrophys. J., 218,
    p.521.

ON THE DETECTION OF THE PERIODIC OSCILLATIONS OF THE INCLINATION OF THE FREQUENCY SPECTRUM OF S-COMPONENT OF SOLAR RADIO EMISSION

A.R. Abbasov, L.F. Golubeva, L.B. Tzirulnik
Institute of Cosmic Research of Natural Resources, Shemakha
Astrophysical Observatory, Azerbaijan SSR, USSR

The values of radio fluxes from some emitting areas were employed at wavelengths λ = 3.2cm and 8cm for 1969-1970 (Solar Activity Charts). 18 groups of spots were selected which were responsible for the longest series of radio observations. Flux differences $\Delta F_i = F_i(\lambda_1) - F_i(\lambda_2)$ were composed for each spot group where $F_i(\lambda_1)$ and $F_i(\lambda_2)$ are radio fluxes at a wavelength of 3.2cm and 8cm respectively for the i-th spot group. Next, another set of differences $\Delta^2 F_{ij} = \Delta F_i(t_j + 1) - \Delta F_i(t_j)$ was constructed where t_j is the j-th moment of observation. All subsequent calculations were carried out with $\Delta^2 F_{ij}$ values.

Let us assume that we have a certain oscillation of the frequency spectrum inclination typical for emitting areas. The existing comparatively short series of obseravtions for each emitting area constitute individual portions of this general periodic function. The least square method was selected for making these portions coincident with the general curve. In the first approximation, this general periodic function was considered as a purely harmonic oscillation with an unknown circular frequency ω. To find this frequency a function of the form

$$\Delta(\omega) = \sum_{i=1}^{18} \Delta_i(\omega)$$

was written where

$$\Delta_i(\omega) = min\Delta_i(\omega, A_i, B_i),$$

$$min\Delta_i(\omega, A_i, B_i) = \frac{\sum_{j=1}^{N_i}\left[y_{ij} - (A_i cos\omega t_j + B_i sin\omega t_j)\right]}{\sum_{j=1}^{N_i} y^2_{ij}}$$

where N_i = number of observations in the i-th spot and $y_{ij} = \Delta^2 F_{ij}$, i.e.

97

W. Fricke and G. Teleki (eds.), Sun and Planetary System, 97–98.

spectrum inclination variation rate in the i-th spot at the j-th time
moment. The values $A_i(\omega)$ and $B_i(\omega)$ were found for each ω from the condition of minimum for the function $\Delta_i(\omega, \Delta_i, B_i)$. The frequency ω was
varied with a step of $\delta\omega = 0.1$.
Four minimum at frequencies of o.7, 1.3, 2.1 and 2.8 can be clearly
defined.
To check our results, a harmonic with a frequency of $\omega = 2.1$ (frequency
of principal minimum) was subtracted from the initial data with corresponding A_j and B_j parameters. Thereupon, the function $\Delta(\omega)$ was again
calculated for the reminder.
The subtraction of the harmonic with a maximum amplitude increased seve-
ralfold the depth of the remaining three minima while shifting slightly
their positions. This is a testimony of four actual periods existing in
all 18 active regions. To improve the values for these periods, the problem was formulated as follows: find local minimum of the function
$\Delta(\omega_1,\omega_2,\omega_3,\omega_4)$ near the frequencies $\omega_1=0.7$, $\omega_2=1.3$, $\omega_3=2.1$, $\omega_4=2.8$
determined earlier.

Here $\quad \Delta(\omega_1,\omega_2,\omega_3,\omega_4) = \displaystyle\sum_{i=1}^{18} \Delta_i(\omega_1,\omega_2,\omega_3,\omega_4),$

$$\Delta_i(\omega_1,\omega_2,\omega_3,\omega_4) = \frac{\displaystyle\sum_{j=1}^{N_i}\left[y_{ij} - \sum_{k=1}^{4}(A_{ik}\cos_k t_j + B_{ik}\sin_k t_j)\right]^2}{\displaystyle\sum_{j=1}^{N_i} y^2_{ij}}$$

As a result we obtained the following frequencies: $\omega=0.6$, 1.4, 2.1 and
2.8. These values suggest that the last three frequencies are obertones
of the main frequency, $\omega=0.7$ which is distorted by the presence of
another lower frequency $\omega<0.7$. On this basis the local minimum was found
for the function $\Delta(\omega_0,\omega_1,\omega_2,\omega_3,\omega_4)$ in the region $\omega_0=0.3$, $\omega_1=0.7$, $\omega_2=1.4$,
$\omega_3=2.1$, $\omega_4=2.8$. It corresponded the following frequences: $\omega_0=0.5$, $\omega_1=0.7$,
$\omega_2=1.4$, $\omega_3=2.1$, $\omega_4=2.8$. It appears then that we have found a certain
oscillation with a period of 9 days and its three overtones. One of possible explanations of this period may be the difference (due either to
emission mechanism or source geometry) in the directivity of the emission
of active areas on transition from the solar centre to the limb (Abbasov
et al., 1967). A simultaneous effect of these factors may be the reason
for abovementioned oscillations.

Our thanks to Prof. A.P.Molchanov for helpful discussions.

REFERENCE

Abbasov,A.R., Akhmedov,S.B., Grebinsky,A.S., Molchanov,A.P.:
        1967, Astron.Zh., 44. p. 1326.

SOME PHOTOSPHERIC CHARACTERISTICS OF THE SUN AS A STAR

E.A. Gurtovenko
Main Astronomical Observatory, Ukrainian
Academy of Sciences, Kiev, USSR

Many non- classical solar features are common among stars. Those features influence spectral lines, and for stellar spectra diagnostic one must know how to discern that influence from other peculiarities of stellar spectra. But one can learn a lot about stellar atmospheres by extending what one knows about the Sun.

We discuss here two quantities: turbulence and damping constant. In 1975-1976 the velocity amplitude of the total motion field seemed to be known right. It was deduced from profiles of weak lines (Figure 1, Gurtovenko. Canfield and Backers). But the micro and macrovelocities were known poorly. Lately the suspition arised that the total velocity amplitude might be also incorrect. The point is that the velocity distribution is not Gaussian, and the real motion field make the line wings more extended comparing to Voigt-function. This may result in additional increase of the velocity amplitude with depth. It is very likely that Gaussian velocity distribution formalism acts correctly for line cors. Thus the possibility

Table 1. Microturbulent velocity amplitude (kms^{-1})

Authors	v_{mic}^{rad}	v_{mic}^{tc}
Gurtovenko and Ratnikova,1976	1.0	1.3
Gurtovenko and Sheminova,1979	0.8	1.9
Sheminova,1981	0.9	1.9
Kostik,1981	0.8	1.4
Blackwell et al., 1976	0.8	1.7
Holweger et al.,1978	1.0	1.6
Blackwell et al.,1979	0.8	1.3
Mean value	0.87	1.6

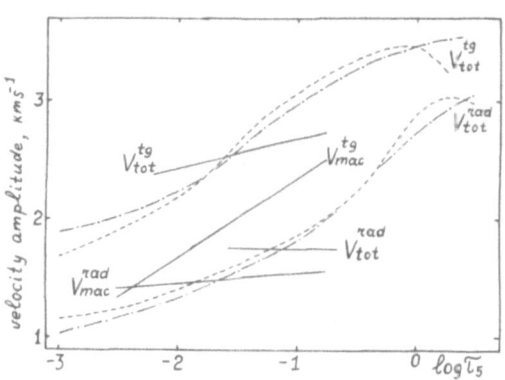

Figure 1. Solar photospheric velocity field components: — — — Gurtovenko, 1975; —·— Canfield and Backers,1976; —— Kostik, 1981.

W. Fricke and G. Teleki (eds.), Sun and Planetary System, 99–100.
Copyright © 1982 by D. Reidel Publishing Company.

for studying the motion field still exist, namely: the microturbulence
should be deduced from equivalent widths and the macroturbulence from
line cors. Table 1 gives the microturbulent velocity amplitude. Those
results are obtained under assumption of a depth-independent model.
Recently Kostik (1981) has found a small increase of both components
with height. In Figure 1 the macroturbulent velocity amplitude is shown.
It has been obtained from the cores of different Fraunhofer lines.
Combining those results with the microturbulence one can calculate the
total velocity amplitude (Figure 1). It differs appreciably from the
results obtained using line profiles.

During the last decade there was a lot of discussions on damping constant
because its observational value occured to be many times larger compared
to the van der Waals damping γ_6. Recently we found that damping constant
derived from wings of moderate lines was systematically overstimated
(Figure 2). This effect is also due to the extra broadening of line wings
by real motion field. Only strong lines give the correct damping constant.
It is a bit too large compared to γ_6. The same real value of damping
constant gives the study of equivalent widths (Table 2). Thus one may
conclude that solar investigations have removed the problem of large
damping constant in stellar photospheres.

One has to keep in mind also that difficulties in the interpretation of
solar Fraunhofer lines will be the same in diagnostic of high stellar
spectra. In particular the description of line profiles by the Voigt
function should be used with the great precaution.

Table 2. Enhancement factor E for
damping constant γ_6

Authors	Method	E
Gurtovenko et al.,1981	Wings of strong lines	1.4
Kostik,1981	Equiv. widths of FeI lines	1.3
"	" NiI "	1.6
"	" CrI "	1.8
"	" TiI "	1.7

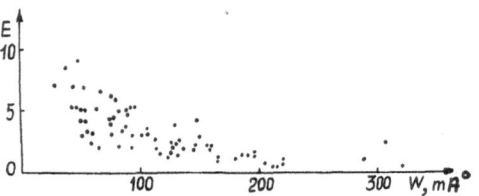

Figure 2. Enhancement factor E for
damping constant γ_6, from
wings of FeI lines.

REFERENCES

Blackwell,D.E., Shallis,M.J.: 1979, M.N.R.Ast.Soc., 186, pp. 673-684.
Canfield,R.C., Beckers,J.M.: 1976, Phys.Mouv.Atm.Stell., pp. 291-331.
Gurtovenko,E.A.: 1975, Solar Phys., 45, pp. 25-33.
Gurtovenko,E.A., Sheminova,V.A.: 1979, Astron.Astrofiz. 38, pp. 29-35.
Holweger,H., Gehlsen,M., Ruland,F.: 1978. A.Astrophys. 70, pp. 537-542.
Kostik,R.I.: 1981, Inst.Teor.Fiziki, prepr. ИТФ-81-20p,Kiev

STARK BROADENING OF HEAVY ION SOLAR LINES

Milan S. Dimitrijević
Institute of Applied Physics, P.O.Box 58,
11071 Beograd, Yugoslavia

Investigation of the Stark broadening of atomic and ionic lines is inpor-
tant for the estimation of the physical conditions in the stellar atmos-
pherae. Stark broadening data are also required for determinatior of the
abundances of elements and for evaluations of the radiative transfer
through stellar interior. In the latter case, data for a great number of
lines are often needed. Recently, a modified version of the Stark broadening theory is published (Dimitrijević and Konjević, 1980), usefull especially for such, large scale calculations with a good average accuracy.
This approach has been tested before for a large number of ionic lines
(Dimitrijević and Konjević, 1980; 1981) of lighter elements and nume-ical
results compare well with experimental data (Dimitrijević and Konjević,
1980). Here, this theory is applied to the most intensive lines of Ti II
and Mn II, observed in the solar spectrum (Moore et al., 1966), 'n order
to test the applicability of the approach to the case of more conplex
transitions and heavier elements.

According to Dimitrijević and Konjević (1980), electron impact line width
(FWHM) can be calculated from the following expression:

$$W_{SEM} = 2(2\pi\hbar/3m)^2 (6m/\pi kT)^{1/2} \sum_{j,j^-=i,f,i;f^-} [\vec{R}^2_{\ell_j,\ell_j+1} \tilde{g}(E/\Delta E_{\ell_j,\ell_j+1}) +$$

$$+ \vec{R}^2_{\ell_j,\ell_j-1} \tilde{g}(E/\Delta E_{\ell_j,\ell_j-1}) + \sum_{j^-} (\vec{R}^2_{jj^-})_{\Delta n\neq 0} \tilde{g}(3kTn_j^3/4Z^2 E_H)] \tag{1}$$

$$\tilde{g}(x) = 0.7 - 1.1/Z + g(x) \tag{2}$$

$$\vec{R}^2_{\ell,\ell^-} = (3n/2Z)^2 [\max(\ell,\ell^-)/(2\ell + 1)] [n^2 - \max{}^2(\ell,\ell^-)] \phi^2 \tag{3}$$

$$\sum_{j^-} (\vec{R}^2_{jj^-})_{\Delta n\neq 0} = (3n_j/2Z)^2 (n_j^2 + 3\ell_j^2 + 3\ell_j + 11)/9 \tag{4}$$

101

W. Fricke and G. Teleki (eds.), Sun and Planetary System, 101–102.
Copyright © 1982 by D. Reidel Publishing Company.

Ion	Transition (mult.no.) wavelength Å	T(K)	W_{SEM} (Å)	W_{SE} (Å)	W_{SC} (Å)
Ti II	$a^4F-z^4G^0$ (1) 3364.8 Å	2500 5000 10000 20000	0.279 0.198 0.140 0.0988	0.262 0.186 0.131 0.0928	
Ti II	$a^4F-z^4D^0$ (5) 3079.4 Å	2500 5000 10000 20000	0.238 0.168 0.119 0.0840	0.230 0.163 0.115 0.0815	0.228 0.164 0.125 0.0996
Mn II	$a^5D-z^5P^0$ (3) 3464.0 Å	2500 5000 10000 20000	0.246 0.174 0.123 0.0870	0.166 0.117 0.0830 0.0587	0.266 0.190 0.144 0.114

$N_e = 10^{17} \text{ cm}^{-3}$

Here, i and f designate initial and final energy levels, n is the effective principal quantum number, (Z-1) is the ionic charge and ϕ is tabulated by Oertel and Shomo (1968). In present calculations, ϕ is also multiplied with corresponding coefficients of fractional parentage (Shore and Menzel, 1965).

Results for two Ti II and one Mn II multiplets are given in the table. Calculations were also performed according to Griem´s (1968) semiempirical (W_{SE}) and semiclassical (W_{SC}) theory (Jones et al., 1971) and obtained results are presented in the table.

The agreement between simple modified semiempirical approach and more sophisticated semiclassical calculations is very encouraging and indicates that this simple method can be used for estimation of electron width for heavier elements.

REFERENCES

Dimitrijević, M.S., and Konjević, N.: 1980, JQSRT 24, pp. 451-459
Dimitrijević, M.S., and Konjević, N.: 1981, in "Spectral line shapes",
          ed. B.Wende, W de Gruyter, Berlin-New York, pp. 211-239
Griem, H.R.: 1968, Phys.Rev. 165, pp. 258-266
Jones, W.W., Benett, S.M., and Griem, H.R.: Univ. of Maryland Techn.
          Rep. No 1971-128, College Park, Maryland
Moore, C.E., Minnaert, M.G.J., and Houtgast, J.: 1966, The Solar Spec-
          trum 2935Å to 8770Å, NBS Monograph 61, U.S. Govt.Print.
          Office, Washington D.C.
Oertel, G.K., and Shomo, L.P.: Astrophys. J. Suppl. Series 1968, 16,
          pp. 175-191
Shore, B.W., and Menzel, D.H.: 1965, Astrophys. J. Suppl. Series 12,
          pp. 187-214

THE ANALYSIS AND OBSERVATION OF THE NEUTRAL
SPECTRUM OF POTASSIUM IN THE SOLAR ATMOSPHERE

N.G. Stchukina
Main Astronomical Observatory, Ukrain Academy of Sciences,
Kiev, USSR

The solar line spectrum carries a weath of information about the
atmospheric structure. But a review of the last papers clearly indicates
that the importance of nLTE effects on the solar line spectrum must be
considered before we can use the lines of some chemical elements for
atmosphere structure diagnostics. We evaluate here the departures from
LTE populations for KI levels and synthesize the profile of the resonance
line 7699A with the usual assumptions of an one-dimensional plane parallel mean atmosphere (HSRA) and complete redistribution of photons in
frequency within each line. Than we investigate the efects of some parameters and photoionization mean intensity approximations on departure
coefficients, source function and profile of the resonance line 7699A.
Those parameters are: multilevel structure, collisional excitation and
ionization cross-sections, photoionization cross-sections. Mean intersity
approximations in bound-free (bf) continuum are following: photospheric
and chromospheric radiative temperature models, exact linearized solution
and two-level approach (Feautrier, 1964).

The nLTE calculations consist of simultaneous and consistent solution of
linearized equations of radiative transfer, statistical equilibrium and
particle conservation for seven level KI atom model by method of Auer and
Mihalas (1969). To compare the theoretical profile of 7699A with observations we have used to centre-to-limb spectrograms which were observed
with high-resolution spectrograph in Kiev. The other details of calculation are published elsewhere (Stchukina, 1981).

The results are summarized as follows:
Photoionization mean intensity uncertainties arising from the choice of
the methods of mean intensity calculation in bf continuum produce appreciably stronger changes in departure coefficients, source function
and profile of 7699A comparing to the influence of photoionization and
collisional cross-sections uncertainties and multilevel structure.
Changing the cross-sections with a factor of the order of 2 to 10 practically does not influence the departure coefficients, source function and
centre-to-limb variation of 7699A. The best fitted profiles of 7699A for
some approximations of mean intensity and two approximations of cross-

W. Fricke and G. Teleki (eds.), Sun and Planetary System, 103–104.

sections (hydrogenic and quantum defect) are shown in figure 1. The comparison of the exact linearized solution for KI bf mean intensity with approximation by Planck function with modeled radiative temperature shows that the last approximation is valid only for deeper layers of atmosphere from $\tau_5 \geqslant 0.2$. It fails for higher layers. So radiative temperature models of KI bf continuum represent the line spectrum of KI incorrectly and can lead to misleading information about the abundance (Reza, Muller, 1975). The two-level approach approximation is fairly correct and within few presents is the same as exact linearized solution.

Departures form LTE occur for considered levels in all depth above $\tau_5=1$. They are especially large in the chromosphere and amount there to the values of order 10^3 for ground level, 10^2 - for $4^2P_{1/2}$ $\sqrt{10}$ - for $4^2P_{3/2}$.

Synthetic profiles computed in the nLTE-assumption reproduce the observed centre-to-limb observations much more better than the LTE computations.

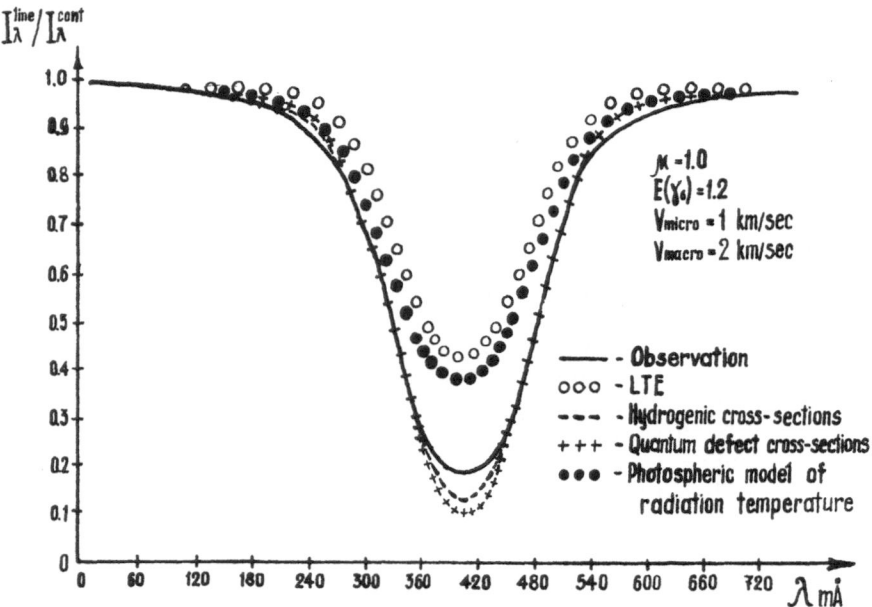

Figure 1. Observed and synthetic LTE profiles of the KI 7699A line at
         μ = 1.0

REFERENCES:

Auer, L.H., Mihalas, D.: 1969, Astrophys. J., 158, pp.641-645.
Feautrier, M.P.: 1964, C.R.Acad.Sci., 258, 3189-3191.
Reza, R., Muller, E.: 1975, Solar Phys., 43, pp. 15-31.
Stchukina, N.G.: 1981, Solar Data, in print.

MESOTURBULENCE IN THE SOLAR ATMOSPHERE

R.I. Kostik
Main Astronomical Observatory of the
Ukrainian Academy of Sciences,KIEV-127, USSR

Solar spectral lines are observed to be broader and stronger than expected from purely thermal and pressure broadening. Classically, the broadening and strengthening are attributed to a macro- and microvelocity field. Such separation of the motions, however, does not take into account the physical nature of the turbulence and has a formal character.

In the seventies there were developed the methods which took into account the distribution of the turbulent elements by sizes (Auvergne et al., 1973). Two stochastic models have been adopted for discussing meso-turbulent line formation: the Uhlenbeck-Ornstein process (UOP) and Kubo-Anderson process (KAP). The equation describing the line profile is stohastic one and depends on correlation length l as well as the mean square turbulent velocity ξ.

In this contribution we shall apply the mesoturbulence formalism (KAP) to derive the correlation lengths in the region of formation of weak and medium-strong solar lines.

In previous paper (Kostik, 1982) we have determined the variation of micro- and macroturbulent velocity with height. Recently (Sedlmayr, 1980; Kostik, 1982) it has been shown that relation

$$\xi^2 = v_{mic}^2 + v_{mac}^2$$

holds with considerable accuracy for solar and different stellar atmospheres. It is also known that the abundance A of a given chemical element determined from the central intensity of the line depend on ξ and l. Therefore if the values A, ξ are known, the correlation length can be easily found by their variation until the abundance of a given element found from the central intensity will be equal to its initial value.

To find the value l using described method we have used the central intensity (Delbouille et al., 1973) of 8 iron lines ($\lambda\lambda$ 4232, 4347, 4445, 5225, 5247, 5250, 5956, 6280) with Oxford oscillator strengths (Black-well et al., 1979a, b), HOLMUL model (Holweger and Muller, 1974),

W. Fricke and G. Teleki (eds.), Sun and Planetary System, 105–106.

$\gamma = 1.3\ \gamma_6$ and $\log A(Fe) = 7.59$ (Kostik, 1982). The average value of correlation length thus determined is equal $l = 570$km. Recently Sedlmayr (1980) derived empirical relations between micro-macroturbulent velocity and the mesoturbulence parameters l, ξ. UOP has been adopted. Having such relation and knowing the velocity of micro- and macroturbulent motions (Kostik, 1981) we found on the average: $l = 680$km (from relations v_{mac}/v_{tot}), $l = 530$km (from relations v_{mic}/v_{tot}). Different methods give close results ($l = 590$km), in good agreement with the data derived by de Jager and Vermue (1977, $l = 600$km), and Sedlmayr (1980, $l = 530 - 640$km). At last we have calculated the mean depth of formation of each line and have obtained the dependence $l(\log\tau)$. As can be seen from Figure 1, the scater of the values l has the systematic trend: the correlation length decreases with height.

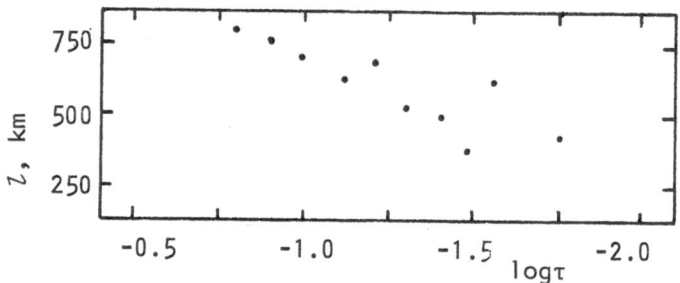

Figure 1. Variation of the correlation length in the solar atmosphere.

REFERENCES

Auvergne,M., Frisch,H., Frisch,U., and Froeschle,Ch.:
              1973, Astron. Astrophys. 29, pp. 93-102.
Blackwell,D.E., Petford,A.D., and Shallis,M.J.:
              1979a, Mon. Not. Roy. Astron. Soc. 186, pp. 657-668.
Blackwell,D.E., Ibbetson,P.A., Petford,A.D., and Shallis,M.J:
              1979b, Mon. Not. Roy. Astron. Soc. 186, pp. 633-650.
De Jager,C. and Vermue,J.: Solar Phys. 54, pp. 313-317.
Delbouille,L., Neven,L., and Roland,C.: 1973, Photometric Atlas of the
              Solar Spectrum from λ3000 to λ10000, Liège.
Holweger, H. and Muller,E.A.: 1974, Solar Phys. 39, pp. 19-30.
Kostik,R.I.: 1982, Solar Phys. 78, in press.
Sedlmayr,E.: 1980, Lect. Notes Phys. 114, pp. 195-210.

VOIGT PROFILE FOR THE CO EMISSION LINES IN THE SOLAR SUNSPOT SPECTRUM

C. J. MACRIS, B. Ch. PETROPOULOS
Research Center for Astronomy and Applied Mathematics,
Academy of Athens

The CO $(A^1\Pi - X^1\Sigma)$ transition was identified by Goldberg et al (1965) in the solar spectrum. Parkinson et al (1969) have used the above transition and determined a solar temperature minimum $T_e = 4000°K$. Jordan et al (1978) have reported the detection of emission lines from CO $(A^1\Pi - X^1\Sigma)$ at 1518A° to 1550A°, in the U.V. spectrum of a sunspot. In a further analysis which was carried out by these authors with a rocket of the Naval Research Laboratory, and a high resolution spectrograph (0.03A°) at 1978 have measured the absolute intensity of some lines. We can remark that in order to construct the B.C.A. model Gingerich et al (1968, 1971) have taken into account that the solar minimum temperature is $T_e = 4660°K$. This temperature is very close to the measured by Parkinson (1969). The concentration of CO obtained with this model and the above temperature is $N = 1.90 \times 10^9 cm^{-3}$. In the present work we have compared the intensities of the lines measured by Jordan et al (1978), to the synthetic profile of the same lines computed with a Voigt function for different values of N and T_e, to determine more correctly N and T_e. The synthetic profile of the vibration rotation line $(u'' = 0, u' = 1)$ for $(J'' = 2, J' = 4)$, has been computed from 1520A° to 1580A° for different concentrations and temperatures of CO, using the Whiting's et al (1969) programme. The data of Table 1 have been used for this computation Herzberg (1950).

Table   1

	$W_e(cm^{-1})$	$W_e X_e(cm^{-1})$	B_e	a_e	$r_e(A°)$	$Too(cm^{-1})$
CO $(X^1\Sigma)$	2170.21	13.641	1.9313	0.01748	1.281	
CO $(A^1\Pi)$	1515.61	17.2501	1.6116	0.2229	1.2351	65074.8

We have also used as data Franck-Condon factors $(q_{0.1} = 0.26)$, computed by the R.K.R. method (Petropoulos et al,1971) and the bipolar electronic moment $R_e = 0.765$ computed in a previus work (Petropoulos, 1974). In figure 1 we give the computed intensities (Voigt profiles) for the $u'' = 0$, $u' = 1$, $J'' = 2$, $J' = 4$ transition for $N = 2.03 \times 10^9 cm^{-3}$ and $T_e = 3800°K$, 4000°K, 4500°K, and for $N = 5.5 \times 10^9 cm^{-3}$ and $T_e = 9500°K$. We can find that the measured intensity of Jordan et al (1978) for the $(u'' = 0, u' = 1, J'' = 2$,

W. Fricke and G. Teleki (eds.), Sun and Planetary System, 107–108.
Copyright © 1982 by D. Reidel Publishing Company.

J' = 4) line is in good agreement with the computed for the concentration
N = 2.03x10^9 cm^{-3} and the electronic temperature 3900°K $\leqslant$ T$_e$ $\leqslant$ 4000°K. This

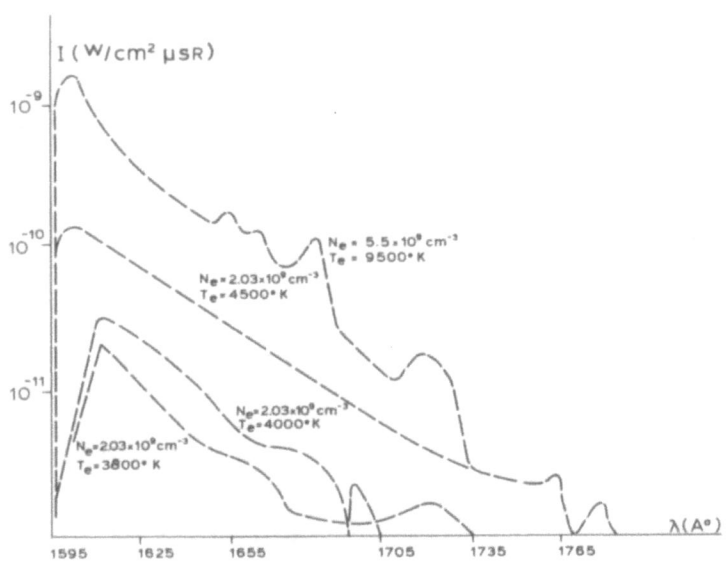

Figure 1.

value is in good agreement with the solar minimum temperature measured by
other methods. Berger et al (1972) found 3500°K having used the 5μ spectrum of CO. Shine et al (1975) measured T$_e$ = 4350°K by the CaII lines. Nelson (1978) taking into account the solar granulation found for his model
T$_e$ = 4145°K.

References
Berger R., Lena P., 1972, AA, 20 , 111.
Goldberg L. J., Noyes R. W., Parkinson W. H., Roeves E. M., Whithbroe G. L.,
        1968, Science, 162, 95.
Gingerich O. J., de Jager C., 1968, Solar Phys., 3, 5.
Gingerich O. J., Noyes R. W., Kalkofen W., Cuny Y., 1971, Solar Phys.18,347.
Herzberg G., Spectra of Diatomic Molecules Vol. II,Van North Co., 1950.
Jordan C., Bartoe J. D. F., Brueckner G. E., Nicolas K. R., Sandlin G. D.,
        Van Hooser M. E., CO Emission Lines in the Solar Atmosphere, M. N.
        R. Astr. Soc., 1979, 187, 473.
Nelson C. R., 1978, Solar Phys., 60, 5.
Parkinson W. H., Reeves E. M., 1969, Solar Phys., 10, 342.
Petropoulos B., Botter R., Proc. of the First Europ. Astron. Meeting, 1971
        (Ed. J. Xanthakis), Vol. I, 134, 1972.
Petropoulos B., Thèse de Doctorat d'Etat, Paris (VI), 1974.
Shine R. A., Milkey R. W., Michalas D., Ap. J.1975, 201, 224.
Whiting E. E., Arnold J. O., Lyle G. C., NASA TN D-5088, 1969.

SOLAR RADIO GRANULATION AT MICROWAVES AND ITS OPTICAL IDENTIFICATION

G.B.Gelfreikh
Pulkovo Observatory, Leningrad, USSR
V.M.Bogod, A.N.Korzhavin
Special Astrophysical Observatory,Leningrad, USSR
E.V.Kononovich, O.B.Smirnova, S.V.Startsev
Sternberg Astronomical Institute, Moscow, USSR
V.V.Piotrovich
Astronomical Observatory of Leningrad Univ., USSR

1. INTRODUCTION. The nature of granulation or supergranulation structure at different levels of the solar atmosphere is important for solving such significant problems of solar physics as underphotospheric convection scales or the solar corona heating and problems of solar activity as well. Radio observations have provided a new method of investigation of the chromospheric network structure both at the chromospheric and chromosphere-corona transition region levels. This so called "radio granulation" is accessible only to a number of large radio telescopes of the world and has already been discussed in a good number of publications (e.g. see all references). This paper deals with the results obtained by using the RATAN-600 in the 1.35-4.0 cm wavelength range.
Some special methods of registration are used to increase the contrast, such as compensation of the signal with a noise generator, scanning the antenna beam, polarization method including, etc. All the study of radio granulation was carried out at the North sector of the RATAN, with aperture of 400m x 80 m (the resolving power at shortest wavelength = 1.35 cm being 9x50 arc sec). The main parameters of radio granules found from these observations follow.

2. DISTRIBUTION OVER THE DISK. All the wavelength range 1 to 4 cm reveals the similar radio granules pattern, as far as their relative contrast is concerned. Taking into account the size of the diagram beam of the antenna and the number of observed details on a record we estimate the full number of granules on the solar sphere to be about 4×10^3. This value is of the same order as the full number of supergranules. At the same time it is an order of magnitude less than the number of the bright X-ray points. No significant varia-

W. Fricke and G. Teleki (eds.), Sun and Planetary System, 109–112.

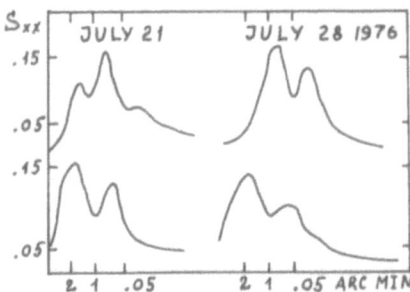

Figure 1. Power spectra of radio granulation (bottom) at = 1,35 cm and optical chromospheric network in K Ca II (top)

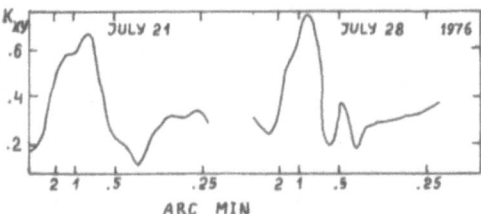

Figure 2. Coherence spectra K_{xy} of optical and radio scans of the Sun with an effective beam of 9 x 50 arc sec

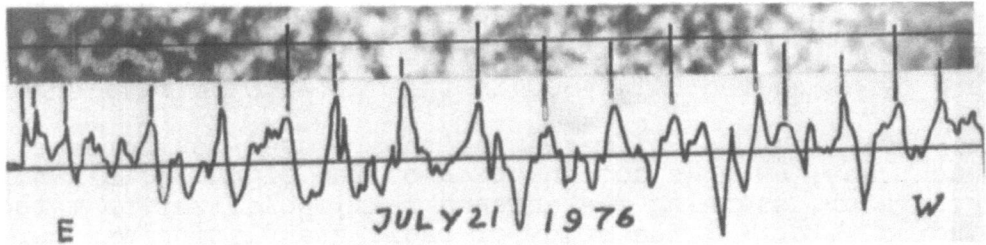

Figure 3. Identification of radiogranules at a radio scan made through the center of the Sun at =1,35 cm with resolution 9x50 arc sec with an optical map of the chromospheric network in K Ca II line

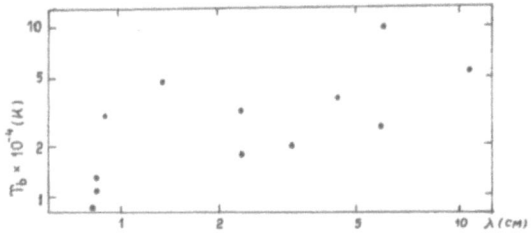

Figure 4. Spectrum of brightness temperatures of radio granules.

tion of concentration of the radio granules is observed
when the Sun is scanned ±10' higher or lower of the solar
disk center.

3. OPTICAL IDENTIFICATION. The most evident and direct
identification was made at =1.35 cm using HPBW equal to
9x50 arc sec. Simultaneous optical and radio observations
have shown the identity of bright features in radio granu-
lations with bright features of the chromospheric network
photographed in the K Ca II line. It must be emphasized
that while each radio granule has a corresponding bright
feature in optical image, not all optical bright elements
have adequate structural elements as the radio,scan. So the
optical and radio wavelength ranges are essentially different sources of information about the physical structure
of supergranules. The power spectra of both radio and optical maps (the latter averaged over the diagram pattern of
the RATAN-600) are shown in Figure 1 together with the coherence function (Figure 2). The best correlation is found
at the scale of 40 to 50 arc sec (K_{xy} = 0.7-0.8). This is
an additional evidence in favour of supergranulation nature
of radio granules.

4. LIFE TIME. The records of the Sun made with the RATAN-
600 at different wavelengths differ in time by some tens of
seconds. Their good identity in radio granulation structure
shows that the life time of the radio granules is not less
than a minute and their origin has nothing to do with the
Earth's atmosphere. On the other hand when comparing the
records for two sequential days the identity is fully lost.
So the life time of radio granules is much less than 24
hours. Crosscorrelation of an optical and a radio map taken
at different moments shows that during an hour's interval correlation coefficient is not reduced significantly
(r~0.7). So the probable life time of radio granulation is
similar to that of the chromospheric network, i.e. 10 hours
order.

5. BRIGHTNESS TEMPERATURE. In the wavelength range of 2 to
4 cm the flux densities are found in the range of 60 to
$300x10^{-26}$ w/m^{2}Hz. As their typical size does not usually
exceed the narrow horizontal section of the beam and much
smaller than the vertical one it is inevitable to use a
certain model to find the value of the brightness temperature. The most probable value lies within the range 5 to
$12x10^3$K. However, the possibility that the real size of the
radio granules is still lower resulting in much higher
brightness temperature cannot be excluded.

6. POLARIZATION. Measurements of circular polarization at the wavelengths from 2.3 to 4 cm with resolution from 14'' x 1.4 to 24'' x2.4 consequently have shown that most radio granules are randomly polarized resulting in magnetic field strength below 3 G (averaged over the diagram). However, some individual features of radio granulation have measurable circular polarization, which amounts to (10-18) x 10^{-26} w/m^2 Hz. They may prove to have quite different e.g. radio manifestation of X-ray bright points or ephemeric active regions. Their magnetic field strength amounts to some tens of G at least but may be higher if their real size is smaller.

7. CONCLUSION. The radio granulation at centimeters is mainly the manifestation of supergranulation of the Sun and has many common features with calcium chromospheric network. At the same time it is not simply a duble of an optical image of the chromospheric network but has different relative brightness distribution. It is quite probable that some features of the radio granulation (e.g.those with high polarization) are due to ephemeric active regions, bright X-ray points etc. That may be the reason of an apparent contradiction of the results obtained by Kundu et al. (1979) – identification with chromospheric network, and Marsh et al. (1980) – identification with bright X-ray points.

REFERENCES

Bogod,V.M. 1978, Soobsch.Spets.Astrofiz.Obs.,№23, p.22.
Bogod, V.M., Korol'kov D.V. 1975, Pis'ma v Astron.Zhurn., 1, №10, p.25.
Efanov, V.A., Moiseev, I.G., Severniy, A.B. 1975, Izv. Krymskoi Astrofiz.Obs., 13, p.121.
Hobbs, R.W. et al. 1974, Sol.Phys., 36, p.369.
Gelfreikh, G.B. et al. 1977, Pis'ma v Astron.Zhurn., 3, №12, p.550.
Fürst, E., Hirth, W. 1974, Astron.Astrophys., 36, p.183.
Kisljakov, A.G. et al. 1975, Pis'ma v Astron.Zhurn., 1, №4, p. 24.
Kundu, M.R. et al. 1979, Astrophys.J., 234, p.1122.
Marsh, K.A., Hurford, G.T., Zirin H. 1980, Astrophys.J., 236, p.1017.

A THEORETICAL MODEL OF THE SOLAR CHROMOSPHERE OVER A SUNSPOT

M. Marik
Astronomical Department of the Lorand
Eötvös University, Budapest, Hungary

1. THE EMISSION COEFFICIENT IN THE CHROMOSPHERE

The emission coefficient of the solar matter in the chromosphere has been calculated by Cox and Daltabuit (1971):

$$\varepsilon_e = 10^{-23} f(T) n_e n_H \ \text{erg cm}^{-3} \text{sec}^{-1}, \tag{1}$$

where $f(T)$ is a numerically calculated function, n_e and n_H are the electron and the hydrogen densities.

According to Marik (1975):

$$\varepsilon_e = 1.33 \times 10^{24} \ F(P,T) \varrho^2, \tag{2}$$

where $F(P,T)$ is a numerically given function, ϱ is the density.

2. THE ABSORPTION COEFFICIENT

The emission coefficient for magneto-acoustic waves, generated in the convective zone of the Sun, have been calculated by Kulsrud (1955). Osterbrock (1961) has suggested that the magneto-acoustic waves transform into weak shock waves in the layer of $V_A \approx V_S$. /V_A is the Alfvén velocity and V_S is the sound velocity./ The chromosphere is heated by these weak shock waves according to Osterbrock (1961), Pikelner and Lifshits (1964), Marik (1966) and others. The absorption coefficient for the weak magneto-acoustic shock waves may be written as follows from Marik (1966):

$$\varepsilon_a = \frac{1}{8} \ 2^{1/2} t_0^{-1} \gamma^{-5/4} P^{-5/4} \varrho^{3/4} F_0^{3/2} x$$

$$x \left(1 + \frac{1}{8} \ 2^{1/2} F_0^{1/2} t_0^{-1} \int_0^h \gamma^{-5/4} P^{-5/4} \varrho^{3/4} dh \right)^{-3}, \tag{3}$$

113

W. Fricke and G. Teleki (eds.), Sun and Planetary System, 113–114.

where t_0 is the the transit time of the shock waves, γ is the ratio of specific heats, P is the gas pressure and F_0 is the energy flux of the shock waves at the level h = 0.

3. THE RADIATIVE EQUILIBRIUM

Let us suppose that the chromosphere is heated by the absorption of weak magneto-acoustic shock waves. In this case $\varepsilon_a = \varepsilon_e$, or combining Equations (2) and (3), and using the Equation of state and the Equation of the hydrostatic equilibrium, we get a differential Equation for P:

$$P'' = -\frac{(P')^2}{P}x$$

$$x \; \frac{5 + 4F(P,P')^{-1}\frac{\partial F}{\partial P}P - 12a(-PP')^{1/6}F(P,P')^{1/3}}{5 + 4\frac{\partial F}{\partial P'}F(P,P')^{-1}P'} \tag{4}$$

The model has been calculated by Equation (4) numerically, using the "ODRA 1304" computer of the Lorand Eötvös University. For the transit time of shock waves, t_0, depending on the magnetic field strenght, 10, 30, 50, ..., 290 seconds have been taken successively.

Figure 1 shows the temperature T(h) as a function of height in the chromosphere for t_0 = 10, 70, 150 and 290 seconds.

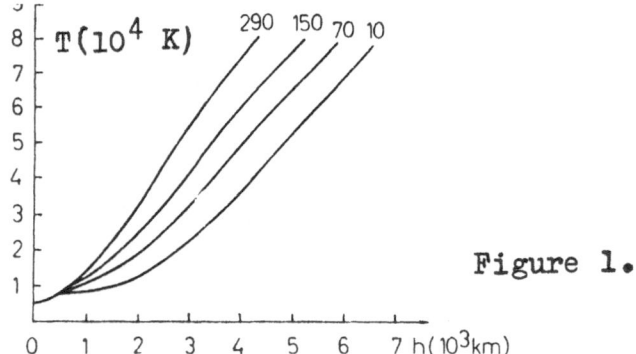

Figure 1.

REFERENCES

Cox, D.; Daltabuit, E.: 1971, Astrophys. J. 167, 113.
Kulsrud, R.: 1955, Astrophys. J. 121, 461.
Marik, M.: 1966, Astron. Zh. 43, 400.
Marik, M.: 1967, Astron. Zh. 44, 336.
Marik, M.: 1975, Bull. Astron. Inst. Czechsl. 26, 317.
Osterbrock, D.: 1961, Astrophys. J. 134, 347.
Pikelner, S; Lifshitz, M.: 1964, Astron. Zh. 41, 1007.

SPACE-BORN AND GROUND-BASED OBSERVATIONS OF A SOLAR ACTIVE REGION AND A FLARE

F. Chiuderi Drago
Arcetri Observatory - Florence, Italy

The Active Region (AR) 2490 has been observed from June 10 to 12, 1930 by several powerful instruments operating either on the ground (Optical and Radio) and on Solar Maximum Mission Satellite (SMM). The atmospheric layer observed by the used instrumentations, their spectral coverage, and the observers involved in this joint research are shown in Table I.

Table I

	Instrumentation		Atmosph. layer		Observers
Radio	VLA[1] (USA)	20	Corona		Felli, Lang, Willson
	" "	2	Tr. Reg., Corona		" " "
	WSRT[2] (NL)	6	" " "		Palagi, Shibasaki, Slottje
	Effelsberg (D)	11	" " "		Fürst, Hirth
	" "	2.8	" " "		" "
	" "	0.9	Chrom., Tr. Reg.		" "
	Helsinki (SF)	0.8	" " "		Urpo
Optical	BBSO[3] (USA)		Chromosphere		Zirin, Tang
	SPO[4] (USA)		"		Falciani
	" "		Photosphere		"
	Kislovodsk (USSR)		Photosphere		Gnevyshev
XUV	SMM (UVSP)[5]	CIV	Tr. Reg.	$(T \sim 10^5)$	UVSP Team (Bruner)
	SMM (XRP)[6]	OVIII	Corona	$(T \sim 3 \cdot 10^6)$	XRP Team (Antonucci)
	"	NeIX	"		"
	"	MgXI	"	$(T \sim 6 \cdot 10^6)$	"
	"	SiXIII	"	$(T \sim 10^7)$	"

(1) Very Large Array; (2) Westerbork Synthesis Radio Telescope; (3) Big Bear Solar Observatory; (4) Sacramento Peak Observatory; (5) Ultraviolet Spectrometer and Polarimeter: Marshall Sp. Flight Centre; (6) X-Ray Polychromator: Lockeed Res. Lab., Mullar Sp. Sci. Lab., Rutherford Appleton Lab.

W. Fricke and G. Teleki (eds.), Sun and Planetary System, 115–118.
Copyright © 1982 by D. Reidel Publishing Company.

1. A.R. OBSERVATIONS

The main results referring to the AR observations are described by Chiuderi Drago et al. (in press), therefore they will be only briefly summarized here.

The A.R. shows, at λ = 6 and 20 cm a double structure not noticed at shorter radiowavelengths. One of the two components (the one observed at all radiowavelengths) overlies the H_α plage and the UV and X-ray regions of enhanced emission. The other one, observed only at λ = 6 and 20 cm, overlies a group of sunspots, all of north polarity. The former one is interpreted as due to thermal free-free emission, while in the latter one the main contribution to the opacity of the corona must be due to the gyroresonance absorption. In fact no X-ray emission in any one of the XRP Channels is observed in this region.

From the maximum observed X-ray intensities, in the three low energy XRP Channels, the following parameters for the "thermal" component are deduced:

$$T_e \approx 4.2 \ 10^6 \ ^\circ K$$

$$\int_{T=T_e} N^2 dh \approx 6.10^{28} \ cm^{-5} \ .$$

Using these data in the computation of the radio brightness temperature, a good agreement with observations is obtained.

In the "non-thermal" component a magnetic field B = 600 G. at a level where the temperature is $T_e \approx 2.10^6$ is deduced from the observed brightness temperature and from the assumption that the observed frequency of 5.10^9 Hz equals the third harmonic of the gyrofrequency. This high value of the magnetic field at a "coronal level" can be justified by the assumption of a thinner transition region above the sunspot, which is suggested by the lack of emission in the CIV line (its intensity is in fact inversely proportional to the temperature gradient in the transition region) and by the absence of a radial displacement of the 6 cm emission with respect to the underlying spots.

2. FLARE OBSERVATIONS

A class 1 flare in A.R. 2490 was observed between June 10 at ~2325 UT and June 11 at ~0030 UT by most of the instrument listed in Table I except the European ones and S.P.O.. The flare was also observed in the two lowermost channels (3.5 ÷ 8 KeV) of the Hard X-ray Imaging Spectrometer (HXIS) on board SMM.

The morphology of the flare, observed in H_α and CIV line (Chromosphere and transition region), is very similar: it presents two bright regions, associated with sunspots of opposite polarity and separated by $\approx$ 2'. In the four XRP Channels, corresponding to OVIII, Ne IX, Mg XI and SI XIII lines, in which the flare was observed, it presents a loop shape connecting the above mentioned spots of opposite polarity. A comparison among the various observations shows that the H_α and CIV emission comes from the two feet of the X-ray loop, while the radio maxi-

mum emission, is located on top of it. (Figure 1 and 2). These results
agree with previous observations made by Marsh and Hurford (1980) and

June 10

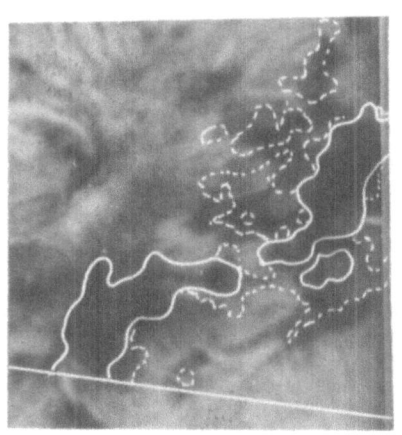 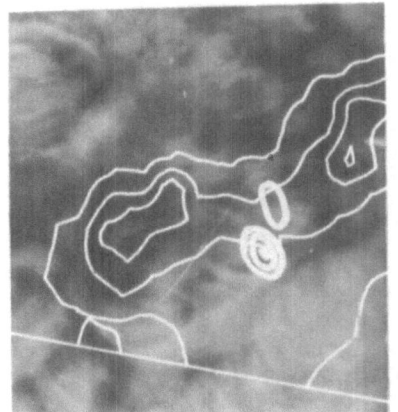

Figure 1. a) Comparison between Hα (BBSO negative picture) and CIV
             (UVSP) line intensity contour (2339 UT).
          b) Isophotes of NeIX line intensity (XRP) and λ = 2 cm bright-
             ness temperature (VLA, thick lines) superimposed to an Hα
             negative picture (BBSO) (2342-2343 UT).

June 11

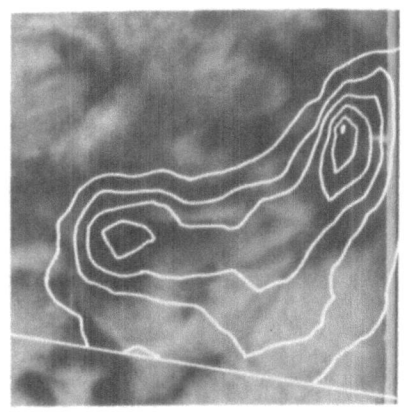 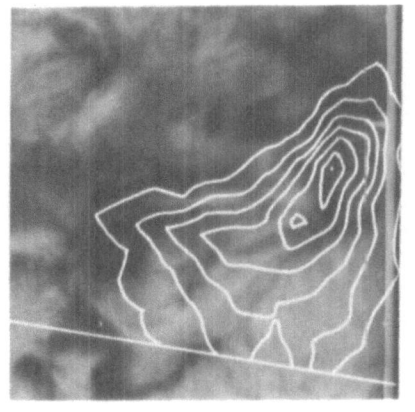

Figure 2. NeIX (a) and SiXIII (b) isophotes (XRP), observed at the fla-
          re maximum (0003 UT), superimposed to a negative Hα picture
          (BBSO) taken at the same time.

Marsh et al. (1981). The CIV line profile shows, during the rising part
of the flare, a considerable broadening and a blu-shift indicating an

upward motion in one of the feet. From the line intensity relative to
the four XRP Channels observed in the same position and at same time of
the VLA burst (Figure 1b) the following parameters for the coronal plasma are deduced:

$$T_e \simeq 7.10^6$$

$$\int N^2 \, dh \simeq 3.10^{29}$$

$$T = T_e$$

Putting these parameters in the radio transfer equation at $\lambda = 2$ cm
a brightness temperature of only $10^{5\,\circ}K$ is deduced while the observed one
is of the order of 10^7 K. This strong discrepancy indicates that another
mechanism, as for instance the gyrosynchrotron absorption, must be present to increase the opacity of the hot coronal plasma.
Although all people listed in Table I have widely cooperated to this research, I am particularly grateful to E. Antonucci, E. Bruner, F. Tang
and R. Willson for supplying the data shown in the figures.

REFERENCES

Chiuderi Drago, F., Bandiera, R., Falciani, R., Antonucci, E., Lang, K.,
        Willson, R., Shibasaki, K., and Slottje, C.:(Submitted to
        Solar Physics).

Marsh, K.A., and Hurford, G.J.: 1980, Ap.J. 240, L111.

Marsh, K.A., Hurford, G.J., Zirin, H., Dennis, B.R., Frost, K.J., and
        Orwig, L.E.: 1981, BBSO Reprint 200.

THE BEGINNING OF OBSERVATIONS OF LARGE-SCALE SOLAR MAGNETIC FIELDS AT THE SAYAN OBSERVATORY: INSTRUMENT, PLANS, PRELIMINARY RESULTS

V.M. Grigoryev, V.S. Peshcherov, M.L. Demidov
Sayan Solar Observatory, SibIZMIR, Irkutsk-33, USSR

Until recently, the investigations on large-scale fields have been limited to analyzing the distribution of the magnetic field radial component. The Stanford group was the first to initiate the most comprehensive study of the poloidal and toroidal component of a mean magnetic field in the photosphere (Duvall et al., 1979). Further progress in the study of the nature of the large-scale magnetic field in the Sun will be connected with investigations of the structure of a toroidal component of the magnetic field and of the nonaxially-symmetric velocity field in the photosphere. At the Sayan Solar Observatory we constructed a telescope and a system of measurement of large-scale magnetic fields and of the large-scale field of line-of-sight velocities in the Sun's photosphere (Grigoryev et al., 1981). The instrument permits the following programs of synoptic observations of large-scale structures: 1. Magnetograms of a large-scale magnetic field with a 3' resolution and 0.1-0.2 Gs sensitivity. Each magnetogram is taken within about 20 minutes. 2. Solar disk magnetograms as half-tone images of the magnetic field distribution with 15 Gs sensitivity and an 8x8" resolution. Each magnetogram is taken within 5 minutes. 3. Measurement of a mean magnetic field of the Sun as a star with about 0.1 Gs sensitivity. The time of measurement is 10 to 20 minutes. Sunlight is directed into the telescope by a 30 cm Jensch-coelostat. A doublet objective (18cm with a 5m focal length) forms a tructs solar image of 4.7 cm. The position of the objective and solar image with respect to the spectrograph slit is determined by the observational program. When a mean magnetic field of the Sun as a star is measured, the solar image is at a distance of about 1.3 m from the spectrograph slit. This produces on the spectrograph slit a region which is illuminated by all parts of the solar image. In observations of large-scale magnetic fields with a 3' resolution solar image is at a distance of 12 cm in front of the spectrograph slit. In this case the slit is illuminated by all parts of the solar image region 3' in diameter. In the mode of taking half-tone magnetograms, solar image is produced on the spectrograph slit. Solar image is scanned quickly with the aid of a two-coordinate scanner and the signal is recorded with a fascimile apparatus as a half-tone black-and-white picture. Behind the spectrograph slit is an electro-optical analyzer of polarization, consisting of a DKDP crystal

W. Fricke and G. Teleki (eds.), Sun and Planetary System, 119–122.
Copyright © 1982 by D. Reidel Publishing Company.

and a polarizer. The analyzer provides amplitude modulation of the luminous flux, whose magnitude depends on the magnetic field strength.
The photometer uses one photomultiplier, in contrast to standard magnetograph configurations. The following approach has been adopted to control
the magnetic signal zero point. One measurement is made in a standard
manner, while in the other, a plate with a $\lambda/2$ phase shift is introduce
into the luminous flux from the Sun in front of the coelostat. In this
case, in the incident beam the circular polarization of σ-components of
the magnetosensitive line is changed, thereby reversing the signal phases.
The phase (sign) of the magnetic zero signal remain unaltered. As a $\lambda/2$
phase plate (30 cm in diameter) we employ a polypropylene film of 0.15
mm thickness.

Construction work on a system of measurement of line-of-sight velocities
is near completion at the Sayan Observatory. The principle of the
instrument consists in a differential measurement of the spectral line
displacement on some place of the Sun's surface with respect to the spectral line, produced by all parts of solar image. Two 30 cm Jensch-
coelostats direct the solar light into two doublet objectives. One
objective projects the solar image on the spectrograph slit or at a
distance of 12 cm in front of the slit, if 3' resolution measurements
are needed. The other objective produces solar image at a distance of
1.3 m in front of the spectrograph slit and creates on the slit a region
illuminated by all parts of solar image. By means of two prisms and a
calcite plate, the two beams enter the slit and have a mutually orthogonal linear polarization.

The quantity of observations at the Sayan Observatory is still small
because since September 1980, the instrument has been adjusted and
individual observations were made. With the aim to develop a technique
and programs of processing the observations we availed ourselves of some
results of the observations made at Stanford in the period from February
1979 through October 1980. As part of this study we obtained some
interesting results. The method of separating out the toroidal component
is similar to that used by Duvall et al. (1979). The toroidal magnetic
field covers the entire surface of the Sun. The toroidal field in the S
and N hemispheres has opposite predominant directions. The direction
(from W to E in the N-hemisphere and from E to W in the S-hemisphere)
corresponds to subphotospheric flux tubes producing active regions in
the cycle 21. The toroidal field strength distribution is not uniform
both in longitude and latitude. The hills of the toroidal magnetic field
do not coincide with those of the radial field and there is a tendency
for them to displace towards the radial field zero-line region. We stress
the fact of the existence of a toroidal field not only on Hall boundaries
of polarity separation of a large-scale radial field but also on anti -
Hall boundaries. We have produced a synoptic map of the slope angle
contained by the field vector and the solar radius. The directions of the
field vectors show clearly on Hall boundaries an arch-like structure of
the field whereas on anti - Hall boundaries there is an "anti - arch"
structure of the field. The magnetic field there looks like an inverted
arch which is open in upper atmospheric layers and closed in photospheric

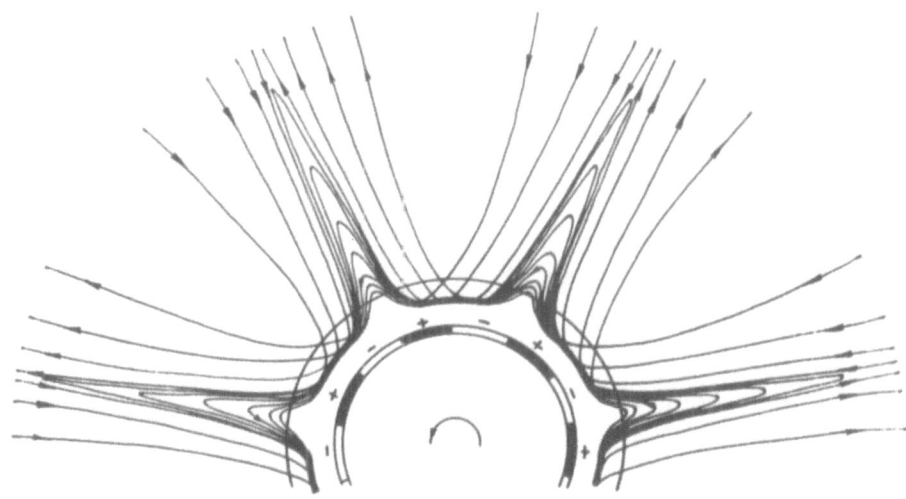

Figure 1. A schematic of the structure of a large-scale magnetic field
in the Sun. Transition from (-) to (+) is the Hall boundary,
while that from (+) to (-), the anti-Hall boundary.

and subphotospheric layers. The latitude distribution of the toroidal
field has a maximum in both hemispheres at ±15-20° latitude. The mean
field strength in the region of the maxima is 1-1.5 Gs. At the beginning
of the cycle, the mean field strength in these zones is 0.1 Gs. The
toroidal field strength maxima at the beginning of the cycle were
arranged at 30-40° latitude in the N-hemisphere and at 30-70° in the S-
hemisphere (Duvall et al., 1979). At latitudes above 40° in both hemispheres there appears a region of a weak toroidal field (0.5-0.1 Gs) with
the reversed direction as distinct from lower latitudes. There is a
marked tendency for the area of opposite toroidal field regions to
increase at higher latitudes during the polar field reversal.

The toroidal field observed in the photosphere is associated with a
global subphotospheric toroidal magnetic field of the Sun, as suggested
in the model of Babcock. The emergent magnetic flux tubes are observed
as a toroidal component of the field in the region of Hall boundaries
of the radial field polarity separation. The toroidal field in the region
of anti-Hall boundaries is indicative of some subphotospheric tubes of
magnetic flux, which connect the leading and following parts of adjacent
bipolar magnetic regions. Since there is a poleward motion of the bipolar
magnetic regions' follower parts and an equatorward motion of the leaders,
this thus leads to a turn of subphotospheric toroidal tubes and to the
formation of a new toroidal field of inverse polarity. A weak (∿0.1 Gs)
toroidal field of inverse polarity observed at high latitudes, possibly
because a new poloidal field due to differential rotation produces a new
toroidal field of the next activity cycle, 22.

The obtained results of the analysis of the toroidal large-scale magnetic field structure of the Sun have led us to an understanding of the general structure of large-scale fields, which are portrayed in Figure 1 (see the next page).

REFERENCES

Duvall, T.L., Scherrer, P.H., Svalgaard, L., and Wilcox, M.: 1979, Solar
        Phys., 61, p.233.
Grigoryev, V.M., Osak, B.F., Kobanov, N.I., Klochek, N.V., Maslov, I.L.,
and Shtol, M.F.: 1981, Issledovaniya po geomagnetizmu, aeronomii i fizike
        Solntsa, 37, p. 147.

THE SOLAR COMPLEX OF SibIZMIR: OBSERVATORIES, INSTRUMENTS AND THE MAIN RESULTS OF THE INVESTIGATIONS

G.Ya.Smolkov
SibIZMIR, Irkutsk-33, P.O.Box 4, 664033, USSR

Regular solar observations were begun at the Irkutsk magnetic ionospheric station (IMIS) in 1956 in connection with the IGY. For that purpose, a chromospheric-photospheric telescope was installed and 1.5 m radio telescope was constructed with the efforts of staff members. In 1959-1960 it was possible to initiate a program of the creation of a complex of instruments to enable simultaneous observations of all manifestations of solar activity accessible to ground-based methods. The different requirements to astronomical climate and the observing mode, the specificity of scientific problems have led to the creation of the following observatories and instruments (Smolkov and Stepanov, 1979): 1. IMIS Radio Astronomical Observatory: regular observations of solar radio emission and cosmic ray intensity variations, mock-up finalization and testing of the Siberian Solar Radio Telescope (SSRT), now under construction. This observatory is equipped with a radio polarimeter, an eight-element radio interferometer and a ten-element compound interferometer operating at about 5 cm wavelengths. These interferometers have a 4' and 1.'5 arc resolution, respectively. 2. Sayan Mountain Observatory (Smolkov, 1966): regular measurements of sunspot magnetic fields, cosmic ray intensity variations and telluric currents, study of large-scale and local magnetic fields and velocity fields of solar features, physical conditions in active regions and their neighbourhood in the photosphere, chromosphere and corona (outside the solar limb), corona's rotation and oscillatory regimes in the Sun's atmosphere. Two horizontal solar telescopes and two Coudé coronographs are installed at 2012 m above sea level. Their characteristics and equipment are given in (Stepanov et al., 1981). 3. Radio Astronomical Observatory Tunkinskaya: research on fluctuations and short-period oscillations of radio emission from active regions with a small-baseline 3.5 cm interferometer, construction work on an SSRT radio heliograph designed to study the structure and evolution of coronal active regions on the solar disk (Gelfreikh et al., 1966). An SSRT-incorporated 256-element radio inter-

123

W. Fricke and G. Teleki (eds.), Sun and Planetary System, 123–124.
Copyright © 1982 by D. Reidel Publishing Company.

ferometer consisting of two rows with 128 antennas in each
in the W-E and N-S. The mirrors are 2.5 m in diameter. The
antennas are connected via a waveguide system arranged as
a stage-parallel scheme, to a multichannel receiver. A complex of automation systems is designed to monitor the status of all SSRT systems, to control their operation, to de-
determine operating modes and the action program, to col-
lect, process and display data in real time and to produce
maps of solar radio images in the non-polarized and circu-
larly polarized emissions. Radio image is produced by frequency scanning the Sun in vertical direction as the Sun
intersects the interferometer's fan-shaped beam lobe. The
SSRT operates at about 5 cm wavelength. The two-dimensional resolution is 20x20" arc in a symmetric cross scheme
and 10x20" arc in a T-scheme. 4. Baikal Astrophysical Observatory: regular observations of photospheric and chromospheric features with a photoheliograph and a chromospheric
telescope (Banin et al., 1981), construction of a large solar vacuum telescope (LSVT) designed for solar feature fine structure studies (Stepanov et al., 1979). LSVT performance and equipment are described in (Stepanov et al.,1981;
1979).

In the course of 1979-1981 the SibIZMIR has co-or-
dinated the observations at Soviet solar observatories on
the SMY, with the help of the Sayan Observatory and IMIS
Radio Astronomical Observatory. Parallel with conventional
methods, our astrophysicists apply extensively the methods
of plasma physics, magnetohydrodynamics, and laboratory
simulation. The results on solar activity studies carried
out at SibIZMIR during 1961-80 are outlined in (Stepanov
et al., 1981).

REFERENCES

Banin,V.G.,Skomorovsky,V.I.,Klevtsov,Yu.A., and Trifonov
    V.D.: 1981, Soln. dannye (in press).
Gelfreikh,G.B.,Korolkov,D.V.,Smolkov,G.Ya.,and Treskov,T.
    A.: 1966,"Results of observations and investigations
    during IQSY", Moscow, "Nauka", 4, p.168.
Smolkov,G.Ya.: 1966, Issled. po geomagn. i aeron., Moscow,
    "Nauka", p. 189.
Smolkov,G.Ya., and Stepanov, V.E.: 1979, Solar Instruments,
    Irkutsk, p. 5.
Stepanov,V.E., Banin,V.G. et al.: 1979, Novaya tekhnika v
    astronomii, Leningrad, "Nauka", p.42.
Stepanov,V.E.,Smolkov,G.Ya.,Kuklin,G.V., and Kovalenko, V.
    A.: 1981, Issled. po geomagn., aeron. i fiz. Solntsa,
    Moscow, "Nauka", 56, p.76.

MOTIONS IN SUNSPOTS LIKE TORSIONAL OSCILLATIONS

S.I. Gopasyuk
Crimean Astrophysical Observatory, USSR

To study the rotation of sunspots, we selected the observational data for one group, namely for group which crossed the central meridian in October 4, 1974. This group was located in the southern hemisphere of the Sun ($\phi \approx -6^0$). It consisted of two large sunspots: preceding N-polarity spot and following S-polarity spot were separated by $\Delta L \approx 9^0$ in longitude. The magnetic field strength of both sunspots was approximately the same, i.e. about 2500G.

Line-of-sight velocities, longitudinal component of the magnetic field and brightness in the line cores were registered with the double magnetograph of the Crimean Astrophysical Observatory at the Solar Tower Telescope in $\lambda5250\text{Å}$ FeI and $\lambda5123,7\text{Å}$ FeI spectral lines, simultaneously. The observations were carried out during four days, from October 3 to October 6, 1974. Photoheliograms and Hα - pictures were also obtained.

On the basis of the line-of-sight velocities data, we have obtained all components of the velocity vector (Gopasyuk, 1977). The azimuthal component of the velocity and its variations with the radius of the spot, are shown in Figure 1 a,b.

The azimuthal component of the velocity shows, that first, gas in both sunspots and the sunpots themselves rotate counterclock wise. This phenomenon led to a clockwise twist of chromospheric structures round the sunspots (see Figure 2a). Then, in October 4-5 the direction of gas rotation in both spots changed into opposite (see Figure 1). In the last two days both spots were rotating clockwise, hence, untwisting the spiral of the magnetic field (see Figure 2b), which was twisted before. A thorough examination of the photoheliograms confirmed these features of the sunspot rotation.

Magnetic fields of the preceding and following sunspots had opposite polarities. Both spots were connected by magnetic lines in the chromosphere.

Therefore, it seems possible to conclude, that a pair of sunspots was

125

W. Fricke and G. Teleki (eds.), Sun and Planetary System, 125–126.

formed by a single magnetic flux tube where torsional oscillations
occured. The period of observed oscillations was approximatly 6 days.

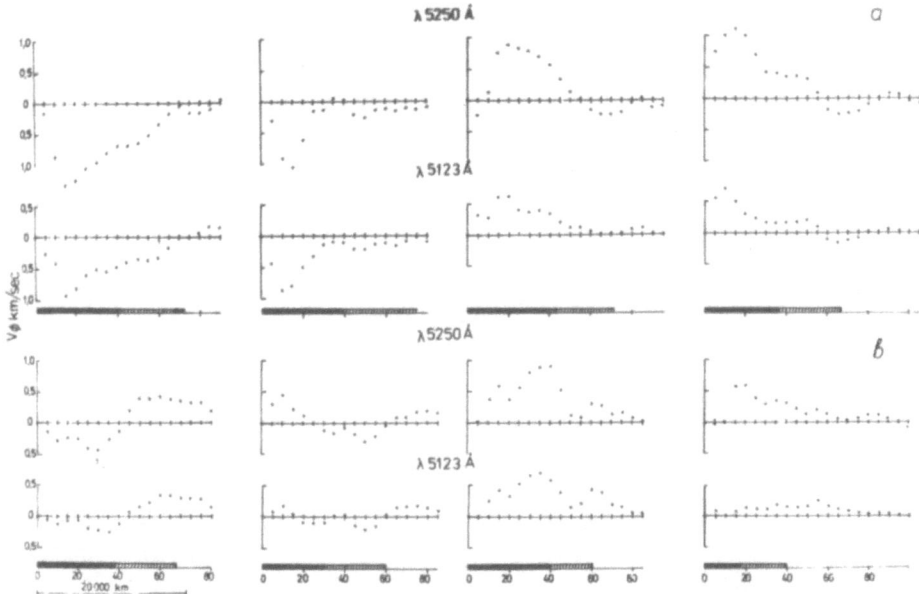

Figure 1. The azimuthal $\overline{V}\phi(r)$ component of the velocity variations with
the radius of preceding (a) and following (b) spots.

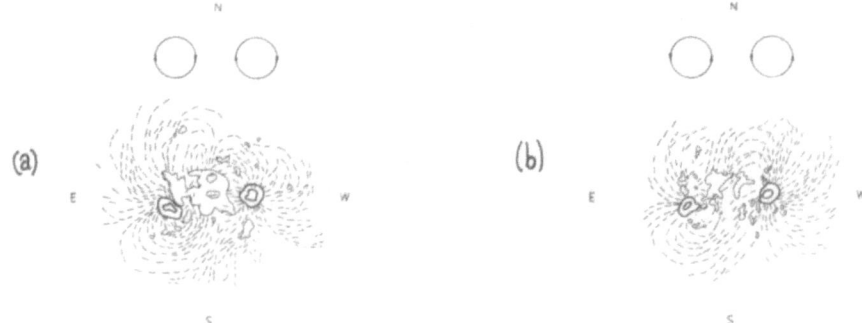

Figure 2. The tracings of Hα pictures showing the structures of the active
region on October 2 (a) and October 6 (b). Thick solid lines are
the boundary of umbra and penumbra of the spots. Thin solid
lines are the boundaries of the plage. Dashed lines show the
filaments of fine structure. The direction of sunspots rotation
is shown by an arrow on the circle.

REFERENCE

Gopasyuk,S.I.: Izv. Krims. Astrophys. Obs., 1977, 57, 107.

SUNSPOTS PROPER MOTIONS AND POSITIONS OF H-ALPHA FILAMENTS IN THE ACTIVE REGION OF AUGUST 1979

Z.B. Korobova
Astronomical Institute of the Uzbek
Academy of Sciences, Tashkent, USSR

Evidence is increasing that the distribution of the longitudinal and transverse components of the magnetic field in an active region is determined by the motions of the main sunspots (Gopasyuk and Moreton, 1967; Kuklin, 1968; Kalman, 1976; Ikhsanov and Shchegoleva, 1980). At the same time the comparison of sunspots trajectories with the position of the line of polarity reversal show that the latter somewhat influences the sunspots proper motions (Gesztelyi, 1977; Korobova, 1974).

We have attempted to compare the trajectories of the umbrae in McMath region 16208 1979 in the period August 10 - August 17 with the configuration of the stable filaments in the region, for filaments are known to mark the dividing line between magnetic polarities (McIntosh, 1972). The Carrington coordinates of the umbrae and filament borders were determined on Tashkent full-disc white light heliograms and IZMIRAN H-alpha films.

The sunspot group consisted of two close spots. The preceding spot of "delta" configuration changed quite a lot and showed a tendency to decay. The following one was a regular preceding polarity spot. The shapes and trajectories of the umbrae are shown in the Figure superpcsed with contours of the filaments on August 14. The preceding polarity is marked with numbers, the following one with a letter. The small numbers indicate the dates of the first and last observation.

Until August 13 all the umbrae revealed an anticlockwise rotation. Later umbrae 2a, 2b persisted in this motion, while umbrae 8a, 8b stopped moving. In addition to the rotation the umbrae 5, 6, 7, 9 north of the group show the poleward drift. But on August 14 umbrae 6 turned south-westward. New umbrae emerging after August 12 moved westward. Umbra A which occupied the gulf of the following polarity revealed a slow motion. The filaments bordered the region of preceding polarity. Between August 10 and 17 the area increased in size due to the migration of its high-latitude border line poleward and eastward.

It is seen at a glance that the umbral trajectories and the filaments

W. Fricke and G. Teleki (eds.), Sun and Planetary System, 127–128.

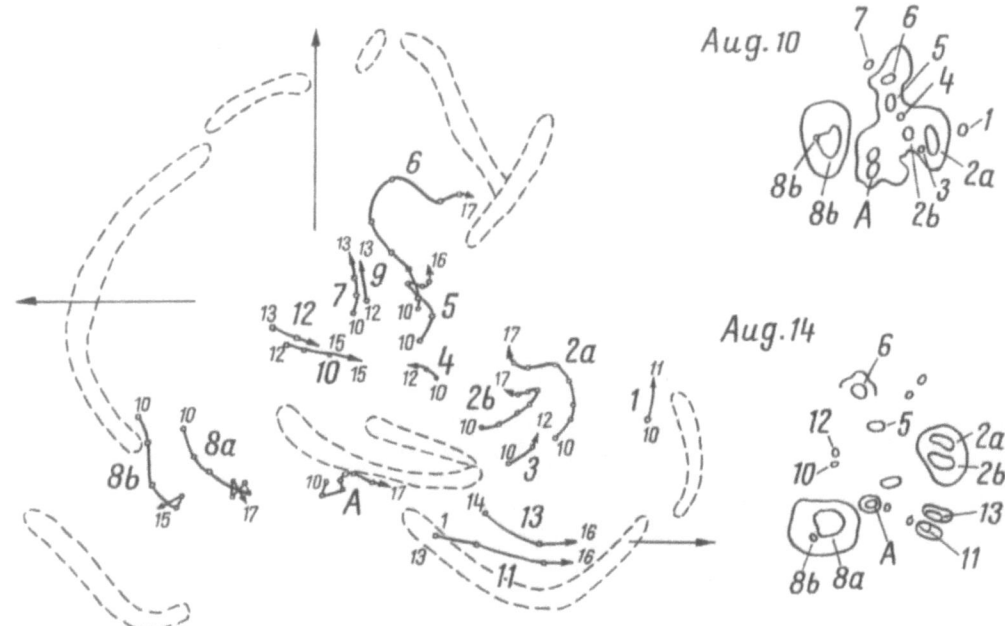

contours are of a single shape. Persistent umbral motion is evident if
the drift follows the zero line or if an umbra is far from the line.
Umbrae approaching nearly at right angles stop or turn off. These
motions were unable to alter the position of the zero line. New spots
moved westward together with the zero line. In the region of the
latitude drift both the umbrae and the zero line moved in the same
direction.

REFERENCES

Gesztelyi,L.: 1977, Publ. Debrecen Heliophys. Obs., pp. 93-109.
Gopasyk,S.I. and Moreton,G.E.: 1967, Proc. Astron. Soc. Austr.I pp. 1-3.
Ikhsanov,R.N. and Shchegoleva,G.P.: 1980, Izv. Glav. Astr. Obs. 198
                                    pp. 39-49.
Kalman,B.: 1976, Izv. Krymsk. Astrofiz. Obs. 60, pp. 60-69.
Korobova, Z.B.: 1974, Soln. Dann. 4, pp. 76-79.
Kuklin,G.V.: 1968, Proc. IAU Symp. 35, pp. 211-213.
McIntosh,P.S.: 1972, in P.S. McIntosh and M.Dryer (eds.), Progress in
                    Astronautics and Aeronautics 30, Cambridge, Massachusetts,
                    pp. 65-92.

ON THE UMBRAL DOTS

I. Sattarov
Astronomical Institute of the Uzbek
Academy of Sciences, Tashkent, USSR

In these note we want to attract attention to some peculiarities of the unipolar sunspot's fine structure. For the study we use the photographs and spectrograms of the unipolar sunspot obtained at Tashkent with a solar horizontal telescope (D = 44cm).

Figure 1 shows that in east and west of the penumbra the penumbral filaments wedges into the umbra in the shape of gulf. The regions of the umbra near the gulf (see prints on Figure 2) contain some more bright umbral dots that other ones. In the regions the dots form chains following the direction of the penumbral filaments wedged in the umbra. One can see tha likeness between the prints on Figure 2. First of all this concerns the chains of dots stretched from east and west of the central section of the umbra and adjoined the dots cluster in its center.

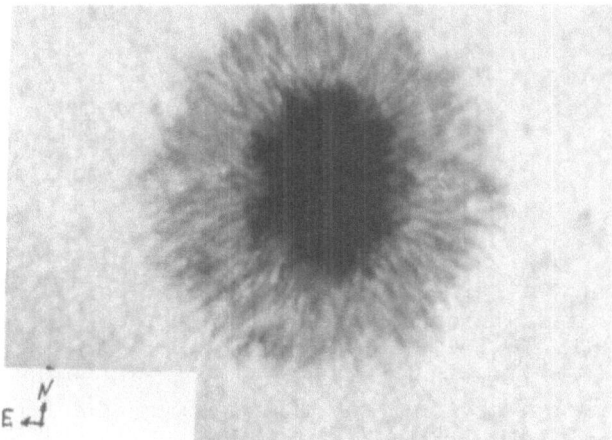

Figure 1. A print from the frame obtained through the broadband glass filter (λ = 6500Å, Δλ = 200Å) at Tashkent on August 8, 1981, at UT 0828 (heliographic coordinates of the unipolar sunspot on August 8, 1981: 10°N, 13°E).

W. Fricke and G. Teleki (eds.), Sun and Planetary System, 129–130.

On the next day photographs the chains are absent. Therefore they live
more than two hours but no longer than a day.

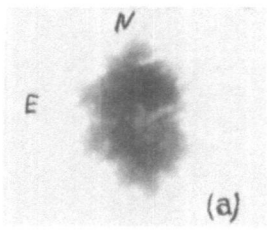 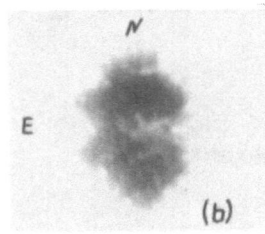

Figure 2. Two prints of the
unipolar sunspot's
umbra:
a) for UT 0650
b) for UT 0828

The bright dot near the eastern central gulf (Figure 2a) increases its
brightness and joints the gulf (Figure 2b). Such processes are observed
in other parts of the umbra. A reverse process, i.e., a darkening of
penumbral boundary parts and an inclusion of it to the umbra is also
observed. The interconnection between the shining of dots and processes
of penumbral filaments wedging into the umbra was noted earlier (Sattarov,
1981). Apparently these processes secure a constancy of a magnetic current. The south part of the umbra on Figure 2 is some more bright than
north one and is covered with umbral dot-net. The north part includes a
small (2".5 × 2".0) and the most dark region. It is attacked from all sides
by the light details and assumed a shaggy shape. The unipolar sunspot
with two dark regions was studied earlier (Sattarov, 1980). It was shown
that the spectral line FeI λ6302.5Å has doublet Zeeman splitting in the
region of the dots cluster and triplet Zeeman splitting in the dark
regions. The same picture one can see at the Figure 3.

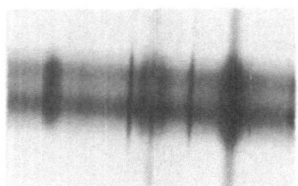

Figure 3. A part of the central section spectrum
with the line FeI λ6302.5Å of the umbra.
On the dots cluster which is in the
umbral center the line shows doublet
splitting.

The aforecited data can not serve as basis for assertion that the
magnetic active lines show the doublet splitting in the light details
of the umbra and the triplet splitting in the dark one. In fact the
light details are present everywhere in the umbra except in its small
region. For the present one may only say that the magnetic field is
longitudinal in the dots cluster.

REFERENCES:

Sattarov,I.: 1980, Sov. Astron.J., 58, 610
Sattarov,I.: 1981, in Morphologija i cyclichnost solnechnoi activnosti,
            ed. V.P. Shcheglov, FAN, Tashkent.

TURBULENT VELOCITY FIELDS IN QUIESCENT PROMINENCES

Eberhart Jensen and Oddbjørn Engvold
Institute of Theoretical Astrophysics,
University of Oslo

The velocity fields of prominences are derived from measurements of CaII K-line shifts in high resolution spectrograms. The turbulent character of the velocity field has been reported earlier (Jensen, 1982). That work was based on spectra of 4 large prominences observed at Oslo Solar Observatory. A much larger set of data is now beeing analysed. These spectrograms are recorded with the main spectrograph at the tower telescope at the Sacramento Peak Observatory. Details of observations are given in an earlier paper (Engvold, 1978). We shall here present and discuss the results obtained from nine quiescent prominences observed during 1973-74. The spectrograms have been analysed by means of the rapid scanning, computer controlled microphotometer of Institute of Theoretical Astrophysics, University of Oslo. The digitally recorded displacements within a given velocity-interval, giving the distribution of radial velocities, was denoted by N(u). In well-observed cases altogether 3-4000 profiles were obtained. Gaussian distributions were then fitted to the observed velocity-distributions by a least square procedure, giving a dependence of the form; $N(u) \propto \exp(-(u-u_o)^2/\alpha^2)$. Here u_o represent the combined rotational and bulk radial velocities, while α is a characteristic velocity parameter, describing the "turbulent exitation" in the prominence. In only one of the prominences investigated, did the velocity-distribution seem completely random and could not be represented by a Gaussian. In all the other cases the correlation-coefficient, r^2, came out in excess of 0.9. This result indicates that the velocity-field, at least to a first approximation, may be assumed to be isotropic and stationary.
The values of the parameter α show considerable scatter, ranging from 2.4 to 16.0 km s^{-1}, with 6.3 km s^{-1} as an average. In all cases except one, we have used observations from the whole prominence to determine α. In the one exeption the prominence consisted of two parts, an upper and a lower, with the slit passing through both. The upper part was more active than the lower and the analysis gave respectively α-values of 8.1 and 4.8 km s^{-1}. In one case, which has not been reduced yet, the prominence shows a marked arc-structure. In the spectra where the slit passes through the "Sauge-Füsse" the velocity-fluctuations are seen to be reduced. Possible variations of α between different parts of promi-

131

W. Fricke and G. Teleki (eds.), Sun and Planetary System, 131–132.
Copyright © 1982 by D. Reidel Publishing Company.

nences will be investigated (cf.Hirayama. 1964). From the distribution
$N(u)$ we may further construct the spectrum of the turbulent velocity
field as the product, $u\,N(u)$. The maximum of this spectrum is at a
velocity $\alpha/\sqrt{2}$. If it is assumed that the temperature in quiescent prominences is in the interval 6000 K < T < 9000 K, the limits for the velocity of sound become, 8.1 km s^{-1} < C_s < 9.9 km s^{-1}. It turns out that
only in a few cases is the spectrum entirely in the sub-sonic range.
In the majority of cases the high velocity tail of the distribution is
definitely supersonic. For the well-observed prominence which gave the
highest value of α in our sample, a substantial part (more than 70%)
of the energy spectrum is in the supersonic range. Actually, the turbulent pressure and the gas-pressure become comparable in this case. If
equipartition between magnetic and kinetic energy density is maintained
as in an Alfven-mode, the mean fluctuations in the magnetic field
should show a linear dependence upon our parameter α.
The tentative conclusions we may draw from our limited material is that
in most cases the velocity-field may be approximated by a Gaussian distribution, with a high correlation. The "most probable velocity", α,
varies considerably from prominence to prominence, and may change with
position within the prominence. The turbulent pressure cannot be
ignored in models of quiescent prominences, nor in discussions on
support mechanisms. In most cases part of the velocity-distribution is
in the supersonic regime. For this reason the theory of incompressible
MHD turbulence cannot in general be applied. The substantial intensity-
fluctuations observed on filtergrams over time-scales of minutes
(Engvold, 1981), are probably connected with compressibility effects.
The presence of a turbulent velocity-field also has consequences for
the values of the transport-coefficients to be used when the energy-
budget is to be evaluated. The thermal conductivity in particular, may
be orders of magnitude higher than the classical value.
The present type of analysis with special emphasis on time evolution
will be applied to disk filaments. An extensive set of spectral observations of filaments in HeI 10830Å has recently been obtained by
Engvold and Keil (1981). They used a 100 x 100 CCD chip in the main
spectrograph of the SPO vacuum tower telescope. The digitally recorded
data have not yet been fully analysed. A cursory inspection of these
data seems to show a fairly small range of velocities; between 1.3 and
2.5 km s^{-1} in filaments. The reason for the smaller velocity dispersion
in filaments as compared to prominences at the limb is as yet uncertain.
It is premature to speculate whether the effect is real or due to instrument or seeing.

References

Engvold, O.: 1978, Solar Phys. 56, 87
Engvold, O.: 1981, Solar Phys. 70, 315
Engvold, O. and Keil, S.: 1981, In preparation
Hirayama, T.: 1964, Publ. Astron.Soc. Japan 16, 104
Jensen, E.: 1982, Solar Phys. 77, 109.

DETERMINATION OF THE ELECTRON DENSITY IN
A QUIESCENT PROMINENCE

Gigolashvili M.Sh.
Abastumani Astrophysical Observatory, Georgia, USSR

As far back as 1930 it was proved by Struve and Elvey that the lines in stellar spectra are broadened due to Stark effect on account of electrically charged particles surrounding the atoms. The Stark effect action on the hydrogen lines was subsequently developed by de Jager (1952), Ivanov-Kholodny (1955, 1958, 1959) and others, but there is no crowning theory which would describe the influence of all disturbing particles.

Nevertheless, there are two rather precise techniques of determining the electron density based on the Stark effect action on the hydrogen lines.

We can determine the electron density by increase of the line widths with the line number. This method relies on the comparison of the observed broadening of the Balmer lines due to the Stark effect with their theoretical values. As it was shown by Hirayama, 1963, the inter-molecular Stark effect causes the broadening of the ultraviolet hydrogen lines with the wave-length shorter than H_{15} only. Therefore, in the works using the first terms of the Balmer series, the values of the electron density are overestimated.

Ther is another possibility: we can determine the electron density by the number of the last resolved Balmer series line. As it is known, the distance between the Balmer series lines decreases with approaching the series limit, but due to the Stark effect, the hydrogen line width increases with the rising of its number. Therefore, the lines merge for a certain value of the main quantum number. The Inglis-Teller formula (1939), connecting the electron density with the main quantum number, was consequently refined by other authors (Ivanov-Kholodny et al., 1960; Ivanov-Kholodny and Nikolsky, 1961; Kuroshka and Maslennikova, 1970).

The electron density of prominences can be determined by the total emission in the continuum. The continuum beyond the Balmer series limit is chiefly due to recombination of protons with electrons, and the emission intensity depends on the electron density and temperature.

W. Fricke and G. Teleki (eds.), Sun and Planetary System, 133–134.

The present paper aims at determining the electron density of the quiescent prominence of October 15, 1969, observed at Abastumani Observatory by means of three different methods. The short-wave spectral region beyond the Balmer series (up to 3200Å) were taken for four different regions of the prominence. The brightest continuum and the greatest number of the Balmer series members were observed in the knot situated at the height of 32". The Balmer lines up to H_{36} are visually traced on the spectrograms.

To ascertain physical conditions and nature of prominences it is necessary to co-ordinate experimental data with theoretical conceptions.

The simplest theoretical method of determining n was suggested by Redman and Suemoto, 1954. Applying such a method, we calculated the theoretical values of the hydrogen line half-widths. The comparison of our modified theoretical curves of the main-quantum-number-dependent half-widths of the Balmer series lines with the observed values of total half-widths gives that

$$n_e = 3 \cdot 10^{10} \text{ cm}^{-3} \quad \text{and} \quad T_e = 5500^0 \text{K}.$$

The electron density determined by the number of the extreme resolved Balmer series line is

$$n_e = 7.4 \cdot 10^{10} \text{ cm}^{-3}$$

and by the total emission of the continuous spectrum beyond the Balmer series limit is

$$n_e \approx 5 \cdot 10^{10} \text{ cm}^{-3}.$$

The values of the electron density obtaind by means of different methods agree well.

REFERENCES:

Hirayama,T.: 1963, Pub. Astron. Soc. Japan 15, pp. 122-144.
Ivanov-Kholodny,G.S.: 1955, Izv. Krymsk. Astrofiz. Obs. 15, pp. 69-94.
Ivanov-Kholodny,G.S.: 1958, Izv. Krymsk. Astrofiz. Obs. 18, pp. 109-135.
Ivanov-Kholodny,G.S.: 1959, Astron. Zh. 36, pp. 589-600.
Ivanov-Kholodny,G.S., Nikolsky,G.M. and Guliaev,R.A.: 1960, Astron. Zh. 37, pp. 799-811.
Ivanov-Kholodny,G.S. and Nikolsky,G.M.: 1961, Astron. Zh. 38,pp. 455-462
Inglis,D. and Teller,E.: 1939. Astrophys.J.90, pp. 439-448.
Jager de,C.: 1952, Rech. Astron. Obs. Utrecht 13, pp. 1-93.
Kurochka,L.N. and Maslennikova,L.B.: 1970, Solar Phys. 11, p. 33.
Redman,R.O. and Suemoto,Z.: 1954, Month. Not. 114, pp. 524-539.

A POLARIZATION INVESTIGATION OF PROMINENCES

Ts.S. Khetsuriani
Abastumani Astrophysical Observatory,
Georgian Academy of Sciences, USSR

A theoretically possible value of the polarization in various spectral lines is determined by quantum mechanical parameters of corresponding transitions (Chandrasekar, 1953). From this standpoint the resonance line of CaI $\lambda4227$ ($4^1S^1- 4^1P^0$) is very convenient for polarization measurements as the polarization predicted theoretically for this line is high ($\sim100\%$). In D_3 He ($\lambda5876$ $3^3D - 2^3P$) line consisting of three components the theoretical polarization in the whole line does not exceed 24%. But in the long wave component of the multiplet $\lambda5875.989$ whose intensity is ten times lower than that of the primary line the polarization is about 100%. Therefore based on the data obtained with high spectral resolution the polarization investigations of these lines can be performed using the method of photographic photometry. Particularly for high prominences, as it is known due to Baumbach's (1938) calculations, the polarization of prominence radiation increases with the distance from the solar limb.

The observational material was obtained on the horizontal solar telescope (diametre of solar image 16 cm). The dispersion was 0.7 Å/mm. The Wollaston prism, placed in front of the film served as an analyser.

Out of the observed prominences one, with such position angle the radius vector of which with the entrance slit made the angle of $\alpha = 45^0$, was chosen. In such a position the instrumental polarization is almost brought to zero as it is known that the plane of the instrumental polarization is directed along the lines of the grating, i.e. along the entrance slit (Nikolsky et al., 1972). In such a case the observed value of the polarization almost equals a true value, if admitted that the polarization plane of prominence radiation is radial (though it is known that there are some deviations, but they are small (Leroy, 1977).

The prominence of August 19, 1968 was such a rare position. A very bright knot was observed in the prominence. The knot radiated in a number of metallic lines including the CaI $\lambda4227$. The action of the CaI $\lambda4227$ absorption line was allowed for. The results of measurements are given in the table.

W. Fricke and G. Teleki (eds.), Sun and Planetary System, 135–136.
Copyright © 1982 by D. Reidel Publishing Company.

α^0	Exp. in min.	h''	P%	P_i%	α^0	Exp. in min.	h''	P%	P_i%
+45	15	30	7.0	-2.0					
		32	9.0	-1.5					
		34	10.0	1.5	+45	5	34	8.0	0
		38	11.0	-2.5			38	6.0	0
		42	8.0	-1.5			42	11.5	-2.5
		46	12.0	0					
		50	10.0	0					
-45	15	30	8.0	2.5					
		32	8.6	2.5					
		34	10.0	2.4	-45	5	34	10.0	0
		38	12.0	-1.5			38	9.0	0.8
		42	10.0	0			42	12.0	1.5

For studying the polarization in the He D_3 we measured the spectral material in the long wave component. In particular the height of 40'', on the average, the polarization was about 5% instead of 25% predicted theoretically.

The value of the ratio P/P_{theor} represents a factor of depolarization and it shows the presence of a depolarizing non-radial magnetic field. If assumed that non-magnetic depolarization effects are negligible (Sahal-Brechot et.al., 1977), the strength of the magnetic field can be estimated . In all cases considered the field was rather strong, but no more than 20 gauss.

REFERENCES

Baumbach,S.: 1938, Astron. Nachr. 267, p. 272.
Chandrasekar,S.: 1953, Transfer of Radial Energy, ed. M., p. 58.
Leroy,J.L.: 1977, Rep. from the Obs. of Lund, 12, p. 95.
Nikolsky,G.M. and Khetsuriani,Ts.S.: 1972, Solnechnie Dannie, 5, p. 91.
Sahal-Brechot,S., Bomier,V. and Leroy,J.L.: 1977, Astr. and Astrophys.
                                                           59, p. 223.

THE FORMATION OF SOLAR PROMINENCES

Claudio Chiuderi
Istituto di Astronomia, Università di Firenze, Italy.

The formation of solar prominences is an interesting example of development of cool condensations that are not due to gravitational forces. It is widely believed that the physical processes operating in this case may also be relevant in different astrophysical systems, like the supernova remnants or the nuclei of active galaxies. The basic mechanism that drives the formation of filamentary structures is the thermal instability, that can be physically described as the tendency of a radiating plasma to react to a cooling condensation by an increase in the radiative output. This in turn produces a further cooling, until a state is reached where this tendency is reversed. In an optically thin plasma of solar composition the thermally unstable behaviour can occur for temperatures roughly in the range $4 \times 10^4 \lesssim T \lesssim 10^7$ K. Thermal conduction, on the other hand, strongly opposes the formation of regions of different temperatures. We thus conclude that situations in which thermal conduction is inhibited are those in which most likely prominences are formed. There are two further physical parameters that influence the possibility of prominence formation and the rate at which they form. These are, of course, gravity and magnetic fields. The weight of the filament must be supported against gravitational forces and this can be done by the magnetic field. This latter has also the effect of channelling the heat flux along field lines and thus to act as an effective thermal insulator for perturbations across $\underset{\sim}{B}$.

A complete treatment of the prominence formation problem therefore involves the solution of a time-dependent set of non-linear equations taking into account all the above mentioned effects. This can only be done numerically. To get some preliminary insight into the problem, it is however useful to consider a series of approximate, limiting cases. The standard approach to any stability problem is to treat linear stability first. The linear stage of thermal instability has been treated long ago by Field (1965). A realistic radiative loss term was not available to him and the magnetic field was taken as uniform. A linear calculation with a correct radiative loss term and a non-homogeneous, sheared magnetic field was performed by Chiuderi and Van Hoven (1979). They showed that in the absence of gravity, thin, elongated structures

W. Fricke and G. Teleki (eds.), Sun and Planetary System, 137–138.

are formed at the location where $\underline{k} \cdot \underline{B} = 0$, i.e. where the perturbation is perpendicular to the field lines. The linear growth time is of the order of the radiative cooling time or 10^3 s in typical coronal conditions. The same calculations have been recently repeated with different field configurations and have shown the same general behaviour, independently of the shearness of the field.

Extensive non-linear calculations have been performed numerically by Hildner (1974), who however did not include the effects of thermal conduction and did not follow the developement of the condensation up to the saturated stage. This has been recently done by Antia and Chiuderi (1981), who have adopted the following scheme. Gravity and magnetic field have been neglected altogether to simplify the problem and to focus on the physical mechanism rather than on the support problem. The basic state is assumed to be uniform. The fastest linearly unstable perturbation is then chosen as the initial stage of the numerical non-linear calculation. The dynamics is then followed until the system reaches a new equilibrium. The actual possibility of finding an equilibrium depends on the initial and boundary conditions imposed on the problem. We have adopted three different sets of boundary conditions representing a fixed slab with no mass or heat flux from the boundaries, or a fixed slab with no mass flux from the boundaries in contact with a thermostat, or finally a fixed mass of plasma free to expand or contract but with boundaries kept at constant pressure and temperature.

From these calculations we are able to conclude that a homogeneous system, linearly unstable to the thermal instability, evolves toward a stable inhomogeneous state where thin cool condensations have formed. This conclusions are likely to be strengthened by the inclusion of magnetic fields, presently neglected.

References

Antia, H.M. and Chiuderi, C.: 1981, in preparation.
Chiuderi, C. and Van Hoven, G.: 1979, Astrophys. J. Lett. 232, L69.
Field, G.B.: 1965, Astrophys. J. 142, 531.
Hildner, E.: 1974, Solar Phys. 35, 123.

RADIATION FROM SURFACE WAVE PACKET AND TYPE II SOLAR
RADIATION EMISSION

V.M. Čadež

Institute of Physics, Studentski trg 12/V, Yu-11001 Beograd

We discuss a possibility of generating the electromagnetic
radiation by a potential surface wave packet in an inhomogeneous plasma
under physical conditions similar to those existing in the solar corona
regions where the type II radio emission originates.

The type II solar bursts are relatively rare but outstanding
events, lasting about 10 min. and are associated with large flares.
They appear several minutes after the onset of the flare and experience
a slow frequency drift of the emitted radiation from higher to lower
ranges, within the interval of the order 10-100 MHz. It is generally
accepted that this type of radio bursts is located at the shock wave
moving away from the region of the flare into the corona, **stretching** the
magnetic field lines at the same time. Charged particle flows along the
magnetic field lines are assumed in the domain of the shock wave discontinuity as a result of eruptive processes in adjacent active regions.
Such particle flows are of finite length in space.

We shall therefore consider the stability characteristics of
an idealized plasma geometry similar to the described conditions at
the region of the type II solar burst. This means that the plasma density is assumed to rise sharply over a short distance $\underline{a}$ (shock wave
front) from some low value, practically vacuum (region outside the
shock wave). Inside the transition layer $\underline{a}$, we assume a particle stream,
parallel to a homogeneous magnetic field $\vec{B}_o$ and perpendicular to the
plasma density gradient. The flow velocity profile may vary in space
inside the layer, i.e. $u=u(x)$, and need not be relativistic. The flow
is also of some final length L. Such plasma configuration becomes unstable if certain resonant conditions for the flow are satisfied and as
the result of the instability a surface wave mode is generated. To see
this we start from known equation for the electric field potential of
small electrostatic perturbations applied to the system:

W. Fricke and G. Teleki (eds.), Sun and Planetary System, 139–140.

$$\frac{d}{dx}\left(\epsilon_1\frac{d\phi}{dx}\right) - k_{\shortparallel}^2\left(\epsilon_1\sin^2\theta + \epsilon_3\cos^2\theta - \frac{\sin\theta}{k_{\shortparallel}}\frac{d\epsilon_2}{dx}\right)\phi = 0 \tag{1}$$

where the dielectric tensor components have their standard form:

$$\epsilon_1 = 1 - \sum_\alpha \frac{\omega_{p\alpha}^2}{\omega^{*2} - \Omega_\alpha^2}, \quad \epsilon_2 = \sum_\alpha \frac{\Omega_\alpha}{\omega^*}\frac{\omega_{p\alpha}^2}{\omega^{*2} - \Omega_\alpha^2}, \quad \epsilon_3 = 1 - \sum_\alpha \frac{\omega_{p\alpha}^2}{\omega^{*2}}, \quad \omega^* = \omega - \vec{k}_{\shortparallel}\cdot\vec{u}$$

and θ is the angle between vectors $\vec{B}_0$ and $\vec{k}_{\shortparallel}$.

Solutions of (1) in constant plasma density regions exponentially fall off from the transition region. These oscillations are therefore localized to the region of density increase and are called surface waves. Their frequency spectrum follows from the related dispersion equation and consists of two modes, the high frequency electron mode ω_H and the low frequency ion mode ω_L /see references/

$$\omega_H = \frac{1}{2}\left[\left(2\omega_{pe}^2 + 2\Omega_e^2 - \sin^2\theta\right)^{1/2} - \sin\theta\right], \quad \omega_L = \left[\frac{\Omega_e^2\omega_{pi}^2\cos^2\theta}{\Omega_e^2\cos^2\theta + \omega_{pe}^2}\right]^{1/2} \tag{2}$$

For our purposes we shall consider the high frequency mode only. If a nonrelativistic particle flow is present inside the transition region then the dispersion equation has complex solutions for perturbation frequencies. The imaginary parts of frequencies i.e. the instability growth rates exist if:

$$\omega_H - \vec{k}_{\shortparallel}\cdot\vec{u}(x_{r_1}) = -\left(\omega_{pe}^2 + \Omega_e^2\right)^{1/2} \quad \text{or} \quad \omega_H - \vec{k}_{\shortparallel}\cdot\vec{u}(x_{r_2}) = 0 \tag{3}$$

The flow thus resonantly excites potential ($\omega/k \ll c$) surface waves at $x = x_{r_1}$, x_{r_2} and when the flow terminates the excited waves will soon decay due to nonlinear effects and dissipative mechanisms of the medium. In our model we take that the flow of finite length L suports the existance of a wave packet of the same length L whose group velocity is $u(x_r)$. In the frame of reference, comoving with the wave packet and for the first resonant condition (3), we shall observe a system of linear oscillators oscillating at frequency ($\omega_{pe} + \Omega_e$)$^{1/2}$ (the upper hybrid frequency). They will radiate as a dipole of length L at this frequency. The second resonant condition (3) will give a stationary charge distribution and no radiation.

The emitted type of radiation would thus come from an effective dipole oscillating at local upper hybrid frequency and moving at speed u (x_{r_1}).

REFERENCES

Čadež, V., Vuković, S.: 1975, Proc. of XII ICPIG Eindhoven, pp. 298.

Jovanović, D., Vuković, S., Čadež, V.: 1976, Proc. of SPIG-76
                                                    Dubrovnik, pp 459-462.

MAGNETOHYDRODYNAMICS AND THERMODYNAMICS OF CORONAL LOOPS

G. Einaudi
Scuola Normale Superiore, 56100 PISA, Italy

Recent X-ray and EUV observations strongly suggest a deep connection between magnetic structures and thermal properties of coronal loops. This means that in order to explain the observations we need to include the energy balance in the set of equations describing the possible equilibria. In other words we need to solve together the equation

$$\vec{\nabla} p = \frac{1}{4\pi} (\vec{\nabla} \times \vec{B}) \times \vec{B} + \rho \vec{g} \qquad (1)$$

expressing the balance of the forces, and

$$E_c = \vec{\nabla} \cdot \chi_0 T^{5/2} (\vec{e}_B \cdot \vec{\nabla}) T \vec{e}_B = E_R (p,T) - E_H (p,T) \qquad (2)$$

in which the coronal heating (E_H), the coronal radiative losses (E_R) and the effects of thermal conduction (E_c) are considered. The solution of Eqs.(1) and (2) is a very difficult problem and therefore requires a certain number of assumptions. The most frequently used is to treat separately the two aspects of what is a single problem, by discussing in turn the thermodynamics and the magnetohydrodynamics of loops. In the first case the only magnetic effect included is the channelling of the heat flux along the magnetic field, disregarding the three-dimensional structure of the magnetic field. In the second case the energy equation used is the adiabatic one, which means to neglect the unhomogeneity of the thermodynamic quantities.

A two-dimensional model can be however worked out by making a less dramatic simplification, i.e. that of neglecting the effect of gravity. The meaning of this assumption can be deduced from Eq.(1) neglecting for a moment the magnetic force. In this case we can estimate a scale height for the pressure, when the temperature is kept constant, as follows: $dp/dz = - p\mu g/RT$ $p = p_0 \exp (- z/h)$; $h = RT/\mu g$.
When the dimensions of the system are $\ll$ h, the gravity can be neglected because the variations of the pressure are due mainly to the magnetic forces. With $g = 2.74 \times 10^4$ cm sec^{-2} and $2 \times 10^4 < T < 2 \times 10^6$, which are typical values for solar corona, h varies in the range 10^8–10^{10} cm. This means that gravity does not have a tremendous influence on the structure of coronal loops and in first approximation can be neglected. Eqs.(1) and (2) then become easier, because the pressure gradients are perpendicular to the magnetic lines of force so that Eq.(2) can be

W. Fricke and G. Teleki (eds.), Sun and Planetary System, 141–142.
Copyright © 1982 by D. Reidel Publishing Company.

solved along a line of force, the only coupling between the two
equations being the relationship between the length of each line and the
value of the pressure on it. In order to model a loop of length L_o, we
choose a cylindrical geometry and a magnetic field of the form $\vec{B} \equiv$
$(0, B_\theta(r), B_z(r))$. Assuming $J_\parallel = \alpha B$, with α constant, it can be shown[1]
that the length of a line at a certain radius, whose expression is:

$$L(r) = L_o (1 + B_\theta^2 / B_z^2)^{1/2} \qquad (3)$$

is independent of the structure of the pressure. It follows that, with
this assumption Eqs.(1) and (2) completely decouple. On the other hand
the form of the magnetic field contains all the features which seem to
be present in coronal structures, so that the model is realistic enough.
We can then solve Eq.(2) along each line by giving the value of the
pressure and computing the length and the temperature profile of this
line[2]. Comparing the computed length with Eq.(3) we can deduce the
radial structure of the temperature. When the magnetic field presents
some shear, which means $dL/dr > 0$, we must for consistency choose a radial
pressure profile with a minimum at the center of the loop and a temperature of similar profile. This model can easily explain some observations[3,4] which show the existence of loops in many lines (CII, OIV, OVI,
NeVII ...) suggesting a temperature structure in the radial direction of
the loop with a minimum of $\sim 2 \times 10^5$ °K at the center and a maximum of
$\sim 2 \times 10^6$ °K at the border of the loop. In fact in this model the radial
temperature gradients are created along the loop by the shear of the
magnetic field and at every height all the temperatures in the range
$\sim 2 \cdot 10^5 - \sim 2 \cdot 10^6$ are present. Of course at different heights we find the
same temperature at different radii and this effect can in principle be
checked in future observations provided the spatial resolution increases. The only unsatisfactory point is that we have the same pressure
gradients at the base of the loop as at the top. This fact could be
ascribed to the neglection of gravity: its proper inclusion introduces
pressure gradients along the lines of force and can solve this problem[5].

Numerical calculations, whose results can be well approximated by
using the scaling laws given in Ref.2, show that a temperature jump
between the center and the border of the loop of 10 at the top and a
pressure jump of 20 are consistent with a length of magnetic lines
increasing by a factor 10. A magnetic configuration of this kind satisfies the sufficient stability conditions given in Ref.6, provided the
loop is not too thin.

An interesting feature of this model is the possibility of relating
observable quantities as temperature, pressure and length of the loop
with topological properties of the magnetic field, which is one of the
big unknowns in the physics of coronal loops.

References:
1) Chiuderi,C.,Giachetti,R.,and Van Hoven,G.:1977,Solar Phys.,54,107.
2) Chiuderi,C.,Einaudi,G.,and Torricelli-Ciamponi,G.:1981,Astron.As-
    trophys.97,27.
3) Foukal,P.V.:1975,Solar Phys.,43,327.
4) Levine,R.H.,and Withbroe,G.L.:1977,Solar Phys.,51,83.
5) Einaudi,G.,Torricelli-Ciamponi,G.,and Chiuderi,C.: in preparation.
6) Einaudi,G.,and Van Hoven,G.:1981,Phys.Fluids,24,1092.

TEMPERATURE AND STEADY FLOWS IN SLENDER MAGNETIC TUBES

ELISABETH RIBES
D.A.S.O.P., Observatoire de Meudon
92190 MEUDON, FRANCE

A theoretical study of steady flows in slender magnetic tubes ("faculae") has been made with a frozen-field assumption (Unno and Ribes 1979 ; Webb, 1980 ; Ribes and Semel, 1982).

The main observational characteristics to be considered are the following :
- a magnetic field strength of 1 to 2 Kilogauss, near $\tau_{5000}=1$
- a downdraft increasing with depth and exhibiting a fairly large velocity gradient in the inner photosphere (Giovanelli and Slaughter, 1978).
- an excess temperature of $1000°K$ or more, in the upper photosphere as deduced from the line weakenings in faculae (Chapman, 1977).

In the low photosphere, the thermal structure of the unresolved magnetic fields is not well known since the interpretation of the continuum contrast depends very much on the assumed width of the magnetic tube. So, a large excess temperature in the faculae throughout the photosphere cannot be disregarded a priori (Koutchmy, 1977).
However, the question arises whether a strong enhancement of the temperature due to the rapid increase of the downdraft in the inner layers can be maintained. To answer, let's compare the radiative cooling time with the dynamical time, that is the time required by the matter to cross through one scale-height. For a typical magnetic tube (B ~ 1.2 Kilogauss, u ~ 2 Kms^{-1}, $\rho_{in} \sim 2\ 10^{-7}$ g cm^{-3}, Hp (pressure scale height) ~ 100 Km ; lateral width of the tube $\lesssim 300$ Km, $T_{in} \gtrsim 6500°K$, at $\tau_{5000}=1$), the radiative cooling time ($t_{rad} \sim 1$ second) is much smaller than the dynamical time ($t_{dyn} \sim 50s$). Since the energy at the bottom of the magnetic tube is radiated away very rapidly, the input of the thermal energy transported down by the flow from the upper levels should be large enough to maintain an excess brightness throughout the tube. An estimate of the ratio between the excess brightness and the thermal energy transport per unit time gives some upper limit for the facular heating:

W. Fricke and G. Teleki (eds.), Sun and Planetary System, 143–144.
Copyright © 1982 by D. Reidel Publishing Company.

at $\tau_{5000}=1$, ΔT could be of the order of a few hundred degrees, probably less than $1000°K$.

Thus, a Bernoulli solution exhibiting an excess Temperature increasing downwards, with a maximum value of $2400°K$ localized near $\tau_{5000}=1$ (Webb, 1980) is in agreement with the observed flow profile but seems inconsistent thermodynamically.

Possible alternatives to reduce the excess temperature will unavoidably affect the dynamical regime of the flow. Either the field strengh at the bottom of the tube is decreased, and the flow speed is reduced as well. Or, we assume that the observations of Giovanelli and Slaughter are averages over the unresolved structures. In this case, much higher velocities could be present leading to a moderate temperature excess in the inner photosphere (Ribes and Semel, 1982). In the extreme case, the flow could even be sonic when entering into the convection zone, with ΔT null (Unno and Ribes, 1979).

The thermodynamics puts severe constraints on the dynamics in a magnetic tube. The photospheric heating of a facula can be explained by two types of consistent steady-flow solutions :
- an accelerating (and rather high speed) downdraft with a high field strengh (B $\sim$ 1.3 Kilogauss)
- or a downdraft decelerating with depth, associated with a moderate field strengh (B $\lesssim$ 800 gauss).

Both situations can exist in the unresolved small-scale magnetic fields, keeping in mind that the interpretation of the observations is strongly model-dependent and, therefore, not unique. However, the first type of solutions seems the most attractive possibility : a rapid increase with depth of the flow speed reduces the magnetic scale-height. It means that the magnetic flux tube will fan out more rapidly than in the case of a low field and low velocity flow (Ribes and Semel, 1980), or in the static case (Webb, 1980) This is supported by recent observations of Giovanelli (1980).

REFERENCES

CHAPMAN G.A. : 1977, Ap J supplement series, 33, pp 35-42.

GIOVANELLI R.G., SLAUGHTER C. : 1978, Solar Phys. 57, pp 255-260.

GIOVANELLI R.G. : 1980, Solar Phys. 68, pp 43-69.

KOUTCHMY S : 1977, Astron. Astrophys. 61, pp. 397-404.

RIBES E. and SEMEL M. : 1980, Proceedings of the Japan-France Seminar on Solar Physics, pp 129-141

RIBES E., SEMEL M. : 1982, in preparation

UNNO W., RIBES E. : 1979, Astron. Astrophys. 73, pp. 314-321.

WEBB A.R. : 1980, PH.D. THESIS, University of St Andrews.

GRAVITO-MAGNETIC EXPLANATION OF THE TEMPERATURE OF STELLAR

CORONAE AND OF PLANETARY VAN ALLEN BELTS

E. Woyk (Chvojková)

Astronomical Institute, Czechoslovak Academy of Sciences,
12023 Praha 2, Czechoslovakia

Charged particles spiralling along magnetic field lines surpass larger altitude distances in the mean time between two subsequent collisions if they spiral towards levels of smaller pitch angles i. (Then the velocity component parallel to the magnetic field grows: $v_{\parallel} \sim \cos i$.)

Thus, if particles alternatively spiraled up and down after subsequent collisions, always beginning with $i=60°$ as shown in Fig.2, a diffusion towards smaller pitch angles would arise. Since the pitch angle i of superescape particles diminishes with altitude r but grows if $v < v_{esc}/\sqrt{3}$ - see Fig.1 - a separation of the high-energy plasma from the deeper cooler rest arises. (Note: In a dipole-like H, each particle path is defined by an $E_{(1)} \equiv (v_{esc}/v)^2_{(1)}$ and the real path between two collisions is always part of an $E_{(1)}$-path, defined by the path formula Woyk 1979.)

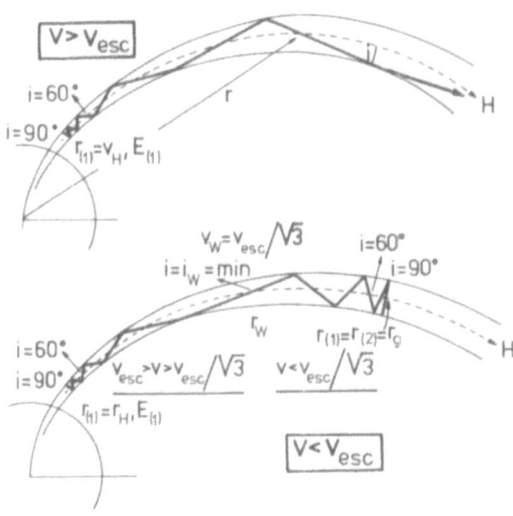

Fig.1. Two particle paths defined by two different $E_{(1)}$.
 Above: Superescape particles spiral along magnetic field lines gradually diminishing their pitch angles i with the distance from the gravity centre, r.
 Below: Underescape ones possess two path ends, two levels $r_{(1)}$ at which $i=90°$. The upper one, $r_{(1)}=r_g$, is the highest altitude a particle of given velocity and path radius can attain in the gravity field. i is minimum at a level r_W at which $v=v_{esc}/\sqrt{3}$ ($E_W=3$). Slow particles spiral above their r_W.

W. Fricke and G. Teleki (eds.), Sun and Planetary System, 145–146.
Copyright © 1982 by D. Reidel Publishing Company.

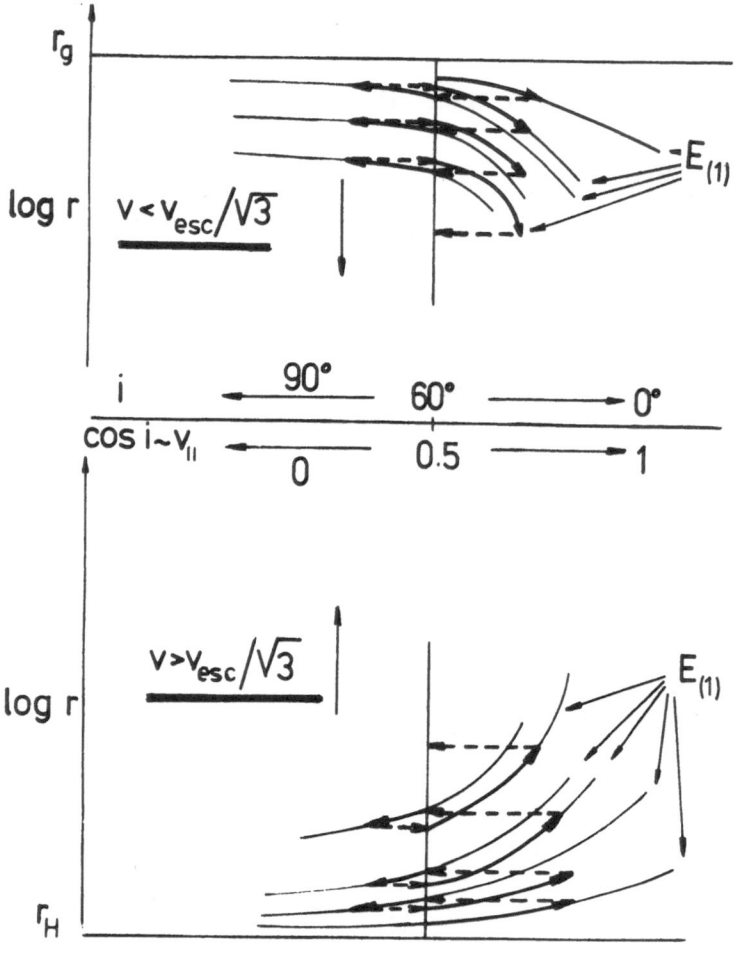

r_g

$\log r$

$v < v_{esc}/\sqrt{3}$

$E_{(1)}$

i $\xleftarrow{\quad 90° \quad}$ 60° $\xrightarrow{\quad 0° \quad}$

$\cos i \sim v_{\parallel}$ $\xleftarrow{\quad 0 \quad}$ 0.5 $\xrightarrow{\quad 1 \quad}$

$\log r$

$v > v_{esc}/\sqrt{3}$

$E_{(1)}$

r_H

Figure 2. Altitude variation i(r) when particles alternatively spiral up and down after subsequent collisions, always beginning with i=60°, in the mean.

Above: Slow underescape particles (v<$v_{esc}/\sqrt{3}$ of Fig.1) diffuse down slowly but surely - since i decreases downwards there.

Below:For the same reason superescape particles diffuse upwards.

Note: Each path is composed of a series of segments of different $E_{(1)}$-curves defined from the velocity and direction, E and i, occurring after the collision (i.e.E and i=60°)

Owing to the separation of the high-energy plasma from the slower deeper rest,shown in Fig.2,a top trapping arises which explains the origin of the temperature of stellar coranae and planetary radiative belts.

The temperature of the solar corona rises from its bottom to a maximum comparable to the escape velocity v_{esc}. (The number of slow particles is already insignificant there while the number of collisions of the superescape ones is insignificant too for contributing to a sizeble temperature increase.) The temperature decreases above this maximum owing to the decrease of the escape velocity with r.

The corresponding formulae as well as many details are given by Woyk 1979. Also several surprising phenomena have been deduced there which arise when the magnetic field becomes deformed by an outer impact. Woyk 1981.

Woyk E., 1979, Bull.Astron.Institut.Czechoslovakia, 30, 341-348.
Woyk E., 1981, Adv.Space Res, (COSPAR), 1, 167.

SECTION II

ASTRONOMICAL, GEOPHYSICAL AND GEODETIC PROBLEMS RELATED TO THE EARTH

ASTRONOMICAL AND GEOPHYSICAL PROBLEMS OF THE EARTH'S ROTATION

J. Kovalevsky
CERGA,Grasse,France

Ya. S. Yatskiv
Main Astronomical Observatory,
Ukrainian Academy of Sciences,
Kiev, USSR

Abstract. Following a general description of the motion of the Earth about its center of mass, the main features of the new 1980 IAU nutation series are described, and the definition of the new Celestial Ephemeris Pole is justified. The use of excitation functions for polar motion and UT1 is explained. Then, the main characteristics of the short period components of the polar motion and UT1 are described, together with their geophysical implications. In connection with the long periodic terms of polar motion and UT1, the problems of core-mantle coupling, of the secular drift of the mean pole and of the non-tidal Earth acceleration are discussed.

I. INTRODUCTION

The four components of the Earth's rotation have a very long history. The precession was discovered by Hipparchus during the second century B.C. although, of course, it was not related to Earth's rotation. The main term of nutation was discovered and explained by Bradley in 1746. Similarly, if Euler has concluded in 1765 that the Earth's pole of rotation could experience a ten month oscillation with respect to the crust, the wobble was discovered by Chandler in 1891. It had a 14 months period and Newcomb showed the next year that it was the Eulerian wobble lengthened by the Earth's departure from rigidity. Finally, the existence of the Earth's deceleration was established by Darwin during the 19th century, while its seasonal component was discovered by Stoyko in 1937.

Nowadays, the techniques used to observe various aspects of the Earth's rotation have reached a remarkable accuracy, and the interest of scientists in them has also grown considerably : a number of important questions have arisen about their causes and consequences in the domains of geophysics, oceanography, atmospheric sciences and astronomy. Let us refer to a recent very valuable book by Lambeck (1980) that discusses in detail their geophysical implications.

W. Fricke and G. Teleki (eds.), Sun and Planetary System, 149–162.

II. DESCRIPTION OF THE EARTH'S ROTATION

Under the general designation of Earth's rotation, are included a number of different aspects that are generally separated for conceptual or for observational reasons. One may define the rotation of the Earth as the time dependent relation that exists between a terrestrial frame with its origin at the center of mass of the Earth and associated with its visible crust (0, XYZ) and a quasi-inertial (or fixed) celestial reference frame (0, xyz)

$$(0, \; XYZ) \Longrightarrow (0, \; xyz) \tag{1}$$

Ideally, both reference systems should be defined and realized independently of the motions of the Earth. However, this is not presently the case. The celestial reference system depends heavily on the Earth's equator (Kovalevsky, 1981) and it is not possible, in practice, to construct a terrestrial reference frame without using some fiducial points outside the Earth, so that the Earth's rotation has to be used in defining and maintaining it (Mueller, 1981).

Let us, however, ignore this reservation and let us assume, for the time being that there exist perfect reference systems. If we also assume as a first approximation, that the Earth is rigid and isolated in space, it is a well known result that the instantaneous axis of rotation $O\omega$ undergoes two independant free motions.

a) a near diurnal motion around the axis of angular momentum OH (see fig. 1). Its actual amplitude is of the order of 0".001.

b) an Eulerian wobble in the Earth fixed frame, with a period of 305 days around the principal axis of inertia $0 \; \zeta$. The amplitude of the equivalent real term is of the order of 0".3.

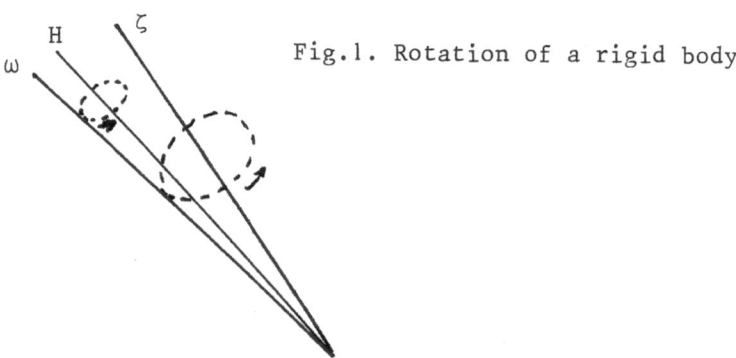

Fig.1. Rotation of a rigid body.

This very simplified approach to the dynamics of the Earth's rotation already shows that it is convenient to introduce a third -intermediate- system of coordinates, which, in this case, would be defined by $O\omega$. But the actual situation is much more complex, since one has to take into account the torques exerted by the Moon, the Sun and the planets. Furthermore, the Earth is not a rigid body and is composed of parts having very different structures and mechanical properties.

Let us call (O, ABC) the intermediate coordinate system, OC being its near polar axis and OA being always at the intersection of the corresponding equator with the ecliptic. The orientation of the Earth in space will be described as the sum of two components.

a) The motion of the intermediary system with respect to the celestial reference frame :

$$(O, ABC) \Longrightarrow (O, xyz) \tag{2}$$

It is an astronomical usage to divide this motion into its very long periodic component, the precession, while all the other terms are included in the nutation. Awkwardly enough, the present celestial reference system is defined in such a way that its origin is at the intersection of the Earth's equator with the ecliptic. The motion of the ecliptic due to the perturbations of the Earth's orbit by other planets introduces a motion of the origin of the axis which is called planetary precession. It has nothing to do with Earth's rotation, but has to be present in the expressions for precession. To distinguish it from the general precession, the motion of OA with respect to a fixed equinox is called the luni-solar precession.

b) The motion of the intermediate system with respect to the Earth-fixed system :

$$(O, ABC) \Longrightarrow (O, XYZ) \tag{3}$$

Obviously (1) is the composition of (2) and (3). This motion is again split into two components.

- The motion of OC with respect to the Earth fixed system is called the polar motion. In some sense, C may be considered as the instantaneous pole of rotation. But one must notice at this stage that if we replace OC by any arbitrary OC', the motion of OC with respect to OC' will be removed from the nutation and included in the polar motion. There is, indeed, a part of arbitrariness in the respective definitions of nutation and polar motion. This property will be used below.

- The motion of OAC with respect to the Earth-fixed system is the sidereal time of the prime meridian. However, it is commonly replaced by Universal time UT1, which is computed using the ratio R of the mean sidereal day to the mean solar day. Its value in the new IAU system of astronomical constants is (P.A. Wayman, 1980) :

$$R = 0,997\ 269\ 566\ 388 - 0.586\ 10^{-10}\ T$$

where T is expressed in julian centuries elapsed since 1900 January 0, 12 h.

As we have already said, a major difficulty in applying these definitions lies in the actual definition of the terrestrial reference system. This has been discussed at length by Guinot (1979), Kovalevsky
(1979) and Yatskiv (1981a).This definition implies that such a system,
called "Conventional terrestrial coordinate system" is realized by the
coordinates of a number of stations as obtained through observations
made during some interval of time and later maintained through continuous observations performed in the same stations. This implies that
there are several different systems (and hence OXYZ frames) depending
upon the technique of observation.

Some systems are based on conventional longitudes and latitudes
of astronomical stations. The most currently used are ;

MPO : Mean pole origin defined by the mean latitudes of observa-
      tories at any moment.
CIO : the conventional international origin defined as the mean
      pole as observed by the International Latitude Service (ILS)
      in 1900-1905.
BIH 1968 system : based on CIO, but referred to the 1968 epoch and
      fixed using conventional coordinates of the stations partici-
      pating in the BIH.
UT (SU) : the reference system adopted by the USSR time service.

Other systems are based on positions of stations defined by their
geocentric trirectangular coordinates. They are used in satellite work.
Let us quote two such systems :

NWL 9Z-2 : it is defined by 20 Doppler stations of the TRANSIT
      system and is adjusted to the BIH 1968 system at the initial
      epoch.
IASOM-LLA 80 : it is defined by 11 laser stations ranging on LAGEOS.
      The pole is the BIH 1968 pole, while one station (Greenbelt)
      defines the longitude origin.

III. NUTATION AND PRECESSION

While the nutation theory that is still used in the Ephemerides,
due to Woolard (1953), assumes a rigid Earth, the model that has been
retained as the basis of the 1980 IAU nutation series (Seidelmann et al.
1982) includes a much better description of the Earth. The model is due
to Gilbert and Dziewonski (1975) and the theory of nutation itself, to
Wahr (1981). The model includes a solid inner core embedded in a fluid
core and a distribution of elastic properties of the Earth's interior
that was deduced from the discussion of a large set of seismological
data.

It is to be noted that the coefficients of the nutation series are
not very sensitive to different Earth models presently available so that
the accuracy of astronomical observations does not permit to distinguish

among them. The choice made is essentially based on its geophysical
likehood. We shall not describe the series, but make only two remarks.

a) The actual Earth models introduce a correction to the amplitudes
α_0 of the corresponding term of period P_0 in the rigid Earth nutation
that has the form :

$$\frac{a}{\alpha_0} = 1 + \frac{1}{P_0} \ f(\frac{1}{P_0}, \ model) \tag{4}$$

This effect is quite important for short period terms (more than 10 %
for semi monthly terms) but is very small for long periodic terms and
vanishes for the precession. The expressions for the precession have
been finalized by Lieske et al. (1977) with a superabundant accuracy.

b) The existence of a compressed ellipsoidal liquid core induces
in the motion of the instantaneous axis of rotation a new free oscillation. In the terrestrial reference frame, it is a small quasi-
diurnal wobble, but a nutation with an amplitude and a period 460 times
larger is also created. Investigations made to identify it in the observations have not been successful (Capitaine, 1975) and, in a recent
review, Yatskiv (1980a) has given an upper bound to this nutation of
0.''01.

The existence of quasi-diurnal wobbles raises the problem of observability of the instantaneous axis of rotation. For instance, astronomical determinations of the pole using upper and lower transits of
stars reach only the center of these wobbles. This is one of the reasons put forward by Atkinson (1975) who advocated that precisely the
center of the wobble be adopted as the pole C of the intermediate system. By construction, there is no nearly diurnal motion of OC with
respect to either celestial or non rotating terrestrial systems. This
point has also the advantage of being the mean axis of the surface of
the Earth as defined by Tisserand (1891). If we decompose the motion
of every point on the Earth's surface into a mean rigid rotation and
a residual deformation, OC moves as it is prescribed by the mean rotation... This pole is used to refer the new nutation theory (Seidelmann
et al, 1982) and is designated as Celestial Ephemeris Pole. This choice
rejects all the purely geophysical effects on polar motion, while the
nutation is produced only by external torques. This is why we may concern ourselves in geophysical phenomena by considering only UT 1 and
the polar motion.

IV. TOOLS FOR THE STUDYING POLAR MOTION AND VARIATIONS OF UT1

In the terrestrial coordinate system, the instantaneous rotation
vector of the Earth can be expressed as

$$\vec{\omega} = \Omega \ (\vec{m} + \vec{i_3} \ m_3) \tag{5}$$

where Ω is the mean angular velocity of the Earth's rotation, $\vec{i_3}$ is the

unit vector of OZ and $\vec{m} = m_1 + i\ m_2$ represents the angular deviation
of Oω from OZ as viewed on the complex plane OXY. The quantity m_3 is
the difference between the mean and the actual rates of rotation of
the Earth. In order to study the dynamics of the Earth's rotation,
a modified excitation function for the polar motion $\vec{\psi} = \psi_1 + i\psi_2$ and
another excitation function for UT-1, ψ_3 are introduced. If we call

$$F_o = f_o + i\ \alpha\ /\ 2\ M$$

the complex Chandler frequency where α^{-1} is the damping time, the
Liouville equations are :

$$\dot{\vec{m}} = 2\ i\ \pi\ f_o\ \vec{m} - 2\ i\ \pi\ F_o\ \vec{\psi}\ ;\quad \dot{m}_3 = \psi_3 \tag{6}$$

while the expressions for the excitation functions are :

$$\vec{\psi} = \frac{1}{2\pi f_o A}\ (\Omega\ \vec{c} - i\dot{\vec{c}} + \vec{h} + \frac{i\vec{L}}{\Omega} - \frac{i\dot{\vec{h}}}{\Omega})$$

$$\psi_3 = -\frac{1}{C}\ (c_{33} + \frac{h_3}{\Omega} - \frac{L_3}{\Omega})$$

where A and C are the principal moments of inertia of the Earth ;
$\vec{c} = c_{13} + ic_{23}$ represents the non-diagonal components of the tensor
of inertia ; $\vec{h} = h_1 + ih_2$ and h_3 are the components of the relative
angular momentum ; $\vec{L} = L_1 + iL_2$ and L_3 are the components of the external forces and c_{33} is the deviation of the polar moment of inertia
from its equilibrium value.

It is known (Yatskiv, 1980 b) that if H (f) is the frequency response of the systems, the relations between the spectral density at
input Sψ (f) and at output, Sm (f) is

$$Sm\ (f) = /H\ (f)/^2\ S\psi\ (f) \tag{7}$$

Three types of problems can be solved.

a) The direct problem. Using geophysical observations and models
of geophysical processes, compute $\vec{\psi}$ and ψ_3 and deduce the polar
motion and variations of UT 1.

b) The inverse problem. From the observed UT 1 and polar motion,
determine the components ψ_1 ψ_2 and ψ_3 of the excitation function
which may be used to evaluate some characteristics of the geophysical
phenomena producing the observed variations.

c) Knowing the excitation functions and the observed m_1, m_2 and
m_3, determine the frequency response H (f). But presently, we have no
good qualitative knowledge of ψ_1, ψ_2 and ψ_3. The problem cannot be
rigourously solved and one can only seek for a theoretical model of
the wobble and attempt to determine its damping time α^{-1} and its
frequency f° by comparison with observations.

V. PERIODICAL TERMS IN POLAR MOTIONS

Many factors contribute to the Chandler wobble and its pertubations, The "solid Earth" effects are the most interesting to study, but they are partly hidden by seasonal, mainly atmospheric effects. It is therefore necessary to free the observed polar motion of its most easily modelled components. This is the case of the seasonal oscillations, mainly due to a geographical redistribution of masses associated with meteorological causes, producing a 0".1 annual and a 0".01 semi-annual terms.

The atmospheric excitation functions have been studied by Sidorenkov (1973), Wilson and Hanbrich (1976) and others. A satisfactory agreement was obtained between computed and observed terms. In evaluating the meteorological excitation function, one must also consider the interaction between the air-pressure and the sea level. The usual approach is to assume a static or inverted barometric response of the oceans to a variable air mass. The increase of atmospheric pressure over the entire ocean depresses the sea surface locally by approximately 1 cm for every millibar of pressure. Snow load, changes in underground water distribution and shifts in ocean masses induced by wind stresses should also be taken into account, but they are not well modelled.

The main characteristics of the remaining part of the polar motion is a 14 months period referred to a Chandler wobble. Its amplitude has varied irregularly since 1900 and has an aspect of a beat phenomenon. Figure 2 shows the power spectrum of the Chandler wobble for three intervals of time. The third interval shows a characteristic double peak structure. One can see that :

a) The spectrum is quite broad suggesting a damping mechanism.

b) The variability of estimates based on different subsets of data suggests a non-stationnary process (Pedersen and Rochester, 1972).

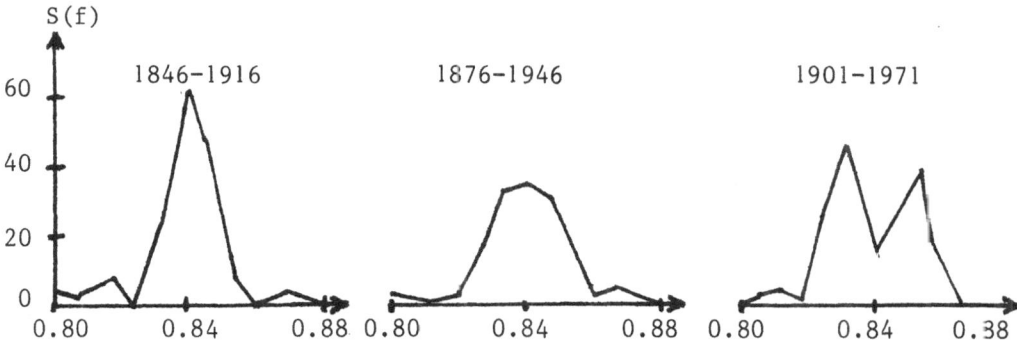

Fig.2. Power spectrum of the Chandler wobble during three periods. The common ordinates S(f) are expressed in $(0".1)^2/c.p.y.$

c) The statistical properties of the noise in observations should be known in order to determine the parameters of Chandler wobble.

Three models have been proposed for describing it : a damped oscillator model, a two component model and a time variable model. In an analysis of the 1846-1971 data, Yatskiv (1974) has concluded that the two component model is not adequate. It seems that the simple damped oscillator model with a frequency wobble response for positive polarisation of the excitation pole given by :

$$H^2(t) = \frac{f_o^2}{(\alpha/2M)^2 + (f - f_o)^2} = \frac{Sm^+(f)}{S\psi^+(f)}$$

is plausible. Solving by a maximum likehood method the following estimations were found :

$$f_o^{-1} = 433.5 \overset{+}{_-} 2.1 \text{ days}$$

$$Q = 40 \text{ to } 60$$

$$S_\psi^+(t_o) = 10^{-15} \text{ rad}^2/\text{cpy}$$

However, using other estimation processes, other authors have recently found values of Q ranging from 60 to 600 (see Graber, 1976 or Wilson and Vicente, 1980).

Another important problem is the value of the Eulerian period. Various Earth models including an inner solid core and an outer fluid core and various laws of mantle anelasticity give values of the Chandler period comprised between 405 and 411 days (Lambeck, 1980). The polar tide, induced by the variation of the position of the pole, produces a non linear coupling whose main effect is to increase the Chandlerian period by 27.4 days (Dahlen, 1976). Recently, Daillet (1981) found a strong correlation between the ellipticity of the Chandler wobble and the enhancement of the polar tide. Finally, the observed and computed periods agree quite well.

The geophysical aspects of the polar motion are fully discussed by Lambeck (1980). Let us however make a few remarks about the excitation mechanism and the sink for energy of the Chandler wobble that are still subject of much controversy.

a) The evidence of significant coherence between the polar motion and atmospheric pressure observations exists (Wilson and Hanbrich, 1976 ; Lambeck, 1980). However the spectrum of the atmospheric excitation function ψ_A is approximately flat over all frequencies with a power of 2.2 10^{-16} rad/c.p.y between 0.67 and 1 c.p.y, while the computed excitation function ψ_o deduced from astronomical observations is five times larger. This effect is therefore probably insufficient and other non seasonal fluctuations in hydrological and oceanic excitations may also contribute. However, no quantitative results have yet been obtained.

b) Certain authors like O'Connell and Dziewonski (1976) believe that earthquakes may excite the Chandler wobble, while others, like Dahlen (1973) or Kanamori (1977) are of the opposite opinion. The basis of disagreement appears to lie in the estimation of moments of earth-quakes and whether significant mass shifts occur immediately before or after the quake.

c) There are evidences of the occurence of aseismic slips. If they occur during a short interval of time in comparison with the Chandler period, they could provide an important excitation mechanism.

d) As for the dissipation of the wobble energy, the most plausible sink is the oceans, although the observational evidence for this re-mains marginal.

VI - SHORT PERIOD IRREGULARITIES OF UT 1

The first kind of irregularities that should be removed from UT 1 before analysing it, are the tidal effects due to the Sun and the Moon. They have been recently recomputed by Yoder et al (1981) inclu-ding all terms larger than 2.10^{-6}. If then one analyses the irregula-rities of UT 1, it appears that the motions of the atmosphere and, in a lesser extent, the motions of the oceanic waters have a direct effect on the Earth's angular momentum and, consequently, on its rate of ro-tation. The major importance of these effects hides the possible contri-butions of the "Solid Earth" to the variations of UT 1, at least for the shorter periods that are considered here. Therefore, it is essential to study first the atmospheric effects so as to be able to remove them.

The seasonal and annual terms are quite well modelled now. The relative amplitude of the annual term of the length of the day is found to be $0.412 \ 10^{-8}$ with a consistency of 1 % between three authors (Lam-beck and Cazenave, 1973 ; Frostman et al., 1967 ; Fliegel and Hawkins, 1967) and the phase is $205° \pm 2°$. This is to be compared with the va-lues $0.402 \ 10^{-8}$ and $188°$ found for the excitation function. The semi-annual term has an amplitude of the order of $0.36 \ 10^{-8}$ and for the biennial term it is $0.097 \ 10^{-8}$. Furthermore, conclusive results were obtained by Siderenkov (1969) on the non-seasonal atmospheric excita-tion function.

Thanks to the improvement of the observational data, new shorter period terms are being discovered. A 50-55 days period has been found by Feissel and Gambis (1980) and meteorological causes seem also pro-bable. Actually, the presently well monitored atmospheric motions (GARP campaigns and meteorological satellites)provide good dayly estimates of the global angular momentum. A very good agreement exists with the variations of the length of the day as obtained by the B.I.H., especially since 1979 (fig. 3). In the mean, more than 75 % are thus explained, although some discrepancies still exist. They may be caused by oceanic effects that are not yet well modelled or by the ice transport and melting. It is only when all these effects will be satisfactorily modelled that one should look for specific excitation events that would have their origin in the Earth's interior.

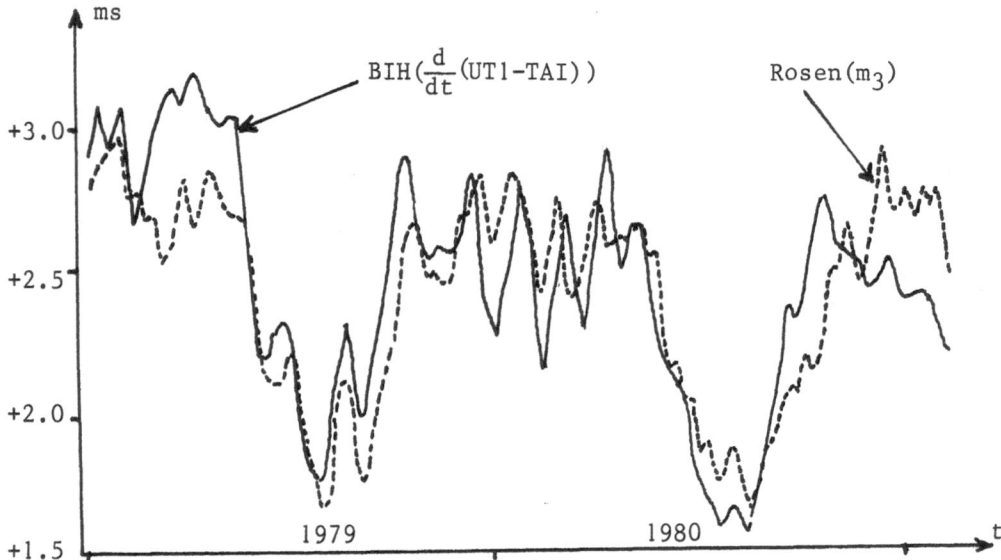

Fig.3. Comparison of the variations of the length of day obtained by BIH and m_3 obtained by R.D. Rosen and D.A. Salstein (ERT Tech. Rep. A, 345-T1,1981) by courtesy of Martine Feissel.

VII. LONG TERM AND SECULAR VARIATIONS

Long term variations of UT 1 and polar motion have been investigated by many authors. Many attempts have been made, for instance, to correlate them with solar activity. No convincing results were ever obtained because, if such effects exist, it brings only a minor contribution to the global phenomena and cannot be separated of other possible mechanisms. Among them, core mantle dynamical or electromagnetic coupling are probably significant. Some correlations seem to exist, for instance, between the UT 1 rate and the total magnetic field with a time lag of some 15 years. One may explain this as follows : while a core mantle reaction produces immediate changes of the Earth's magnetic field, the propagation of the angular momentum through the anelastic mantle takes a long time to propagate and be transmitted to the crust (Le Mouel and Courtillot, 1981).

If we turn now towards secular effects, several new problems arise on which we would like to comment.

For many years, now, it was noticed that the pole as determined by

I.L.S. has a drift. The estimated rate is 0".0035 per year and the
direction of motion is 80°W. It is difficult to ascertain now whether
this is an apparent phenomena (due to star catalogue errors or some
local effects) or whether it is real.

It can however be shown qualitatively, that local effects in I.L.S.
stations could be mainly responsible for it. Let us consider the yearly
mean values of polar coordinates given by I.L.S. and B.I.H. for the
period 1962-1979. The local effects are certainly less important in
the B.I.H. system where the large number of stations smooth them out.
Figure 4 shows the X and Y coordinates of I.L.S.-B.I.H. results, the
contribution of the variations of the mean latitude of Ukiah to the

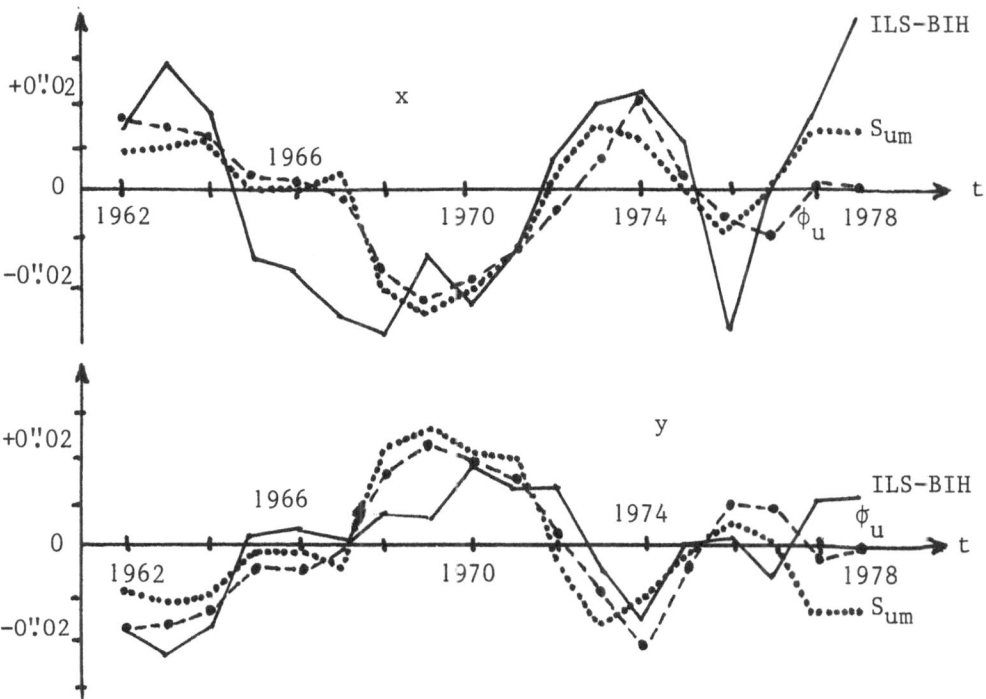

Fig.4. Comparison of ILS-BIH in x and y coordinates with the variations
of local vertical S_{um} and the mean latitude of Ukiah ϕ_u

coordinates and the variations of the angle Sum between the vertical
of Ukiah and the vertical of a mean observatory composed of Pulkovo,
Kazan, Poltava and Kitab. The latter does not depend upon the polar
motion. The existence of similarities between these three curves shows
that the secular and low frequency deviations of C.I.O. with respect
to the mean pole of epoch or to the B.I.H. system is mainly due to the
local non-polar effects of the Ukiah station.

This applies in particular to the 24 year component found in the
secular drift. No correlation was found between the coordinates of the
mean pole derived from two groups of stations (Mizusawa, Carloforte,

Ukiah and Pulkovo, Greenwich, Washington), confirming again the analysis
of Ukiah data. We conclude that this 24 year component is also due to
the slow local variations of the mean latitude of the stations.

If it is possible to remove all the local and star catalogue errors
from the S.I.L. data, and that a secular drift is shown to be real,
large scale mass displacements should be looked for. Dickman (1977) and
Lambeck (1980) have shown that the shift in latitude due to the motion
of lithospheric plates contributes unsignificantly (10 %) to the observed secular polar motion. The effects of deglaciation have also been
investigated and give again a much smaller polar drift. However recently,
Nakiboglu and Lambeck (1980) have shown that the observed drift can be
accounted for by the late Pleistocene (180 to 60 centuries ago) deglaciation concomitant with a sea level change and a mantle rebound if the
Newtonian viscosity of the mantle averages about 10^{23} p.. This was also
supported by Sidorenkov (1980).

Whatever are the origin and the characteristics of the secular drift,
it is very important that the actual polar motion in the years 1900-1960
be well established. All the astrometric observations used in dynamics
of the solar system or in kinematics of the Galaxy rely on I.L.S. pole.
Any error in it affects the theory of planets, the determination of the
equinox, etc... (Kovalevsky, 1982). If the future HIPPARCOS catalogue
can be used to improve the errors of the I.L.S. star catalogues and if
other type of errors, like local vertical drifts, can be modelled in
order to correct the position of the pole, all earlier astronomical
observations will then be subject to corrections.

The secular variations of UT 1 are also very important to determine, in particular in order to find what portion of the secular acceleration of the Earth has a non tidal origin. This quantity is a function of the following parameters :

    - The value of the mean deceleration of the Earth that can be ob-
tained using old eclipse observations ;
    - The value of the acceleration of the Moon on its orbit for which
discrepancies of 20 % still remain between the results of different
methods : Lunar laser (Ferrari et al., 1980), artificial satellites
(Cazenave and Daillet, 1981) or solar eclipses (Muller, 1976) ;
    - A possible variation of the constant of gravitation (Van Flandern,
1975).
As shown by Lambeck (1979), a non zero dG/dt considerably affects the
non tidal acceleration. Furthermore, the dispersion of results of the
difference E.T. and T.A. is still large, so that its interpretation
is not yet easy. This problem remains open and needs more observations
and discussions of the lunar motion as well as a modelling of possible
causes of a non tidal acceleration of the Earth.

VIII. CONCLUSION

Many geophysical and astronomical investigations depend upon precise unbiased determinations of various components of Earth's rotation. The accuracy of these determinations does not depend only upon the precision of individual observations, but also upon their consistency over long periods of time. It is necessary, in the future to produce continuously such a set of consistent observations referred to well defined celestial and terrestrial reference frames. The forthcoming MERIT campaign should help to select those observational techniques that will provide them and to provide a new operational reference terrestrial system that could be accurately maintained during a sufficiently long time so as not to introduce in the results spurious trends that might bias their scientific interpretation.

REFERENCES
Atkinson, R. d'E., 1975: Mon Not. R. Astr. Soc., 71, p. 381.
Capitaine, N., 1975 : Geophys. J. 43, p. 573.
Cazenave, A. and Daillet, S., 1981, J. of Geophys. Res., 86, p. 1659.
Dahlen, F.A., 1973 : Geophys. J., 32, p. 203.
Dahlen, F.A., 1976 : Geophys. J., 46, p. 363.
Daillet, S., 1981 : Geophys. J., 65, p. 407.
Dickman, S.R., 1977, Geophys. J., 51, p. 229.
Feissel, M. and Gambis, D., 1980 : Compte-Rendus Ac. Sc. Paris,
    B 291, p. 271.
Ferrari, A.J., Sinclair, W.S., Sjogren, W.L., Williams, J.G. and Yoder,
    C.F., 1980, J. of Geophys. Res., 85, p. 3939.
Fliegel, H.F. and Hawkins, T.P., 1967 : Astron. J., 72, p. 544.
Frostman, T.O., Martin, D.W. and Schwerdtfeger, W., 1967 : J. of
    Geophys. Res., 72, p. 5065.
Gilbert, F. and Dziewonski, A.M., 1975 : Phil. Trans. R. Soc. London,
    A 278, p. 187.
Graber, M.A., 1976 : Geophys. J., 46, p. 75.
Guinot, B., 1979 : in IAU symp. 82, Cadix, p. 7.
Kanamori, H., 1977 : J. of Geophys. Res., 82, p. 2981.
Kovalevsky, J., 1979 : in IAU symp. 82, Cadix, p. 151.
Kovalevsky, J., 1981 : in IAU coll. 56, Warsaw, p. 77.
Kovalevsky, J., 1982, "Hipparcos and the dynamics of the Solar System",
    Celestial Mechanics, 26, p. 213.
Lambeck, K., 1979, in "The Earth, its origin, structure and evolution",
    M.W. Mc Ethinny ed., Academic Press, London, p. 59.
Lambeck, K., 1980 : "The Earth's variable rotation : geophysical causes
    and consequences", Cambridge University Press, Cambridge.
Lambeck, K. and Cazenave, A., 1973 : Geophys. J., 32, p. 79 and 38,
    p. 49.
Le Mouel, J.L. and Courtillot, V., 1981: "Core motions, electromagnetic
    core-mantle coupling and variations in the Earth's interior : new
    constraints", Phys. Earth and Planet. Int. (in press).
Lieske, J.H., Lederle, T.,Fricke, W. and Morando, B., 1977 : Astr. and
    Astroph. 58, p. 1.
Mueller, I.I., 1981 : in IAU coll. 56, Warsaw, p. 1.

Muller, P.M., 1976 : J.P.L. Report SP 43-36, Pasadena.

Nakiboglu, S.M. and Lambeck, K., 1980, Geophys. J., 62, p.49.

O'Connell, R.J. and Dziewonski, A.M., 1976 : Nature, 262, p.259.

Pedersen, G.P.H. and Rochester, M.G., 1972 : in IAU symp. 48, Morioka, p.33.

Sidorenkov, N.S., 1969 : Soviet Astron., 12, p.706.

Sidorenkov, N.S., 1973 : Izv. Atmos. Ocean. Phys., 9, p.339.

Sidorenkov, N.S., 1980 : Meteorology and Hydrology, 1, p.52.

Seidelmann, P.K., Abalakin, V.K., Kinoshita, H., Kovalevsky, J., Murray, C.A., Smith, M.L., Vicente, R.O., Williams, J.G. and Yatskiv, Ya.S., 1982 : "1980 IAU theory of nutation", Celest. Mech., 27, 79.

Tisserand, F., 1891 : "Traité de Mécanique Céleste", Gauthier Villars ed., vol. 2, p.506.

Van Flandern, T.C., 1975 : Monthly Not. R. Astr. Soc., 170, p. 333.

Wahr, J., 1981 : Geophys. J. 64, p. 705.

Wayman, P.A., 1980 : Transactions of the International Astronomical Union, vol. XVII B, p. 70.

Wilson, C.R. and Hanbrich, R.A., 1976 : Geophys. J. 46, pp. 705 and 745.

Wilson, C.R. and Vicente, R.O., 1980 : Geophys. J. 62, p. 605.

Woolard, E.W., 1953 : Astron. Pap. Am. Ephem. and Naut. Alm., 15, p.3.

Yatskiv, Ya. S., 1974 : in Proc. 2nd Int. Symp. Geodesy and Physics of the Earth, Potsdam, p.143.

Yatskiv, Ya. S., 1980a : in IAU symp. 78, Kiev, p. 59.

Yatskiv, Ya. S., 1980b : in "Geodinamika i Astrometrija", Naoukova Dumka ed., Kiev, p.63.

Yatskiv, Ya. S., 1981 : in IAU coll. 56, Warsaw, p. 155.

Yoder, C.F., Williams, J.G. and Parke, M.E., 1981 : J. of Geophys. Res., 86, p. 881.

A NOTE ON THE INITIAL RESULTS AND FUTURE PLANS OF PROJECT MERIT

G. A. Wilkins
Royal Greenwich Observatory
Herstmonceux Castle, Hailsham
East Sussex, BN27 1RP, UK

Chairman of IAU/IUGG Joint Working Group on
the Rotation of the Earth

Project MERIT is a special programme of international collaboration to Monitor Earth-Rotation and to Intercompare the Techniques of
observation and analyses. Its origin, objectives and programme have
been briefly described by Wilkins (1980a) and a more extensive review
of the project and of the techniques to be used has been published
(Wilkins, ed., 1980b). A short campaign of observations was held during
the period 1980 August to October and the data were analysed by many
groups. The operational aspects of the project were discussed at the
MERIT Workshop held at Grasse on 1981 May 19-21, and some preliminary
results were presented on May 22 at IAU Colloquium No 63 (Wilkins,
1981; Calame, ed., 1982). A full report on the Short Campaign and
the Workshop is in preparation (Wilkins and Feissel, eds., 1981); it
contains reports by the principal coordinators for each technique, a
summary of the discussions at the Workshop, and lists of the values
of universal time (UT1-UTC) and of the coordinates of the pole that
were obtained by each participating group. These results are discussed
by M. Feissel elsewhere in this volume. Information about the progress
of the Project and related activities is issued from time to time in
the MERIT Newsletter.

The Short Campaign provided valuable experience of the operational
arrangements that will be required during the Main Campaign, which
will take place during the period 1983 September 1 to 1984 October 30,
and also produced high-quality data for scientific analysis. The
campaign provided a powerful stimulus to the faster development of the
new techniques and led to improvements in the quality and speed of
communication of the results from the classical techniques. Perhaps
even more importantly, it has led to much more cooperation and inter-
change of information between the various groups engaged in the observations and in the analysis of the data.

Between now and the Main Campaign a great deal of effort will be
required to ensure its success. For the new techniques of laser
ranging and radio interferometry it will be necessary to build new

W. Fricke and G. Teleki (eds.), Sun and Planetary System, 163–164.
Copyright © 1982 by D. Reidel Publishing Company.

equipment or upgrade current equipment, and then bring it to the point where it will operate regularly throughout the period and provide data of higher precision than that now available. In addition, the techniques for the communication of data between the observing stations and the analysis centres must be improved, and the computer programs for the analyses must be modified and extended so that the final results provide an accurate record of the variations in the rate of rotation of the Earth and of the motion of the pole of the rotation of the Earth with respect to the pole of figure. A Working Group, under the chairmanship of W. Melbourne, is developing a standard set of parameters and models so that the results obtained by the analysis centres may be easily compared and combined with each other. It will also be necessary to develop a self-consistent catalogue of station coordinates and this will provide a firm basis for a new terrestrial reference system for use in astronomy, geodesy and geophysics. All the data and results obtained during the main campaign will be made available for further analysis and for comparison with other astronomical, geophysical and meteorological datasets.

The experience and results obtained during the main campaign of Project MERIT is expected to provide a sound basis for recommendations to the IAU and IUGG about the future international earth-rotation service. The aim is that the new service should provide, in an economical manner, data that are more precise and more frequent than is now possible. Such data are required for practical purposes and for use in studies of the various phenomena that affect, or are affected by, the rotation of the Earth. The success of the Project will be assured if it continues to receive the excellent support that it has been given by many institutions throughout the world during the past few years.

REFERENCES

Calame, O., ed., 1982. High-precision Earth-rotation and Earth-Moon dynamics: lunar distances and related observations. D. Reidel Publishing Company.

Wilkins, G. A., 1980a. A note on the origin, objectives and programme of Project MERIT. In "Reference Coordinate Systems for Earth Dynamics", E. M. Gaposchkin and B. Kolaczek (eds.), 275-6.

Wilkins, G. A., ed., 1980b. A review of the techniques to be used during Project Merit to monitor the rotation of the Earth. Published by the Royal Greenwich Observatory and by the Institüt für Angewandte Geodäsie, Frankfurt.

Wilkins, G. A., 1981. Report on the Merit Workshop. CSTG Bulletin No 3, 35-40.

Wilkins, G. A., and M. Feissel, eds., 1981. Project MERIT: Report on the short campaign with observations and results on earth-rotation during 1980 August-October. To be published by the Royal Greenwich Observatory.

STEPS TOWARDS THE DETERMINATION OF EARTH ROTATION PARAMETERS NOT DEPENDENT ON THE OBSERVATION TECHNIQUE

Martine Feissel
Bureau International de l'Heure

The rotation of the Earth is currently measured by classical astrometry, Doppler and laser satellite tracking, laser ranging to the Moon, and radio interferometry. Several years long time series of pole coordinates and/or UT are available from most of these techniques. The various series are intercompared and their stability in the time frame of years to days is estimated.

INTRODUCTION

In 1978-79, IAU and IUGG initiated the international program MERIT with the aim of intercomparing the measurement techniques used for the determination of the Earth rotation parameters (ERP). Two intensive periods of observations organized by MERIT are the 1980 three months preliminary campaign and a main campaign, 1983 September 1-1984 October 31. However, the various available techniques are already currently used and have produced series of results covering several years. These results are widely available, in particular through their publication in a special section of the Annual Report of the Bureau International de l'Heure (BIH).

The BIH has included in its current solution of ERP the satellite Doppler tracking data in 1972 and laser ranging to Lageos and connected elements radio-interferometry data in 1981. In 1979 the BIH introduced a calibration of the annual and semi annual errors of their initial system of ERP, which had been established on classical astrometry data in 1966-67. This calibration was based on the comparison with satellite Doppler tracking and Lunar Laser Ranging.

The numerous global geophysical phenomena that are acting on the Earth's rotation and which can be studied through analysis of the ERP cover a large range of temporal frequencies. It is therefore necessary

W. Fricke and G. Teleki (eds.), Sun and Planetary System, 165–172.

that series of ERP be accurate in all parts of the spectrum. On the other hand, each of the techniques used nowadays may be characterized by a frequency domain in which it is accurate. Theoretical analysis of the techniques and intercomparison of independent series allow to identify the systematic errors as well as the type and level of noise that are present in the series of ERP. A further step is a combination of techniques which takes into account the conclusions of such studies. Examples are the introduction of the 1979 BIH System and a combined solution at five days intervals in 1980, using the data of four different techniques (Feissel, 1982).

THE COMPARED SERIES

The compared series are as in Table 1. They, or a reference to them, can be found in BIH (1981). All of the mentionned institutes participated in the MERIT short campaign in 1980, together with other institutes (Wilkins and Feissel, 1981).

Intercomparisons are performed in order to evaluate accuracies in the long term (one year to 20 years), in the medium term (one month to one year), and in the short term (one day to one month).

Table 1. Series of compared ERP, 1962-1980

Technique	Program or Institut	Years	ERP x y UT lod	Reference
Classical Astrometry (AST)	ILS	1962-78	1 1 0 0	Yumi and Yokoyama
	IPMS	1962-78	1 1 1 0	(1980)-A.R. of IPMS
	BIH	1962-80	1 1 1 0	Annual Reports of BIH
Satellite Doppler tracking (DOP)	DMA	1972-80	1 1 0 0	See BIH (1981)
	MEDOC	1977-80	1 1 0 0	See BIH (1981)
Laser Ranging to LAGEOS (SLR)	IASOM	1976-80	1 1 0 1	Schutz et al (1981)
	GSFC	1976-78	1 1 0 1	See BIH (1980)
	SAO	1980	1 1 0 1	See BIH (1981)
Lunar Laser Ranging (LLR)	EROLD	1971-80	0 0 1 0	Calame (1981)
	JPL	1971-80	0 0 1 0	Fliegel et al (1982)
	MIT	1971-80	0 0 1 0	Langley et al (1981)
Very Long Base Interferometry (VLBI)	JPL	1971-80	1 1 1 0	See BIH (1981)
	NGS	1980	1 1 1 0	See Wilkins and Feissel (1982)
Connected Elements Radio Interf. (CERI)	USNO	1979-80	1 1 1 0	See BIH (1981)

LONG TERM STABILITY

For the pole coordinates, classical astrometry provides the long term reference. The ILS programme with five permanent stations will come to an end in the eighties. For the period 1890-1968 pole coordinates were also computed from latitude observations made in several observatories (Fedorov et al., 1972). Since 1962, they are regularly computed from time and latitude observations of a worldwide network of stations. The longest homogeneous series obtained by a new technique is the one by DMA (DOP). The tie between the older data and the present and future ones can then be established. Figure 1 shows the path of the mean pole from 1963 to 1980 as measured by ILS, IPMS and BIH (AST) and by DMA (DOP).

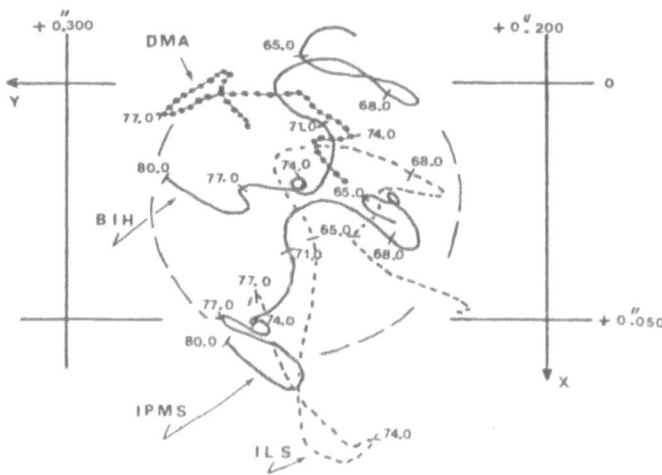

Figure 1. AST and DOP. Path of the mean pole determined every 0.06y from 1963 to 1980. The radius of the circle is one metre.

For the period 1975-1980, more series are available. Table 2 gives the Allan variance of the yearly mean differences between series.

Table 2. Pole position 1975-1980. Square root Allan pair variance of the yearly mean differences between independent series.

Difference	$\sigma(2, \tau = 1\ \text{year})$	
	x	y
SLR (IASOM) - DOP (DMA, Sat. 67)	0".0031	0".0063
SLR (IASOM) - AST (BIH)	0".0045	0".0056
DOP (DMA, Sat. 67) - AST (BIH)	0".0031	0".0063

For universal time, classical astrometry provides the main long term reference. However, homogeneous series of LLR and sparse determinations by VLBI are now available ; they go back to 1971. The differences of these series with AST are not easy to interpret, as each of them can show long term effects of inaccurate reference frame realization. However, their comparison over 10-15 years suggests that their spurious drifts are under 0.0002 s/yr, while some few years oscillations have less than 0.003 s total amplitude.

MEDIUM TERM STABILITY

As the Earth's rotation irregularities have strong components with periods 1.19 and 1 year which should be rigourously related to their geophysical causes, it is of primary interest to evaluate the accuracy of the different techniques at these frequencies. Table 3 gives the rms differences of the amplitude of these terms as determined from different series.

Table 3. Root-mean-square differences of amplitudes at 1.19 and 1 year periods. (units : 0".001 for x, y, 0.0001s for UT)

Series	Years	Period 1.19y		Period 1 year		
		x	y	x	y	UT
IPMS, BIH(AST) (parallel reduction)	1967-78	6	10	9	9	13
AST(BIH), SLR(IASOM) DOP(DMA, sat. 67)	1975-80	9	9	9	6	-

In the higher frequencies (1 c/year to 1 c/month), the role of inaccuracies of the reference frames realization usually becomes less critical and the observational errors become more important and difficult to take into account, as they are often correlated. In the case of pole position, Figure 2 shows the smoothed residuals to a common reference in 1980.

a) Combined series

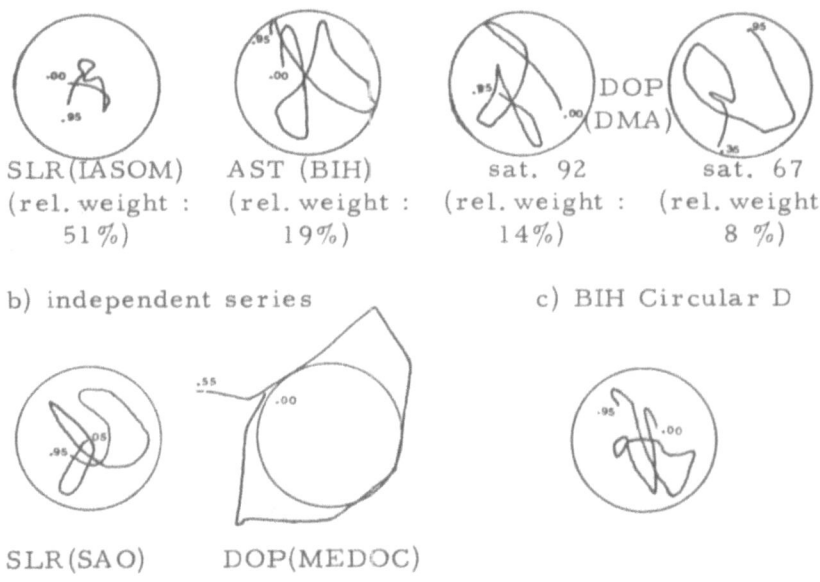

SLR (IASOM)	AST (BIH)	sat. 92	sat. 67
(rel. weight : 51%)	(rel. weight : 19%)	(rel. weight : 14%)	(rel. weight : 8%)

b) independent series

c) BIH Circular D

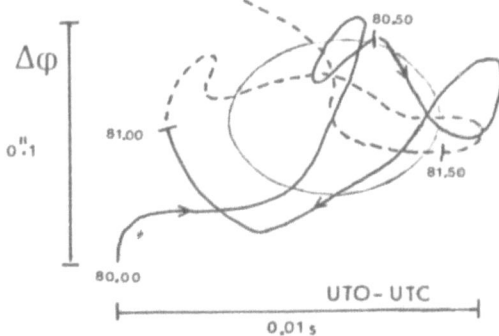

SLR (SAO)   DOP(MEDOC)

Figure 2. Medium term noise in pole positions determinations. Residuals to a combined solution in 1980. Circles have a radius of 1 metre.

Figure 3 shows the smoothed residuals of CERI with respect to the same combined solution in 1980/81. In this case, the medium term stability is not well assured, probably due to a lack of accuracy in some modeled corrections that are necessary in the reduction of the observations.

Figure 3. CERI (USNO). Smoothed residuals to a combined solution in 1980/81. The length of the semi-axes of the ellipse is 1 metre.

As for UT, recurrent oscillations with a pseudo period of about 55 days and an amplitude of 0.001 to 0.005 s have recently been discovered (Feissel and Gambis, 1980). This motion is detected by CERI (when corrected for longer term effects), LLR (Langley et al, 1981 b), and AST, which implies that the lack of accuracy of these techniques around the 55 days period is not larger than 0.001s. The determinations of the duration of the day (lod) by SLR are referred to the modelized motion of the satellite orbit/node. Apart for long term errors, some errors are still present in the medium term : as an example, the 55 days oscillations are not well detected by this technique (the amplitude is found about 50 % too small).

SHORT TERM

The pole coordinates are usually determined at 2 to 5 days interval. When at least three independent series are available, it is possible to obtain an estimation of their stability by use of the Allan variance (Gray and Allan, 1974). The short term stability of SLR, DOP and AST has been evaluated from the results available from 1980 Sep. 8 to 1981 May 20 (Table 4).

Table 4. Pole position. Short term stability

Series	Square root of Allan pair variance (unit:0".001)					
	τ = 5 d		τ = 10 d		τ = 20 d	
	x	y	x	y	x	y
SLR (IASOM)	9	7	6	5	6	2
DOP (DMA, Sat. 92)	22	15	14	12	9	6
DOP (DMA, Sat. 67)	22	19	-	--	-	--
AST (BIH)	13	13	11	9	10	8

The short term stability of AST for UT and the one of CERI have been evaluated by comparison to a smoothing of UT1-UTC which cuts off periodicities longer than 1 c/20 days (Table 5).

Table 5. AST(UT) and CERI. Short term stability (1980)

Series	Parameter	time resolution	rms residual to smoothing
AST (BIH)	UT1*	5 d	$0^s.0007s$
CERI (USNO)	UT0	3 d	0.0011 s
	latitude	3 d	0".011

*UT1-UTC is corrected for the effect of zonal tides with periods < 122d.

It is now possible to go further in the separation of short term and long term effects in the Earth's rotation irregularities : an external information is now provided by direct measurements of the angular momentum of the atmosphere (Rosen and Salstein, 1981). Such data are available since 1976 at 0. 5 d interval. Their comparison with the ERP are presently in process. Figure 4 shows the measurements of both types that are available during a part of MERIT 1980 Campaign.

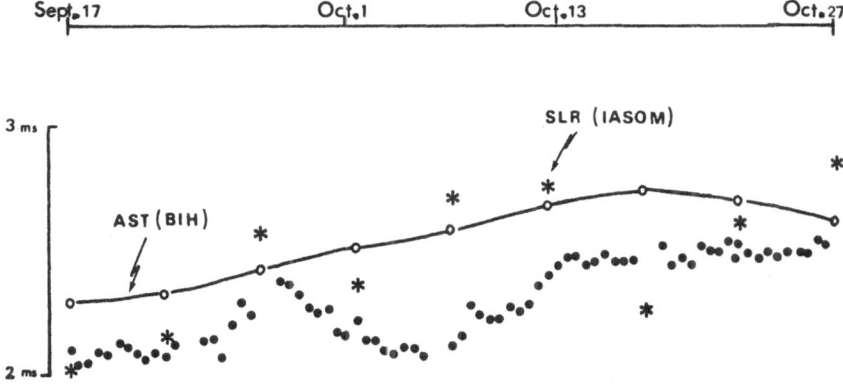

Figure 4. Measurements of the duration of the day, and variations expected from the influence of the atmosphere (•).

CONCLUSION

Although some of the techniques that are used for measuring the ERP have not yet reached their ultimate accuracy and time resolution, it is already possible to identify the frequency domains in which each of them is able to bring precise and accurate information on the Earth s rotation. The combined solution that is already performable (and performed at the BIH) has presently the accuracy characteristics summarized in Table 6.

Table 6. ERP combined solution. Accuracy according to time frame

Time frame	Pole position (AST, DOP, SLR, CERI)	UT1 - UTC (AST, LLR, CERI)
1 - 20 years	0''007 (25 cm)	0. 0020s (100 cm on the equator)
1 year	0''009 (30 cm)	0. 0015s (70 cm on the equator)
1 month - 1 year	0''009 (30 cm)	0. 0015s (70 cm on the equator)
5 days - 1 month	0''005 (15 cm)	0. 0008s (40 cm on the equator)

REFERENCES

- BIH, 1980 : Annual Report for 1979, Paris

- BIH, 1981 : Annual Report for 1980, Paris

- Calame, O., and Guinot, B., 1979 : Ann. Rep. of the BIH for 1978, D-27

- Calame, O., 1981 : see (BIH, 1981)

- Fedorov, E. P. et al, 1972 : The motion of the pole of the Earth from 1890.0 to 1969.0, Naukova Dumka, Kiev

- Feissel, M. and Gambis, D., 1980 : C.R. Acad. Sci. Paris, 291, B 271.

- Feissel, M., 1982 : IAU Coll. 63, 3.

- Fliegel, H.F., Dickey, J.O., Williams, J.G., 1982 : IAU Coll. 63, 53.

- Gray, J.E., and Allan, D.W., 1974 : 28th Ann. Symp. on Frequency Control

- Langley, R.B., King, R.W., and Shapiro, I.I., 1981a: see (BIH, 1981)

- Langley, R.B., King, R.W., Shapiro, I.I., Rosen, R.D., and Salstein, D.A., 1981 b : Nature (in press)

- Rosen, R.D. and Salstein, D.A., 1981 : ERT Techn. Rep. A 345-TI

- Schutz, B.E., Tapley, B.D., and Eanes, R.J., 1981 : see (BIH, 1981)

- Wilkins, G. and Feissel, M., (eds), 1981 : Project MERIT : Report on the short campaign with observations and results on earth-rotation during 1980 August-October to be published by Royal Greenwich Observatory

- Yumi, S., and Yokoyama, K., 1980 : Results of the International Latitude Service in a homogeneous system, Mizusawa.

SECULAR AND DECADE FLUCTUATIONS IN THE EARTH'S ROTATION: 700 BC - AD 1978.

L.V.Morrison, Royal Greenwich Observatory, UK.
F.R. Stephenson, University of Liverpool, UK.

ABSTRACT: Secular and decade fluctuations in the Earth's rotation in the period 700 BC - AD 1978 are deduced from timings of lunar eclipses and occultations on the assumption that the Moon's tidal acceleration is $-26''/cy^2$. Besides the tidal deceleration of the Earth, it is found that there is an accelerative component which implies a fractional decrease in the moment of inertia of $8.4 \pm 1.0 \times 10^{-11}/yr$.

1. INTRODUCTION

Before the introduction of the international atomic time-scale (TAI) in 1955, the determination of changes in the Earth's rotation over decades or longer periods has been largely dependent on observations of the Moon's position derived from timings of occultations of stars and eclipses of the Sun. The principle by which these changes are determined is as follows. The time-argument of the lunar ephemeris is regarded as a uniform measure of time and the instants of occurrence of occultations and eclipses are calculated on that time-scale (known as dynamical time, TD). The observations, on the other hand, are timed on the universal time-scale (UT) which is directly related to the variable period of rotation of the Earth. The difference, TD-UT = $\Delta T(t)$, is the cumulative discrepancy in time at epoch $\underline{t}$, and the first derivative is the difference between the observed rate of rotation and the adopted standard which is equivalent to a mean solar day of 86400 SI seconds.

2. TIDAL ACCELERATION OF THE MOON

The results derived by this method for changes in the Earth's rotation will be erroneous unless the gravitational theory of the Moon's motion is complete. In particular, an allowance must be made for the empirical tidal acceleration in mean longitude. Recent determinations of the tidal acceleration ($\dot{n}$) obtained by independent methods are listed in Table 1.

173

W. Fricke and G. Teleki (eds.), Sun and Planetary System, 173–178.

Table 1   Tidal acceleration of the Moon

Method	Author	$\dot{n}$ ($''$/cy^2)
Lunar laser ranging	Ferrari et al (1980)	-23.8 ± 3.1
Lunar occultations	Van Flandern (1981)	-21.4 ± 2.6
Artificial satellites	Cazenave & Daillet (1981)	-25.1 ± 2.9
Meridian-circle obsns.	Oesterwinter & Cohen (1972)	-38 ± 8
Transists of Mercury	Morrison & Ward (1975)	-26 ± 2
Total solar eclipses	Muller (1976)	-30 ± 3

In the following derivations of changes in the Earth's rotation the value of $-26''$/cy^2 found by Morrison & Ward (1975) has been adopted for the Moon's tidal acceleration.

3. DECADE FLUCTUATIONS 1620-1978

The values of ΔT in the period 1620-1954 were derived from an analysis of 40000 timings of lunar occultations as described in Morrison (1979) and Morrison & Stephenson (1981). A smooth curve has been fitted through the observations and this is shown in Fig.1(a). The uncertainty in the position of the curve varies from $\pm$ 10 s at 1650 to $\pm$ 0.05 s at 1950.

The overall shape of the curve is parabolic, which corresponds to a long-term deceleration in the Earth's rate of rotation. The parabola shown as a dashed line has the equation $\Delta T = -15 + 32.5 (\underline{t} - 0.1)^2$ seconds, where the coefficient 32.5 s/cy^2 is obtained in the next section from an analysis of ancient Babylonian lunar eclipses, and $\underline{t}$ is measured in centuries from AD 1800. The constants -15 s and -0.1 cy were chosen to secure an overall fit to the ΔT curve: they have no physical significance.

The first derivative of ΔT is shown in Fig.1(b). The left ordinate measures the excess in milliseconds in the length of the mean solar day compared to the standard day of 86400 SI seconds. The right ordinate measures the change in angular velocity relative to the adopted standard of 1.00273 78119 06 (2π/86400) rad/s (see Explanatory Supplement 1961, p.76).

The data before 1800 indicate that the length of the day increased by 3 ms between 1650 and 1800, but they are too uncertain to draw any conclusions about possible decade fluctuations. Around 1900, however, the standard error of the annual points through which the curve in Fig.1(b) has been drawn is about 0.2 ms and thus the fluctuations are undoubtedly real. The long-term deceleration in Fig.1(a) corresponds to an increase in the length of the day of 1.78 ms/cy, which is shown as a dashed line in Fig.1(b). The decade fluctuations are usually attributed to torques arising from core-mantle coupling. By contrast, the annual variation due to the couple between the atmosphere and the mantle produces an amplitude of $\sim$ 0.4 ms in the length of the day.

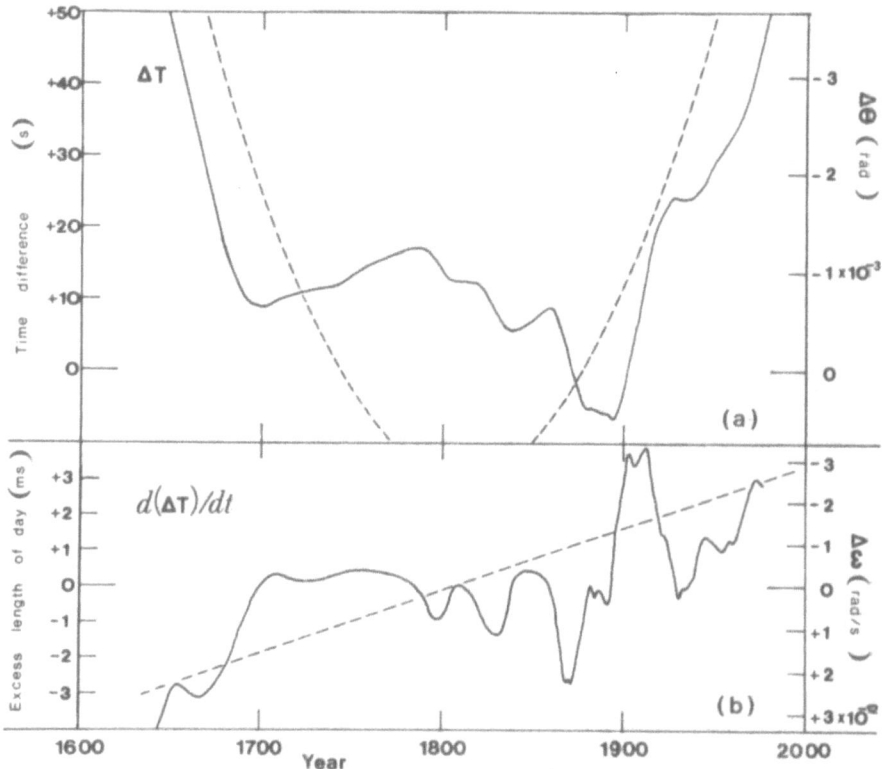

Figure 1.   Smoothed ΔT curve (a) and its first derivative (b)
derived mainly from timings of lunar occultations.
The dashed parabola in (a) and straight line in (b)
indicate the secular deceleration in the Earth's
rotation

4.   SECULAR DECELERATION

It is clear from Fig. 1 that the secular deceleration cannot be
determined from the data since AD 1600 because the comparatively large
fluctuations occurring over decades distort the result. For this reason
several investigators (see eg Muller, 1976) have analysed the recorded
occurrences of total solar eclipses in the ancient world, but these have
the disadvantage that very few are of high reliability. Furthermore,
the simultaneous solution for the lunar acceleration and the Earth's
rotational deceleration gives unreliable results because of the high
correlation between them. A new source of data is required.

Ancient Babylonian timings of lunar eclipses observed between 700 and
50 BC have survived on clay tablets which are preserved in the British
Museum. These observations record to the nearest u̯s (4 minutes) the

interval of time between the rising or setting of the Sun or Moon and the
beginning, end or middle of a lunar eclipse. In order to minimize the
errors due to clock drift of the ancient Babylonian timing devices, this
analysis is restricted to 17 observations measuring a time-interval of
20 uš or less. The UT of the rising or setting phenomenon and the TD
(dynamical time) of each lunar eclipse can be calculated from the date on
the tablet and a knowledge of Babylonian chronology. A lunar tidal acceleration of $-26''/cy^2$ and an increase of 9 ± 2 per cent in the radius of the
umbra were used in these calculations. The 9 per cent increment allows
for the Earth's atmosphere and the deep penumbral shadow which is mistaken
for the umbra in observations made with the unaided eye.

The UT of the eclipse is the algebraic sum of the UT of the rising or
setting phenomenon and the observed time-interval. The difference TD−UT=ΔT
is listed in Table 2 for 13 of the 17 lunar eclipses. Three of the observations were rejected because the eclipses were less than 0.5 magnitude
and one reached mid-eclipse very soon after moonrise.

Table 2 Analysis of timed Babylonian lunar eclipses

Year	ΔT (hours)	$t^2(cy^2)$	$c(s/cy^2)$
−684	5.21 ± 0.26	617	30.4 ± 1.4
−536	5.14 ± 0.13	546	33.9 ± 0.8
−482	4.71 ± 0.09	521	32.6 ± 0.6
−464	4.44 ± 0.19	513	31.2 ± 1.2
−420	4.22 ± 0.16	493	30.8 ± 1.1
−406	4.38 ± 0.13	487	32.4 ± 0.9
−405	4.49 ± 0.13	486	33.3 ± 0.9
−352	4.11 ± 0.07	463	32.0 ± 0.5
−239	3.87 ± 0.06	416	33.3 ± 0.5
−211	3.18 ± 0.17	404	28.3 ± 1.4
−153	3.43 ± 0.07	381	32.4 ± 0.6
−142	3.39 ± 0.08	377	32.4 ± 0.7
−79	2.96 ± 0.17	353	30.2 ± 1.6

The standard errors for ΔT comprise a 13 per cent error in the elapsed
time, a ± 2 min error for rounding to the nearest uš, a 2 per cent error
in the umbral radius and a ± 1 min eror in the time of rising and setting.
These errors were estimated in a separate investigation of the Babylonian
observations. Assuming that the decade fluctuations are smoothed out over
a period of 2500 years, the values of ΔT can be represented by a parabola
whose vertex is close to the epoch AD 1800, as shown in Fig. 1(a). The
results for the coefficient of t^2, where t is measured in centuries from
AD 1800, are listed under c. The mean value of c and standard error are
+32.5 ± 0.3 s/cy^2. The error of ± $2''/cy^2$ in the adopted value of ṅ is
dominant and leads to an error of ∓ $2s/cy^2$ in the mean value of c.

5. GEOPHYSICAL IMPLICATIONS

From the consideration of conservation of angular momentum in the Earth-Moon system, the expected tidal acceleration $\dot{\omega}_T$ in the Earth's rotation is found to be (see eg Morrison, 1978)

$$\dot{\omega}_T/\omega = 1.16\dot{n} + 1.1 \times 10^{-11}/\text{yr}.$$

Substituting $\dot{n} = -26 \pm 2$ "/cy^2, we find,

$$\dot{\omega}_T/\omega = -29.0 \pm 2.3 \times 10^{-11}/\text{yr}.$$

The actual measured acceleration is given by $\dot{\omega} = -2c$; thus with $c = +32.5 \mp 2.0$ s/cy^2 from above,

$$\dot{\omega}/\omega = -20.6 \pm 1.3 \times 10^{-11}/\text{yr}.$$

The residual, non-tidal acceleration is given by

$$\dot{\omega}/\omega - \dot{\omega}_T/\omega = +8.4 \pm 1.0 \times 10^{-11}/\text{yr}.$$

The errors in $\dot{\omega}/\omega$ and $\dot{\omega}_T/\omega$ are both principally dependent on the error in $\dot{n}$, and since they enter with the same sign they tend to cancel, leaving a final error of only $\pm 1.0 \times 10^{-11}/\text{yr}$.

This value of $+8.4 \pm 1.0 \times 10^{-11}/\text{yr}$ is clear evidence of a long-term, non-tidal accelerative component in the Earth's rotation which probably arises from a fractional change of the opposite amount in the moment of inertia. This is the average value over 2500 years and it is not necessarily constant over the intervening period. Further work reeds to be done on the medieval observations to establish whether there are fluctuations in the non-tidal component on a time-scale longer than decades.

REFERENCES

Cazenave, A., and Daillet, S., 1981. J. Geophys. R., 86, 1659.

Explanatory Supplement to the Astronomical Ephemeris and the American Ephemeris and Nautical Almanac, 1974, HMSO, London.

Ferrari, A.J., Sinclair, W.S., Sjogren, W.L., Williams, J.G. and Yoder, C.F., 1980. J. Geophys. R., 85, 3939.

Morrison, L.V., 1978. in Tidal Friction and the Earth's Rotation, eds. Brosche and Sundermann, pp 22-27, Springer-Verlag, Berlin.

Morrison, L.V., 1979. Geophys. J.R. astr. Soc., 58, 349.

Morrison, L.V., and Stephenson, F.R., 1981. in Reference Coordinate
    Systems for Earth Dynamics, eds. Gaposchkin and Kołaczek,
    pp 181-185, D. Reidel, Dordrecht.

Morrison, L.V., and Ward, C.G., 1975. Mon. Not. R. astr. Soc., 173, 183.

Muller, P.M., 1976. Determination of the Cosmological rate of change of
    G and the Tidal Accelerations of the Earth and Moon from Ancient
    and Modern Astronomical Data. Spec. Publ. 43-36, Jet Propul. Lab,
    Pasadena. Calif.

Oesterwinter, C. and Cohen, C.J., 1972. Celestial Mech. 5, 317.

Van Flandern, T.C., 1981. Is the Gravitational Constant Changing?,
    pre-print.

OCEANIC TIDES AND THE ROTATION OF THE EARTH

Peter Brosche
Observatorium Hoher List der Universitäts-
Sternwarte Bonn, D-5568 Daun, F.R.Germany

ABSTRACT
Periodic effects of oceanic tides are near to the limits of
modern observational techniques. The largest effects are
due to the secular transfer of angular momentum and energy
between the Earth's rotation and the lunar orbit. This
interaction is mediated by oceanic tides. Because of the
very complex resonance structure of the oceans, the changes
due to continental drift give rise to considerable variations
of the torque not only within the 100 million time scale
but even within 1-10 million years. These variations are
about factors 2 or 3; if they appear in the sense of diminishing torques, they might settle the enigma of a narrow
Earth-Moon-system at a time not far enough in the past.

1. INTRODUCTION

Oceanic tides are easily observable in coastal waters only.
Therefore our knowledge of oceanic tides in general is
based on computer models which are tested by the observations
at the continental coasts and at deep sea islands. These
models evaluate the hydrodynamical equations in a simplified
version. The simplifications are allowed by the circumstances
in the oceans and dictated by the limitations in computer
time. The results represent fairly well the quantities of
oceanographic interest: instantaneous values of tidal elevations and current velocities. Difficulties arise with an
adequate representation of small scale processes, especially
if the determination of time averages is wanted.Since this
is the case for most astronomical applications, the models
have to be used with an unusual high degree of accuracy. The
limitations are defined by the mesh size of the discretisa-
tion (usually a few degrees on the earth's surface). In case
of the more direct use of the hydrodynamical equations (HN-
models), also the time step (about 1/100 of the tidal period)

179

W. Fricke and G. Teleki (eds.), Sun and Planetary System, 179–184.

is of importance. With such models, the application of a
more realistic quadratic bottom friction law is possible.
The assumption that elevation and velocity vary sinusoi-
dally with the period of the special tide under consideration leads to the so-called Laplace tidal equations (LTE)
where the time dependence is omitted.

2. PERIODICAL EFFECTS

For treating such effects, the use of LTE models is certainly
adequate. However, since the measurements in astronomy and
geosciences usually refer to the crust of the earth, the
total tidal variations of the crust are needed. They depend
not only on the oceans but even more on the tides of the
solid earth. For the latter, effects of the mantle and the
fluid core have to be treated separately (Yoder et al., 1981).
The combination of effects from the solid earth and the
oceans is by no means a simple addition, since both are interacting: the determination of oceanic tides has to include the
effects of elastic yielding of the mantle, loading and self-
attraction. Since external torques can be neglected for
tidal periods, the change of the moments of inertia due to
the elastic properties of the different parts of the earth
causes reciprocal variations in the angular velocities.
Yoder et al. (1981) arrive at a value $k/C = 0.94\pm0.04$ where
k is the effective Love number and C the dimensionless polar
moment of inertia. This result is based on the earth model
1066 of Gilbert and Dzienowski (1975) and on an ocean
correction of +0.04 in k/C by Dahlen (Agnew and Farrell,
1978). In deriving the oceanic part, the assumption of an
equilibrium response of the oceans at the periods involved
(> 5 days) is made, especially at the important fortnightly
and monthly periods. The validity of this assumption has
been questioned (Wunsch, 1967; Agnew and Farrell, 1978). If
we accept it at the moment, we can conclude that the oceans
contribute $\approx4\%$ to the UT variations; that is ≈0.3ms in case
of the maximum amplitudes at 14^d and 28^d. For the essential
short periods of ≈0.5 and 1 day the ocean tides cannot be
considered anymore as equilibrium tides and the complicated
models mentioned above have to be used. Yoder et al. (1981)
obtained from different models amplitudes in UT up to
serveral 0.1ms. This relies on the assumption "that the
ocean tides' harmonic components act as if they are tied to
the mantle"; this means that the varying storage of angular
momentum in the water motion is neglected. As we found from
our 4° model of the world ocean, the variation in the storage
term is about three times larger than the effect on the polar
moment of inertia. The corresponding change in UT has a range
of 0.05ms and is therefore at the VLBI limit. Also the phase
is completely different from the one of the moment of inertia

(see Fig. 1).
Through non-diagonal parts in the moment of inertia tensor, the oceanic tides produce latitude variations of a few $0\overset{..}{.}0001$
Finally,the variations in water elevation cause a periodic change in the barycenter of the water with respect to the barycenter of the whole earth. The barycenter of the solid earth has to follow a corresponding path 180° out of phase with an amplitude of ≈ 1cm (Yoder et al. 1981, and unpublished work of ourselves).
To sum up, periodic effects of oceanic tides are either too small to be observed at present, or observable only as part of a sum of effects, or swamped by larger unmodelled effects.

3.SECULAR EFFECTS

3.1 Present Oceans

Observational techniques with time scales ranging from ≈ 10 years until ≈ 2000 years have revealed a secular deceleration in the mean motion of the moon ($\approx 25"$/century^2) and in the rotation of the earth (increase of the day ≈ 2ms/century) (Calame and Mulholland, 1978; Morrison, 1978; Stephenson, 1978 and 1981; Cazanave, 1981). The first means an increase in the orbital angular momentum of the moon whereas the latter corresponds to a decrease of the rotational angular momentum of the earth. While the observed quantities for the torques on the two bodies agreee roughly (expect for the sign), it is not yet clear wether there is a significant deficit in the balance. It has been shown that the tides of the solid earth are responsible for at most a few percent of the observed torques (Zschau, 1978 and 1981). The old idea of Kant (1754) that the tides of the oceans are the most important process has remained valid: order-of-magnitude arguments supported it but it is the task of the numerical models of today to represent the observed values. The main contribution stems from the M_2 tide, and very different models of ocean tides agree within 10% or 20% on the torque between the moon and the oceans (Lambeck, 1980, tables 10.9 and 10.10).Our own result with a 4°-model of Zahel is a torque

$$L = 4.9 \cdot 10^{23} \text{ dyn} \cdot \text{cm} \qquad\qquad \text{or}$$
$$d \; E_{rot}(\text{earth})/dt = - 3.6 \cdot 10^{19} \text{ erg/s}$$

The interactions of the oceans and the solid earth are, however, much more difficult to model. After integrating several hundred tidal periods, we obtained at best an order of magnitude agreement only. So far the theoretical problem cannot be considered as fully solved. On the other hand, the assumption that the water cannot be a sink or source of angular momentum is in the long run certainly valid (while we found it invalid in the foregoing chapter); therefore we can

attribute the torque between moon and oceans to the solid
earth also.

3.2 Old Oceans

The growth increments of corals, mussels and other animals
show periodic pattern which have been interpreted as due to
the solar day, the synodic month and the year (Scrutton, 1978).
Those data seem to indicate that the tidal torque had the
same order of magnitude as today for the last 500 million
years.Theoretical computations for such epochs cannot be
based, of course, on present day ocean models. Therefore, we
have determined the M_2-tide and its average torque for re-
constructions of old oceans configurations, taking into
account the effects of continental drift (Brosche and Sündermann, 1977; Sündermann and Brosche, 1978; Brosche, 1981;
Krohn, Brosche and Sündermann, 1981; Krohn and Sündermann,
1981). These models contain two depth steps only, one for the
shelf areas and one for the deep sea areas. Comparing such a
schematized model for the present with the detailed model one
finds a ratio 1:1.6 of the resulting torques. This is probably due to an underrepresentation of the shelf influence
in the schematic models. It is doubtful whether one can apply
the ratio above in general as a correction factor. Hence the
following table summarizes our results for the M_2-tide within
geological timescales without any correction.

Epoch	Time (million years)	Torque ($- 10^{23}$ dyn cm)
Present time	0	3.1
Upper Cretaceous	70	3.1
Upper Permian	240	2.0
Middle Silurian	420	3.6
Ordovician	450	4.3

Since the astronomical parameters (amplitude and frequency
of the tidal force) change only a few percent within the
above time span, the variation of a factor 2 is mainly caused
by the changing ability of the oceans to oscillate with the
exciting force. In other words, it is of great influence on
the resulting torque whether or not (and how precisely) the
M_2 frequency coincides with an eigenfrequency of the oceans.
But the strength of that influence is not a priori predictable
since the oscillation systems belonging to nearby resonance
frequencies may be quite different. In any case, the resonance
frequencies given by several authors (see, e.g., Platzmann
et al., 1981) are so near to each other that we must expect
considerable torque variations already within time scales of
1 to 10 million years. In order to examine this in a first
crude example, we shifted America in small steps up to $\approx 10^{\circ}$

east and west (Brosche and Hövel, 1981). The resulting torque
curve has a maximum a few million years ahead, which is about
the twofold of the broad level of the last 20 million years.
Consequently, realistic caculations for the history of the
earth-moon-system require a very dense sequence of oceanic
tide models. The presently measured values might be not a
representative average even for such a short time interval
as for 10 million years. The effects of the resonance are
not smoothed out even for the very long time scale of several
10^9 years. As was shown by Webb (1980,1981), the very schematic
model of a hemispheric ocean averaged over all possible pole
positions exhibits still some resonance maxima if it is integrated backwards in time. Oceanic tides can also affect the
orientation of the Earth's angular momentum vector. The K_1
tidal forces produce the normal precession. Acting on the
nonsymmetrical K_1 tide elevations of the oceans, they can
lead to an 'anomalous precession' in a direction deviating
from the vernal equinox. As very preliminary results we have
obtained 0."004 per century towards $\alpha \approx 3^h 40^m$.

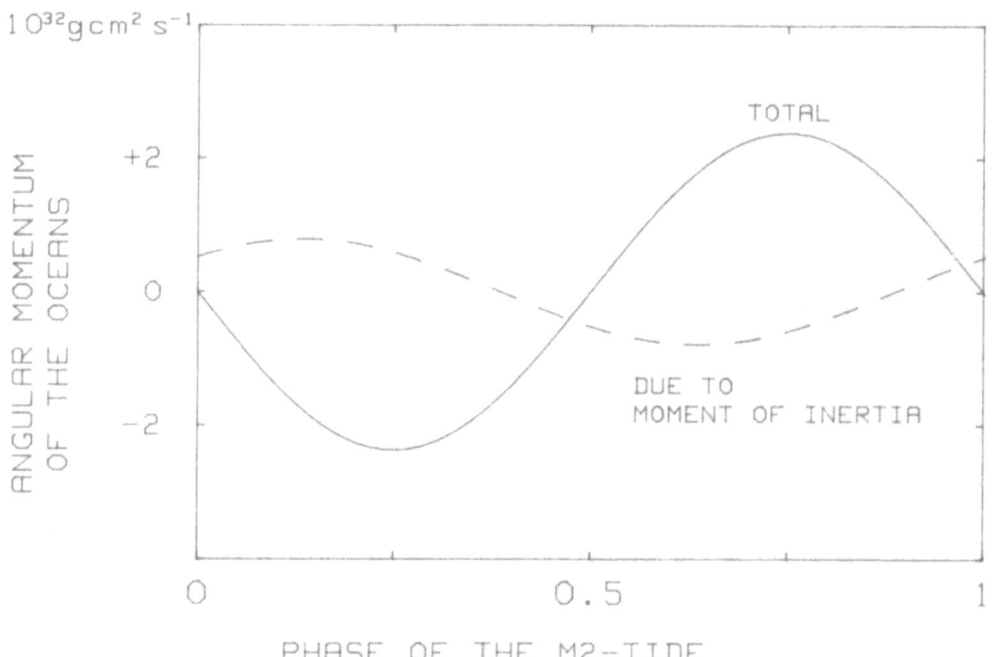

Fig. 1. The total variation of the angular momentum
content of the oceans (solid line) and that part of the
variation which is attributable to changes in the polar
moment of inertia (broken line). Both curves refer to
the M_2-tide.

REFERENCES

Agnew, D.C., and Farrell, W.E., 1978, Geophys. J.
      R. astr. Soc., 55, 171
Brosche,P., 1981, Mitt. Astron. Gesellschaft, 51, 81
Brosche,P., and Hövel, W., 1981, BII
Brosche,P., and Sündermann, J., 1977, in:
      Scientific Applications of Lunar Laser
      Ranging (ed. J.D. Mulholland), Reidel,
      Dordrecht Boston, p. 133
Calame, O., and Mulholland, J.D., 1978, BI, p. 43
Cazenave, A., 1981, BII
Gilbert, F., and Dziewonski, A.M., 1975, Phil. Trans. R.
      Soc. London, Ser. A., 278, 187
Kant, I., 1754, Wöchentliche Königsbergische Frag-
      und Anzeigungs-Nachrichten vom 8. und 15.6.1754
Krohn, J., Brosche, P., and Sündermann, J., 1981,
      Geologische Rundschau 70, 64
Krohn, J., and Sündermann, J., 1981, BII
Lambeck, K., 1980, The Earth's Variable Rotation:
      Geophysical Causes and Consequences, Cambridge
      Univ. Press, London et alibi
Morrison, L.V., 1978, BI, p. 22
Platzman, G.W., Curtis, G.A., Hansen, K.S., and Slater, R.D.,
      1981, J. Phys. Oceanography 11, 579
Scrutton, C.T., 1978, BI, p. 154
Stephenson, F.R., 1978, BI, p. 5
Stephenson, F.R., 1981, BII
Sündermann, J., and Brosche, P., 1978, BI, p. 125
Webb, D.J., 1980, Geophys. J.R. astr. Soc., 61, 573
Webb, D.J., 1981, BII
Wunsch, C., 1967, Rev. Geophys. Space Phys. 5, 447
Yoder, CH.F., Williams, J.G., and Parke, M.E., 1981
      J. Geophys. Res. 86, B2, 881
Zschau, J., 1978, BI, p. 62
Zschau, J., 1981, BII

BI and page number = P. Brosche, J. Sündermann (eds.),
                     "Tidal Friction and the Earth's
                     Rotation" (Proceedings of a workshop
                     held in Bielefeld 1977), Springer-
                     Verlag, Berlin-Heidelberg-New-York 1978

BII   = Workshop on the sampe topic, held in
                     Bielefeld 1981, to be published by
                     Springer

EFFECTS OF THE NON-RIGIDITY OF THE EARTH DERIVED FROM ASTRONOMICAL OBSERVATIONS

Nicole CAPITAINE
Observatoire de Paris and Bureau International de l'Heure
FRANCE

ABSTRACT

The non rigidity of the Earth affects the astronomical observations in three ways : the tidal deflexion of the local vertical of the observer, the non coincidence of the instantaneous rotation axis with the adopted one corresponding to a rigid Earth and periodic variations of the rotation rate. The principal tidal and nutational effects have been derived from latitude and time data of the Paris astrolabe from 1956.6 to 1979.0. Some of them have also been derived from the z-term of latitude, w-term of UT, or UT1-TAI as computed by the BIH. These results are given here and a comparison between the local and global determinations is made.

INTRODUCTION

The lunisolar tidal force is responsible for the precession and nutations of the Earth's axis of rotation ; the amplitudes of these nutations are dependent on the Earth's elasticity and on the existence of a fluid core. Because of the non rigidity of the Earth this lunisolar tidal force is also responsible for oceanic and body tides giving rise to the deflexion of the local vertical and for periodic variations of the Earth's angular rate.

The tidal deflexion of the local vertical affects each astronomical observation and then the observations from a single station are suitable for deriving the amplitude of this effect.

The deviation of the instantaneous rotation axis with the adopted one corresponding to a rigid Earth affects globally the reduction of the observations. It can then be studied through the observations of a single station or through the z-term of latitude and w-term of UT as computed by the BIH in the determination of the Earth rotation parameters

W. Fricke and G. Teleki (eds.), Sun and Planetary System, 185–188.

The periodic tidal variations of the Earth rotation rate can be derived from global determination of UT1-TAI as the BIH one but also, with a lower precision, from local determination of UT1-TAI.

In the first part of this paper we give new determinations of the lunar tidal deflexion of the vertical from the Paris astrolabe observations (1956.6 to 1979.0). They are the extension of the previous determinations (Debarbat 1967, Capitaine et al 1979, 1980) to both latitude and time data and to both diurnal and semi-diurnal terms of this variation, with a more suitable treatment.

In the second part of this paper we give a comparison between the principal and fortnightly coefficients of nutation as derived from the Paris latitude data (Capitaine 1980 and this paper) and from the z and w - BIH data (Capitaine & Xiao 1981).

In the third part of this paper we give the results for the coefficient of the short term tidal variations in UT1, as expressed by Yoder et al (1981), obtained from the Paris astrolabe observations in time and from the BIH UT1-TAI determinations.

1. LUNAR TIDAL DEFLEXION OF THE VERTICAL

The coefficients Λ_D and Λ_{SD} of the diurnal and semi-diurnal variations in latitude and time observations, due to the tidal deflexion of the local vertical, have been derived from the Paris astrolabe observations, according to expansions given by Melchior (1978).

Λ is theoretically the combination $1+k-\ell$ of the Love number k and the Shida number ℓ.

The latitude and time residuals used for this purpose have been derived from a Vondrak's smoothing among the 6754 data of each series, obtained by the usual method from 1956.6 to 1979.0, corrected for group corrections and referred to TAI.

A complementary correction has been applied in order to refer these residuals to Wahr's (1981) fortnightly coefficients of nutation.

The least squares results for the Λ coefficient and phase lag τ are given below.

1.1. latitude residuals

Coefficient of the diurnal term :

$\Lambda_D = 1.36 \pm 1.34$, $\tau = 25°$ (whereas Λ_D is 11.32 ± 1.34 when referring to Woolard's (1953) coefficients of nutation)

Coefficient of the semi-diurnal term :
$\Lambda_{SD} = 1.06 \pm 0.15$, $\tau = 8°$

and $\Lambda_{SD} = 1.43 \overset{+}{_{-}} 0.15$, $\mathcal{T} = 27°$ when using a preliminary correction for the indirect oceanic effect in Paris (Souriau 1979).

1.2. Time residuals

Coefficient of the diurnal term:
$\Lambda_D = 0.96 \overset{+}{_{-}} 0.28$, $\mathcal{T} = 8°$ (whereas Λ_D is $2.33 \overset{+}{_{-}} 0.28$ when referring to Woolard's coefficients of nutation)

Coefficient of the semi-diurnal term:
$\Lambda_{SD} = 0.99 \overset{+}{_{-}} 0.13$, $\mathcal{T} = 23°$

These values of Λ are consistent with other computed and theoretical values, the computed ones being included between 0.40 and 1.70 and the theoretical ones between 1.20 and 1.22 (Melchior 1978).

The correction for the oceanic effect has to be more completely considered.

2. NUTATIONAL EFFECT

2.1. Principal term of nutation

The coefficients of this term, respectively in longitude and obliquity, as previously computed (Capitaine 1980) from latitude and time group differences of the Paris astrolabe are, when referred to the Celestial Ephemeris Pole and 1971.0 :
$N_P \sin\epsilon = -6''843 \pm 0''006$, $\Omega_P = 9''212 \overset{+}{_{-}} 0''006$ from latitude group-differences

$N_P \sin\epsilon = -6''832 \overset{+}{_{-}} 0''008$, $\Omega_P = 9''206 \overset{+}{_{-}} 0''008$ from time group-differences.

A similar determination made from the BIH data every 0.05 year from 1962.0 to 1980.0 (Capitaine & Xiao 1981) has given :
$N_P \sin\epsilon = -6''843 \pm 0''002$, $\Omega_P = 9''210 \overset{+}{_{-}} 0''002$ from the z term of latitude.
$N_P \sin\epsilon = -6''841 \pm 0''006$, $\Omega_P = 9''210 \pm 0''005$ from the w term of UT.

2.2. Fortnightly term of nutation

The coefficients of this term derived from the latitude residuals of the Paris astrolabe computed in 1., without nutational corrections but with tidal corrections are :
$N_f \sin\epsilon = -0''089 \pm 0''002$, $\Omega_f = 0''100 \pm 0''002$.

A similar determination made from the BIH-z-data every 5-days from 1967.0 to 1980.0 has given (Capitaine & Xiao 1981) :
$N_f \sin\epsilon = -0''091 \overset{+}{_{-}} 0''001$, $\Omega_f = 0''098 \overset{+}{_{-}} 0''001$

These determinations are quite consistent with the theoretical Wahr's coefficients $(N_P \sin\epsilon = -6''8406, \Omega_P = 9''2022), (N_f \sin\epsilon = -0''0905, \Omega_f = 0''0977)$

and the uncertainties of the two kinds of determination are of the same order of magnitude.

3. PERIODIC TIDAL VARIATION OF ANGULAR RATE

The coefficients $(k/C)_{Mf}$ and $(k/C)_{Mm}$ of the fortnightly and monthly periodic terms in UT1 have been derived from UT residuals, according to expansions given by Yoder et al (1981). The least squares results obtained from the Paris time residuals, as computed in 1, are :

$$(k/C)_{Mf} = 1.06 \pm 0.17 \qquad \text{or} \qquad k_{Mf} = 0.35 \pm 0.06$$

$$(k/C)_{Mm} = 0.90 \pm 0.16 \qquad \text{or} \qquad k_{Mm} = 0.30 \pm 0.06$$

The least squares results obtained from the BIH UT residuals every 5 days from 1967.0 to 1980.0, are (Capitaine & Xiao 1981) :

$$(k/C)_{Mf} = 0.90 \pm 0.08 \qquad \text{or} \qquad k_{Mf} = 0.30 \pm 0.03$$

$$(k/C)_{Mm} = 0.88 \pm 0.08 \qquad \text{or} \qquad k_{Mm} = 0.29 \pm 0.03$$

A non permanent peak at 13.81 d appearing in spectral analyses of the UT residuals, corrected for M_f and M_m tidal terms, can perturb (as mentioned by Guinot (1981)) the determination of $(k/C)_{Mf}$.

The two kinds of determination of these coefficients are nearly equivalent and consistent with the theoretical value ($k/C = 0.94$) but the noise level in spectral analyses is lower for global UT data than for local one.

CONCLUSION

Some parameters of the non rigidity of the Earth can be derived from astronomical observations with a wholly independence from the other geophysical methods. Long series of homogeneous data or global determination provide equivalent results.

REFERENCES

Capitaine, N., 1980, in "Nutation and the Earth's Rotation", IAU Symp. 78. E. P. Fedorov, M. L. Smith & P. L. Bender ed.
Capitaine, N., Chollet, F., Débarbat, S., 1979, Proc. 8th International Symp. on Earth Tides, ed P. Melchior & A. Bonatz.
Capitaine, N., Chollet, F., Débarbat, S., 1980, 4th International Symp. "Geodesy and Physics of the Earth", Karl Max Stadt, Mai 1980.
Capitaine, N., Xiao, N. Y., 1981 Geophys. J. R. astr. Soc., in press.
Capitaine, N., Xiao, N. Y., 1981, in preparation.
Débarbat, S., 1967, Bull. Astr. Paris, Serie 3, 2, 541.
Guinot, B., 1981, Personal Communication.
Melchior, P., 1978, The Tides of the Planet Earth, Pergamon Press, Oxford.
Souriau, M., 1979, Geophys. J. R. astr. Soc., 57, 585.
Wahr, J. M., 1981, Geophys J. R. astr. Soc. 64, 705.
Woolard, E. W., 1953, Astr. Pap. Amer. Ephem. Naut. Almanach 15, 1.
Yoder, C., Williams, J., Parke, M. E., 1981, Geophys. Res. 86, 881-891.

CORRELATION BETWEEN SOLAR ACTIVITY AND UNIVERSAL TIME VARIATIONS

Djurović Dragutin
Department of Astronomy, University of Belgrade
Yugoslavia

In the difference of universal times UT2 and TAI, the geomagnetic indice Ap and the zonal component of the global atmospheric angular momentum AAM common cyclic variations of 3 and 4 months exist. The first one was identified by Belocerkovsky 1963, the second one-by Djurović 1970, both in universal time UT2. Its amplitudes are, approximately, 2-3 milliseconds (ms). The 4-month term is also observed in the pole coordinates and the upper stratosphere zonal circulation (Belmont et al. 1974).

Beside the two mentioned cyclic variations, in UT2-TAI and AAM the common cyclic variation of 55 days exists. It is not identified in the spectrum of Ap.

By using the known relation between the length of day (l.o.d.) and AAM variation, the amplitudes of 55-day, 3 and 4-month cycles in l.o.d. were computed. The computed values are close to the corresponding observed ones (the agreement is 90--95%).

In the spectrum of Ap the peaks for period P=28.6 days and P=6 months are emphasized. They are explained by the Sun's rotation and its apparent annual motion, respectively.

In the range of mean-period variations (from 1 year up to 1 decade) the peaks for P=3.3 and P=6.6 years exist in the spectra of Ap and UT2-TAI. The AAM is not analysed in the mentioned range of P.

Since the phases of assumed 3,4-month, 3.3 and 6.6-year variations of Ap are close to the phases of corresponding terms in UT2-TAI, the author supposes that solar activity affects the angular velocity of the Earth's rotation. It seems that 3 and 4-month terms represent indirect effects of solar activity. Probably, an unknown mechanism causes AAM variation.

189

W. Fricke and G. Teleki (eds.), Sun and Planetary System, 189–190.
Copyright © 1982 by D. Reidel Publishing Company.

"Solid" Earth response is opposite angular momentum change.

Since the observed irregular changes of geomagnetic field intensity (geomagnetic storms) can have a great amplitude and a duration of few days, it is assumed that they could be correlated with irregular fluctuations of UT2-TAI. Taking into account that the phase difference between them is possible, the coefficient of correlation R is computed.In all cases R is small (Rmax=0.002).From this analysis it follows that irregular fluctuations of UT2-TAI are independent from solar activity.

The results of the above investigations in a more complete form are presented in the following papers:

D.Djurović:Solar activity and Earth's rotation,Astron.Astrophys.100,156-158 (1981),

D.Djurović:Short-period geomagnetic variations and Earth's rotation,Astron.Astrophys.,submitted (1981),

D.Djurović:Common short-period cyclic variations of the atmospheric angular momentum and the length of day,Astron.Astrophys.,submitted (1981).

ON INVESTIGATIONS OF MEAN LATITUDE VARIATIONS

B.Kołaczek
Planetary Geodesy Department, Space Research Centre,
Polish Academy of Sciences, Warsaw, Poland
G.Teleki
Astronomical Observatory, Belgrade, Yugoslavia

ABSTRACT: Nonpolar variations of mean latitude are discussed. Some examples of variations of mean latitude differences of stations, located in the vicinity or along a common meridian, are presented and discussed.

1. INTRODUCTION

Variations of latitude and UTO, the source data for the polar motion determinations, are caused not only by free and forced nutations of the Earth's axis of rotation but also by local geophysical phenomena and other nonpolar effects like: atmospheric influences, including anomalous refraction, crust motion and variations of local vertical, instrumental and personal errors, declination and proper motion errors and errors caused by the adopted reduction procedures and fundamental constants.

Such nonpolar variations of latitude detected by analysis of different sets of latitude data are reviewed in brief in the paper. Now, nonpolar latitude variations can not be removed from latitude data, but analyses of these phenomena could enable some preventive actions diminishing their influences. Analyses of mean latitude variations, especially in the case of location of several stations or instruments in a near vicinity or along a common meridian is a good method for detection and study of nonpolar and local effects.

2. SHORT REVIEW OF NONPOLAR LATITUDE VARIATIONS

2.1. Atmosphere influences including refraction anomalies. Influences of refraction anomalies on latitude variations have been studied deeply by Teleki (1976,1977,1978) and others (Takagi and Goto, 1979). Atmospheric influences are variable in time and size, and it is very hard to eliminate them from latitude data. Instability of annual terms of systematic corrections R determined by BIH, for every instrument and every year, in order to preserve the BIH system, indicates a sensitivity of astrometric methods on local meteorological parameters. Several latitude stations like Herstmonceux, Paris, San Fernando, Calgary, Shirley-Ottawa have more stable annual terms. Their observational conditions ought to be

W. Fricke and G. Teleki (eds.), Sun and Planetary System, 191–194.
Copyright © 1982 by D. Reidel Publishing Company.

studied. Thermal protection of the instrument and the pavilion at the
Belgrade Observatory (Milovanović et al. 1980) as well as thermal
insulation of astrolabe pavilions (Débarbat, Chollet, private commu-
nications) have given good results. So, some preventive actions like
careful site selection, constructions od adequate pavilions (Høg, 1978)
with a thermal and wind protection of instruments are needed in order to
diminish these effects.

2.2. Crust motions, variations of local verticals and mean latitude variations.

Parameters of plate tectonic motions were determined from
latitude and UT variations by Arrur and Mueller (1971), Proverbio and
Quesada (1970,1973) and Feissel (1974). The results are in good agreement with values obtained from geophysical investigations only. Besides
global motions of plate tectonics, there are local crustal motions like
vertical uplift, for instance in Finland, crust motions in seismical
active regions etc., which can cause secular or irregular variations of
latitude and mean latitude (e.g. Melchior, 1957; Lobanova et al. 1977).
Local secular variations of mean latitude of order of $0\overset{''}{.}001$ per year of
some IPMS stations were determined by Yumi (1968). Nonpolar variations
of differences of Orlov' mean latitudes of stations, located on a common
meridian, were found by Filipov (1956), Teleki (1969), Kołaczek et al.
(1977), Galas (1976). These irregular variations of mean latitude reach
to 0.1 second of arc per year in maximum.

Nonpolar variations of zenithal arcs between pairs of station verticals
characterizing the vertical motions, were determined and studied by
Fedorov et al. (1972), Mironov and Korsun (1974), Mironov (1974). Secular
variations of these arcs reach the magnitude of 0.01 second of arc per
year in the case of the arc between Ukiah and Kitab. There are irregular
variations of zenithal arcs of the order of a few hundredth of an arc
second and annual variations of the order of a tenth of an arc second.

Variations of constant A-terms of the BIH systematic corrections R show
high correlations with variations of Orlov's mean latitudes (Kołaczek
and Teleki, 1981) and characterize well local or regional effects in latitude data. Variations of the constant A-terms of stations show some regional similarities. There is a common increase of mean latitudes of
stations in the West Europe $(\lambda<1^h)$, except Paris, in the years 1975-1980,
Fig. 1A, and of some stations in the central Europe $(1^h<\lambda<2^h)$ in the
years 1971-1973, which is followed by the latitude decrease in the next
years, Fig. 1B. There are a common decrease of mean latitude of Irkutsk
and Blagoveschensk, the similar latitude variations in the case of a few
instruments of Mizusawa at the end of seventhies and others. Regional
character of some of these variations implies possibility of connections
with some crust motions. Analysis of variations of BIH systematic corrections R on the background of these data of stations located in the vicinity allows the detection of some instrumental anomalies like the mean
latitude decrease of Potsdam, the large mean latitude variations of
Washington and of the astrolabe in Mizusawa, the unusual large annual
term of Carloforte, etc. Instability of BIH systematic corrections R,
and specially their annual terms, ought to be carefully studied by every

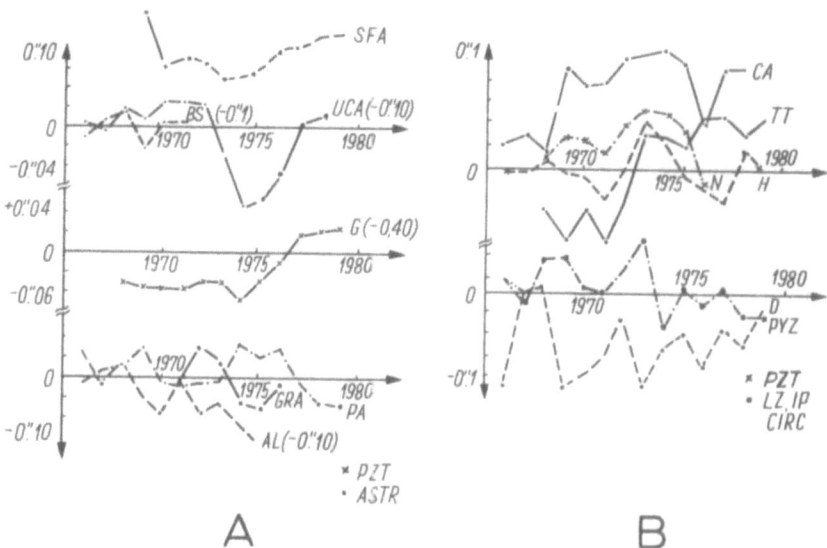

Figure 1. Variations of the BIH A-terms of latitude stations in Europe.

station in order to detect the main sources of these phenomena and undertaken appropriate preventive actions. More attention ought to be paid
also to regional distribution of stations participing in polar motion
determinations.

2.3. Instrumental and personal errors. Systematic instrumental errors
and their variabilities affect strongly latitude data, and probably they
are responsible for the most of variations of BIH systematic corrections
R. Systematic instrumental differences equal to 0.04 second of arc were
found in the case of two instruments located at Mizusawa by Sugawa and
Kitago (1970). Their annual variations were found by Kalmykov (1966) and
Mironov (1974). Personal errors of latitude determined by visual zenith
telescope of the order 0.1 second of arc were found by Ishii (1973) and
Dukwicz-Łatka et al. (1970).

It is difficult to separate instrumental influences from the other ones.
Participation of a large number of instruments in polar motion services
and careful investigations of parameters of instruments are necessary for
diminishing influence of systematic instrumental errors on latitude variations.

2.4. Errors of star coordinates, fundamental constants and errors introduced by the adopted method of computations. The accuracy of star coordinates is not satisfactory for the need of polar motion determinations
and of latitude variation investigations. Local declination systems
worked out by stations have systematic errors, and every change of an
observational program introduces systematic errors to determined latitudes. Proper motion errors can introduce secular variations of latitudes.
Space astrometry catalogue will improve this situation. Nowdays, every

station ought to preserve a homogenous declination system by determination of declination of new stars introduced to a program in a previous
system.

More homogenous set of latitude data can be obtained by introducing in
practice homogenous methods of computations of apparent declinations,
refraction corrections, instrumental corrections as well as final adjustment and smoothing of latitude data.

Computation methods of mean latitude values need also some considerations.
Many different filters are used for computations of mean latitudes, but
variations of parameters of main periodical terms of latitude variations
as well as the existence of other ones, insufficiently known cause systematic differences.

2.5. Conclusions. Improvement of polar motion determinations by astrometric methods can be achieved by undertaking all possible preventing
actions ensuring the best observational conditions at stations, careful
investigations of local latitude and UT0 variations and by the future
space astrometric cataloque. Such improvement is important not only for
the future but also for polar motion determinations in the past based
on the unique set of astrometric data.

REFERENCES

Arrur,M.G. and Mueller,I.I.:1971. J.Geoph.Res., 76,p.2071.
Dukowich-Latka,M. and Pieczynski,L.:1970,Seminarium "Badania Ruchu
                              Wirowege Ziemi", Warszawa,p.71.
Fedorov,E.P.,Korsun,A.A. and Mironow,N.T.:1972,Proc.IAU Symp.No.48,p.78.
Feissel,M.:1974,Geophys. J.Roy.Astr.Soc., 38,p.21.
Filipov,A.B.:1956, Trudy Poltav. Gravim. Obs. 6,p.1.
Galas,R.:1976, Proc. 3rd Symp.Geod. and Phys.Earth, Weimar, p.569.
Høg, E.: 1978, Proc.IAU Colloq. No 48, p.229.
Ishii,H.: 1973, Proc.Int.Latit.Obs.Mizusawa, 13, p.202.
Kalmykov,A.M.: 1966, Trudy Tashkent Astr.Obs., 11, 12, p.49.
Kołaczek,B., Galas,R., Barlik,M.: 1980, Proc.Symposium No 78, p.211
Kołaczek,B. and Teleki,G.: 1981, Proc.IAU Colloq. No 56. p.161.
Lobanova,V.V., Urasina,I.A. and Chudinov,N.N.:1977.Astr.Cirk.951, p.6.
Melchior,P.J.: 1957,Comm.Obs.R. Belgique, 130, p.225.
Milovanović,V.,Teleki.G. and Grujić, R:1980,Publ.Astr.Obs.Sarajevo, 1.
Mironov,N.T.: 1974, Proc.IAU Colloq. No 26, p. 79.
Mironov,N.T. and Korsun, A.A.: 1974,Proc. 2nd Symposium Geod. and Phys.
                              of the Earth, Potsdam, p. 173.
Proverbio, E. and Quesada,V.: 1970, Bull.Geod. 112, p. 187.
Proverbio, E. and Quesada,V.: 1973, Ann.Geofisica, 26, No 2-3.
Sugawa, C. and Kitago,H.: 1970, Proc.Int.Latit.Obs.Mizusawa,10,p. 18.
Takagi,S., Goto,Y.: 1979, Proc. IAU Symposium No 89, p. 119.
Teleki,G.: 1969, Bull.Obs.Astr.Belgrade, 27, No 2, p. 58.
Teleki,G.: 1976, Wiss.Z.Techn.Univers.Dresden, 25, p. 913.
Teleki,G.: 1977, Bull.Obs.Astr.Belgrade, 128, p. 19.
Teleki,G.: 1978, Bull.Obs.Astr.Belgrade, 129, p. 1.
Yumi, S.: 1968, Proc. IAU Colloq. No 1, p. 193.

The work is supported in part by the Smithsonian Inst.Grant No.FR-6-50015.

CLOSING ERRORS OF THE BELGRADE LATITUDE
OBSERVATIONS AND TEMPERATURE INFLUENCES

M. Djokić
Astronomical Observatory Belgrade, Yugoslavia

1. In the period 1949.0-1980.0 the Belgrade Latitude Service operated in succession two observational programs. Under the earlier program (the old program) the observations were made in the interval from 1949.0 to 1961.0 and under the second (the new program) from 1960.0 to 1980.0. Relevant information on these observing programs is given in the papers of Ševarlić (1961) and Ševarlić and Teleki (1959). During 1960 the observations were made of both programs concurrently in order to make comparison of the results obtained separately by each one of them.

An analysis of the series of latitude observations in the interval from 1960.0 to 1980.0 (Djokić, 1980) showed that there existed a correlation between: latitude differences (Dϕ) between consecutive subgroups, observed on the same night and the temperature differences ΔT_i, ΔT_v and ΔT_z respectively. ΔT_i are the differences of the mean instrument temperatures measured during the observation of consecutive subgroups. ΔT_v are the differences of the mean temperature inside of the pavilion and ΔT_z are the differences of the mean temperature outside of the pavilion. The last two temerature differences are also related to the time of observation of the relevant subgroups.

All of these differences, those in latitude from subgroups, as well as those in telescope temperatures, temperatures inside pavilion as well as those outside pavilion exhibit an annual periodicity, with the maximum in summer and minima in the winter. This is typical of the continental latitude stations (Sugawa, 1954).

The temperature insulation of the instrument and its pavilion, provided in 1969 (Milovanović et al., 1981) resulted in mutual accordance of the temperature differences ΔT_i, ΔT_v and ΔT_z. A diminishing of the amplitude of the curve, representing latitude differences (Dϕ) has also been stated along with the change in its pattern. The latter change can also be a consequences of the diminishing of the inclination measured by the Talcott levels.

195

W. Fricke and G. Teleki (eds.), Sun and Planetary System, 195–196.
Copyright © 1982 by D. Reidel Publishing Company.

Further studies revealed that there existed a correlation between the
annual closing errors (R) and the temperature differences of the type
$\Delta T_i - \Delta T_v$ and $\Delta T_i - \Delta T_z$. This is an indication of the room refraction
effects on the closing errors. However, these are second order effects.

2. In order to check the possibility of using, on equal terms, the
latitudes obtained by the observations under both programs an analogous
analysis of the latitude differences $(D\phi)$ obtained from the observations
under the old program as well as one of the corresponding temperature
differences $(\Delta T_i, \Delta T_v$ and $\Delta T_z)$ in the interval from 1949.0 to 1961.0 has
been carred out.

From the mean monthly values of latitude differences $(D\phi)$ along with the
temperature differences $(\Delta T_i, \Delta T_v$ and $\Delta T_z)$ the following harmonic formulae
are derived:

$$D_\phi = -0\overset{.}{.}02 + 0\overset{.}{.}08 \cos (t - 150\overset{\circ}{.}7)$$

$$\Delta T_i = +1\overset{\circ}{.}01 + 0\overset{\circ}{.}49 \cos (t - 162\overset{\circ}{.}7)$$

$$\Delta T_v = +0\overset{\circ}{.}53 + 0\overset{\circ}{.}29 \cos (t - 166\overset{\circ}{.}2)$$

$$\Delta T_z = +0\overset{\circ}{.}55 + 0\overset{\circ}{.}30 \cos (t - 162\overset{\circ}{.}9)$$

On the basis of the data representing the closing errors (R) and the
mean annual temperature differences $\Delta T_i - \Delta T_v$ and $\Delta T_i - \Delta T_z$ the correlation
coefficients are determined and given in Table 1.

Table 1.   Correlation coefficients between the closing errors and the
           mean annual temperature differences.

The observational period 1949.0-1961.0/r	R, $(\Delta T_i - \Delta T_v)$	R, $(\Delta T_i - \Delta T_z)$
	- 0.70	- 0.73

3. On the basis of the available data the conclusion can be drawn that
the temperature influences on the latitude differences (D_ϕ) in the groups
of the old program are confirmed as well as the existence of the correlation between the closing erros (R), deduced from the latitude
observations under this program and the corresponding temperature
differences.

REFERENCES:

Djokić,M., 1980: Bull.Obs.Astron. Hvar
Milovanović,V., Teleki,G., Grujić,R.,1981:Publ.Obs.Astron.Sarajevo,1,131.
Ševarlić,B. and Teleki,G., 1959: Bull.Obs.Astron.Belgrade, 112-113, 19.
Ševarlić, B., 1961: Publ.Obs.Astron. Belgrade, 8.
Sugawa, C., 1954: Publ.Astron.Soc.Japan, 6, 3.

REFERENCE SYSTEMS LINKAGE FROM SPACE

P.Farinella[*], A.Milani[**], A.M.Nobili[**], F.Sacerdote[**].
[*] Osservatorio Astronomico di Brera, Merate, Italia
[**] Istituto di Matematica,Università di Pisa, Italia

The HIPPARCOS satellite will be launched by ESA in 1985, carrying on board an astrometric telescope of very high accuracy (1). The expected output of the mission is a catalogue of angular positions, proper motions and parallaxes of $\sim 10^5$ stars. This will provide a stellar reference system with an internal accuracy of about 0".002. Then the necessity arises of linking this reference system to the others bound to different bodies, such as the Earth (classical astrometry and satellite geodesy), minor planets (they will be also observed by HIPPARCOS), extragalactic radio-sources (VLBI). These links must have the maximum achievable accuracy compared to the internal consistency of various reference systems.

We propose to use HIPPARCOS itself for this purpose (2). The idea is that the HIPPARCOS telescope could look at an Earth-bound laser beacon which would appear to it as a star-like point source. (The seeing is negligible because the scattering medium is near the source). When one tube of the telescope looks at the dark zone of the Earth the other one will look at the sky and in the field of view of 54'x54' there will be, as a mean, a star of apparent magnitude $m \sim 9$. A 10mW laser with 10^{-3} rad beam divergence will appear to HIPPARCOS as a star of about the same magnitude. In this way it should be possible to measure the angular position of the laser beacon with respect to the stellar field with the same accuracy predicted for the astrometric mission. The accuracy of a single measurement is the same expected for the stars (the beacon is observed only one time per great circle but the observation time is longer); moreover, a given beacon can be observed at about the same frequency as a star. Obviously, the larger is the number of available Earth-bound laser beacons, the larger will be the amount of resulting data. If the satellite position could also be determined with a comparable accuracy (i.e., 40cm in all directions) with respect to Earth-bound stations, this would allow a direct link between the Earth-fixed system and the stellar one defined by HIPPARCOS. In order to determine very accura-

W. Fricke and G. Teleki (eds.), Sun and Planetary System, 197–198.

tely the satellite position a quasi-static laser triangulation is required and the spacecraft must have two laser retroflector arrays on board.
With current laser technology (range accuracy ~20cm, i.e. longitude
accuracy ~ 2m - due to the bad geometry) it will be possible to achieve
an accuracy of about 0".01. Some laser stations (involved in the SIRIO2-
LASSO mission) are planning to improve the tracking accuracy by about
one order of magnitude: these improved stations could be able to measure
the HIPPARCOS'position with an accuracy comparable to the internal accuracy of its measurements.

Since Earth-fixed systems will be linked to the reference system
defined by extragalactic radio-sources (by very accurate VLBI techniques)
this method would allow an indirect link between stellar and radio reference systems (provided that the positions of VLBI stations, laser
stations and laser beacons sites are determined with a comparable accuracy). One of the original goals of the HIPPARCOS mission, combined with
the Space Telescope project, was just to link these two systems. We
think that an independent way of linking the two systems is interesting
by itself. Moreover, the problem of matching accurately the extragalactic
radio-sources with their optical counterparts could be bypassed.

In order to carry out with the present HIPPARCOS configuration this
experiment, more attention must be devoted to the study of the Earth's
straylight effect when the beacon is observed: a very large background
noise prevents the angular measurements between the beacons and the
stars.

REFERENCES
(1)   "HIPPARCOS Spacecraft Design and Development",(1979) ESA (ESTEC),
      PF 616, 5 december 1979.
(2)   A.Milani,B.Bertotti,P.Farinella,A.M.Nobili and F.Sacerdote: 1981,
      "A Proposal for the Geophysical Use of HIPPARCOS: Preliminary
      Studies and Discussions" Gruppo di Meccanica Spaziale, Internal
      Report n.10/81.

HIGH PRECISION TRACKING OF GEOSYNCHRONOUS
SATELLITES : OCEANOGRAPHIC APPLICATIONS

P. Farinella[*], A. Milani[**] and A.M. Nobili[**]

* Osservatorio Astronomico di Brera, Merate (Como), Italy

** Ist. Matematico "L. Tonelli", Università di Pisa, Pisa, Italy

The sea surface topography (SST) is defined as the radial departure of the sea surface from the geoid (averaged for tides and wave motion), and is produced by the interplay of oceanic currents (subjected to the Coriolis acceleration) with the frictional forces due to winds and bottom topography (Cook, 1981). The SST has a long-wavelength amplitude of the order of one meter, with likely seasonal variations of ± 20 cm. A world-wide map of SST should be very important to understand the global current patterns in the oceans ; it could be derived by satellite altimetry, provided the geoid were accurately known. As shown by Mather et al. (1978), who have referred the altimetry data obtained by GEOS 3 to the GEM 9 geoid model, the major error source in the low-degree SST determinations by altimetry is at present the uncertainty of the corresponding geopotential harmonics, which limits the attainable results essentially to the second-degree zonal harmonic (whose resulting amplitude is 46 ± 6 cm).

A significant improvement of this technique requires the refinement of the low-degree geopotential at least to 2 parts in 10^9 (that is a geoid accuracy of the order of 1 cm). For this purpose, an important role can be played by the tracking of geosynchronous satellites, whose peculiar orbital behaviour (with a long-term longitude drift) is mainly determined by the resonant low-degree terms of the geopotential (with even $(\ell-m)$, e.g., C_{22} and S_{22}). If a precise tracking method, like laser tracking or optical "astrometry" of the satellite (Anselmo et al., 1982), is coupled to a refined dynamical model of the perturbations (the most critical one for a high satellite is the solar radiation pressure), a longitude accuracy of the order of 100 m is not difficult to achieve. Over a 100-days orbital arc, corresponding to a longitude drift of a few degrees, this means that the resonant longitude acceleration can be measured up to a 10^{-4} relative accuracy. A comparable accuracy of the involved geopotential coefficients (i.e., in the 10^{-10} to 10^{-9} range) is attainable provided a

199

wide longitude coverage of the tracking data is available. This method
can be applied in several different ways, involving different levels of
cost and organizational effort : (1) analysis of the tracking data of a
single satellite (our group is presently planning to use the range data
of SIRIO 2, the first geosynchronous satellite with laser retroreflectors
which will be launched by ESA in 1982 - see Bertotti et al., 1979);
(2) analysis of data from many satellites positioned at different longitudes, for instance by a campaign of "astrometric" optical observations
using 1 meter-sized telescopes and determining the angular coordinates of
a number of telecommunication satellites with accuracies of the order of
1 arcsec (Anselmo et al., 1982); (3) launch of an ad hoc geophysical satellite in synchronous orbit, equipped with an array of laser retroreflectors and minimizing problems due to radiation pressure (no despun antenna,
nearly spherical shape, small area-to-mass ratio, etc.).

   This latter solution would provide the possibility of determining
the resonant geopotential terms below the 10^{-10} accuracy level, allowing
at the same time a direct measurement of the seasonal or long-term SST
variations, simply by monitoring their contribution to the Earth's gravity
field (the SST contribution to the geoid is about 1/10 of the local SST
height). This method implies that either the geopotential is recovered
from data not too spread in time (i.e., from arcs not longer than ~ 1
month), or a variable fraction of the geopotential terms, depending on
some parameters to be solved for, must be included in the force model.
In a similar way, other changes of the resonant geopotential coefficients
due to different causes (e.g., slow mass displacements in the Earth's
interior - see Wagner, 1973) could be detected by observing their dynamical effects on a synchronous orbit.

   The authors thank C.A. Wagner for several helpful discussions and
suggestions. This work is partially supported by CNR.

REFERENCES

Anselmo,L.,Bertotti,B.,Farinella,P.,Milani,A.,Nobili,A.M.,Sacerdote,F.:
     1982, "Comparative Study of the Planets" (Vulcano, Italy),
     A. Coradini and M. Fulchignoni (eds.), pp. 195-202.

Bertotti,B.,Bevilacqua,R.,Farinella,P.,Gianni,P.,Milani,A.,Nobili,A.M.:
     1979, Int.Rep.Osservatorio Astronomico di Brera no. 8/80.

Cook,A.H.: 1981, Q.Jl.R.Astr.Soc. 22, pp.125-132.

Mather,R.S.,Lerch,F.J.,Rizos,C.,Masters,E.G.,Hirsch,B.: 1978, presented
     at the International Symposium on "The Use of Artificial Satellites
     for Geodesy and Geodynamics" (Lagonissi, Greece).

Wagner,C.A.: 1973, J.G.R. 78, pp.470-475.

OPTICAL TRACKING OF SYNCHRONOUS SATELLITES FOR GEOPHYSICAL PURPOSES

S.Catalano[*]; P.Farinella[**]; A.Milani,A.M.Nobili[***]

*Istituto di Astronomia,Università di Catania (Italia)
** Osservatorio Astronomico di Merate (Italia)
*** Istituto Matematico "L.Tonelli", Univ.di Pisa (Italia)

The tracking of synchronous satellites, combined with an orbit propagation model of comparable accuracy, allows the determination of the
resonant geopotential coefficients (ℓ-m even) of low degree ℓ (mainly
C_{22} and S_{22},which correspond to the Earth's equator ellipticity). The
radio tracking techniques used in the sixties with the first telecommunication satellites (Wagner 1965;1966) were not very accurate,so that
the resulting relative accuracy in C_{22},S_{22} was not better than few per-
cents.In the present situation, the coefficients relevant for synchronous orbits derived within the existing global models show rather large
discrepancies going from 10^{-3} to 10^{-1} of their values. Much more accurate tracking techniques are now available (LASER, Doppler twin band),
but only the synchronous satellite SIRIO2, scheduled for launch by ESA
in 1982,will have a laser retroflector array on board; no synchronous
satellite has - nor is planned to have - the required payload on board
to allow Doppler twin-band tracking. The ground laser stations tracking
SIRIO2 will provide ranging data with an uncertainty in the radial direction less than 1 m. Due to the bad geometry of the range measurements,
the corresponding uncertainty in longitude will be about 10 times larger,
providing, in principle, a relative accuracy in C_{22},S_{22} of the order of
10^{-5}. Unfortunately the results which could be obtained by using LASER
data are seriously limited by radiation pressure perturbations (Anselmo et al.,1981),which are very difficult to model (unless ad hoc spherical or very expensive drag-free satellites are used). The bad modelling of the radiation pressure perturbation causes the longitude
uncertainties to grow up to 100÷1000m, so that the resulting relative
accuracy in C_{22},S_{22} will not be anyway better than 10^{-3}÷10^{-4}.

For this reason we have analyzed again the possibility of performing optical tracking of geosynchronous satellites (no ad hoc device on
board is required!)(Milani and Nobili,1980). First of all it is necessa-

W. Fricke and G. Teleki (eds.), Sun and Planetary System, 201–202.

ry to estimate the optical magnitude in diffused light, because only in very peculiar situations (near the equinoxes) it is possible to receive the sunlight directly reflected from the satellite, and in these situations it is several orders of magnitude brighter. It is generally a faint object (for SIRIO2 we predict a magnitude $m \sim 15 \div 16$), moving with respect to the stars at a speed of about 15"/sec, so that the telescope must track the satellite in order to allow the impressing of its image on the plate. Then the stars appear on the same plate as linear trails. A large enough number of luminous enough catalogue (AGK3, SAO catalogues) stars is necessary in order to obtain good angular measurements. Moreover, unless the star trails are marked in some way, only the N-S angular coordinate is determined. The recovery of the E-W coordinate is possible simply by using a very fast shutter or by moving very fast the plate in the N-S direction at accurately recorded times. Some trials at the SAO astronomical station of Agassiz and at the Observatory of Serra la Nave (Catania) have shown that an accuracy of $1 \div 2$" (i.e., $200 \div 400$ m in longitude at geosynchronous distance) is achievable, and therefore optical tracking data could be available with errors of the same order as the unavoidable errors in the orbit propagation due to radiation pressure.

In a world-wide astronomical campaign, involving small-sized and intermediate-sized telescopes, many synchronous satellites at many different longitudes could be optically tracked, giving the possibility to improve the global geoid knowledge.

REFERENCES

Anselmo, L., Farinella, P., Milani, A., and Nobili, A.M.: 1981, Proceedings of the International ESA Symposium on "Spacecraft Flight Dynamics", Darmstadt, 1981, pp. 47-52.

Milani, A., and Nobili, A.M.: 1980, Internal Report n.1/80, Gruppo di Meccanica Spaziale, Università di Pisa.

Wagner, C.A.: 1965, J.Geophys.Res. 70, pp. 1566-1568.

Wagner, C.A.: 1966, J.Geophys.Res. 71, pp. 1703-1711.

GRAVIMETRIC GEOID DETERMINATION FOR THE AREA OF GREECE

D. Arabelos, L.N. Mavridis, I. Tziavos
Department of Geodetic Astronomy, University of Thessaloniki,
Thessaloniki, Greece.

In a previous paper (Arabelos, 1980) a gravimetric geoid determination for the area of Greece ($34^{\circ} \leqslant \varphi \leqslant 42^{\circ}$, $18^{\circ} \leqslant \lambda \leqslant 27^{\circ}$) was presented. This determination was based on the following data:
1) The spherical harmonics model complete to degree and order 20 published by Rapp (1973).
2) 783 terrestrial free-air gravity anomalies in $1^{\circ}x1^{\circ}$ blocks covering the area: $24^{\circ} \leqslant \varphi \leqslant 52^{\circ}$, $8^{\circ} \leqslant \lambda \leqslant 37^{\circ}$, taken from the file of the Institut für Theoretische Geodäsie, Universität Hannover (IfTG/UH).
3) 8925 terrestrial free-air gravity anomalies in 6'x10' blocks contained in the area: $31^{\circ} \leqslant \varphi \leqslant 45^{\circ}$, $14^{\circ} \leqslant \lambda \leqslant 35^{\circ}$ compiled by the first of the authors.

The computations have been carried out by the method of least squares collocation using a grid $0^{\circ}.5x0^{\circ}.5$. The IAG 1975 System was used as a reference system. For an estimation of the accuracy of the geoid thus obtained, a comparison was made between this geoid and the altimetric geoid computed by Rapp (1979) on the basis of GEOS 3 satellite altimetry. Only the 94 altimetric values corresponding to the sea surface and having an accuracy better than ±1 m have been used in this comparison. A rms difference of ±0.8 m was found after a reduction of a 0.7 m bias.

In the present paper an attempt was made to extend and improve the above geoid using the following data:
1) The spherical harmonics model GEM-9 complete to degree and order 20 published by Lerch et al. (1977), instead of the model published by Rapp (1973).
2) 5506 terrestrial free-air gravity anomalies in $1^{\circ}x1^{\circ}$ blocks covering the area: $4^{\circ} \leqslant \varphi \leqslant 74^{\circ}$, $-25^{\circ} \leqslant \lambda \leqslant 65^{\circ}$, and taken from the file of the IfTG/UH.
3) 21440 terrestrial free-air gravity anomalies in 6'x10' blocks contained in the area: $31^{\circ} \leqslant \varphi \leqslant 48^{\circ}$, $0^{\circ} \leqslant \lambda \leqslant 35^{\circ}$. The additional 6'x10' gravity anomalies used in the new determination were obtained as follows:
a) For the area of continental Italy from Bouguer maps published by Ballarin et al. (1972), which were transformed into free-air anomalies with the help of the densities given by Vecchia (1955) and the mean elevations computed by Wenzel et al. (1980) on the basis of the data given

W. Fricke and G. Teleki (eds.), Sun and Planetary System, 203–204.
Copyright © 1982 by D. Reidel Publishing Company.

by Ballarin et al. (1972).b) For the area of the Mediterranean sea with
$\lambda < 14^{\circ}$ from the data obtained by Wenzel (1980) on the basis of the free-
air anomalies maps published by Morelli (1970) and Morelli et al. (1975).
c) For the remaining blocks the values were predicted by the method of
least squares prediction (Arabelos, 1980).

The computations were carried out by the Stokes-A integration method
(Rapp and Rummel, 1975), which is more economical than the method of least
squares collocation, using a grid $0^{\circ}.5x0^{\circ}.5$. GRS 80 was used as a refer-
ence system. The new geoid covers the area: $34^{\circ} \leqslant \varphi \leqslant 42^{\circ}$, $16^{\circ} \leqslant \lambda \leqslant 27^{\circ}$.
The computer programs used were put at our disposal by the IfTG/UH.

For an estimation of the accuracy of the geoid thus obtained a comparison was made between this geoid and the altimetric geoid computed, by
Rapp (1979). Again only the 136 altimetric values corresponding to the sea
surface and having an accuracy better than ±1m were used in the comparison. A rms difference of ±0.95m was found after a reduction of a
0.07 m bias.

From the same comparison a tilting in the NW-SE direction of the new
geoid as compared to the altimetric geoid was found. After a reduction
with the help of a first degree polynomial the rms difference between
the two geoids became equal to ±0.84m. Another reduction with a second
degree polynomial gives a rms difference equal to ±0.7m. The tilting
may have been caused by a systematic difference between the gravity
anomalies data used for Greece and Italy.

References

Arabelos, D.: 1980, Wiss. Arb. der Fachr. Vermessungswesen der Univ.
        Hannover Nr. 98.
Ballarin, S., Palla, B., and Trombetti, C.: 1972, Pubbl. della Commis-
        sione Geodetica Italiana, IV, No. 19.
Lerch, F.J., Klosko, S., Laubscher, R.E., and Wagner, C.A.: 1977, Grav-
        ity Model Improvement Using GEOS 3 (GEM 9 and 10). Goddard Space
        Flight Center, Greenbelt, Maryland.
Morelli, C.: 1970, Boll. Geof. Teor. Appl., XII, 275.
Morelli, C., Pisani, M., Gantar, C.: 1975, Boll. Geof. Teor. Appl. XVII,
        67.
Rapp, R.H.: 1973, Ohio State Univ. Rep. No. 200.
Rapp, R.H., Rummel, R.: 1975, Ohio State Univ. Rep. No. 233.
Rapp, R.H.: 1979, Private communication.
Vecchia, O.: 1955, Carta della densita media in Italia sino all livello
        del mare. Pubbl. della Commissione Geodetica Italiana, III, No. 9.
Wenzel, H.-G., Weber, G., Arabelos, D.: 1980, Private communication.
Wenzel, H.-G.: 1980, Private communication.

COINCIDENCE OF SOME MAGNETIC AND GRAVITY FIELD CHARACTERISTICS

G. Barta
Department of Geophysics, Eötvös University,
Budapest, Hungary

ABSTRACT: A 50-years' period can be found in the magnetic secula-
variation and in several features of the Earth's rotation as well; its
relation to mass movements is very likely. The symmetry centre of the
magnetic s.v. vectors is near India. Presumably the eccentricity of the
magnetic dipole towards Australia is a consequence of mass asymetry
which must be apparent in the Earth' figure, too.
The geoid is the sum of two rotation-symmetrical figures with axes towards Australia and India; the connection of the two fields is therefore
very probable. Subtracting the best-fit theoretical surface from the
measured one we obtain a residual figure which is linked to surface for-
mations.

It has been known long since that the Earth's magnetic dipole - now being
eccentric towards the Marshall islands - has been drifting west-wards at
a velocity of 0°2/year. If beside the magnetic secular variation the
eccentricity of the magnetic dipole is also assumed to originate from
some kind of mass asymmetry in the "background", then this must be
apparent in the Earth's figure, too. When these investigations began -
about 20 years ago - we had only certain forms of hypotheses about the
triaxiality of the Earth and a general belief in geodesy was that the
equatorial major axis of the Earth pointed towards Australia. Although
this direction coincided with the direction of eccentricity of the
magnetic dipole, this conception was not more than a guess owing to the
inaccuracy of primary measurement data.

The first reliable geoid heights were computed from the perturbations of
satellite orbits 15 years ago. At that time we made an attempt to calculate the equatorial ellipticity from the equatorial section of the
1966 geoid approximating it with a zonal spherical harmonic the axis of
which was directed towards Australia. The calculations gave the surprising results that a spherical harmonic of this type could not fit well
the equatorial section. Varying the axis of approximation we determined
the direction of best-fitting ellipse and, subtracting it from the
measured data, obtained a characteristic antisymmetry in the residual
map. Thus we came to the conclusion that the equatorial section could be
approximated not by one, but two zonal harmonics with axes nearly at

205

W. Fricke and G. Teleki (eds.), Sun and Planetary System, 205–208.
Copyright © 1982 by D. Reidel Publishing Company.

right angles to each other. This approximation is really very good.
Assuming that the two reflection-symmetric figures obtained in the
approximation were rotation-symmetric, we drew a map of the combined
body. Surprisingly this combination reproduced the six well-known
anomalies of geoid, i.e. the origin of the six large geoid anomalies can
not be six separate, independent mass inhomogeneities, but the geoid is
the sum of two great effects. The pairs of anomalies in the northern and
southern temperate zones are the antipodal superpositions of the two main
effects. The axes of the best-fitting component figures are not far away
from those mentioned above (Australia, Pakistan) (Figure 1).

It is very demonstrative that this geoid map has been computed solely
from the equatorial geoid heights, in other words the anomaly pattern of
temperate zones is implied in the equatorial data. If a separate density
inhomogeneity is attributed to every geoid anomaly, these inhomogeneities
should satisfy this very peculiar condition the probability of which is
extremely small.

It is also worth mentioning that according to a schematic calculation,
if the inner core is eccentric, and of high density, the level surface
becomes egg-shaped, peaked towards the direction of eccentricity. In our
case the zonal harmonic approximation gives an egg-shaped surface with
its axis and peak pointing towards Australia. It is interesting that the
rotation-symmetric component the axis of which is directed towards India
contains no even spherical harmonics, so it has no ellipticity. The
rotation symmetry is represented by harmonics of order 3,5 and 7. In this
stage of investigation, since we use only the equatorial data, the lines
connecting the oceanic anomaly pairs of temperate zones are prependicular
to the plane of equator. In reality these lines are not perpendicular to
the equator, instead they are inclined to each other northward. This suggests that the mass inhomogeneity lying behind the geoidal figure is
located north of the equatorial plane. And really, the magnetic dipole
is shifted north of this plane.

In the next stage of research - to explain this distortion of anomalies-
we had to abandon the equator as a plane of approximation and search for
a better fit. This generalization of computations made the formulae
extremely complicated. The best approximation was difficult to determine
because the surficial "inhomogeneities" of topography have also an
influence on the geoidal figure, so not only the two rotation-symmetric
effects but the irregular surficial mass inhomogeneities are also present
in geoids like SE III. Setting out from the consideration that the
influence of the surficial source bodies diminishes with the altitude
more rapidly than that of the global sources, moving away from the Earth's
surface we can separate the different types of anomalies. To achieve this
separation we have calculated the geoid for 1000, 2000, 3000, 6000, 10000,
20000 and 50000 km altitudes with the well-known method. The pictures at
20000 and 50000 km heights are rather schematic and more than the
equatorial ellipticity can not be observed. Between 5000 and 10000 km
altitudes the effect of the Indian source is clearly discernible although
the maps are still schematic. Their schematic pattern means that the

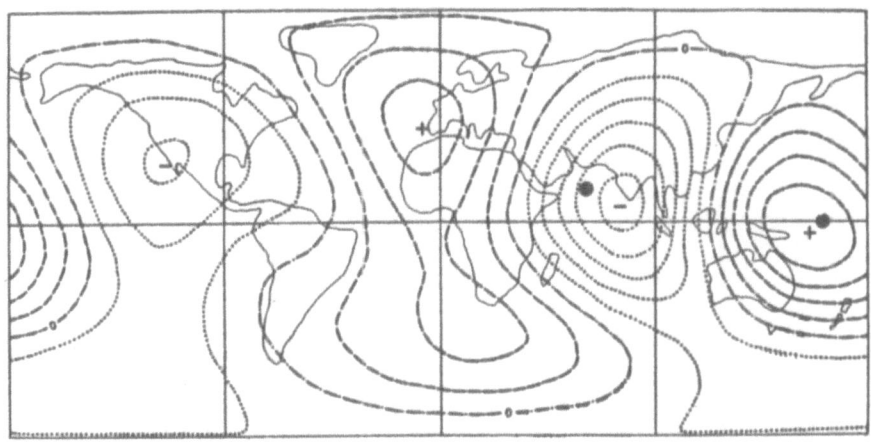

Figure 1. The geoid at a heigh of 6000 km with surficial intersecticn
points of the axes of approximating spherical harmonics

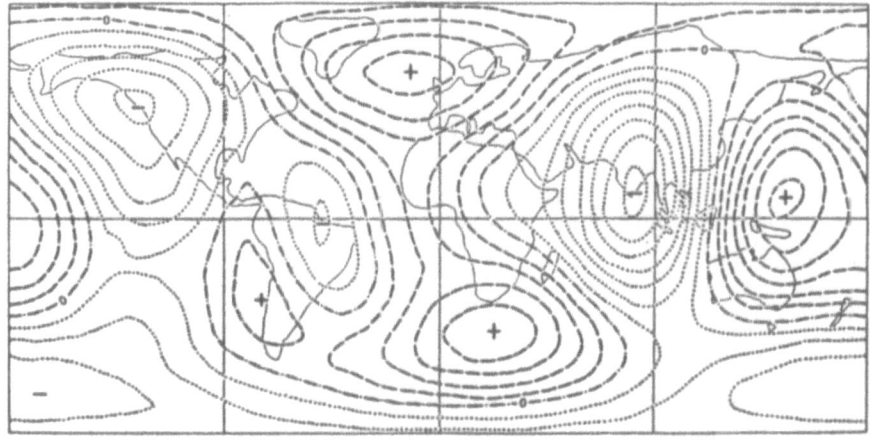

Figure 2. Approximating geoid figure on the Earth's surface.

effect of surficial disturbances is negligibly small at his height.
Therefore the data of the map corresponding to 6000km altitude were
chosen for the purposes of further research (Fig. 1). During the
calculations a mathematical method was found which determined the axial
directions of the two approximating rotation-symmetric figures quite
uniquely. The calculations can be carried out in two different ways:
either from the data along the main circle given by the two axes or from
the complete map of data measured all over the world. If the source
body is a composition of rotation-symmetric figures, the two definition
will give the same set of spherical harmonic coefficients when and only
when the two axes of approximation are properly chosen. In this way we
determined the best fit directions excluding the effect of surficial
inhomogeneities. Then the geoid so obtained was extrapolated back to the
Earth's surface (Fig. 2). Finally the results - supposed to reflect the
effect of deep seated sources - was subtracted from the measured geoid
surface, thus giving the effect of the surficial sources.

The residual map reveals a correlation between the anomaly pattern and
the relief of topography. The rows of positive anomalies of the residual
picture usually coincide with the big mountain ranges of the Earth and
similarly the system of 0 lines with the mid-oceanic ridges. Furthermore
the zonal asymmetry of geoid (its paper-shaped figure) can be attributed
only to the deep-seated sources. The surficial sources have no zonal
harmonic components. The polar ellipticity of the Earth may also be
imagined as the sum of two parts corresponding to our conception. The
major term is due to the Earth's rotation, and about 0.2 percent is attributed to the internal mass inhomogeneities.

The connection of magnetic and geoidal anomalies leads to the suggestion
that - just like the magnetic field - the gravity field might also under-
goe a secular change. The magnitude of this change would be very small,
at most several times 10 μgal a year .   No physical process is at
present suggested.

REFERENCE

Barta,G., 1973: Nature, 243, 5403, pp. 156-158.

PRECESSION AND NUTATION INFLUENCE ON THE HARMONIC COEFFICIENTS OF A
PLANETARY POTENTIAL

Oumlensky, V., Shkodrov, V.
Section of Astronomy
Bulgarian Academy of Sciences,
Sofia, Bulgaria

Abstract. In this paper an analysis of the precession and nutation influence on the coefficients of spherical harmonics is made. The greatest influence on the accuracy of the zonal coefficients is exerted by the deviation of the z-axis of the inertial system with respect to the instantaneous one. In this case the relative error is proportional to the order of
the harmonic coefficient. The x-axis deviation does not exert any influence.

In the classical form the differential equations of motion are valid only
in a reference system, whose axes are fixed in the space. In order to be
precise, when we solve kinematical problems, the perturbation function
must be also written in the same system. But having in mind certain
consideration, the harmonic coefficients in the development of the potential are expressed in a reference system, rigidly fixed in the Earth.
From the above said it follows, that these coefficients must be transformed in coefficients related to a fixed reference system. One method to
perform such a transformation is examined in this paper.
Let us denote the motionless reference system by $S\{Oxyz\}$. Let its origin
be the mass centre of the planet, and the plane Oxy coincide with mean
equator in the epoch T. The reference system, fixed with respect to the
Earth, will be denoted $S_0\{x_0y_0z_0\}$. The origin of the system is also at
the mass centre O, and its z-axis coincides with the rotational axis of
the planet. Besides S and S_0 we shall introduce a motionless reference
system $S'\{x'y'z'\}$, whose main plane $Ox'y'$ coincides with instantaneous
equator. We shall define the mutual positions of S and S' by means of the
three Euler angles $\hat{\psi},\hat{\theta},\hat{\phi}$, assuming, that they depend on the precession
and nutation and that they are small angles.
To connect the elements of the orbit, along which the body moves, with
expression of the potential V in the system S ($V \in S$), we shall introduce
one more reference system $S''\{Ox''y''z''\}$, so that $Ox''y''$ coincides with the
orbital plane, Oz'' being oriented towards the orbit's pole, and Ox''
towards the pericentre. The position of S'' in relation to S is determined
by the Euler angles α,β,γ, as in (Shkodrov, 1980).
Proccedings from (Shkodrov, 1981), for potential we can write

W. Fricke and G. Teleki (eds.), Sun and Planetary System, 209–212.

$$V = \frac{GM}{r} \sum_{n=0}^{\infty} (\frac{a}{r})^n N_n(\psi), \tag{1}$$

where G=gravitational constant, M= planet's mass, a=its equatorial radius, r=radius-vector of the point, in which we determine the potential and $N_n(\Psi)$ is given by the expression

$$N_n(\psi) = \int_{(T)} (\frac{\rho}{a})^n P_n(\cos\psi) \frac{dm}{M}. \tag{2}$$

In (2) ρ=the radius-vector of the elementary mass dm, $P_n(\cos\psi)$= the Legendre polynomial with the angle between ρ and r as an argument. To apply the generalized theorem of the spherical functions in (2) (Shkodrov, 1981), we shall assume, that $\rho \in S'$ in which the planet potential harmonical coefficients are known, i.e.

$$P_n(\cos\psi) = \frac{1}{2n+1} \sum_{(mm'm'')} (-1)^{m''} i^{m'+m''} Y_{nm''}(\frac{\pi}{2},v) D_{mm''}^{n*}(\Omega,I,\omega)$$
$$\times D_{mm'}^n (\hat{\psi},\hat{\theta},\dot{\phi}) Y_{nm'}(\theta',\lambda'), \tag{3}$$

where Ω is the length of the ascending node of the orbit, I - its inclination, ω - argument of the pericentr, v - true anomaly, $D_{mm'}^n(\alpha,\beta,\gamma)$ - Wigner function, $Y_{nm}(\theta,\lambda)=\bar{P}_n^m(\cos\theta)e^{im\lambda}$, ($\bar{P}_n^m(\cos\theta)$ - normalized associated Legendre functions), $i=\sqrt{-1}$, and complexly conjungated quantities are denoted by (*).
The final form of the perturbation function R in the system S is obtained from (1) and (3)

$$R = \sum_{(nm)} R_{nm}, \tag{4}$$

where
$$R_{nm} = \frac{GMa^n}{r^{n+1}} \sum_{(m'k)} (-1)^{n-2k} i^{m'+n-2k} \beta_{nn-2k} A_{nm'}^* D_{mn-2k}^{n*}(\Omega,I,\omega+v)$$
$$\times D_{mm'}^n(\hat{\psi},\hat{\theta},\dot{\phi}). \tag{5}$$

In (5) $A_{nm'}^* \in S'$ and β_{nn-2k} is denoted

$$\beta_{nn-2k} = (-1)^{n-k} [\frac{\Gamma(n-k+1/2)\Gamma(k+1/2)(2n+1)}{\pi\Gamma(n-k+1)\Gamma(k+1)}]^{1/2}. \tag{6}$$

When m=0, we obtain

$$R_{no} = R_n = \frac{GMa^n}{r^{n+1}} \sum_k i^{n-2k} \beta_{nn-2k} \frac{\alpha_n^*}{\sqrt{2n+1}} Y_{nn-2k}(I,\omega+v), \tag{7}$$

where

$$\alpha_n^* = \sum_{m'} (-\hat{i})^{m'} (2n+1)^{-1/2} A_{nm'}^* Y_{nm'}^* (\hat{\theta}, \hat{\phi}),$$ (8)

or

$$\alpha_n^* = -\bar{J}_n P_n (\cos\hat{\theta}) + \sum_{m'}' (-i)^{m'} A_{nm'}^* Y_{nm'}^* (\hat{\theta}, \hat{\phi}) (2n+1)^{-1}/2 ,$$ (9)

where $\sum'$ indicated a sum by the index m' without the value $m'=0$. Evidently, the last expression gives a relation between the zonal coefficients α_n^* in the reference system S' and harmonical coefficients $A_{nm'}^*$ in the system S'.
The difference $\delta\alpha_n^* = \alpha_n^* - (-\bar{J}_n)$ between the zonal coefficients $\alpha_n^* \in S$ and $(-\bar{J}_n) \in S'$ is

$$\delta\alpha_n^* = \bar{J}_n - \bar{J}_n P_n (\cos\hat{\theta}) + \sum_{m'}' (-i)^m A_{nm'}^* Y_{nm'}^* (\hat{\theta}, \hat{\phi}) (2n+1)^{-1/2}.$$ (10)

Let us determine the character of this difference as well as the influence of the variables $\hat{\theta}$ and $\hat{\phi}$ on it. For the purpose we shall develop the right-hand side of (10) in Taylor series, as a function of two variables around the point $\hat{\theta}=0$, $\hat{\phi}=0$ and we shall limit ourselves to the terms of the first order, i.e.

$$\frac{\delta\alpha_n^*}{\bar{J}_n} = [1 - P_n(\cos\hat{\theta}) + \sum_{m'}' (-i)^{m'} \frac{A_{nm'}^*}{\bar{J}_n \sqrt{2n+1}} Y_{nm'}^*(\hat{\theta}, \hat{\phi})]_{\hat{\theta}=0 \ \hat{\phi}=0}$$

$$+ [-\frac{dP_n(\cos\hat{\theta})}{d\hat{\theta}} + \sum_{m'}' (-i)^{m'} \frac{A_{nm'}^*}{\bar{J}_n \sqrt{2n+1}} \frac{\partial Y_{nm'}^*(\hat{\theta}, \hat{\phi})}{\partial\hat{\theta}}]_{\hat{\theta}=0 \ \hat{\phi}=0} \cdot \hat{\theta}$$ (11)

$$+ [\sum_{m'}' (-i)^{m'} \frac{A_{nm'}^*}{\bar{J}_n \sqrt{2n+1}} \frac{\partial Y_{nm'}^*(\hat{\theta}, \hat{\phi})}{\partial\hat{\phi}}]_{\hat{\theta}=0 \ \hat{\phi}=0} \hat{\phi}.$$

But since

$$[P_n(\cos\hat{\theta})]_{\hat{\theta}=0} = 1, \ Y_{nm'}^*(0,0) = 0 \ (m' \neq 0)$$ (12)

and

$$\frac{\partial Y_{nm'}^*(\hat{\theta}, \hat{\phi})}{\partial\hat{\phi}} = -im' Y_{nm'}^*(\hat{\theta}, \hat{\phi}), \ 2 \frac{\partial Y_{nm'}^*(\hat{\theta}, \hat{\phi})}{\partial\hat{\theta}} = -\sqrt{n(n+1)+m(1-m)}$$ (13)

$$\times Y_{nm-1}^*(\hat{\theta}, \hat{\phi}) e^{-i\hat{\phi}} + \sqrt{n(n+1)-m(m-1)} Y_{nm+1}^*(\hat{\theta}, \hat{\phi}) e^{i\hat{\phi}}$$

instead of (11), we obtain

$$\delta\alpha_n^* / \bar{J}_n = \hat{\theta} i \sqrt{n(n+1)} (A_{n1}^* - A_{n1}) / 2\bar{J}_n$$ (14)

or

$$| \delta\alpha_n^* / \bar{J}_n | = \hat{\theta} \cdot \sqrt{n(n+1)} \, | \bar{S}_{n1} / \bar{J}_n | \tag{15}$$

where use is made of the known relation

$$\bar{S}_{n1} = (A_{n1}^* - A_{n1}) i/2 . \tag{16}$$

If we set in (15)

$$| \bar{S}_{n1} / \bar{J}_n | = 1 \qquad (n \neq 2),$$

we obtain the following simple expression for the relative value of the difference between the zonal coefficients in S and S'.

$$| \delta\alpha_n^* / \bar{J}_n | = \hat{\theta} \sqrt{n(n+1)} \approx n\hat{\theta} \qquad (n \neq 2). \tag{17}$$

When n=2

$$| \delta\alpha_2^* / \bar{J}_n | = 0, \tag{18}$$

as $S_{21} = 0$.

The expression (17) shows, that the relative value of the difference is proportional to the zonal coefficients order n, the most essential influence being exerted by the angle $\hat{\theta}$, between the z-axis of the reference system S and S'. The precession and nutation influence of the $\delta\alpha_n^* / \bar{J}_n$, through the variable ϕ is by one order smaller than that of $\hat{\theta}$. Precession and nutation do not affect $\delta\alpha_n^* / \bar{J}_n$ through the variable ψ. From (18) it follows, that precession and nutation in this approximation do not lead to a difference between α_2^* and $(-\bar{J}_2)$.

The qualitative analysis of $\delta\alpha_n^*$ shows that the order of differences between $\alpha_n^* \in S$ and $(-\bar{J}_n \in S')$ is equal to this of the third digit after the decimal point of the corresponding coefficient (in a system 1.10^{-6}). These results are obtained assuming that $(-\bar{J}_n) \in S'$. If in the investigation we introduce a rotating together with the Earth reference system S_0 and the convention $(-\bar{J}_n) \in S_0$, then the coefficients α_n^* will be explicit functions of the sidereal time s, too. We can show, that the absolute value of the relative difference has a maximum when $s = \mathrm{arctg}(-S_{n1}/C_{n1})$ and is equal to 0 when $s = \mathrm{arctg}(C_{n1}/S_{n1})$.

The quantitative estimates we have made permit to determine the magnitude of the time interval, in which the precession and nutation do not cause noticeable difference between the values of the zonal coefficients in the reference systems S and S' (respectively S_0).

REFERENCES

Shkodrov, V.G.: 1980, Compt. rend. Acad. bulg. Sci. 33, 8, p. 1025.
Shkodrov, V.G.: 1981, Compt. rend. Acad. bulg. Sci. 34, 5, p. 605.

AEROSOL STRATIFICATION OF THE STRATOSPHERE AND MESOSPHERE

G.G. Mateshvili
Abastumani Astrophysical Observatory of the
Georgian Academy of Sciences, USSR

At present, the atmospheric pollution by the aerosols of terrestrial and cosmic origin changing solar radiation influx represents an actual problem. Since already forty years, at Abastumani Astrophysical Observatory in Georgia (USSR) the upper layers of atmosphere are studied by the twilight sounding. This method has given the possibility to determine the aerosol distribution in the middle atmosphere, for altitudes from 20 to 120 km. In order to obtain the parameters characterizing the atmospheric aerosol, the twilight sky brightness was measured by photometers of various designs, and a wide spectral range was included, from 400 to 740 nm.

The following quantities were considered: logarithmic brightness gradient $d\ln B(z)/dH$ (where $B(z)$ is the twilight sky brightness and H is the effective altitude of scattering), the scattering ratio and the atmospheric turbidity S, calculated from this ratio.

The altitudinal dependences of logarithmic brightness gradient and of turbidity allowed us to reveal the following properties of atmosphere:

1. All the curves display pronounced maxima, giving the evidence of aerosol clouds or layers.

2. Aerosol layers are situated at the altitudes of 25, 37 to 45, 55 to 60, 70 and 85 to 90 km.

3. Altitudes of maximum aerosol concentrations are rather stable. Their deviations from average values do not exeed 3 km.

From altitudinal dependences we have obtained it may be seen that in stratopause region the values of turbidity vary from 0.1 to 2 or to 3, then they grow: at the mesopause the turbidity reaches, on the average, the values range from 6 to 12. It is supposed that the mesosphere contains a background aerosol with concentrations rising from stratopause to mesopause. On the background of the aerosol, due to continuous influx of the matter from space, regions of higher concentrations are observed.

213

W. Fricke and G. Teleki (eds.), Sun and Planetary System, 213–214.
Copyright © 1982 by D. Reidel Publishing Company.

Periods were observed, when the turbidity within the mesosphere region
rises.

Our conclusions are as follows:
a) there is a background   aerosol in the mesosphere;
b) the accumulations of aerosol are observed as layers with maximum
   concentrations in regions of 50 to 60, 60 to 70, 80 to 90 km. The
   arrangement of aerosol within the layers is irregular, its horizontal
   component has a scrappy, cloudy pattern.

The origin of the upper atmosphere aerosol is the meteoric matter: this
term means all the contributions to the atmosphere from the space, i.e.
sporadic and shower meteors, comet dust, etc. This leads to effects of two
kinds. There are directly incoming particles, as well as the products of
their fission, ablation and vaporization. It seems that the secondary
particles, arising at the recondensations of meteoric vapours, constitute
the principal share of mesospheric aerosol. Considering that their sizes
are much lesser than those of primary particles and therefore, the
probable lifetime at the high altitudes is longer, too, one may consider
them as responsible for the scattering at such altitudes.

As to the stratospheric aerosol, in the shapes obtained by us there is
an evidence of a layer at the altitude of 25 km which was identified as
the Yunge layer. Above this layer, at the altitudes of 37 to 45 km, in
most of our shapes there is an evidence of a rather stable atmospheric
layer. Such stratospheric pattern was observed especially clearly in
August and September 1980, and we relate this fact to a complicated
cosmic and geophysical situation during that period. On the one hand,
there was a maximum of activity of the Perseides meteor shower. On the
other hand, the mighty eruptions of the Mayon, Hekla and Etna volcanoes
were observed in 1980. Besides, the paroxysm of the St Helens Volcano in
the Cascade range (North America), situated at above the same latitude
as Abastumani is, was continuating during the whole year. That is why,
owing to such feeding, the stratosphere layers were especially pronounced.
Sporadic observations of aerosol at the altitudes of 37 to 45 km by the
other observers and by the other methods convinced us that the stratosphere contains, apart from the Yunge layer, another layer, whose
thickness is genetically related with the Yunge layer.

INTERPLANETARY MAGNETIC FIELD AND SOLAR CYCLE MODULATION OF THE TROPO-SPHERIC CIRCULATION OVER WEST MEDITERRANEAN

BASIL P. TRITAKIS
Research Center for Astronomy and Applied Mathematics
Academy of Athens

Etesians are North-South direction winds which flow over the Aegean Sea every summer from May to October. Repapis et al. (1978) showed that a thermal low over Iraq in conjunction with high pressure centers over Central Europe-Balcans is responsible for the construction of the Etesians background. Previous papers (Carapiperis, 1960; Xanthak·s, 1975; Tritakis, 1981) pointed out that Etesians have a close correlation with the solar cycle, while positive interplanetary magnetic field (I.M.F.) days look to control the frequency distribution of these winds within the solar rotation. The Etesians interannual variability looks smooth during the period of their occurrence (May-October) except June when significant irregularities are noticed. Acutally, for 29 Junes of the period 1947-1975, where valid data are available, we notice statistically significant peaks as well as deep hollows of the Etesians and I.M.F. distributions. This is very important because June, from the meteorological point of view, is the time when the Etesians background is mainly constructed while, from the astronomical aspect, it is the time when the earth crosses the solar equator (6th June) and is located on the solstice (21st June). Figure 1a shows the Etesians frequency distribution for all Junes of the period 1947-1975. We notice statistically significant peaks on the 6th (the earth has heliolatitude 0°) and the 26th-29th June. In the opposite, the I.M.F. distribution doesn't show any significant variability (Fig. 1b). However, if we have in mind the different way where solar activity controls the I.M.F. during the epochs of extrema and intermediate, detected in previous papers (Tritakis, 1979, 1980, 1981) a close relationship can be found between Etesians and I.M.F. predominant peaks. Figures 2 and 3 are similar to Figure 1 but they represent data of two different phases of the solar cycles they cover the period 1947-1975. Namely, Figure 2 shows the frequency distribution of the Etesians and I.M.F. for all Junes of the epoch of extrema. This epoch (15 years) comprises the years of maximum and minimum of the solar cycles as well as the preceding and the following year. Figure 3 is the same with Figure 2 but it concerns to the epoch of intermediate. This epoch comprises years they don't belong to the epoch of extrema (14 years). From Figures 2 and 3 it is evident that the main peaks of Figure 1a correspond to different phases of the

W. Fricke and G. Teleki (eds.), Sun and Planetary System, 215–216.
Copyright © 1982 by D. Reidel Publishing Company.

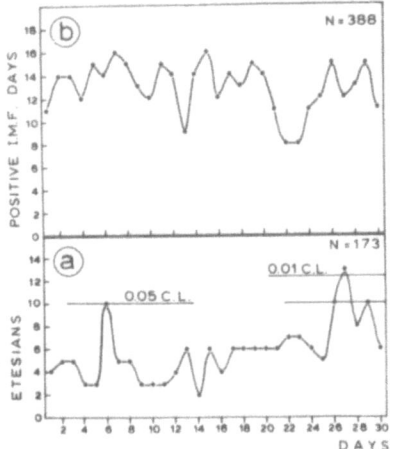

Figure 1. Etesians and I.M.F.
positive days for the Junes
of the period 1947-1975.

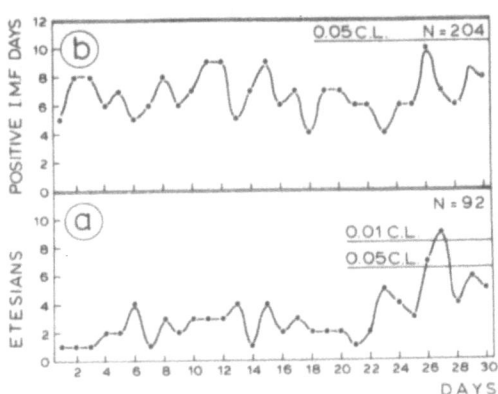

Figure 2. Etesians and I.M.F. positive
days for the Junes of the epoch of
extrema.

solar cycle. Actually, the epoch of
extrema is responsible for the peak
of the 27th while the epoch of the
intermediate causes the peak of the
6th of June. It is noteworthy that
the Etesians significant peaks of
Figures 2a and 3a correspond to significant peaks of the positive I.M.F.
days (Figures 2b and 3b). In a previous paper (Tritakis, 1981) we have
discussed the possible interpretation
of the correspondence between significant peaks of both Etesians and
I.M.F. positive days.

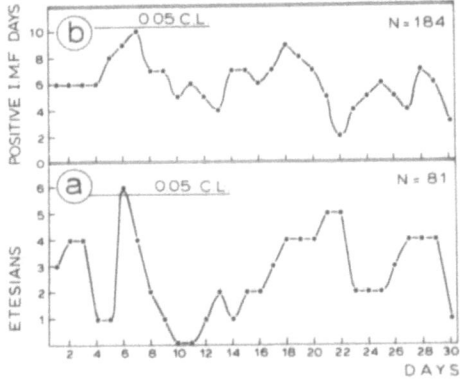

Figure 3. Etesians and I.M.F.
positive days for the Junes of
the epoch of intermediate.

References
Carapiperis, L.:   1960, Geoph. Pura e Appl. 46.
Repapis et al.:   1978, Praktika Academy of Athens, 52, pp. 572-606.
Tritakis, B.:   1979, Sol. Phys. 63, pp. 207-215.
Tritakis, B.:   1980, Astrophys. Space Sci. 66, pp. 385-390.
Tritakis, B.:   1981, J. Atmos. Terr. Phys. (to be appeared).
Xanthakis, J.:   1975, In Memoriam Demetrios Eginitis, Athens, Greece,
                 pp. 305-317.

ON THE EVOLUTION OF NEARLY CIRCULAR ORBITS
OF SATELLITES WITH CRITICAL INCLINATION

A.S. Sochilina
Institute for Theoretical Astronomy, Leningrad, USSR

The present paper is dealing with an investigation of "Navstar" type orbits including the influence of the critical inclination, the principal terms of luni-solar perturbations and resonance terms of the geopotential. The parameters of a "Navstar" type orbit are the following: the eccentricity e = 0.005, the inclination i = 63°.4, the period of revolution is commensurable with the Earth's rotation as 2:1.

A qualitative solution of the critical problem for the satellites moving along the nearly circular orbits has been given by A.Jupp (1979). The results of this work may be applied to the "Navstar" type satellite. The system of two non-linear equations for the variables $h = e \sin\omega$ and $k = e \cos\omega$ (ω = argument of perigee) on which the critical inclination has prominent influence can be reduced to the linear form in the case of high orbits.

But for such orbit it is necessary to take into account luni-solar perturbations. The significant part of these perturbations when the inclination is equal 63°.4 contains the terms depending on ω and e. Therefore the final equation for h and k can be written in the following form:

$$dh/dt = Ak, \qquad dk/dt = Ah + S \tag{1}$$

where A reflects the influence of the Moon and Sun ($A = 8.53 \cdot 10^{-5}$) and S - the odd order harmonics of the geopotential ($S = -1.01 \cdot 10^{-6}$). The solution of (1) is expressed in terms of hyperbolic functions:

$$h = h_0 ch\, At + Ak_0 sh\, At + (S/A)(ch\, At - 1),$$

$$k = k_0 ch\, At + Ah_0 sh\, At + (S/A)\, sh\, At, \tag{2}$$

where h_0, k_0 = initial values of a and k at the moment t = 0.

It has been shown earlier (Sochilina, 1982) that the resonance perturbations arising from the geopotential cause considerable changes not

W. Fricke and G. Teleki (eds.), Sun and Planetary System, 217–218.

only in λ_N (longitude of stroboscopic mean node) but also in ω if the orbit is nearly circular independently of its inclination. Introducing the variables h and k and also adopting the GEM - 10B model equation (9) (Sochilina, 1982) for "Navstar" type satellites can be written as follow:

$$
\begin{aligned}
d^2\lambda_N/dt^2 = \Big[&-(0.6196 - 32.02h - 21.85k) \sin 2\lambda_N + \\
&+ (0.9111 - 55.76h + 12.55) \cos 2\lambda_N - \\
&- 0.2871 \sin 4\lambda_N - 0.4253 \cos 4\lambda_N \Big] \cdot 10^6
\end{aligned}
\tag{3}
$$

The equation for h and k will have the following form

$$dh/dt = 1.333 \cos(2\lambda_N + 0.5203) \cdot 10^6$$

$$dk/dt = -3.401 \sin(2\lambda_N + 0.5203) \cdot 10^6$$

The results of numerical integration (3) and (4) has shown that for "Navstar" type orbits the resonance perturbations play the principal role. The evolution of these orbits can be represented in the following way: a satellite placed into equilibrium point leaves it and moves in a spiral trajectory in the phase plane $(2\dot{\lambda}_N, 2\lambda_N)$. The character of evolution strongly depends on initial ω_0. The perigee of orbit approaches the values of 120^0-180^0 independently of the initial value of ω_0 in 2000 days. The eccentricity begins slowly to increase. The luni-solar perturbations and the critical inclination affect only the variation of ω for the initial $\omega_0 = 0^0$.

The computations have been made with different initial values ω, e, λ_N, on intervals from 4500 up to 8000 days.

REFERENCES:

Gedeon,G.S., 1969, Celes. Mech., 1, pp. 167-189.
Jupp,A.H., 1979, Celes. Mech., 21, pp. 361-393 .
Sochilina,A.S., 1982, Celes. Mech. 26, 337.

SECTION III

PHYSICS OF PLANETS, MINOR PLANETS, SATELLITES

AND INTERPLANETARY MEDIUM

EXCITATION IN THE EARLY SOLAR NEBULA - NEW EXPERIMENTAL FINDINGS

G. Arrhenius, M.J. Corrigan, R.W. Fitzgerald, and C. Schimmel
Scripps Institution of Oceanography, La Jolla, CA 92037

ABSTRACT

Inferences about the formation of primordial matter in our solar system rest on analysis of the earliest preserved materials in meteorites, of the structure of the solar system today, and of matter in evolving stellar systems elsewhere.

The isotope distribution in meteorites suggests that molecular excitation processes similar to those observed today in circumstellar regions and dark interstellar clouds were operating in the early solar nebula. Laboratory model experiments together with these observations give evidence on the thermal state of the source medium from which refractory meteoritic dust formed. They indicate that resonance excitation of the broad isotopic bands of molecules such as $^{12}C^{16}O$, MgO, O_2, AlO and OH by strong UV line sources such as H Lyα, Mg II, Hβ and Ca II may induce selective reactions resulting in the anomalous isotopic composition of oxygen and possibly other elements in refractory oxide condensates in meteorites.

The temperature of the grains condensing from this medium can be determined from the interdiffusion of elements between metal grains in contact with each other; the results of such analyses illustrate the large temperature differential between condensing dust and the surrounding source plasma. The metal diffusion couples mostly consist of platinum or platinum metal alloys in contact with nickel iron, encased in refractory oxide grains. These consist of minerals such as magnesium aluminate (spinel) and calcium aluminum silicates (melilite and pyroxene). The metal interdiffusion shows that they have formed at temperatures $\leq$ 1000 K; this is less than or about one half of the temperature surmised from consideration of thermodynamic rather than thermal radiation equilibrium.

W. Fricke and G. Teleki (eds.), Sun and Planetary System, 221–232.
Copyright © 1982 by D. Reidel Publishing Company.

NATURE OF THE SPACE MEDIUM AND ORIGIN OF THE SOLAR SYSTEM

The space medium, wherever it has been studied experimentally in detail, is found to have a highly complex filamentary structure with large changes over short distances in number density, chemical composition and thermal and electromagnetic parameters. This inhomogeneity is due to the pervasive magnetic fields and electric currents characteristics of matter in motion (see e.g. Alfven 1980). The resulting photochemical excitation typically leads to large differences between rotational, vibrational, translational, neutral, ion and electron temperatures. High internal molecular temperatures coupled with low kinetic temperatures give rise to extensive isotope effects maintained in steady state chemical reaction cycles or by isotopically selective resonance excitation.

There is no reason to believe that the situation was qualitatively different in the past, and particularly not during the formation of our solar system. If we wish to reconstruct those formative conditions it would thus seem necessary to rely on the processes known to control the circumstellar medium today, and to identify preserved features in the present day solar system which bear witness of the primordial conditions. An attempt to bring available information together in this way has been made by Alfven and Arrhenius (1976). One of the few potential sources of information on the chemical state of the medium from which solids formed and on the properties of primordial condensates is provided by meteorites, particularly of the carbonaceous type which show the least effect of subsequent alteration.

The way in which gas and plasma temperatures couple to solid particles determines the thermal state and growth conditions in condensing grains. We wish to reconstruct these conditions for meteoritic grains using isotopic anomalies as a means for understanding the state of the source medium from which they formed, and metal diffusion as a built in solid state thermometer for evaluating the temperature of the grains themselves.

A fundamental requirement for condensation and growth to proceed is that energy is removed from the system, primarily by radiative cooling of the grains since solids are effective IR emitters. Radiative cooling can only take place if the medium surrounding the condensate is optically thin. The rate of condensation in contracting dense regions of protostellar clouds should consequently be inversely related to the optical depth of the locale. This has important consequences for the temperature differential between the condensing grains and the source medium; the solid particles characteristically maintain steady state temperatures an order of magnitude below the kinetic temperature of the medium. Accordingly, in the $\approx 7000°$ cool plasma in nova shells, dust is observed to emit at $\approx 1000°$ K (Geisel et al. 1970; Clayton and Wickramasinghe 1976); ice and dust particles embedded in the Jovian magnetospheric plasma at several million degrees assume temperatures below

200 K (cf., Fig. 1). The radiation balance determining the temperature differential between grains and source medium at condensation has been calculated (Lehnert 1970, Arrhenius, 1971, De and Arrhenius 1979).

Considering this temperature differential it is impossible to determine "condensation temperatures" even within an order of magnitude from assumed thermodynamic equilibria, and without distinguishing between the effects of isobaric cooling and isothermal compression. The only way to reconstruction of actual grain temperatures would seem to be by measurement of such phenomena in the grains which are sensitive to temperature or thermal exposure , i.e., the length of time which they have spent at specific temperatures. An opportunity for such measurements is offered by the presence in carbonaceous meteorites of microscopic spherules of platinum metals, some of them consisting of practically pure platinum, overgrown with or in intimate contact with nickel iron. Since these metals are soluble and diffuse freely in each other at a rate determined by temperature and composition, determinations of metallic interdiffusion in these couples give upper limits for the temperature and for the duration of thermal exposure. The metal

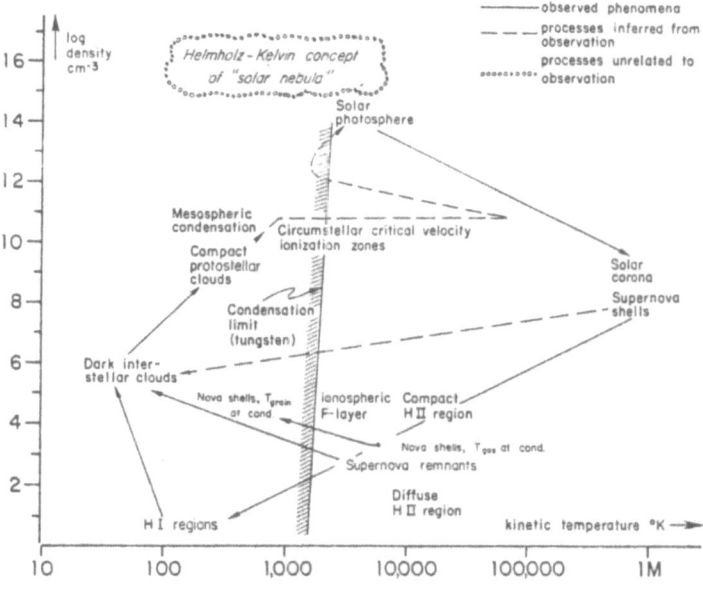

Figure 1. The astrophysical P-T setting for chemical reactions, isotope fractionation and condensation in interstellar and protostellar clouds with embedded or surrounding UV line radiation sources (from Arrhenius and Raub, 1978).

couples are in turn encased inside crystals of refractory silicate and oxide minerals which are the carriers of selective oxygen isotope anomalies (Clayton et al 1973, 1976). These host crystals, important as indicators of the state of excitation of the source medium can therefore not have received any thermal exposure more extensive than that determined for the metal couples.

The picture that emerges from these studies is an initial condensation of the most refractory oxides and silicates at a grain temperature which could hardly have exceeded about 1000° K and which could have been much lower. The excitation temperature (in contrast to the kinetic temperature) of the source medium, would have been sufficiently high to achieve the observed isotope fractionations in nitrogen, carbon and hydrogen (not necessarily in the same region and at the same time), and at least the mass dependent effects in oxygen and magnesium in carbonaceous meteorites.

NEW DATA FROM TERNARY SYSTEMS

If individual diffusion profiles are established for each element of a binary alloy such as Fe Ni into a metal such as Pt, the difference in their diffusion rates conveys additional information. This makes it possible to determine not only the thermal exposure, i.e. the continued effects of time and temperature, but also to define the maximum effective temperature to which the system has been exposed.

The temperature dependence of the intermetallic diffusion rate for a specific concentration, represented by the interdiffusion coefficient D_c, is typical for chemical reaction rates in that it follows the expression

$$D_c = (D_o)_c \exp - (Q_c/RT)$$

The pre-exponential factor $(D_o)_c$ and the activation energy Q_c, which is related to the bond enthalapy, can be determined by graphing $\ln D_c$ vs $1/T$.

Diffusion in an n-component system is described by an extension of Fick's first and second laws. A composition dependent interdiffusion coefficient D can be determined using a solution, developed by Matano (1933) and extended to n component systems (Kirkaldy and Lane 1966)

$$\int_{C_{i_o}}^{C_i} x dC_i = -2t \sum_{j=1}^{n-1} D_{ij}^n \ dC_j/dx$$

where dC_i/dx and dC_j/dx denote the concentration gradients of components i and j at a specific composition, n is the chosen dependent component, x the distance in the direction of diffusion and t is the diffusion time.

For the ternary system of Pt-Ni-Fe this reduces to

$$\int_{C_{Ni_0}}^{C_{Ni}} x dC = -2t \left[D_{NiNi}^{Pt}/dC_{Ni}/dx + D_{NiFe}^{Pt} dC_{Fe}/dx \right]$$

$$\int_{C_{Fe_0}}^{C_{Fe}} x dC = -2t \left[D_{FeNi}^{Pt} dC_{Ni}/dx + D_{FeFe}^{Pt} dC_{Fe}/dx \right]$$

Assuming that these ternary diffusion coefficients also exhibit normal reaction rate behavior, then for a given diffusion path evaluation for specific concentrations requires the simultaneous solution of four equations. Since in virtually every system the species are diffusing at different rates it is expected that the solution would yield a unique temperature. This result is of primary importance here.

EXPERIMENTAL AND ANALYTICAL PROCEDURE

The diffusion couple used to model the simple Pt-Ni Fe grains in meteorites consisted of an arc melted Ni_2Fe alloy and pure platinum. They and then sectioned. The diffusion experiments were carried out at temperatures ranging from 1173° to 1573°K, also in argon. The couples were then mounted in epoxy, sectioned and polished in preparation for quantitative analysis.

Eleven samples were annealed; at 1173°K (68.5 and 183 hours), 1223°K (74.5 and 121 hours), 1273° (39.5 hours), 1373° and 1473°K (3.99 and 5.53 hours). The concentration measurements were done on a Cameca electron microprobe, and were carried out in two stages. Each couple was scanned along several traverses while recording the digital output for each of the elements. All results were found to be reproducible within ± 1%. Three of the concentration profiles are presented in Fig. 2, showing the progress of diffusion at three different temperatures and after different lengths of diffusion time.

Two interesting features can be seen. One is the beginning of ordering at 1223 K, manifested by acceleration of the diffusion in the composition region around $Pt_{0.5}Fe_{0.25}Ni_{0.25}$ (Pt[FeNi]).

The other is the rapid increase in the diffusion rate of iron relative to that of nickel, beginning between 1223 and 1373°K. This phenomenon provides a temperature indicator independent of diffusion time. The degree of interdiffusion of nickel in platinum in graph c), for example, can be achieved during 16.2 hours at 1373°K as shown, but also at an infinite number of other combinations of longer times with lower temperatures and vice versa (at higher temperature the times rapidly become unrealistically short). A distinguishing feature is, however, that Fe is seen in this temperature range to diffuse faster than Ni into the Pt, thus fixing the temperature at about 1280°K. On the basis of this and the total diffusion, the time of diffusion can be determined to be around 1 day or less.

In none of the Pt-NiFe diffusion couples observed in carbonaceous meteorites can even incipient diffusion be observed on the scale resolvable with electron microprobe techniques, i.e. distances of the order of 3000Å and concentration differences of the order of 5-10 at percent in the configuration and compositions in question. This confirms the conclusion that these metal couples and their host silicate and oxide minerals have during their joint life together, not been

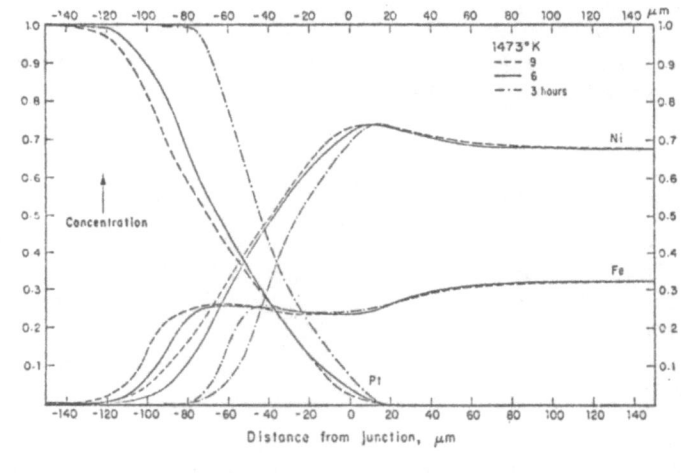

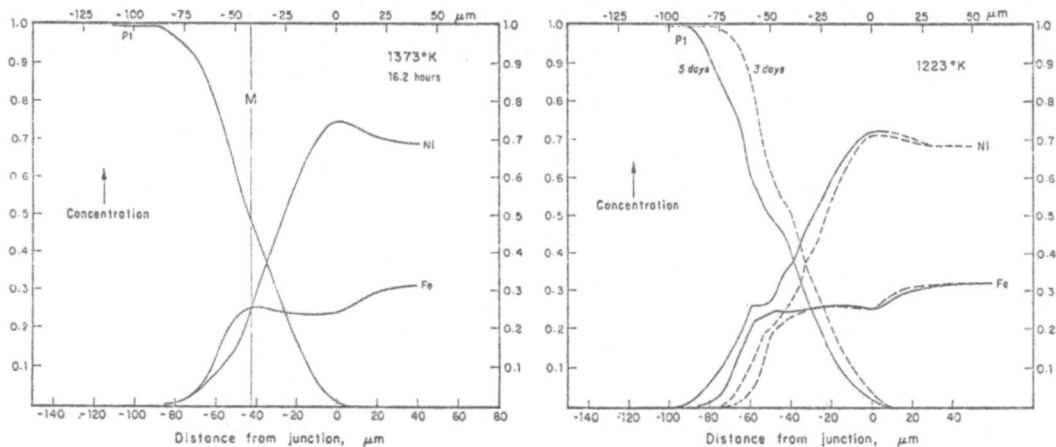

Figure 2. Diffusion profiles for Fe and Ni in Pt, and Pt in Ni$_2$Fe at three different temperatures (separate graphs) and after different lengths of diffusion time (indicated in graphs).

exposed to temperatures exceeding about 1,000 K for more than a few minutes. Such a time period appears too short to permit the formation by condensation of such crystals or precursor glass from the circumsolar medium. Actual growth may of course have taken place and probably did proceed at much lower grain temperatures. This does not exclude kinetic and excitation temperatures of the surrounding medium several orders of magnitude higher than that of embedded grains in thermal steady state (Arrhenius 1978; De and Arrhenius 1979). This points to the importance of finding indicators also of the various thermal parameters of the source medium.

STATE OF EXCITATION OF THE SOURCE MEDIUM

Some of the most productive clues to the state of excitation of the medium from which the solids formed in the early history of the solar system come from chemical isotope effects that can be determined with great precision on meteorite materials. Corresponding present day fractionation effects are observed at even larger scale by microwave spectrometry in dark interstellar clouds (see e.g. Winnewisser et al. 1979).

The proposed relationship between the effects seen in the space medium and those observed in meteorites is most applicable to the widely varying isotopic composition of hydrogen, carbon, nitrogen, oxygen and magnesium and possibly some of the effects observed in occluded noble gases, and in odd isotopes of heavy elements such as barium in carbonaceous meteorites. Polyisotopic elements such as oxygen, magnesium and sulfur are of particular interest in this respect since single isotope enhancement as observed in the laboratory (e.g., Liuti et al. 1966; Vikis 1978; Turro and Kraeutler 1978) and predicted to be effective in the space medium (Haberkorn et al. 1976; Arrhenius et al. 1979b) can generally be verified only if at least two other isotopes exist as reference points. Such enhancement of single isotopes or isotope pairs has been shown or predicted to be caused by various mechanisms of selective excitation and/or transition into reaction complexes or predissociating states (see review in Arrhenius et al. 1979b). Selective resonance excitation by spectral line sources has been observed astronomically for several molecular isotopic species in the space medium (Gahm et al. 1977). Chemical effects of this kind if found to be the likely cause also of nonlinear isotopic anomalies in meteorites, would give useful indications about the physical state of the source medium from which primordial solids formed in our solar system.

An alternative explanation of effects found in meteorites is that not only small anomalies in heavy elements, but also large effects in e.g., ^2H, ^{16}O, ^{15}N and ^{26}Mg, could be of nucleosynthetic rather than chemical origin. Observational proof one way or another is generally not possible as far as meteorites are concerned, since the effects recorded in them are caused by events in the past, and since confirming nucleosynthetic experiments can in most cases not be carried out. In contrast, observations of currently active processes in the interstellar cloud medium provide direct evidence of large scale chemical isotope fractionation in carbon, hydrogen and nitrogen, and with more refined techniques the potential exists for discerning the details of selective fractionation of individual isotopes of oxygen and magnesium (Winnewisser 1979). Selective enhancement of ^{16}O relative to $^{17,18}O$ in CO has been observed in some regions of the interstellar medium and is considered as a puzzling feature if interpreted in terms of nucleosynthetic origin (Penzias 1980).

SINGLE ISOTOPE ENHANCEMENT BY RESONANCE EXCITATION

Several types of chemical fractionation processes have been considered as potentially relevant to the excess of ^{16}O (Arrhenius et al. 1979). One is resonance excitation in the rotational fine structure of a molecule with well separated isotopic vibrational bands. The excitation, to be important, should be caused by a prominent line in interstellar space, the excited molecule must be reasonably abundant and undergo reaction, and the density, unless ionization takes place, must be high enough for collision to occur within the time of radiative decay. Regions of high density behind shock fronts and in stellar atmospheres may be the loci of such conditions.

Selective excitation of ^{18}O and ^{13}C by excitation with a VUV line is well known from laboratory experiments (Liuti et al. 1966; 1969). In space the coincidence of the intense H L line at 1216 Å with the 14th vibrational band of the A-X transition in $^{12}C^{16}O$ has been proposed as a potential cause of fractionation (Arrhenius et al. 1980; Corrigan et al. 1980); other such isotopic concidences may also occur between strong stellar emission lines and abundant interstellar molecules. In the laboratory the reaction of CO* with CO forming CO_2 and C_3O_2 is particularly convenient to study as a model system; in the space environment collisions of the excited molecule with hydrogen species will be by far the most important reactions. The close line matching needed for resonance excitation might intuitively seem to make this an exceptional phenomenon achievable only by tunable sources or by combining emitters and absorbers selected among a large number of species, most of them unlikely to be of importance in nature. What makes the phenomenon common in astrophysical settings, however, (Gahm et al. 1977) is the band structure of the spectra of abundantly occuring molecules such as CO or O_2; with each electronic-vibrational level is associated a large number of rotational levels which may be populated under excitation conditions characteristic of the space medium. This together with the occurrence of several strong emission lines in stellar UV spectra makes coincidences common enough for isotopic selection to occur in a number of important molecular reactions in the space

medium. The coincidence between a $^{12}C^{16}O$ absorption band and the
hydrogen Lyman α emission line is a case in point (Figure 3).

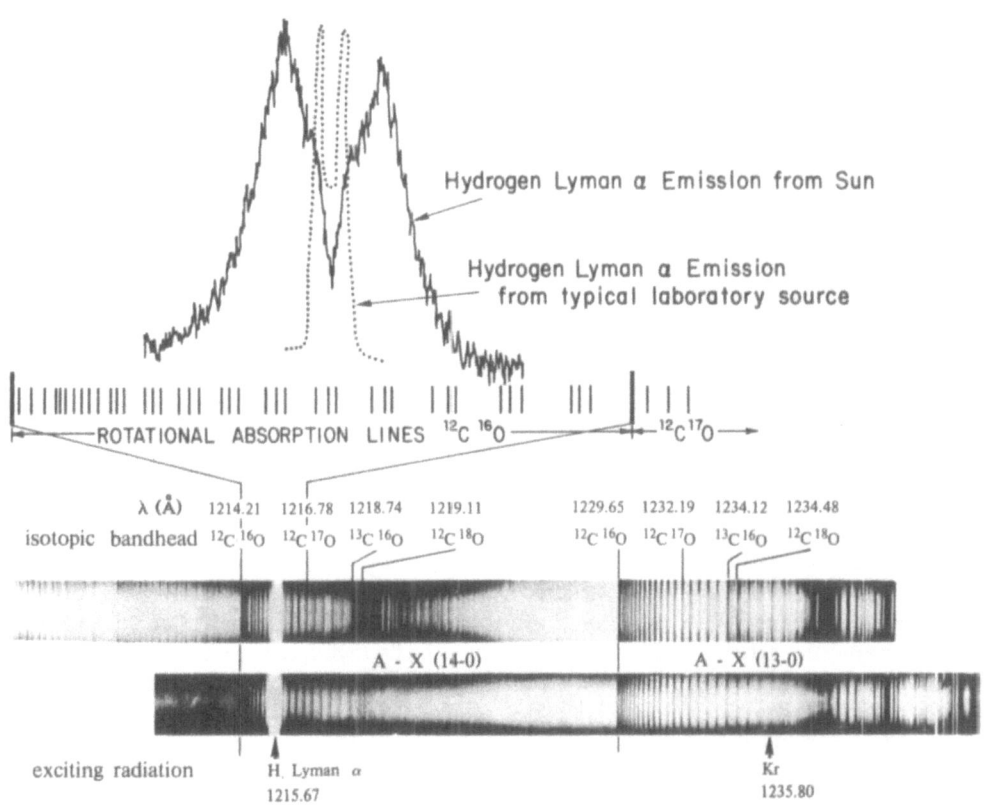

Fig. 3 (Below) Overlap of laboratory Lyman α emission line with rotational lines of CO in the (14-0) vibrational band of the Z-X electronic
transition (from Tilford and Simmons 1972). The figure illustrates
that emission overlaps absorption lines of $^{12}C^{16}O$-isotopic species
only, leading to selective excitation of this species.

(Above) Expansion of scale of $^{12}C^{16}O$ portion of (14-0) band with
schematic rotational absorption lines and with Lyman α emission profiles from laboratory sources and from the Sun. Emission profile width
is to scale relative to location of $^{12}C^{16}O$- and $^{12}C^{17}O$- bandheads.
Absorption lines of other isotopic CO species are off the figure to the
red (adapted from Purcell and Tousey, 1960).

The broadening of the stellar emission profile (with overlap over
about 40 rotational $^{12}C^{16}O$ lines) leads to an order of magnitude
enhancement in a astrophysical settings over the effect achieved in the

laboratory where the narrow profile lamp source can be made to effectively overlap only two such rotational lines.

On the basis of the work by Gahm et al., the circumstellar region
around emission line stars of spectral classes O through G, and the T-
Tauri stars is an interesting astrophysical setting for explaining oxygen isotope effects in meteorites. The Sun is now a G2v star and in its
early evolution it is considered possibly to have passed through a T-
Tauri stage. Other situations where dense protostellar clouds are
irradiated by strong line sources have also to be investigated.

A quantitative treatment is being given elsewhere of the applicability to astrophysical situations of isotope fractionation by the process discussed above. Laboratory experiments, guided by such theoretical considerations provide insights into the reaction mechanisms following the selective isotopic excitation of molecules in space. We
have used the system (C-O)* for such model studies because of our earlier findings of large isotope effects (Arrhenius et al. 1979), the
importance of the system in protostellar cloud chemistry, experimental
convenience and laboratory demonstration of large selective fractionation of ^{18}O (sevenfold) in product CO_2 and C_3O_2 by resonance VUV excitation (Liuti et al. 1966; Vikis 1978). Concurrent fractionation of
^{13}C was also monitored in our investigations.

Selective excitation induced ^{16}O fractionation reactions in the
laboratory are inefficient compared to those in the space medium due to
the low HLyα photon flux and signal to noise ratio in available emission sources--we use a radio frequency excited helium-hydrogen lamp,
separated by a thin MgF_2 window from the liquid nitrogen cooled reaction chamber (Fig. 4).

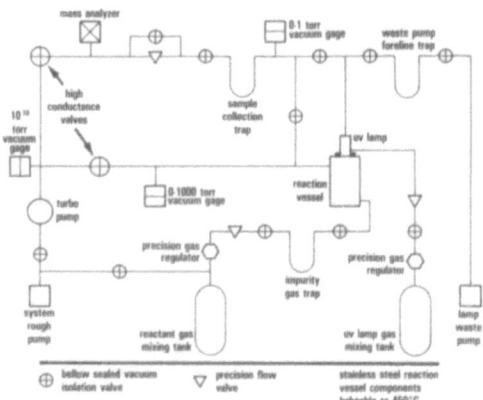

Figure 4. Experimental system for single isotope fractionation experiments by selective isotopic molecular excitation by atomic line emission in the vacuum ultraviolet.

Due to these limitations and the mass spectrometric need for preparing pure oxygen from the produced CO_2 the effects observed so far (Clayton 1981) are small ($\leq$ 1 per mil negative deviation from linear δ values) but believed to be real since they are 2 to 4 times the deviations observed in terrestrial samples (Antarctic snow) by the same author. In further experimental model studies we are aiming at amplifiying the effects by improving the spectral and intensity of the excitation source and investigating corresponding excitation effects in the Schumann-Runge bands of oxygen, where the product O_3 can be used directly for mass spectrometric analysis.

REFERENCES

Alfven, H.: 1980, Cosmic plasma, Reidel, Dordrecht.

Alfven, H. and Arrhenius, G.: 1976, NASA Spec. Publ. SP-345, U. S. Govt. Print. Off.

Arrhenius, G.: 1972, Proc. Nobel. Symp., 21, p. 117.

Arrhenius, G.: 1978, in S. F. Dermott, ed. Origin of the Solar System, Wiley, N. Y.

Arrhenius, G., Corrigan, M. and Fitzgerald, R.: 1980, Lunar Planet. Sci., 11, p. 34.

Arrhenius,G., McCrumb.J. and Friedman,N.: 1979,Ap. Space Sci.,60, p.59.

Arrhenius, G. and Raub, C.: 1978, J. Less Common Metals, 62, p. 417.

Clayton, R, Grossman, L. and Mayeda, T.: 1973, Science, 182, 485.

Clayton, D. and Wickramasinghe, N.: 1976, Ap. Space Sci., 42 p. 463.

Corrigan, M., Fitzgerald, R., Mendis, D. and Arrhenius, G.: 1980, Meteoritics 15, p. 4.

De,B. and Arrhenius,G.: 1979,Adv. Colloid and Interface Sci.,10, p.253.

Gahm, G.: 1977, Astron. Ap. Suppl., 27, p. 277.

Geisel, S., Kleinmann, D. and Low, F.: 1970, Ap. Space Sci. p. L101.

Haberkorn, R., Michel-Beyerle, M. and Michel, K.: 1977, Astron. Ap. 55, p. 315.

Kirkaldy, J. and Lane, J.: 1966, Can. J. Phys., 44, p. 2059.

Lehnert, B.: 1970, Cosmical Electrodynamics, 1, p. 219.

Liuti, G., Dondes, D. and Harteck, P.: 1966, J. Chem. Phys., 44, p. 4052,: 1969, Adv. Chem. Ser. 89, p. 65.

Matano, C.: 1933, Japan J. Phys., 8, p. 109.

Penzias, A.: 1980, Science, 208, p. 663.

Purcell, J. D. and Tousey, R.: 1960, JGR, 65, p. 370.

Tilford, S. G. and Simmons, J. D.: 1972, J. Chem. Phys. Ref. Data. 1, p. 147.

Turro, N. and Kraeutler, B.: 1978 JACS 100, p. 7432.

Vikis, A.: 1978, J. Chem. Phys., 69, p. 697.

Winnewisser, G., Churchill, Co and Walmsley, C.: 1979, Astrophysics of interstellar molecules. In: G. Chantry, ed., Modern Aspects of Microwave Spectroscopy, Acad. Press, N. Y.

PHYSICAL PROCESSES OF RELEVANCE TO THE FORMATION OF THE PLANETARY SYSTEM

H.J. Völk
Max-Planck-Institut für Kernphysik
Postfach 10 39 80
6900 Heidelberg
W. GERMANY

ABSTRACT

The processes mainly reviewed are grain evaporation and condensation, grain-grain coagulation, grain diffusion, sedimentation and
radial drifts, as well as self-gravitational fragmentation of dust
disks. All this is done within the framework of a turbulent protostellar
accretion disk. It is shown that such a model can in principle be consistent with a number of cosmochemical observations. However, steady
turbulent disks probably do not allow effective grain sedimentation due
to losses by radial drifts. An intermittent turbulence is suggested to
allow the formation of large seed bodies onto which the disk material
can accrete to form planetesimals.

1. INTRODUCTION

The purpose of this paper is to discuss the role and the interre-
lation of different physical processes during early planetary formation.
This can best be done in the framework of a model and in the next chapters we shall concentrate on turbulent protostellar disks. An objection
against such an approach could be that then the definition of processes
like for example condensation or turbulent coagulation as being relevant for planetary formation, appears to be connected with the details
of the model. Although this is true to a certain extent there are, of
course, the astronomical and cosmochemical observations which any model
has to account for. Either they can only be understood in terms of very
specific processes or, at least, they favour certain processes at the
expense of others. Therefore we shall review in the next section some
of the observations together with theoretical considerations. Having
identified the relevant processes within this framework we proceed to
a discussion of turbulent protostellar disks and of grains and trace
gases in them in the next sections.

2. OBSERVATIONS AND THEORETICAL CONSIDERATIONS

Let us start with some key cosmochemical results. Measurements in

233

W. Fricke and G. Teleki (eds.), Sun and Planetary System, 233–242.

primitive meteorites have shown chemical fractionations in different forms which are essentially correlated with variations in equilibrium volatility rather than other chemical properties. These abundance patterns have been interpreted as the result of a thermal condensation sequence from a hot gaseous "solar nebula" (Urey, 1954; Larimer, 1967; Larimer and Anders 1967; see e.g. Grossmann and Larimer, 1974, for a recent review). Therefore c o n d e n s a t i o n and - considering that stars form in dense clouds - grain e v a p o r a t i o n ought to be relevant processes.

However, the resulting effects of mixing must be limited. More recently variations in isotopic composition have been found both in bulk (differentiating various meteorite classes from terrestrial and lunar material and from each other) as well as locally between different sub-components (on scales of the order of millimeters) within carbonaceous chondrites. This isotopic diversity, most apparent in oxygen (Clayton et al., 1973; see Begemann, 1980, for a review), is not of a physico-chemical nature alone. It therefore implies true isotopic heterogeneity not only on a planetary scale but also between neighbouring chunks of material in meteorites. Such a heterogeneity can hardly be preserved in a purely gaseous phase, but we know that much of the heavier elements should have entered the protostellar nebula in the form of submicron-sized grains. If one assumes large scale isotopic heterogeneity in this protostellar material and assumes complete evaporation of solids, then recondensation must have occurred in locally separated regions to explain the planetary scale variations. If the macroscopic isotopically anomalous inclusions are due to anomalous grains, then c o a g u l a - t i o n must be an important process and able to produce mm-sized objects locally within the protostellar nebula. Of course, this is not the only argument for coagulation (see below), but it appears indicated to emphasize this process also in the context of isotopic anomalies.

Astronomically, one of the most important facts is that the solar system is essentially flat. This can hardly be considered as accidental but points to a rather isolated, joint formation of sun and planets, without close interference from other stars, unless one explicitly dissociates the formation of the sun from its (later) acquisition of a protoplanetary gas and dust cloud which then settles into a disk configuration. A model of the latter kind is that of Alfvén (1954), expounded in Alfvén and Arrhenius (1976) and will be presented at this conference. Its central physical process is the critical velocity ionization of spherically symmetric accreting collisionfree neutral gases, a plasma physical mechanism that has been widely confirmed in laboratory experiments, as well as recently in an active space experiment by Haerendel et al. (see Petelski, 1981). Yet the question is how convincing the observational evidence for its operation in the context of planetary formation is. In addition some of the accreting material should have finite angular momentum and show a hydrodynamic behaviour. It should have embedded grains that contain most of the matter later to be found in the planets. To grains, however, the critical velocity effect does not

address itself, at least not directly. Leaving therefore aside this
otherwise intriguing effect and, for the time being, also capture theories in general, it is then most plausible that the sun formed as a
single star together with the planets.

Observations also indicate that star formation occurs in high density, cool, neutral clouds. Numerical collapse calculations modelling
this process, rather generally suggest that even a small amount of initial angular momentum prevents star formation without some non-molecu-
lar angular momentum transfer (e.g. Tscharnuter, 1980; Regev and Shaviv,
1981). At the high densities and low temperatures of gravitational
collapse in interstellar clouds, the magnetic field should not be dynamically important, leaving t u r b u l e n c e in the gas a s the
main p r o c e s s o f a n g u l a r m o m e n t u m t r a n s f e r.
Although this is a mechanism invoked ad hoc, it is quite likely to exist
in protostellar disks(v. Weizäcker, 1944, 1948; Lynden-Bell and Pringle
1974; Cameron, 1973). Mainly radiative energy transport should be responsible for the temperature structure of the resulting disk, whereas
radiation pressure effects (e.g. Yorke and Krügel, 1977) should be negligible for solar mass protostars as considered here.

The extensive craterisation of all planetary bodies, observable at
least on all those planets without deep gaseous atmospheres, together
with the present existence of many small solid bodies (the asteroids)
points towards the direct formation of large solid bodies from small
solid bodies. Although the large gas planets might have formed by gravitational fragmentation of a gaseous disk, this formation process is
quite unlikely for their moons, the terrestrial planets and the asteroids whose masses are so much smaller (e.g. Völk, 1981). And indeed,the
giant planets might only be the result of g a s a c c r e t i o n
onto a large solid core.

The direct formation of large solid bodies probably occurs by means
of sedimentation of grains through the gas towards the plane of gravitational symmetry of the disk unless the particles are lost by r a -
d i a l d r i f t s due to gas drag. Subsequent s e l f - g r a -
v i t a t i o n a l f r a g m e n t a t i o n of the resulting dust
disk should lead to planetesimals with sizes in the km-range (Safronov,
1969; Goldreich and Ward, 1973). Finally, due to mutual gravitational
perturbations these planetesimals could collide and remain gravitationally bound in order to a c c u m u l a t e in a late phase into solid planets or satellites on time scales of the order of 10^8 yrs (e.g.
Safronov 1969, 1980; Wetherill, 1978, 1980).

The last process is relatively well understood and has been extensively reviewed in the cited references. In addition, it can be considered rather independently of the previous history of the system. However, this does not hold for the earlier stages of solar system evolution. In the next chapters we shall therefore be concerned with the
question of planetary formation from turbulent protostellar disks and
the physical processes associated with it.

3. TURBULENT PROTOSTELLAR ACCRETION DISKS.

Collapse calculations of protostars rotating approximately at the galactic rate show the appearance of a disk-like structure of scale 100 AU after about one initial free fall time. With turbulent viscosity this optically thick disk evolves, slowly forming a central object, whereas the outer free falling envelope feeds the disk continuously. Close to the central object the temperature is large enough to evaporate all grains (Tscharnuter, private communication). We idealize this configuration schematically in Figure 1: In the center of the proto-

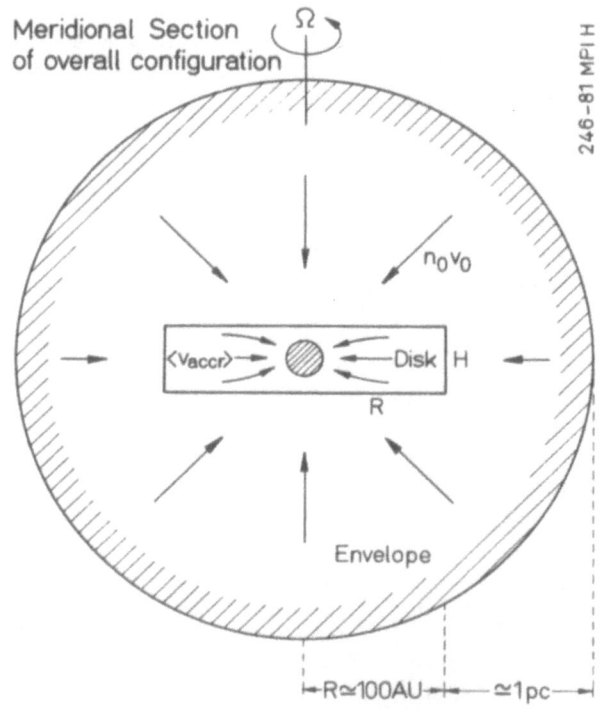

Figure 1:
Schematic picture of protostellar accretion. The particle flux density from the envelope is $n_0 v_0$, its initial size is of the order of 1 pc and its initial rotation rate is Ω. The radial accretion speed is $\langle v_{acc} \rangle$ in the disk.

stellar cloud there exists an optically thick disk of radius $R = O(10^2)$ AU and constant height $H = O(10^{-1} R)$, with turbulent mass transport towards the interior. The free falling envelope feeds this disk at a constant rate until it is exhausted. The density and temperature structure of the disk vary only quasistatically.

For a fixed mass input rate $\dot{M}$ from the envelope over a time τ_e, the time structure of the disk depends, in particular, on the size of the turbulent viscosity $\nu = (\alpha/3) \cdot v_{th} \cdot H$, where $\alpha < 1$ is a strength parameter. For "large" α, a quasisteady state is reached quickly, i.e. for times $t \ll \tau_e$, with a rather low, constant disk mass M_{disk}, the accretion flow linearly increasing the mass $M_*(t)$ of the central object. For

the special case of mass ·input only in the equatorial plane, the steady
state disk surface density Σ is given by

$$\Sigma \underset{\sim}{\sim} \dot{M} \cdot (G \cdot M_*)^{1/2} \cdot r^{-3/2} \cdot \{1 - (R_*/r)^{1/2}\} \cdot (2\pi\alpha v_{th}^2)^{-1} \sim \dot{M}/\alpha \tag{1}$$

cf. Shakura and Sunyaev (1972). Here R_* is the radius of the central
object, and r the equatorial radial distance within the disk. Decreasing α builds up Σ which,however,reaches a steady state only at later
times. A further decrease of the viscosity may not allow a steady state
disk for $t < \tau_e$ at all. Then the mean surface density $\langle\Sigma\rangle$ is given by:

$$\langle\Sigma\rangle = \langle\Sigma_0\rangle + \dot{M} \cdot t \cdot (2\pi R^2)^{-1} \underset{\sim}{\sim} \dot{M} \cdot t \cdot (2\pi R^2)^{-1} \sim t \quad \text{for } t < \tau_e . \tag{2}$$

The formation of macroscopic solid bodies within such an accretion disk will presumably depend considerably on the type of disk which
establishes itself.

4. GRAINS AND TRACE GASES (VAPOURS) IN TURBULENT PROTOSTELLAR DISKS.

The mode of transport of small grains and their vapours relative to the
gas in the disk is turbulent diffusion. It can transport information
about the hot interior where grains evaporate to the cold outer regions
where vapours recondense again (Morfill, 1981). Larger grains have in
addition a systematic radial drift inwards due to friction with the gas
which, in contrast to the grains, is partly pressure supported (Whipple,
1972). Turbulent diffusion acts against sedimentation and self-gravitational fragmentation of the dust component,whereas random velocity
differences due to turbulence as well as systematic radial drift velocity differences due to different particle sizes induce collisions and coagulation of grains.

To illustrate the configuration in the disk equatorial plane with
the temperature decreasing from the hot center down to some ten degrees
in the outer parts,we schematically consider only two types of grain material: a volatile and a refractory component (Figure 2). Then inside
r_v the volatile grain material can only exist as vapour, the remainder
of the grain existing as refractories only, whereas for $r < r_R$ even refractory grains are evaporated. The systematic drift v_d of large grains
tends to enhance the trace component concentrations towards the center.

Let us first consider the vertical structure of the grain component in such a turbulent disk, i.e. the question of sedimentation. After
a short initial phase the vertical scale of the dust component (cf mass
density ρ_d) reaches a quasisteady state which is roughly determined
through

$$\partial/\partial z \{\langle w_d\rangle \rho_d - \kappa \cdot \partial\rho_d/\partial z\} \underset{\sim}{\sim} 0 \tag{3}$$

where $\langle w_d\rangle = -4\pi \cdot \tau_f \cdot g_z$ is the vertical drift speed (in Z-direction)

Equatorial Section
for trace constituents
in disk

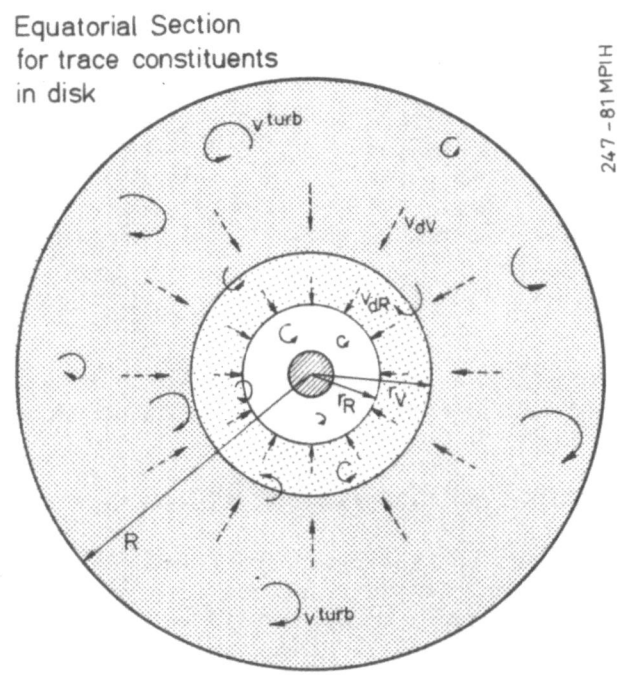

Figure 2: The proto-
stellar accretion disk
view along the rotation
axis. Turbulent gas
motions v^{turb} diffuse
grains and their vapours
across the disk. Within
r_V and r_R, volatile,
and even refractory
grain materials only
exist in vapour form,
respectively. Large
grains have radial drift
velocities v_{dV} and v_{dR}
in the respective re-
gions.

due to the gravitational acceleration g_z, and $\tau_f = (\pi/2)^{1/2} \cdot m_d \cdot (\rho_g$
$\pi \cdot a^2 \cdot v_{th})^{-1} \sim a$ is the grain-gas friction time; the quantities m_d, a,
and ρ_g denote the grain mass, its radius, and the gas mass density, respectively. The diffusion coefficient κ is defined as:

$$\kappa = (1/3) \cdot \ell_t \cdot <\delta v^2> \cdot <\delta v_g^2>^{-1/2} = (\alpha/3) \cdot v_{th} \cdot H \cdot <\delta v^2> \cdot <\delta v_g^2>^{-1} \qquad (4)$$

where the strength parameter α, used before, is given by

$$\alpha = <\delta v_g^2>^{1/2} \cdot \ell_t / (v_{th} \cdot H) < 1 \qquad (5)$$

, ℓ_t being the largest turbulent scale, $<\delta v^2>$ and $<\delta v_g^2>$ denoting the
mean square gas and grain velocities, respectively. For $\tau_f << \ell_t <\delta v_g^2>^{-1/2}$
, i.e. "small" grains, we have $<\delta v^2> = <\delta v_g^2>$ and κ equals the eddy
diffusion coefficient for a trace gas. For the reverse case ("large"
grains)

$$<\delta v^2>/<\delta v_g^2> \sim\!\!\!\!- <\delta v_g^2>^{-1/2} \cdot \ell_t / \tau_f \sim a^{-1} \qquad (6)$$

, and κ becomes very small since the large grains decouple from the
turbulence (e.g. Völk et al., 1980).

The turbulent scale height Z_d of the grains is then $O(\kappa/<w_d>) \sim$

$(\alpha/a) \cdot <\delta v^2>/<\delta vg^2>$ as long as this is smaller than H; otherwise $\bar{z}_d \sim H$. For large grains $Z_d \sim (\alpha/a^3)^{1/2}$.

Having discussed sedimentation we shall assume for the moment that it is very effective and consider then the further evolution of the dust component.

A thin dust disk may initially contract $\sim \exp\{-wt\}$ by selfgravity, on scales $2\pi/k \ll R$, if, according to Goldreich and Ward (1973):

$$w^2 = 2\pi G \cdot k \cdot \Sigma_d - k^2 \cdot <\delta v^2> - 2\Omega[\Omega + d(r\Omega)/dr] > 0. \qquad (7)$$

In the first term on the r.h.s. Σ_d denotes the surface density of the dust component. The second term is stabilizing small scales through turbulent grain pressure $\sim <\delta v^2>$. In the last term, stabilizing large scales by velocity shear, we may neglect the relative azimuthal motions of grains and gas and identify Ω with the gas rotation frequency. An approximately Keplerian disk becomes unstable for $\Sigma_d > \Omega \cdot <\delta v^2>^{1/2}/(\pi \cdot G)$, i.e. for sufficiently large grains, or weak turbulence, cf. eq. (6). Safronov (1969) and Goldreich and Ward (1973) have argued that, indeed, large solid bodies, the planetesimals of scale $2\pi/k$, should ultimately form from the instability, if it operates in the first place.

Within a dense dust disk large bodies might also form due to coagulation on account of differences in radial drift velocity caused by gas drag (Weidenschilling, 1977). This, however, still has efficient sedimentation as a prerequisite.

If we assume the collision mean free path in the gas to be large compared to the grain size, this drift is given by (cf. Morfill and Völk, 1981)

$$v_d \simeq \tau_f \cdot (\partial p_g/\partial r) \cdot \rho_g^{-1} \cdot \{1 + \tau_f^2 \cdot v_\phi^2 /r^2\}^{-1} \qquad (8)$$

, where p_g, v_ϕ, and r denote gas pressure, gas rotation velocity, and radial distance from the central object, respectively. Clearly $v_d \sim a$ for small, and $v_d \sim a^{-1}$ for large particles, the transition between these two regimes given by $v_{d,max} \simeq \tau_f \cdot (\partial p_g/\partial r)/(2\rho_g)$, at $\tau_f \cdot v_\phi = r$.

Whatever the detailed reason for planetesimal formation, the radial drift plays an important role in disks of finite radial extent. Whereas in a disk of effectively infinite radius R, particles may grow to large sizes by turbulent coagulation (Cameron, 1973; Völk, 1981) and then sediment efficiently, this growth is limited for finite R by the drift time $O(R/v_d)$ into the center. In fact, in a turbulent disk, as considered here, the grain lifetime is in addition limited by the diffusion time $t_{diff} = O(R^2/\kappa)$, a particle being lost on average after diffusing towards the center of this time scale.

Thus, in order to see whether sedimentation and planetesimal formation, or at least the local coagulation of mm-sized conglomerates (cf. section 2) is possible, we must estimate the time scales available for coagulation, as well as the resulting particle sizes. Taking $v_{th} =$

10^5 cm sec^{-1}, R = 10^2 AU, and α = 0,1 as nominal values, we obtain first of all

$$t_{diff} = 4,5 \times 10^{12} \ sec \ \{(\alpha/0,1)^{-1}\cdot(R/100)\cdot(v_{th}/10^5)^{-1}\} \tag{9}$$

Since t_{diff} should be smaller than the formation time of the central object, but not by orders of magnitudes, no complete mixing should be expected within the disk on a planetary scale. Still, recondensed vapours (on existing grains, requiring no nucleation) can diffuse out into the disk in the form of thermally and chemically modified grains. This will lead to an obvious fractionation with volatility at increasing radial distances due to the average temperature gradient.

For coagulation, the determining time scale turns out to be that of the radial drift. Using for purposes of estimate in the outer disk: T = T(R)$\cdot$(r/R)$^{-1}$, with T(R) = 80 K, we consider particle growth, starting at distance r with interstellar grains of radius a = $0(10^{-5})$cm, and ending at r_i due to drift over the distance $r-r_i$ = $r\cdot(1-\sigma)$. For turbulence induced coagulation between equal particles (Fig. 3) we obtain for the final radius $a(r,\sigma)$:

$$a(r,\sigma) = 33 \ cm(r/R)^{4/3}\cdot(1-\sigma^2)^{2/3}(M_{disk}/M_{\odot})$$
$$\cdot(R/100)^{-2}\cdot(\rho_s/1)^{-1}(\alpha/0,1)\cdot(Q/1)^{2/3} \tag{10}$$

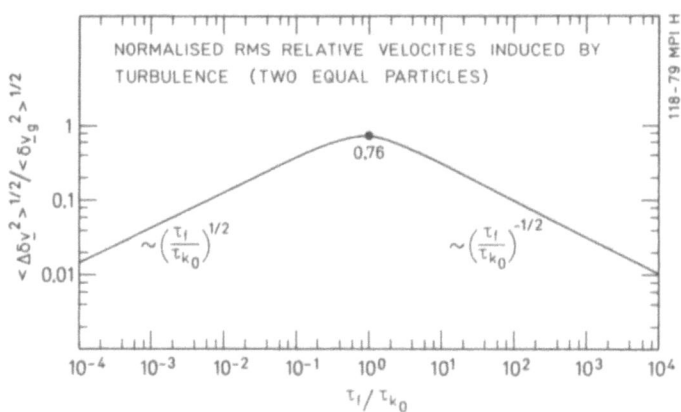

Figure 3: Relative grain velocities for equal particles as a function of dimensionless particle radius. The lifetime $\ell_t<\delta v_g^2>^{-1/2}$ of the largest eddies is denoted here by τ_{k_o}. The other quantities are defined in the text.

, where ρ_s is the grain material density in g cm^{-2} and Q is the sticking probability. For M_{disk} = $3\cdot10^{-2}M_{\odot}$, for example, we get a $\sim 10^{-1}$ cm for the nominal parameters. But, of course, not all particles grow to such a large size. In fact, as shown numerically by Röser (private communication) the value of a above corresponds to $<s^3>^{1/3}$ within a factor of 2, where $(4\pi/3) \cdot \rho_s \cdot <s^3>$ is the grain mass averaged over the resulting distribution of sizes s.

Coagulation due to sweeping up of small grains by large grains leads to

$$a(r,\sigma) = 28 \text{ cm} \cdot (r/R) \cdot (1-\sigma) \cdot (M_{disk}/M_\odot) \cdot (R/100)^{-2} \cdot (\sigma_s/1) \cdot (Q/1) \qquad (11)$$

which is a result very similar to eq. (10).

Thus, even for $\sigma = 0(1)$ sizeable local coagulation is possible to form mm-sized conglomerates. However, we still have

$$\tau_f \cdot <\delta v_g^2>^{1/2} \ell_t^{-1} \underset{\sim}{\sim} 10^{-2} \cdot (r/R)^{4/3} \cdot (1-\sigma^2)^{2/3} \cdot (\alpha/0,1)^2 \cdot (Q/1)^2 << 1 \quad (12)$$

except for $\alpha \underset{\sim}{\sim} 1$. Therefore, even if we allow $\sigma \rightarrow 0$, there is no essential sedimentation possible.

Large particles may be manufactured by recondensation of vapours near the evaporation points. But this needs high vapour densities and should only occur in high density disks ($\alpha << 1$) at the end of accretion from the envelope when few grains remain for heterogeneous nucleation (Morfill and Völk, 1981).

We may, however modify the above turbulent disk model in a speculative way: the turbulence of such disks should not be steady, but rather interm ittent, or patchy (see also Spiegel, 1973). At those spots, where turbulence is temporarily very small, or absent, rapid sedimentation of mm-sized conglomerates should be possible, followed by gravitational fragmentation. This should produce seed bodies which are only slightly perturbed by friction with the gas. They could accrete the other material floating around, turbulently mixed all the time, to form the planetesimals whose existence is considered central for the formation of at least the solid planets.

ACKNOWLEDGEMENTS

The author would like to thank Drs. G. Morfill, J. Ray, and W. Tscharnuter for a number of valuable discussions regarding various aspects of this paper.

REFERENCES

Alfvén, H., 1954, "On the Origin of the Solar System", London: Oxford Univ. Press.
Alfvén, H., and Arrhenius, G., 1976, "Evolution of the Solar System", NASA SP-345, Washington, D.C.: US Government Printing Office.
Begemann, F., 1980, Rep.Prog.Phys. 43, 1309.
Cameron, A.G.W., 1973, Icarus 18, 407.
Clayton, R.N., Grossman, L., and Mayeda, T.K., 1973, Science 182, 485.

Goldreich, P. and Ward, W.R., 1973, Astrophys.J. 183, 1051.

Grossman, L. and Larimer, J.W., 1974, Rev.Geophys. Space Phys. 12, 71.

Larimer, J.W., 1967, Geochim.Cosmochim. Acta 31, 1215.

Larimer, J.W. and Anders, E., 1967, Geochim.Cosmochim. Acta 31, 1239.

Lynden-Bell, D. and Pringle, J.E., 1974, Mon.Not.R.Astr.Soc. 168, 603.

Morfill, G., 1981, submitted to Icarus.

Morfill, G. and Völk, H.J., 1981, submitted to Mon.Not.R.Astr.Soc.

Petelski, E.F., 1981, in "Relation between Laboratory and Space Plasmas" (Ed. H. Kikuchi), p. 23, D. Reidel Publ.Comp. Dordrecht: Holland.

Regev, O. and Shaviv, G., 1981, Astrophys.J. 245, 934.

Safronov, V.S., 1969, "Evolution of the Protoplanetary Cloud and Formation of the Earth and the Planets", Moscow, Nauka Press; NASA TTF-677, 1972.

Safronov, V.S., 1980, in "Early Solar System, Processes and the Present Solar System", p. 58, L XXIII Corso, Soc. Italiana di Fisica, Bologna, Italy.

Shakura, N.I. and Sunyaev, R.A., 1972, Astron.Astrophys. 24, 337.

Spiegel, E.A., 1973, in "On the Origin of the Solar System" (Ed. H. Reeves), Editions du CNRS, Paris, p. 165.

Tscharnuter, W., 1980, IAU-Colloquium on "Stellar Hydrodynamics", Los Alamos (Ed. Stobie).

Urey, H.C., 1954, Astrophys.J., Suppl. 1, 147.

Völk, H.J., 1981, Mitt.Astron.Ges. 51, 63.

Völk, H.J., Jones, F.C., Morfill, G.E., and Röser, S., 1980, Astron. Astrophys. 85, 316.

Weidenschilling, S.J., 1977, Mon.Not.R.Astro.Soc. 180, 57.

von Weizsäcker, C.F., 1944, Z.Astrophysik 22, 319.

von Weizsäcker, C.F., 1948, Z.Naturforschung 3a, 524.

Wetherill, G., 1978, in "Protostars and Planets", (Ed. T. Gehrels), Univ. Arizona Press, p. 565.

Wetherill, G.W., 1980, Ann.Rev.Astron.Astrophys. 18, 77.

Whipple, F.L., 1972, "From Plasma to Planet" (Ed. A. Elvius), Wiley (London), p. 211.

Yorke, H.W. and Krügel, E., 1977, Astron.Astrophys. 54, 183.

MAGNETOSPHERES OF JUPITER AND SATURN

Norman F. Ness
NASA Goddard Space Flight Center
Greenbelt, Maryland 20771   USA

ABSTRACT

Jupiter and Saturn possess significant magnetic fields. The surrounding regions of space in which the forces on charged particles are due primarily to the magnetic field are their magnetospheres. These are similar, in certain respects, to that of Earth: large radiation belts, a compressed dayside field, an extended nightside magnetic tail with imbedded plasma sheet and detached bow shock wave around these obstacles to solar wind flow. However, differences with Earth are significant and relate to the large amount of entrapped low energy plasmas which form a magnetodisk and an Io associated torus at Jupiter while at Saturn there is a Titan torus and a substantial ring current. The interaction of the satellites and rings deep within these magnetospheres leads to many special phenomena tightly coupling the properties of the material forming the satellites and rings with the rapidly corotating planetary magnetospheres. This paper will review recent results obtained by the Voyager 1 and 2 spacecraft.

The two giant planets, Jupiter and Saturn, and their extensive satellite and ring systems have recently been the focus of attention of the USA NASA program in planetary exploration. Beginning in 1973 and culminating 8 years later in 1981, the spacecraft Pioneers 10, 11 and Voyagers 1 and 2 have examined in increasingly higher resolution the characteristics of these unique members of the solar system. Of special interest has been the visual imaging studies, which provided dramatic evidence of 2 mini-solar systems. This paper is concerned with the study of the magnetospheres of these planets: the immediate region of their environment in which the planetary magnetic field is the dominant physical force, and also with the boundary of these magnetospheres, which is formed by the interaction with the solar wind. Our attention will be restricted to the recent results obtained by the Voyagers. For comparison, Table I summarizes the salient features of the spacecraft trajectories, especially as they relate to magnetosphere studies.

W. Fricke and G. Teleki (eds.), Sun and Planetary System, 243–248.

TABLE I - In-situ Studies of Magnetospheres of Giant Planets

Spacecraft/ Planet	Date of Periapsis	Sun-Planet-Spacecraft Angle (Outbound at 100 R_p)	Periapsis (R_p=Planetary Radius)
Voyager I/J	5 Mar 1979	117°	4.89
Voyager II/J	9 July 1979	141°	10.1
Voyager I/S	12 Nov 1980	120°	3.09
Voyager II/S	25 Aug 1981	94°	2.67

During approach to the target planet, all spacecraft are on trajectories which are close to the planet-sun line and so penetrate the magnetospheres and detached bow shock waves near the stagnation point region of solar wind interaction. After closest approach, the departure trajectories vary widely with respect to the sun-planet line (see SPS angle). Also note the significant differences in the periapses distances. Both the angle and the distance are a natural result of the spacecraft being placed in a special trajectory corridor for subsequent objectives, such as the 24 Jan. 1986 encounter of Uranus by Voyager 2.

The basic features of the magnetospheres of Jupiter and Saturn are similar to those of the Earth:
1. A compressed dayside magnetic field region, due to the deflected solar wind, and an extended nightside magnetic tail trailing far behind the planet with an imbedded plasma sheet separating the opposite polarity lobes whose field lines connect directly to the polar cap regions.
2. A detached bow shock wave surrounding the magnetosphere, where the supersonic and super Alfvenic solar wind undergoes an abrupt transition as the solar originating high speed plasma interacts with the obstacle formed by the planetary magnetic field. A trapped radiation environment of charged particles, some of very high energies, and a low energy plasma environment.
3. Planet encircling azimuthal equatorial systems of electrical currents flow in the Jovian (2×10^8 amps) and Saturnian (7×10^6 amps) magnetospheres.

There are striking contrasts with the Earth's magnetosphere:
1. Because the solar wind momentum flux is so reduced at 5 and 10 AU, and because the magnetic moments of these planets are so large, the size of the magnetospheres is enormous, even when scaled by the radius of the parent planet (See Table II).
2. Embedded deep within the magnetospheres of both Jupiter and Saturn are a large number of satellites and also an extended ring system at Saturn. These "moons" are significant as both sinks and sources for the entrapped radiation belts and plasmas, quite unlike Earth's moon, which is too far distant to play any significant role in the terrestrial magnetosphere. Indeed the chemical composition of the Jovian and Saturnian magnetospheres is dominated not by the solar wind

nor atmospheres of the planets but by the characteristics of these moons, especially volcanically active Io and heavily atmosphered Titan. 3. Strong electrodynamic interactions between some of the moons, Io and Titan, with the corotating magnetospheres of Jupiter and Saturn.

Table II - Salient Features of Planetary Magnetospheres

Planet	Radius (km)	Sub-Solar magnetopause (planetary radii)	Dipole Moment (Gauss cm^3)
Earth	6371	10	8×10^{25}
Jupiter	71372	50-100	1.6×10^{30}
Saturn	60330	20	4.6×10^{28}

It is impossible in this short paper to adequately discuss all aspects of these planetary magnetospheres, or to even provide a meaningful list of individual references. Rather, the reader is referred to the collected sets of papers which have appeared in the following journals:
1. 1979, Science, 204, pp.945-1008 (Voyager 1: Jupiter)
2. 1979, Science, 206, pp.925-996 (Voyager 2: Jupiter)
3. 1981, J. of Geophy. Res., 86, to appear, (Voyagers: Jupiter)
4. 1981, Science, 212, pp.159-243 (Voyager 1: Saturn)
5. 1981, Nature, 292, pp.675-755 (Voyager 1: Saturn)
6. 1982, Science, to appear, (Voyager 2: Saturn)
7. 1982, J. of Geophys. Res., 87, to appear, (Voyager 1: Titan)
There are additional articles which have appeared and there are scheduled plans for special meetings and symposia, with associated publications during the next several years. Clearly, the scientific data returned by the Voyagers shall be studied extensively during the next decade by many investigators.

Since 1955, Jupiter has been known to be a strong source of radio emission, and shortly after the 1958 discovery of the terrestrial radiation belts, these non-thermal emissions were interpreted in the framework of a trapped population of charged particles in a planetary magnetic field. Analyses of these decimetric and decametric emissions provided quantitative estimates of the magnetic field which were, in the main, confirmed and elaborated by the Pioneer 10 and 11 flybys. While it was originally hoped that the Voyager spacecraft would augment our study of the main field of Jupiter, their relatively large values of periapses have precluded this. However, a detailed study and model of the "magnetodisk" current system was obtained and future studies of the main field will be possible.

The 9.6° tilt of the dipole axis at Jupiter combined with the rapid rotation rate of 9.92 hours leads to a wobbling of the entire Jovian magnetosphere back and forth. Periodic peaks in particle fluxes are observed by spacecraft at moments when the relatively thin magnetodisk is at the spacecraft location and troughs at those times

when the disk region moves away. The radiation is highly concentrated
around the disk. Close to the planet, the magnetodisk is coincident
with the magnetic equator, but further away the disk approaches the
centrifugal equator as the heavier sulfur and oxygen ions load down the
field.

An important side effect of the entrapped plasma and radiation
belts is that there is a significant distension of magnetic field lines
near the equator. This is illustrated in Figure 1 where the meridian
plane projection of field lines shows how field lines are extended and
how the local field intensity is reduced. The disk current in this
case is modeled by a current system whose intensity decreases with
radial distance and is bounded at the inner limit of 5 R_J and outer at
50 R_J, thickness =5 R_J. An important consequence of the distorted
field geometry is the time shifting of expected satellite absorption
features, which provide a unique independent check on models of the
magnetosphere.

The most unique features of the Jovian magnetosphere are due to Io
and its indigenous volcanic activity. There is a permanent torus
shaped region of plasma originating at Io and composed primarily of
sulphur and oxygen. The ions arise from the electrodynamic interaction
of the magnetosphere of Jupiter co-rotating past Io, which induces a
several hundred thousand volt drop across Io's ionosphere and generates
a pair of Alfven wing currents of $2.8x10^6$ amps with a net power of 10^{12}
watts. By a careful study of the geometry of the Alfven wings, we have
been able to make an independent assessment of the mass density in the
Io torus. The Alfven wings are in fact similar to field aligned
currents, a configuration which can only occur when the mass of the
ions involved approaches zero. There are other species in the
magnetosphere of Jupiter and their entire energization cycle is now the
object of intense study. The modulation of Jovian decametric emissions
by Io has been known since 1964 and a strong interaction long proposed.
We now have quantified these processes. Future studies of the Jovian
magnetosphere can be accomplished by careful ground based observing of
the many species of ions and neutrals in the Jovian magnetosphere.

Nothing was known about Saturn's magnetosphere until 1979. No
ground based studies of radio emissions preceeded the in-situ
observations. There then followed, in 2 years, a quick succession of 3
very successful flybys with Pioneer 11 and Voyagers 1 and 2 (See Table
I). Pioneer 11 established that the planetary field is weaker than
expected, mainly dipolar, and with the dipole axis aligned within 1° of
the rotation axis. Radiation belts were observed in the magnetosphere
with a large number of examples of absorption of various energetic
particles by the innermost moons of Saturn and an almost complete
absence of charged particles inside the region spanned by the rings.

Because there is no tilt of the magnetic dipole axis, there is no
wobbling of the magnetosphere. However, because of the entrapped
plasma and charged particles there is a substantial azimuthal ring

Fig. 1. Voyager 2 model
of Jovian inner mag-
netosphere structure,
showing isointensity
contours and field
lines, stretched out
due to magnetodisk
azimuthal current sys-
tem. Numbers indicate
distance to equatorial
crossing of dipole
field lines originat-
ing at same latitude.

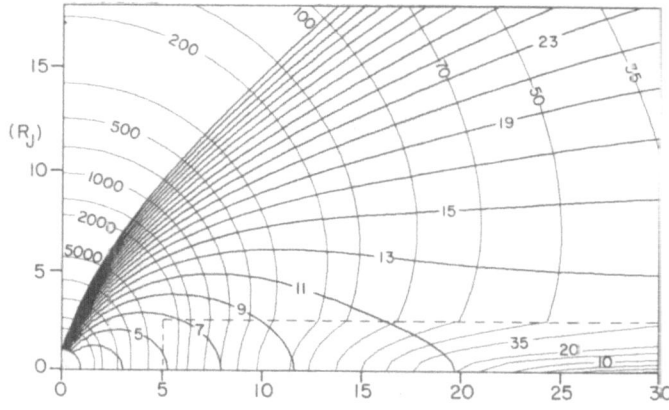

Fig. 2. Voyager 1 model
of Saturnian magneto-
sphere, illustrating
compressed dayside field,
extended nightside tail
and ring current (shaded
areas). Values on field
lines indicate co-lati-
tude of field lines.
Ultraviolet aurorae were
seen in the southern
polar cap region near
latitude 78°-80°, a
range consistent with
the magnetic field
topology.

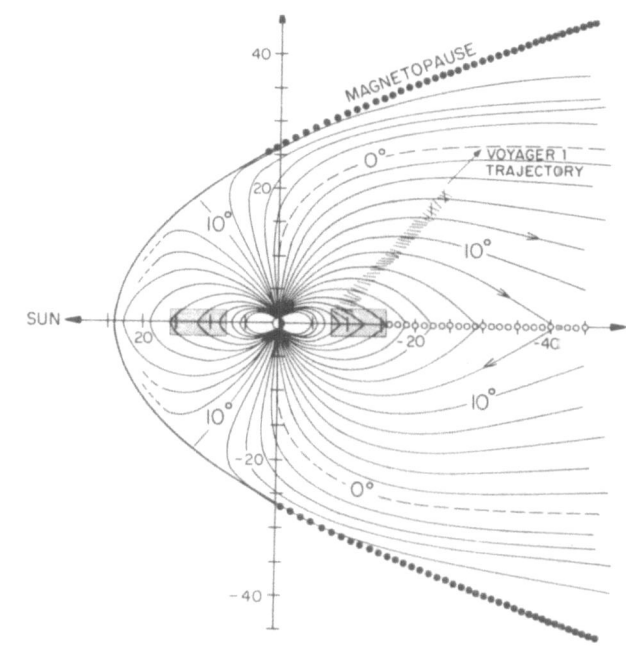

Fig. 3. Schematic diagram
indicating flow of co-
rotating Saturnian mag-
netosphere around Titan
and development of in-
duced bipolar magnetic
tail as field lines are
draped around ionosphere
exosphere of Titan.

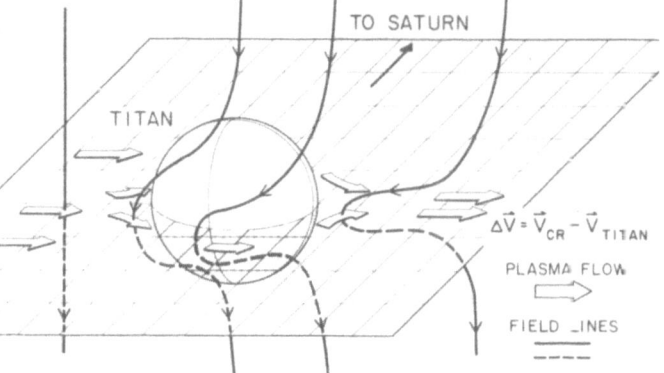

current system. This is illustrated in Figure 2, in a noon-midnight cross section of the Saturnian magnetosphere. The current system extends from 8 to 15 R_s radially, and thus is described as a ring current, rather than a magnetodisk. The innermost boundary of the ring current is almost coincident with the orbit of Rhea. The outermost boundary is slightly inside the average magnetopause position at 20 R_s.

Penetration of the current system was accomplished by both Pioneer 11 and Voyager 1, but the discovery and analysis of the ring current only occurred by careful study of Voyager 1 data. There has been ample evidence for the existence of several lighter ion torii associated with the rings and satellites of Saturn. The impact of highly energetic ions and electrons on the mainly water ice satellites and rings releases substantial hydrogen and oxygen and those are the major ionic constituents of the inner Saturnian magnetosphere.

The satellite Titan is special because of its very dense atmosphere. The close flyby of Titan by Voyager 1 at 7000 km distance from its center should be considered a separate planetary encounter because of the size and nature of Titan and results obtained. Also, Titan occupies a special orbital position being at 20 R_s, which is the average distance of the subsolar magnetopause. Thus, depending upon the intensity of the solar wind, Titan may be located inside or outside of the Saturnian magnetosphere when in the sunward portion of its orbit.

Voyager 1 results were obtained when Titan was located at 1330 local Kronographic time and fortunately the magnetosphere was expanded so that Titan was located well inside the magnetopause. There was no evidence for any substantial magnetic field (dipole moment less than 2×10^{21} Gauss cm^3) but good evidence for a strong electrodynamic interaction. The over taking corotating magnetosphere of Saturn induces only a few thousand volt drop across Titan, and the main feature of the interaction is the development of an induced bipolar magnetic tail and plasma wake region, leading Titan in its orbit (see Figure 4). It has even been suggested that the plasma wake is capable of being preserved for several Saturn rotations so that a plasma plume would develop. Titan contributes a rich source of nitrogen to the Saturnian magnetosphere.

This brief survey has only touched on a few of the unique features of the Saturnian and Jovian magnetospheres. It is now possible to make direct comparisons of many processes common to these and Earth's. For example, the total energy stored in the radiation belts and plasma torii of these three planets appears to be the same fraction of the energy in the planetary magnetic field. Thus, the effectiveness of natural astrophysical magnetic "bottles" is now established, at least for slow rotators. Extension to pulsar magnetospheres and other astrophysical situations may be possible.

A PHYSICAL DETAIL RELEVANT TO THE SAVIĆ-KAŠANIN THEORY OF BEHAVIOUR OF MATERIALS UNDER HIGH PRESSURE

Vladan Čelebonović
Narodna Opservatorija, Gornji Grad, Kalemegdan,
11000 Beograd, Yugoslavia

P.Savić and R.Kašanin have proposed a theory of behaviour of materials under high pressure (Savić,1981 and earlier references given there). Their theory (also known as the Sk theory) can be applied to the explanation of the internal structures of planets and stars. In the present paper, a simple method for the calculation of the internal temperatures of the terrestrial planets is proposed. All the parameters needed for the application of our method can be obtained from the Sk theory.

According to a recent review (Zharkov and Trubitsyn, 1980) planetary temperatures are usually calculated by assuming an equation of state and chemical composition for planetary matter, and then deriving the temperatures. Our approach will be based on considerations of energy.

The energy per unit volume of a solid body can be calculated as (Landau and Lifchitz, 1976):

$$E = \frac{\pi^2 k^4}{10(\hbar\bar{u})^3} T^4 \qquad (\text{erg cm}^{-3}) \tag{1}$$

where $\bar{u}$ denotes the mean velocity of sound in the solid, while the other symbols have their usual meaning. We approximate $\bar{u}$ by the Bohm--Staver formula (Ashcroft and Mermin, 1976):

$$\bar{u}^2 = \frac{1}{3} Z \frac{m}{M} v_F^2 \tag{2}$$

where v_F is the Fermi velocity, m and M are the masses of electrons and ions in the crystal lattice, and Z is the charge of ions. On the other hand, the SK theory gives the following expression for the energy per unit volume of a solid:

$$E = 4,6143 \cdot 10^{-19} \rho^{4/3} (\frac{N_A}{A})^{4/3} \quad (\text{erg cm}^{-3}) \tag{3}$$

W. Fricke and G. Teleki (eds.), Sun and Planetary System, 249–250.

where N_A is the Avogadro number, A is the atomic weight of planetary material, and ρ is the mass density.

By combining (1) to (3) and expressing v_F from solid state theory, we obtain the following:

$$T = 1,422 \cdot 10^5 \left(\frac{\rho}{A}\right)^{7/12} \cdot \left(\frac{m}{M}\right)^{3/8} \cdot Z^{5/8} \tag{4}$$

As a test we have applied equation (4) to the model of Earth calculated by Savić (1981). The results are presented in the following table:

depth (km)	0-39	39-2900	2900-4980	4980-6371
max.dens.(g cm^{-3})	3.0	6.0	12.0	19.74
Z	1	2	2	3
T (K)	700	1610	2410	4160

Our results are reasonably close to those of other authors. For example, Schubert et al (1979) have obtained a value of 1580 K for the present temperature of the Earth's mantle. Similar values are given by Zharkov and Trubi·tsyn (1980).

According to these authors, the temperature of the Earth's core is about 6300 K - notably higher than in our method. However, Baumgardner and Anderson (1980), using a different equation of state, have calculated the value of 4800 K for the same quantity. We think that the differences which we have mentioned are due to two main reasons: the precision of our method, which is limited by uncertainities in the determination of the velocity of sound, and problems with the choice of the equation of state.

References:

Ashcroft,N.W. and Mermin,D.N.: 1976, Solid State Physics, Ed.Holt, Rinehart and Winston, London, p.514.

Baumgardner,J.R. and Anderson,O.L.: 1980, Publ. 2056 of the Inst.of Geophys. and Planet. Phys., UCLA, Los Angeles.

Landau,L.D. and Lifchitz,E.M.: 1976, Statisticheskaya fizika, 1, p.213, Ed. Nauka, Moscow.

Savić,P.: 1981, Adv. Space Res., 1, p. 131.

Schubert,G., Cassen,P. and Young,R.E.: 1979, Icarus,38,p.192.

Zharkov,V.N. and Trubitsyn,V.P.: 1980, Fizika planetnih nedr, ch.1, Ed.Nauka, Moscow.

ON THE LIFETIME OF E-RING GRAINS AND THEIR NATURE

A. Cheng
Rutgers University, New Brunswick, New Jersey
L. J. Lanzerotti
Bell Laboratories, Murray Hill, New Jersey
V. Pirronello
Osservatorio Astrofisico di Catania, Catania, Italy

In the attempt to try to explain the existence of the hydrogen ring atmosphere, found by Weiser et al. (1977) as L_α emission and confirmed by Pioneer 11 (Judge et al.,1980), as due to the erosion of icy surfaces induced by the inpact of the ions of the Saturn magnetosphere in the vicinity of the E ring and of the moons Dione and Tethys, as it is suggested bythe considerable decrease in the ion phase space densities between 4.5 and 8 Saturn radii (Krimigis et al., 1981), a strong constraint on the lifetimes of the grains which belong to the E ring or on their nature has been found.

The E ring is extremely opticallythin and extends from about 210000 to 300000 Km in radius (Smith et al., 1981). Its visibility from Voyager I only in forward scattered light strongly supports observations of Terrile and Tokunaga (1980) which indicate that most of its particles are very small in size.

The rate of erosion of the grains belonging to this E ring is easily deduced if they are composed only of water ice assuming an omnidirectional flux of protons of about 10^7 cm^{-2} sec^{-1} (Krimigis et al., 1980) and an average erosion yield of one H_2O molecule per inpinging ion at htat temperature on the base of new laboratory data (Brown et al., 1978 and Brown et al., 1981).

In this conditions a monolayer of 10^{15} H_2O/cm^2 from a grain will be eroded in 10^8 sec. For a grain thickness of about 10^{-4}cm its lifetime will be $\simeq 10^{12}$ sec, i. e. 3×10^4 years. Such a value, quite short if compared with the solar system age, implies either that the grain population of the E ring is regularly replenished, for instance by meteoric impact on the icy surfaces of nearby satellites followed by the ejection of icy material in the ring plane, or that the E ring grains are mainly composed of silicates or, more generally, of non-icy materials; in such a case, in fact, their lifetime would be much longer because the erosion mechanism would be dominated by the nuclear collision cascade processes (Sigmund, 1969) which give sputtering rates two or three order of magnitude lower than those due to mechanisms related to the electronic energy loss of the inpinging ion, which are responsable for the erosion of frozen volatiles (Brown et al., 1978; Pirronello et al., 1981).

251

References

W.L. Brown, L.J. Lanzerotti,J.M. Poate,W.M. Augustyniak
         Phys. Rev. Letters 40, 1027, 1978
W.L. Brown et al. Nucl. Instr. and Methods (in press), 1981
A.L. Broadfoot et al. Science 212, 206, 1981
D.L. Judge, F.M. Wu, R.W. Carlson Science 207, 431,1980
S.M. Krimigis et al. Science 212, 225, 1981
V. Pirronello, G. Strazzulla, G. Foti, E. Rimini Astron. Astroph.
         96, 267,1981
P. Sigmund Phys. Rev. 184, 383, 1969
A. Smith et al. Science 212, 163, 1981
R. Terrile, A. Tokunaga Bull. Am. Astr. Soc. 12, 701, 1980
H. Weiser, R.C. Vitz, H.W. Moos Science 197, 755, 1977

H$_2$ ENRICHMENT OF INTERPLANETARY MEDIUM

V. Pirronello,G. Strazzulla
Osservatorio Astrofisico di Catania,Italy

G. Foti
Istituto di Struttura della Materia,Catania,Italy

Observational evidences show the presence in the solar system of interplanetary matter.A considerable contribution to the enrichment of dust and gas in the interplanetary medium is given through release by comets under the action of the solar radiation field and of its particle emission (solar wind,impulsive phenomena etc.).

Most of the observed molecules and radicals are belived to be generated by the breaking of few parent molecules under the action of the already mentioned external agents.Among these parent molecules there are mainly hydrates: from the simple H$_2$O,NH$_3$,CH$_4$, to more complex as those observed in interstellar and circumstellar regions.

The relative importance of solar electromagnetic radiation and solar energetic particles in the modification of cometary nuclei is still an open question mainly because of the poor knowledge of the modification that a frozen gas suffers by the interaction with energetic particles.The problem has at least two aspects: the first one concernes with the number of atoms or molecules ejected by a frozen target per impinging ion,the second one concernes with the chemical composition of these erosion yields.

The first and really very important aspect of the problem has been investigated by some authors for frozen volatiles of astrophysical inte rest (Brown et al. 1978,Bøttiger et al.1979,Pirronello et al.1981,a).

The main result of all of these experiments is that the erosion coefficient is two or three order of magnitude greater than that one provided by classical sputtering theory and for this reason the importance of such a phenomena is greatly increased in astrophysical environ ments.

Some new experimental results (Ciavola et al.1981,Brown et al.1981) showed the molecular character of the particles released from ice targets (T~ 10÷100 K) bombarded by energetic (keV-MeV) ions.The analysis of the eroded particles performed by a quadrupole mass spectrometer shows the main components are H$_2$,H$_2$O and O$_2$ molecules.

These results are quite general: we have detected high H$_2$ production also for frozen gas of C$_6$H$_6$ and CH$_4$ bombarded by energetic particles and this seems to be a general property of hydrates.

The role of this process in explaining H$_2$ formation in dense mole-

W. Fricke and G. Teleki (eds.), Sun and Planetary System, 253–254.
Copyright © 1982 by D. Reidel Publishing Company.

cular clouds as a consequence of the interaction of cosmic rays with frozen gases on grains has been already discussed (Pirronello et al.1981,b).
    Also in cometary environments the interaction of solar energetic
particles with cometary ices rich of frozen hydrates results in the release of considerable amount of H_2 molecules.Such a process assumes a relevant role due to the fact that H_2 is only a minor product of the H_2O
photodissociation (Wallis 1973).

    In this way comets lose a considerable amount of hydrogen already
during their life away from the Sun due to cosmic rays bombardment and
contribute to the enrichment of interplanetary medium.When they approach
to the Sun the H_2 production rate increases both because of the increase
in the particle flux and because of the rise in thetemperature of the
cometary ice and then of the erosion yields (Brown et al.1981).

    On the basis of our experimental results we can estimate an H_2 pro-
duction rate at I AU ranging between 10^{26} and 10^{28} particles/sec depending
by the frozen surface effectively exposed (we assume $\sim 10^{18}$ cm that
is one tenth of the value used by Kristofferson and Fredga 1977) to the
solar wind and by the chemical composition of the frozen,the erosion
yields ranging from about I/IO for pure H_2O to about IO for pure NH_3.
    This rate is consistent with an upper limit at I AU of $3\cdot10^{27}$
quoted by Mendis and Ip (1976) for comet Kohoutek.

References:

Bøttiger J.,Davies J.A.,L'Ecuyer J.,Matsunami N.,Ollerhad R.,   1979,in
    Proc.Conf. on Ion Beam Modification of Materials (Contr.Res.Inst.
    for Physics,Budapest) p.I52I
Brown W.L.,Lanzerotti L.J.,Poate J.M.,Augustyniak W.M.,   1978, Phys.
    Rev. Lett. 40,1027
Brown W.L.,Augustyniak W.M.,Simmons E.,Marcantonio K.J.,Lanzerotti L.J.,
    Johnson R.E.,Boring J.W.,Reimann C.T.,Foti G.,Pirronello V.,
    1981, Nucl.Instr.Methods (in press)
Ciavola G.,Foti G., Torrisi L.,Pirronello V.,Strazzulla G.,   1981
    Radiation Effects (in press)
Kristofferson L.,Fredga K.,   1977, Astrophys.Space Sci.,50,105
Mendis D.A.,Ip W.H.,   1976, Astrophys.Space Sci.,39,335
Pirronello V.,Strazzulla G.,Foti G.,Rimini E.,   1981 a,Astron.Astrophys.
    96,267
Pirronello V.,Strazzulla G.,Foti G.,   1981 b,Astron.Astrophys.(in press)
Wallis M.K.,   1973, Astron.Astrophys.,29,29

STUDY OF THE INTERPLANETARY DUST AT HIGH ECLIPTIC LATITUDES : DOPPLER-FIZEAU SHIFTS

Robley R., Bücher A.
Observatoires du Pic du Midi et de Toulouse
1 Av. C. Flammarion 31500 Toulouse, France
Koutchmy S.
Institut d'Astrophysique, 98bis Bd. Arago, 75014 Paris, France

Measurements of the Doppler shifts in the zodiacal light spectrum were already performed by James and Smeethe (1970) and more recently by Fried (1978). The results showed that dust grains of the zodiacal cloud are on prograde orbits around the Sun, but with velocities higher than what was expected : hyperbolic type velocities were found at small heliocentric distances. Using a different method of observation, our purpose is to obtain direct spectra of the light scattered by the dust grains, in order to measure the observed, averaged along the line of sight, Doppler shifts at high ecliptic latitudes. The observations are performed with a spectrograph of conventional Czerny-Turner type and described by Bücher et al. (1981) (entrance slit : 1x80mm - holographic grating : 160x140mm, 2400gr/mm, 1st order - image intensifier : VARO, 2 stages, fiber optic entrance window $\emptyset$=50mm and blue phosphor output). The site of the observations is the Pic du Midi Observatory (altitude 2800m) ; we used a 103aF film directly pressed against the output window, in order to obtain spectra with an exposure time of 10 min ; the instrumental set-up was permanently oriented to the North : azimut 180°, high above horizon 20°. A spectral region has been selected near the triplet b_1, b_2 and b_3 of MgI around 5180 Å, giving absorption lines contrasted enough to be measured in any case with a comparator of Hilger-type.

METHOD OF ANALYSIS, RESULTS AND CONCLUSION.

On each spectrum we managed to superpose on the lower part of the night sky spectrum a calibration spectrum coming from a Rubidium spectral lamp. Additionally the 5199 Å line of NI, produced in the high terrestrial atmosphere, is always present in our spectra ; we decided to use it as a standard of wavelength, because it appears superposed on the entire height of the spectra. Then we deduced the calibration scale of wavelength using a twilight spectrum with the lines : b_1=5184 Å, b_2=5173 Å, b_3=5167 Å of MgI, the 5199 Å line of NI, and Rb_1=5188 Å, Rb_2=5162 Å of Rb, giving a dispersion almost linear in the region of interest. Finally the differences between the apparent b lines and the calibration lines give the possibility to deduce Doppler shifts ; averaged on the set of

W. Fricke and G. Teleki (eds.), Sun and Planetary System, 255–256.
Copyright © 1982 by D. Reidel Publishing Company.

9 differences, then we obtain the value of the shift and its standard
deviation. To make a control, we determine the values of (b_1-b_2), (b_1-b_3)
and (b_2-b_3) from the (NI-b) or (Rb-b) values ; these differences should
remain constant during the whole night of observations. Radial velocities
can be determined knowing that 10 μm on a spectrum correspond to 0.238 Å
or 13.7 km/s at 5200 Å. The correction over radial velocities due to the
orbital Earth's motion is given by $v=30\cos b \times \sin(1-1_\odot)$ km/s, where b and
1 are ecliptic coordinates of the line of sight and $1_\odot$ is to solar ecliptic longitude. The reported results were obtained during the nights of
June 30, July 1, August 25 and 26, 1981. The display of results as a
fonction of U.T. (fig.1a) shows in any case an important red shift corresponding to a radial velocity up to or more than 100 km/s ; the behaviour of the variations seems to be identical for all nights : velocities
reach a nearly zero value at 23h U.T.. Furthermore, if the results are
plotted as a function of the relative ecliptic longitudes $(1_\odot-1)$ (fig.1b),
then the maximum of Doppler red shifts seems to occur near $20°$, taking
into account the fact that all observations were performed off the
galactic plane.
It would be premature to propose an interpretation of these results. If
we note that ecliptic declinaisons of the line of sight are larger than
$43°$, then the concerned interplanetary grains are situated at distances
from the Sun not less than 0.7 A.U., corresponding to an orbital heliocentric velocity of 50 km/s. We are forced to conclude that our measurements correspond to rather higher values, so a quite large part of the
"off-ecliptic" grains should gravitate on hyperbolic orbits.

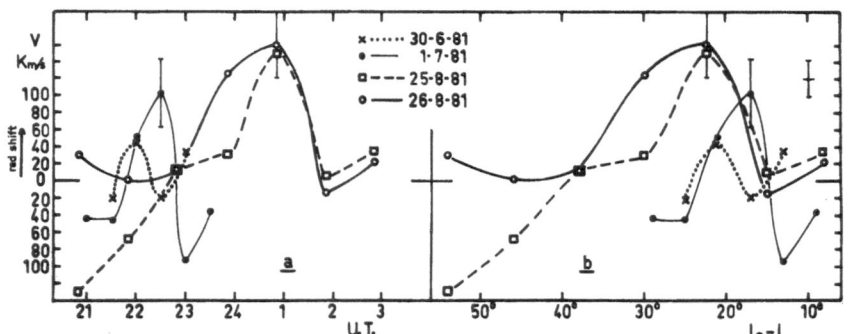

Fig.1 : Doppler shifts for ecliptic declinaisons larger than $43°$: (a)
v.s. U.T.. (b) v.s. differences between both ecliptic declinaisons
(Sun-line of sight).

REFERENCES

James,J.F., Smeethe,M.T. : 1970, Nature 227, pp. 588-589.
Fried,J.W. : 1978, Astron. Astrophys. 68, pp. 259-264.
Bücher,A., Koutchmy,S., Robley,R. : 2e Colloque national français du
Télescope spatial, mars 1981, Orsay.

NEAR INFRARED EMISSION FROM THE SOLAR CORONA

A.Mampaso, C.Sánchez Magro and J.Buitrago
Instituto de Astrofísica de Canarias

During the summers of 1978 and 1979 the F Corona was observed from
ground in the near infrared at 2,2μ and 3,5μ in an attempt to measure
the thermal emission peaks predicted by Peterson (1963) and found experi
mentally by Peterson (1967) and Mc Queen (1968). These measurements made
in solar eclipse time were necessarily short and of limited signal to
noise ratio. However they could find emission peaks at 3,4 Ro , 3,9 Ro ,
8,7 Ro and 9,2 Ro with the dominant peak at 3.9 Ro. The equivalent width
of the peak was of the order of 8 arc. minutes. Different authors have
found difficulties to explain the form and the existence of these peaks
directly related with the depletion of interplanetary dust falling towards the sun Saito et al (1977) Calbert and Beard (1972) Roser and Stau
de (1978). Another authors Mc Queen and Poland (1977), Bohlin (1971)
could find coronal brightness changes justifiable by electron streams
with long persistence time.
We thought in the possibility of observing the solar corona from ground,
after checking the background emission, improving in this way the parameter determination. Emission, colour, time variations and even polarization looked very promising indicators of the radiation mechanisms for
this purpose we prepared a specific instrumentation and observational
techniques. We built a 25 cm. aperture telescope, with vibrating primary
mirror, gold coated optics and newtonian tipe with a focal ratio F/4.
The telescope was fixed refered to the sky and could see just out of the
edge if an occulting screeen 12 m. above the ground. This screen was in
fact the occulting screen. In this way the solar corona was scanned in
front of the telescope using the earth's motion. The measurement started
just when the telescope was in shadow reaching 2 Ro from the solar centre. The chosen aperture was 2,5' , enough to resolve the hypothetical
8´ peaks. The synchrons detections system used controlled the vibrations
of the primary at 15 Hz. In this way we were able to measure variations
of brightness but we could not measure background emission. These emission was measured by D.C. detection systems along the observing period.
Over one hundred of scanns between 2 and 10 Ro were recorded, roughly
60% at 3.5μ, 35% at 2,2μ and the rest at 1.2μ, 1,65μ and 5μ. Of them the
most usefull scanns were done at 3.5μ (filter L). The calibration was do
ne with scanns of α Lyrae.

W. Fricke and G. Teleki (eds.), Sun and Planetary System, 257–258.
Copyright © 1982 by D. Reidel Publishing Company.

Although we could not detect the emission peaks in isolated days, in an attempt to improve the detection peaks limits for the coronal peaks we added all the scanns at the same wavelengh for selected good days. Even so the results were negative after adding more than forty scanns.
In a following step we deconvolved the scanns with the instrumental pro file. The results were always negative. Lastly even been sure of the impossibility of detection of the Peterson peaks, these were added to the deconvolved profile and convolved numerically. The peaks were evident on the artifitial scanns demonstrating the detection possibilities and the advantage of sky chopping technique. Peaks one order of magnitude less than the Peterson Peaks were detected easily, well below the noise level.

We can conclude saying that the coronal emission peaks, if some, must be lower than the found ones by Peterson and Mc Queen. Electron streams must be considered in the existence of these local maxima. Colour, and polarization measurements con be a powerfull tool to distinguish the different mechanisms of coronal infrared emission.

References :

Bohlin, J.D.: 1971, Solar Physics. 18, 450.
Calbert, R. and Beard, D.B.: 1972, Astrophysics.J. 176,497.
Mac Queen, R.M. : 1968, Astrophys. J. 154, 1059.
Mac Queen, R.M; and Poland, A.J. : 1977, Solar Phys. 55, 143.
Peterson, A.W. : 1963, Astrophys. J. 138, 1218
Peterson, A.W. : 1969, Astrophys. J. 155, 1009
Roser, S. and Staude, H.J. : 1978, Astron. Astrophys. 67, 381.
Saito, K.; Poland, A.I.; Munro, R.H. : 1977, Solar Phys. 55, 121.

ASSOCIATIVE IONIZATION AND SODIUM IN THE ATMOSPHERES OF PLANETARY SYSTEM BODIES

V.Vujnović
Institute of Physics of the University
P.O.Box 304, 41001 Zagreb, Yugoslavia

The intensity of a spectral line is consequence of excitation and ionization processes which govern the population of ionization stages and levels. Sodium atom is one of the emitters found in cometary heads, high planetary atmospheres including their satellites (e.g. Jovian magnetosphere and moon Io). Processes studied in the case of sodium atom include photoexcitation and photoionization, the electron impact excitation and ionization, charge exchange and energy transfer in collisions with other atoms and molecules. In the cases when direct ionization from the ground level is not likely, the ionization could be effected through the collisions between heavy particles. In general, collisions betweeen two atoms - one of which is excited at least, lead to the ionization by three channels of chemiionization:

associative ionization (AI) $A^* + B \rightarrow AB^+ + e$

atomic ion formation $A^* + B \rightarrow A + B^+ + e$

ion pair formation $A^* + B \rightarrow A^+ + B^-$.

We will give information on the laboratory data obtained recently for the alkalies, with particular emphasis on sodium, i.e. $A = B = Na$.

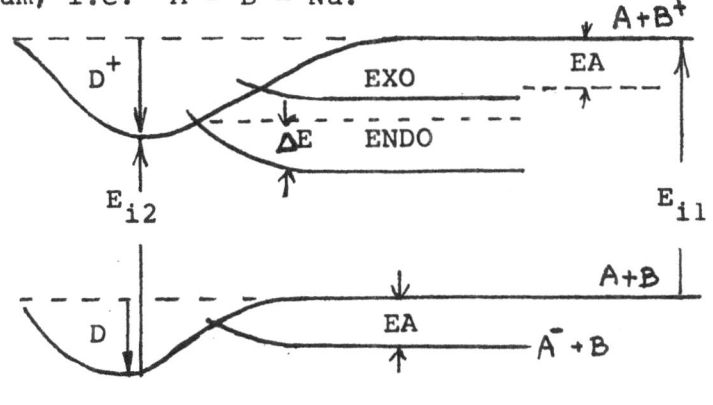

The processes are governed by the energy relations.

W. Fricke and G. Teleki (eds.), Sun and Planetary System, 259–260.
Copyright © 1982 by D. Reidel Publishing Company.

The following table gives data (Klucharev et all 1977,1977,
T$\approx$500 K) for the case when both atoms are excited at the
resonant level (ΔE is estimated energy defect or endothermi-
city, $\langle Qv \rangle$ is in 10^{-13} cm^{3}s^{-1}, $\overline{Q}$ in 10^{-16} cm^2)

INPUT	2Cs(6P)	2Rb(5P)	2K(4P)	2Na(3P)	2Li(2P)
ΔE /eV	0,4	0,2	$\leqslant$0,1	-	0,2-0,3
$\langle Qv \rangle$	2	3,2	9	400	no data
$\overline{Q}$	40	1,6	0,1	5	no data

Resonantly excited alkalies show exothermic AI only for Na.

In the case when two excited atoms collide, the excited levels could be low levels! In the case when only one atom is
excited, AI proceeds only from the high levels and it can
overlap with other chemiionization processes, i.e. ion pair
formation, which is possible with the binding energy of the
level equal to or less than the attachment energy EA. If
this happens, ionization increases. The measurements for the
excited levels which are at the onset of pair formation, made for Cs(9P)+Cs(6S), Rb(8P)+Rb(5S), K(7P)+K(4P) gave for
the ionization rate coefficients 5, 3 and 45 respectively
(in units of 10^{-10}cm^{3}s^{-1}). It is difficult to extrapolate
to sodium, but its rate cannot fall much apart.

In the real case, the levels are not excited selectively as
in experiments. Level mixing also produces the excited levels, with high azimutal quantum numbers. Large ionization
rate coefficients were revealed for the high D-levels, e.g.
$1,6 \times 10^{-9}$cm^{3}s^{-1} for the Rb(6D)+Rb(5S). A source of excited
atoms is again heavy particle collision; for example:
$$Na + N_2^* , \quad Na + Na_2^* , \quad etc.$$
In the rarefied gaseous medium, with radiation from distant
sources, atom-atom collisions become important. Observations
frequently indicate not resolved mechanisms. Sodium auroral
intensity rises in comp-arison to the nightglow; the aurora
illuminated by the Sun has molecular bands of N_2^+ more pronounced than unilluminated aurora. Also,abundance studies ask
for consideration of atom-atom collisions (AI was initially
proposed by Oppenheimer and Dalgarno 1977 in order to account for the interstellar cloud depletion of titanim). It is
therefore worth-while to search for possible effects of AI
and other chemiionization processes during atom-atom symmetric and unsymmetric collisions, in the interplanetary and
interstellar medium.

REFERENCES
A.N.Klucharev,V.Ju.Sepman,V.Vujnović,1977,J.Phys.B **10**,715
A.N.Klucharev,V.Ju.Sepman,V.Vujnović,1977,Opt.Spectrosc.
                                          (USSR) **42**,588
M.Oppenheimer,A.Dalgarno,1977,Astrophys.J. **212**,683

ON PHYSICAL INTERPETATION OF HYPSOMETRIC CHARACTERISTICS
OF THE MOON AND PLANETS

I.V. Gavrilov
Main Astronomical Observatory of the Ukrainian
Academy of Science, Kiev 127, USSR

Hypsometric characteristic of the planets contain important information
on the origin and evolution of these bodies. With the use of earthbased
and space observations, during the past two decades many hypsometric
data have been obtained for the Moon, Mars, Mercury and some other
planets and their satellites. For the purposes of comparative planet-
ology it is very important that hypsometric characteristics of these
bodies were homogeneous from point of view of its physical and geometrical
interpretation.

At present, hypsometric characteristics of planets are analysed using
spherical and sampling functions (see references). But, if observational
data are not complete, spherical harmonics cannot describe the variety of
all forms of relief. Therefore, it is proposed here to use the following
empirical formula:

$$H_i = \sum_{n=0}^{N} \sum_{m=0}^{n} (A_{nm} \cos m\lambda_i + B_{nm} \sin m\lambda_i) \cdot P_{nm}(\beta_i) +$$

$$+ \sum_{k}^{K} \phi_{kl} (\lambda_i, \beta_i) + h_i \qquad (1)$$

where H_i are absolute heights of a surface points with planetographic
longitude λ_i and latitude β_i; A_{nm}, B_{nm} are coefficients of spherical
harmonics; P_{nm} are Legendre functions; N is a limit order of spherical
approximation; ϕ_{kl} are special functions, which describe unique forms of
relief; k is a code of these forms of relief; l is a code of special
functions; h_i are residual heights, which are described statistically;
i the number of points with measured heights.

The first part of this formula is an ordinary approximation of global
forms of planetary surfaces by means of spherical harmonics to a certain
order. It should be used only if observational data cover the whole
surface of a planet.

W. Fricke and G. Teleki (eds.), Sun and Planetary System, 261–262.

The second part of the formula is the sum of approximations for unique and not numerous forms of relief, which can be described by means of special functions.

The third part of the formula describes small relief, which can be analysed by statistical methods.

On the basis of formula (1) we propose the following classification of hypsometric characteristics, which may be physically and geometrically interpreted: parameters of geometrical figure, megarelief, macrorelief and toporelief.

Parameters of figure can be determined if in formula (1) three of the first coefficients of spherical harmonics (case N=2) are computed. Mean radius, position of center and semidiameters of threeaxial ellipsoid, which approximate the surface of a planet, can be obtained from these coefficients.

Megarelief (mean levels of large regions of highlands and maria, linear and circular mountain ranges, other large surface features) can be determined if in formula (1) coefficients of spherical harmonics by $n = 3 \div 6$ are computed. These coefficients describe global forms of megarelief and asymmetry of a planetary surface. For computation of these coefficients mean levels of areas with size about $30^\circ \times 30^\circ$ should be used. Individual forms of megarelief should be described by using special functions from the second part of formula (1).

Macrorelief (mean levels of areas with size about $5^\circ \times 5^\circ$) is determined by rims of craters and other mountain features. Characteristics of macrorelief are described by spherical harmonics with $n = 7 \div 18$.

Toporelief (small surface features, which are smoothed when mega- and macrorelief are determined). If observational data are complete, toporelief can be described by special functions from the second part of formula (1). When observational data is deficient, toporelief should be described statistically by means of the third part of formula (1).

Hypsometric characteristics, which are described briefly above can be determined for planetary litospheres, hydrospheres and some levels of atmospheres. When comparing these characteristics for various planets and its satellites, these aspects should be taken into account.

The principles of classification of hypsometric characteristics, which are described above, are used in Kiev for investigations of the Moon and planets.

REFERENCES:

Bills,B.G., Ferrari,A.J., 1977: Icarus, 31, 244-259.
Chujkova,N.A., 1975: Astron. Zh., 52, 1279-1292.
Gavrilov,I.V., Yanovitskaya,G.T.: 1972: Astrom. i Astrof., 16, 46-52.
Goudas,C.L., 1963: Icarus, 2, 423-439.
Yanovitskaya,G.T, 1975: Astrom. i Astrof., 25, 58-65.

THE LUNAR PHOTOMETRIC CONSTANT IN THE SYSTEM OF TRUE FULL MOON

V.V. Shevchenko
Sternberg State Astronomical Institute, Moscow, USSR

A photometric analysis of 26 lunar phases has been undertaken on the basis of earth-based observations and space survey from the space craft Zond-3, Zond-6, Zond-8 and Apollo-13.

The average brightness distribution over the visible disk of the Moon has been constructed from earth-based data. The analogous distribution of brightness has been obtained for Zond and Apollo photographic data.

The effect of opposition was investigated and the true albedo values have been found. For interval of phase angles from 0 to 2.3 deg the effect of opposition is about 11 per cent and from 0 to 5 deg it is 18 per cent.

The values of the photometric function in the system of the true full Moon were obtained for the angles of incidence from 0 to 80 deg, for the angles of reflection from 0 to 90 deg and for the angles of azimuth from 0 to 180 deg with interval of 10 deg. From these data the spatial scattering indicatrix was derived. It includes 1710 values of the photometric function. The average photometric function reproduces the brightness dependence of mare and highland features under the photometric heterogeneity about 5 per cent and within the limits of unresolved roughnesses from several kilometers to several millimeters.

For numerous series of phase angles α the values of average brightness B_{AV} have been obtained by means of the spatial scattering indicatrix data. Then the values of integrated phase-brightness B was calculated from the equation

$$B = B_{AV} \cdot 1/2 \ (1 + \cos\alpha)$$

The values of integrated brightness B can be connected with measurements of phase stellar magnitude of the Moon m_α by means of the equation of linear regression of the form

$$m_\alpha = r \ (-2.5 \ \lg B) + m_q$$

W. Fricke and G. Teleki (eds.), Sun and Planetary System, 263–264.

The used values m_α is data of photoelectrical measuring made by Rougier (1933) and Nikonova (1949). The values constants r and m_α have been obtained by least-squares process, leading to

$$m_\alpha = 0.987 \ (-2.5 \ lg \ B) \ -12\overset{m}{.}13 \pm 0\overset{m}{.}05,$$

where: m_α is the integral stellar magnitude of true full Moon; Δ_v is the value of correction in visual system obtained by Martynov (1959): $\Delta_v = - 0\overset{m}{.}78$. So, we obtained the new values of the integral stellar magnitude of the full Moon, the phase integral, the light constant of the Moon, the geometrical and spherical albedo and the stellar magnitude of the Moon at the unity distance from the Earth and the Sun:

m = -12.91 ± 0.05   p = 0.147

q = 0.509   A = 0.075

E = 0.449 lux $V(1.0) = + 0\overset{m}{.}02 \pm 0\overset{m}{.}05$

The empirical expression for the lunar phase curve has been found to be of the form

$$\Delta m = + 3.38 \ (\frac{\alpha}{100^0}) - 1.07 \ (\frac{\alpha}{100^0})^2 + 0.99 \ (\frac{\alpha}{100^0})^3$$

REFERENCES

Martynov,D.Ya.: 1959, Astron. Zhurn., 36, p.648.
Nikonova,E.K.: 1949, Izv. Krymskoi Astrofiz. Obs., 4, p. 114.
Rougier,G.: 1933, Ann. de L´Observ. de Strasbourg, 2, p.205.

SURFACE PROPERTIES OF ASTEROIDS

H.J.Schober
Institut für Astronomie
Universität Graz
A-8010 Graz, Austria

ABSTRACT

From spectrophotometry, polarimetry,IR-radiometry and UBV-photo-
metry surface properties of asteroids are derived, such as:diameters,
spectral reflectivity, albedo; based on observable parameters a classification into taxonomic types (Bowell et al.1978) is made and an interpretation in terms of the mineralogy of the surface and meteoritic analogs can be done. From accurate photometry during rotational cycles irregularities on the surfaces are found. A short summary of our present
knowledge of asteroid surfaces is given.

1. FORMATION AND EVOLUTION OF ASTEROIDS

Starting with an outlook to a possible future mission and exploration of asteroids it might be no sacrileg anymore to think about mining
and exploiting a kilometer-sized earthcrossing asteroid. In such a con-
nexion or with respect to understand the origin and evolution of the solar system with asteroids to be considered as a powerful tool, it is of
high interest to know if an asteroid consists out of pure nickel or if
it just shows a surface covered with a thick regolith. The determination
of asteroid masses is not precise enough to derive good densities, roughly 2-3 g cm^{-3},and to provide the physical composition. For the largest
asteroids 1 Ceres,2 Pallas and 4 Vesta densities of only 50% accuracy
were found by Schubart and Matson (1979).

From an asteroid we get the reflected sunlight,altered due to its
surface properties.Considering the very rough schematic view of the formation and evolution of asteroids by Wilkening (1979) we assume that out
of the primitive solar nebula in the asteroid zone first primitive asteroids in the order of 100 km size were formed either by condensation or
by accretion.Without chemical evolution such an object should go through
impact cratering by smaller objects and should end up as a typical C-ob-
ject with surface structures; even a breakup due to a catastrophic collision may occur and would originate a physical asteroid family. Under
higher temperature conditions the asteroid instead could go through the

265

W. Fricke and G. Teleki (eds.), Sun and Planetary System, 265–268.

process of differentiation and metamorphism,heating up and cooling down,
so that different layers symmetrical around the core are formed;the density would vary along the radius up to the surface,but again impact cratering or even fragmentation takes place and a typical S-asteroid with a
dense core could result.Due to craters or a breakup the object would exhibit its deeper layers and would show variations in the surface properties if observed from earth during its rotation.

2. SPECTROPHOTOMETRY AND THE SURFACE

Spectrophotometry of asteroids was done as early as by Bobrovnikoff
(1929) when he stated the high blue reflectance of Pallas contrary to
other more reddish asteroids. The most detailed study of asteroids with
the coverage of the optical spectrum 0.3 - 1.1 µm was made by Gaffey
and McCord (1979) with a 25-filter spectrophotometer. 4 Vesta is unique
among the asteroids with a very deep absorption band near 0.95 µm due
to the mineral olivine-pyroxene. A comparison with the reflectance spectra of meteoritic material and mixed powder such as olivine,pyroxene,
nickel-iron metal in different percentage was made; for each asteroid
the spectral curve was measured and types were defined representing the
possible mineral assemblage,meteoritic analogs and the asteroidal/albe-
do/polarization classification.Chapman and Gaffey (1979) analysed 277
asteroids and presented average spectra for types C (flat,silicates plus
opaque carbons), S (steep and absorption at 0.95 µm,silicates plus metal)
and M (medium, metal).They also tried without result to group spectra
due to composition of C:S material ranging from 1:6 to 6:1.

3. POLARIMETRY, RADIOMETRY, UBV-PHOTOMETRY AND THE SURFACE

Measurements of the partly polarized scattered light on the surface
of asteroids is used to study the broadband reflectivity or the surface
texture itself (Dollfus and Zellner,1979). Using the slope-albedo law of
the polarization curve,the geometric albedo of a large number of asteroids was measured and the bimodality in the albedo was found: low albedo C-objects with flattish spectrum and $p_V < 0.065$,typically 0.035-0.04,
and the moderate albedo S-objects with $p_V \sim 0.065-0.23$,typically 0.14
(Zellner and Gradie,1976; Morrison,1977). The polarization curve is characterized by the negativ polarization P_{min} at small phase angles and
the width of the negative branch up to the inversion angle; both are a
signation of a rough,porous or a particular surface; the slope of the
curve in the linear part near the inversion angle is used to derive po-
larimetric albedos and in consequence diameters.

The measurement of the thermal emission of asteroids in the infrared at 10 and 20 µm, combined with the reflected visual brightness in B
and V was used to derive radiometric albedos and diameters for about 200
asteroids by Morrison and Lebovsky (1979). Within an accuracy of 10% the
results do agree with determinations by polarimetry and stellar occultations. A standard thermal model of an insulating regolith with low conductivity and high thermal inertia was assumed in order to account for
slow cooling of the side not illuminated.

Bowell initiated an UBV-survey in 1975; the most complete list of asteroids was created with magnitudes,colors,types and diameters in the TRIAD-file by Bowell et al.(1979). The taxonomy of asteroids combined albedo,polarization,three spectroscopic parameters and color indices B-V and U-B. The classification was done for about 750 asteroids: 75% dark C-objects, 15% bright S-objects; 10% are M (metal),E (enstatic a-chondrites),R (reddish) and U (unusual,unknown);Degewij and van Houten (1979) extended to D (RD) for dark and reddish,distant asteroids.

4. PHASE-MAGNITUDE RELATION AND THE SURFACE

The observed brightness of an asteroid depends on the geometry and on the scattering properties of the surface. Before other methods were available albedo and diameters were derived from the phase coefficient in the linear part of the magnitude-phase relation.At small phase angles the nonlinear opposition effect is present. Scaltriti and Zappalà (1979) have shown that among the asteroids the opposition effect is similar and that it is not affected by compositional type or albedo.Phase coefficients vary little with 0.02 mag/deg for bright to 0.05 for dark asteroids.

Bowell and Lumme (1979) have presented a new theory describing the radiative transfer in the surface of atmosphereless bodies,taking into account multiple scattering. The multiple scatter factor Q is combined with the photometric albedo at zero-phase,from where diameters are computed. The opposition effect in the phase curve is mainly controlled by the volume density or porosity, the linear slope of the curve is due to the surface roughness. From an analysis of more than 1500 UBV observations it turns out in consequence of the similarity of the opposition effect that asteroid surfaces are strikingly similar in texture with a moderately porous surface. From the different slopes in the linear part it is concluded that they are moderately rough on a scale greater than the wavelength of the light applied; they differ slightly in surface roughness due to type. Bowell and Lumme (1979) have listed constants to be applied for different taxonomic types,if not measured for the individual asteroid, which will influence the future standard system of the asteroid magnitude system.

5. PHOTOMETRY AND SURFACE VARIEGATION DURING ROTATION

Asteroids generally do rotate at a rate of hours up to a few days. During rotation they show lightvariations from a few 0.01 up to 1.5 mag. to be interpreted as the variation of the cross-section with respect to sun and earth and the pole orientation. Therefore the geometric surface properties in terms of flattening of an asteroid body or structures are obtainable. A statistical investigation of B-V obtained by Bowell and Lumme (1979) did not show a general variation at a level of ±0.005. But on the other hand careful measurements of polarization, frequent color measurements during rotation were made,including indications from spectrophotometry made at different rotational phases; results are listed by Degewij et al.(1979) and by Schober and Schroll (1981): out of 49 asteroids ten, among them 3 Juno and 4 Vesta, do show evidence for variations of the integrated surface properties during rotation; this means that either local color paches or albedo spots are present, or that there is a locally changing mixture of S and C-type material on the surface.

Finally it must be mentioned that in the last years lightcurves of asteroids were becomming more and more complex in their structure. Rotational rates therefore often had to be revised, doubled or turned out to come down to half of the value derived before. This mostly was due to misidentifications of similar lightcurve features and asteroids not observed long enough; lightcurves do show humps in the most simple case due to small scale features, but they also exhibit triple extrema in many cases as for 51 Nemausa (Chang and Chang,1963), or for 1580 Betulia (Tedesco et al.,1978),or even the possibility of quadruple extrema as it might be the case for 37 Fides (Schober,1981).Therefore the surface characteristics of asteroids must be regarded under two general aspects: a) the large scale characteristic with significance for the entire body within a classification system – and b) the local irregularities and surface structures in detail, with the possibility that there are in reality e.g.two objects – a binary asteroid or a satellite.

REFERENCES
"Asteroids" ed.Gehrels,T., Univ.of Arizona Press,Tucson, 1979
Bobrovnikoff,N.T.: 1929, Lick Obs.Bull.14, p.18
Bowell,E.,Chapman,C.R.,Gradie,J.C.,Morrison,D. and Zellner,B.: 1978,
       Icarus 35,p.313
Bowell,E.,Gehrels,T. and Zellner,B.: 1979, in"Asteroids",p.1108
Bowell,E. and Lumme,K.: 1979, in "Asteroids", p.132
Chang,Y.C. and Chang,C.: 1963, Acta Astron.Sinica 11,p.139
Chapman,C.R.and Gaffey,M.J.: 1979, in "Asteroids",p.655
Degewij,J. and van Houten,C.J.: 1979, in "Asteroids",p.417
Degewij,J.,Tedesco,E.F.,and Zellner,B.: 1979, Icarus 40,p.346
Dollfus,A.and Zellner,B.: 1979, in "Asteroids",p.170
Gaffey,M.J. and McCord,T.B.: 1979, in "Asteroids",p.688
Morrison,D.: 1977, in "Comets,Asteroids,Meteorites:Interrelation,Evo-
       lution,Origin",Delsemme,A.H.ed.,Univ.Toledo Press,Toledo,p.177
Morrison,D. and Lebovsky,L.A.: 1979, in "Asteroids",p.184
Scaltriti,F. and Zappalà,V.: 1980, Astron.Astrophys.83,p.249
Schober,H.J.:1981, Astron.Astrophys.,in press
Schober,H.J. and Schroll,A.: 1981, this volume
Schubart,J.and Matson,D.L.: 1979, in "Asteroids",p.84
Tedesco,E.,Drummond,J.,Candy,M.,Birch,P.,Nikoloff,I. and Zellner,B.:
       1978, Icarus 35, p.340
Wilkening,L.L.: 1979, in "Asteroids",p.61
Zellner,B.: 1979, in "Asteroids",p.1011
Zellner,B. and Gradie,J.: 1976, Astron.J.81,p.262

EARTH-CROSSING ASTEROIDS: NEW DISCOVERIES

Eleanor F. Helin
Jet Propulsion Laboratory, California Institute of Technology,
Pasadena, CA 91109

A total of 43 Earth-crossing asteroids are now known. Twenty-five were discovered or recovered in the last decade. There were only six numbered Earth-crossing asteroids prior to 1970. Since then, twenty-one have been numbered. The Aten asteroids, a new group of Earth-crossing asteroids, have orbits smaller than that of the Earth. The largest Earth crosser, Hephaistos, has C-type UBV colors and is probably about 10 km in diameter. The smallest, Hathor, has unusual UBV colors and a probable diameter of about 200 m.

INTRODUCTION

Fifty years of astronomical observations have led to the discovery of 43 Earth-crossing asteroids (Table 1). Nearly sixty percent of these were found in the decade 1971-1981 as the result of systematic search and incidental discoveries from observations made for other purposes.

An Earth-crossing asteroid is one whose orbit will intersect the orbit of the Earth as a result of secular perturbations (Shoemaker et al., 1979). Three recognized classes of Earth-crossing asteroids are 1) Atens (a < 1.0 AU, Q > 0.983 AU), 2) Apollos (a ⩾ 1.0 AU, q ⩽ 1.017 AU), and 3) Amors (a ⩾ 1.0 AU, 1.017 AU < q ⩽ 1.3 AU). All four known Atens, all but one of the 27 known Apollos, and 14 of the 29 known Amors are Earth crossers.

HISTORY AND BACKGROUND

The first Earth crosser discovered was the Amor asteroid 887 Alinda found by M. Wolf at Heidelberg in 1918. Apollo was discovered at Heidelberg by K. Reinmuth in 1932, and Amor was discovered by E. Delporte at Uccle in the same year. By the end of the 1930's, six Earth crossers, three Amors and three Apollos, had been discovered; all but Alinda and Amor were lost. Ivar was accidentally rediscovered in 1957, and Apollo and Adonis were recovered in 1973 and 1977,

269

Table 1. Earth-Crossing Asteroids
(Atens, Apollos and Earth-Crossing Amors)

Number	Asteroid Name	Discovery Year	q (AU)	Q (AU)	a (AU)	e	i (Deg)
1566	Icarus	1949	.19	1.97	1.08	.827	23.0
2212	Hephaistos	1978	.36	3.97	2.16	.835	11.9
	1974 MA	1974	.42	3.13	1.78	.762	37.8
2101	Adonis	1936	.44	3.30	1.87	.764	1.4
2340	Hathor	1976	.46	1.22	.84	.450	5.9
2100	Ra-Shalom	1978	.47	1.20	.83	.437	15.8
	1954 XA	1954	.51	1.05	.78	.345	3.9
1864	Daedalus	1971	.56	2.36	1.46	.615	22.1
1865	Cerberus	1971	.58	1.58	1.08	.467	16.1
	Hermes	1937	.62	2.66	1.64	.624	6.2
1981	Midas	1973	.62	2.93	1.78	.650	39.8
	1981 VA	1981	.63	4.22	2.43	.741	21.8
2201	1947 XC/1979 XA	1947	.63	3.72	2.17	.712	2.5
1862	Apollo	1932	.65	2.29	1.47	.560	6.4
	1979 XB	1979	.65	3.88	2.26	.713	24.9
2063	Bacchus	1977	.70	1.45	1.08	.349	9.4
1685	Toro	1948	.77	1.96	1.37	.436	9.4
2062	Aten	1976	.79	1.14	.97	.182	18.9
2135	Aristaeus	1977	.79	2.40	1.60	.503	23.0
2329	Orthos	1976	.82	3.99	2.40	.658	24.4
	6743P-L	1960	.82	2.42	1.62	.493	7.3
1620	Geographos	1951	.83	1.66	1.24	.335	13.3
	1959 LM	1959	.83	1.85	1.34	.379	3.3
	1950 DA	1950	.84	2.53	1.68	.502	12.1
	1973 NA	1973	.88	4.04	2.46	.642	68.1
	1978 CA	1978	.88	1.37	1.12	.215	26.1
1863	Antinous	1948	.89	3.63	2.26	.606	18.4
2102	Tantalus	1975	.91	1.67	1.29	.298	64.0
	6344P-L	1960	.94	4.21	2.58	.635	4.6
	1979 VA	1979	.98	4.29	2.64	.627	2.8
	1978 DA	1978	1.02	3.93	2.48	.587	15.6
	1980 PA	1980	1.04	2.82	1.93	.459	2.2
2061	Anza	1960	1.05	3.48	2.26	.537	3.7
1915	Quetzalcoatl	1953	1.05	3.99	2.52	.583	20.5
	1980 AA	1980	1.05	2.73	1.89	.444	4.2
1917	Cuyo	1966	1.06	3.23	2.15	.505	24.0
1943	Anteros	1973	1.06	1.80	1.43	.256	8.7
1221	Amor	1932	1.08	2.76	1.92	.436	11.9
	1980 WF	1980	1.08	3.38	2.23	.514	6.4
1580	Betulia	1950	1.12	3.27	2.20	.490	52.0
2202	Pele	1972	1.12	3.46	2.29	.510	8.8
1627	Ivar	1929	1.12	2.60	1.86	.397	8.4
887	Alinda	1918	1.15	3.88	2.52	.544	9.1

respectively, through carefully planned search efforts. Hermes, discovered in 1937, remains lost. At the close of the 1950's, fourteen Earth crossers had been found; over half of them were lost. Many of these discoveries were by-products of proper motion surveys and other surveys of the sky.

The discovery in 1971 of Daedalus at Palomar by T. Gehrels led off a period of rapid discovery of Earth crossers. The Palomar Planet-crossing Asteroid Survey (PCAS) was initiated by E. F. Helin and E. M. Shoemaker in 1973; PCAS uses the Palomar 46-cm Schmidt camera to photograph a large area of the sky near opposition each month. This survey was the first systematic program dedicated to searching primarily for Earth crossers. Starting about the same time, C. T. Kowal found a number of Earth crossers with the 122-cm Palomar Schmidt camera, which he regularly uses to conduct a Solar System Survey. During this period, R. West and M. Schuster at the European Southern Observatory in Chile also discovered Earth crossers on plates taken for the Southern Extension Sky Survey. L. and N. S. Chernykh at the Crimea Astrophysical Observatory have reported several fast moving asteroids found during their regular minor planet observations. P. Wild, L. Kohoutek, and A. Mrkos have also begun reporting discoveries of Earth crossers. The Apollo asteroid, 1979 XB, was discovered with the 122-cm U.K. Schmidt by K. Russell at Siding Spring, Australia. The most recent Apollo discovery, 1981 VA, was found with the 122 cm Palomar Schmidt by E. Helin.

There is an observational bias against finding Amor asteroids, primarily because of the comparatively slow motion of many of them near opposition (Helin and Shoemaker, 1979). The discovery of nine Amors in the last two years, most of which are Earth-crossing, is noteworthy. Most of these objects displayed prograde motion on discovery and had daily motion ranging from about 0.25 to 1.5 degrees. E. Bowell and his colleagues at Lowell Observatory, who discovered several of these Amors, have recently become active in asteroid search.

NOTEWORTHY RESULTS OF THE LAST DECADE

The Aten Asteroids. -- In January 1976, 2062 Aten was discovered with the 46-cm Schmidt at Palomar (Helin and Shoemaker, 1977). It is the first asteroid found with an orbit smaller than the Earth's (Figure 1). Aten was also the first asteroid to be thoroughly observed by photoelectric photometry and infrared radiometry on its discovery apparition. It is a relatively high albedo S-type asteroid, about 900 meters in diameter (Morrison et al., 1976).

In October, 1976 another object, now named Hathor, was discovered which has an even smaller orbit. Although Hathor, like Aten, was discovered by the PCAS with the Palomar 46-cm Schmidt camera telescope, there were three other independent discoverers. While Helin was recording Hathor on the 46-cm Schmidt, W. L. Sebok observing on the Pal-

omar 122-cm Schmidt photographed the same field practically simultaneously on October 25, 1976. The same day it was also found by C. T.
Kowal on a film taken 3 days earlier on the 46-cm Schmidt. A fourth
independent discovery was made by N. S. Cherynkh at the Crimea Observatory. Physical observations of Hathor on its discovery apparition
suggest it is only about 200 m diameter. It has unusual UBV colors
somewhat similar to the colors of Icarus (Shoemaker and Helin,
1978a). The multiple discoveries resulted in part from the fact that
Hathor approached the Earth within 1.16 million kilometers on its discovery passage. Among the known Earth crossers, only Hermes in 1937
came closer (about 800,000 kilometers).

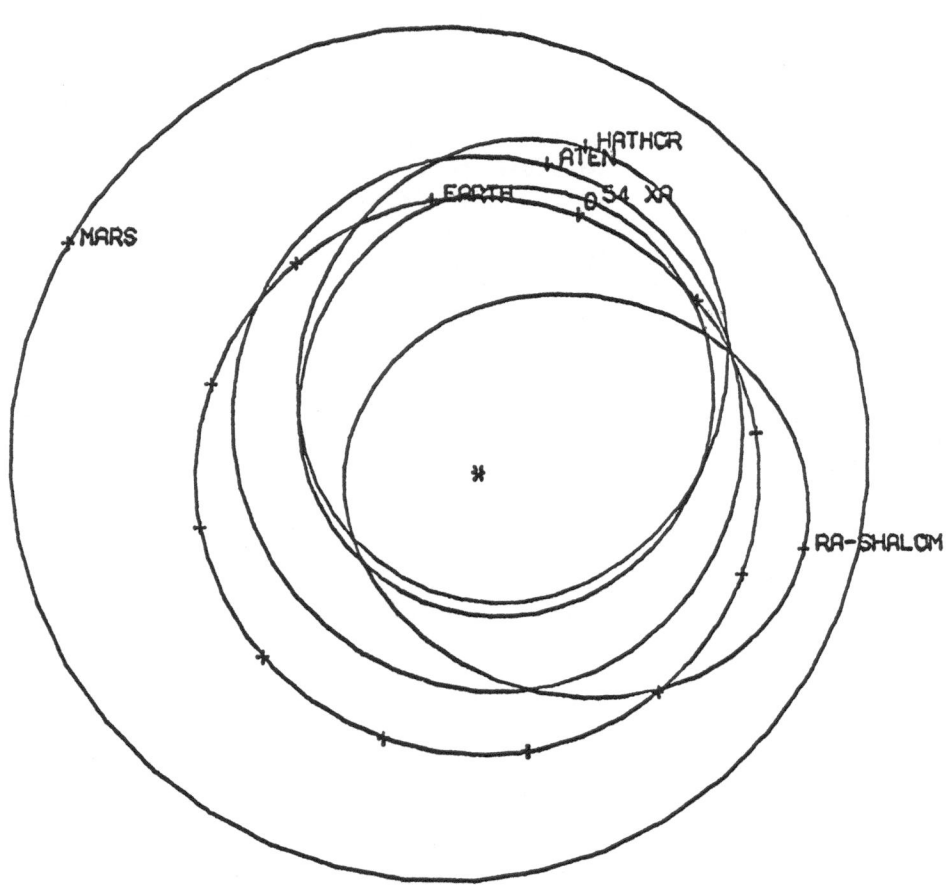

Figure 1.   Orbits of Earth, Mars and four Atens projected on ecliptic
plane.   Atens labelled at aphelion.   Time ticks are 1st day of each
month.

The third Aten discovered was 2100 Ra-Shalom. Physical observations showed Ra-Shalom to be a C-type asteroid with a diameter of 3.4 km (Bowell, 1978; Lebofsky et al., 1979). Apparently its surface is more rocky than a typical main belt asteroid.

The fourth Aten, 1954 XA, was first observed by G. Abell on plates taken for the National Geographic Sky Survey at Palomar in December 1954. Recovery plates were taken six nights later at Palomar and sent to L. Cunningham at Lick Observatory. The measurements were never reported, however. J. G. Williams, on learning of the existence of these plates in 1979, sought them out and J. Gibson found and measured the image on one plate. Using the discovery positions on December 5 combined with recovery image on December 11, Williams computed an Aten-type orbit for 1954 XA (MPC 4823, 1979); as no orbit was available for more than 25 years, 1954 XA is lost.

Accidental rediscoveries. -- Three Earth-crossing asteroids have been accidentally rediscovered in the last fifty years. 1627 Ivar was discovered first by E. Hertzsprung in 1934, five years after the plates had been taken in Johannesburg, South Africa in 1929. It was rediscovered in 1957, 28 years later, by J. Schubart at Sonneberg Observatory. When the orbit was obtained by Schubart from these latter observations, it was found to be identical with Hertzsprung's object.

2100 Ra-Shalom was discovered with the Palomar 46-cm Schmidt on September 10, 1978 by E. F. Helin. Within a week of its discovery it was realized that the asteroid had been seen before. In November 1975, it had been photographed twice by R. West at the European Southern Observatory at La Silla, Chile. Designated 1975 TB, the observations by West were, by themselves, insufficient to calculate an orbit. However, they became crucial when combined with the later observations by Helin and others, in calculating a definitive orbit.

Apollo asteroid 1979 XA discovered by the PCAS program at 0.32 AU from the Earth was found to be identical to 1947 XC discovered by E. Giclas at Lowell Observatory (Marsden, 1979). The asteroid was nearly in the same position in respect to Earth and Sun as it was on initial discovery 32 years earlier. As only two plates were originally obtained by Giclas, the observed arc was insufficient to calculate an orbit adequate to predict future positions. Hence, the PCAS observation of the fourth known Apollo was purely accidental.

Whipple (1973) suggested that the frequency of accidental rediscovery of near-Earth asteroids can be used to make a rough estimate of the total number to the threshold of detection. Wetherill (1976) refined this approach and showed that, under certain circumstances, the frequency of accidental rediscovery may be used to set a rough lower bound to the population. From the theory developed by Wetherill, the three accidental rediscoveries suggest that there are at least several hundred Earth crossers bright enough to be detected by current methods of search.

Asteroids of possible cometary origin. -- Four Earth-crossing asteroids, 6344 P-L, 1973 NA, 2212 Hephaistos and 1979 VA have orbits with
large aphelion distances (Q = 3.97 to 4.29 AU). Hephaistos has an
orbit somewhat similar to that of P/Encke; it has C-type UBV colors
and a probable diameter of about 10 km. The orbit of 1979 VA takes it
closer to Jupiter than any other known Apollo asteroid. It may be
just stable against strong perturbations by encounter with Jupiter.
The reflectance spectrum appears to be consistent with some carbonaceous meteorite assemblages (Helin and Gaffey, 1980). Both the orbit
and spectral reflectance of 1979 VA suggest it may be a recently de-
funct short period comet.

Candidates for rendezvous and sample return missions. -- Earth-
crossing asteroids with low inclinations and relatively low eccentricities are attractive candidates for exploration. They have low
escape velocities, and negligible propulsion is required to land on
them or escape. The most accessible Earth crosser is the Amor asteroid Anteros discovered by J. Gibson in 1973. The ΔV from low Earth
orbit for rendezvous with Anteros, for favorable trajectories, is
about 5.5 km/sec (Niehoff, 1977; Hulkower and Ross, 1981). Other
Earth crossers that are considered good mission candidates are
Bacchus, Geographos, Hathor, Toro, and 1980 PA (Hulkower, 1980; Stancati and Soldner, 1981). These have minimum ΔV to rendezvous of about
6 to 8 km/sec (Shoemaker and Helin, 1978b). More than 100 opportuni-
ties to rendezvous with asteroids by means of ballistic trajectories
have been identified for the remainder of this century (Stancati and
Soldner, 1981).

Populations and collision rates with Earth. -- The discoveries of
Earth-crossing asteroids from various systematic surveys of the sky
have been combined by Helin and Shoemaker (1979) in order to estimate
the populations of these objects. Estimates of the populations of the
Atens, Apollos and Earth-crossing Amors to absolute visual magnitude
18 are given in Table 2.

Table 2.   Estimated Number of Earth-Crossing Asteroids
to Visual Magnitude 18

Atens	~ 100
Apollos	750 ± 300
Earth-crossing Amors	~ 500

Collision rates with the Earth can be estimated from the populations given in Table 2 if it is assumed that the orbits of the dis-

covered Earth crossers are a representative sample of the orbits of the entire population. Shoemaker et al. (1979) found that about three Earth-crossing asteroids equal to or brighter than visual magnitude 18 strike the Earth every million years.

ACKNOWLEDGEMENTS

The author is grateful to D. Bender, N. Hulkower and C. Lee for generating the plot of orbits of Aten asteroids. I also wish to thank my colleague E. M. Shoemaker for critical review of this paper. The research described in this paper was carried out by the Jet Propulsion Laboratory, California Institute of Technology, under contract with the National Aeronautics and Space Administration.

REFERENCES
Bowell, E. 1978. 1978 RA, Internat. Astronom. Union Circ. No. 3244.

Helin, E. F., and Gaffey, M. J., 1980. 1979 VA, A possible carbonaceous asteroid, Meteoritics, 4, pp. 299.

Helin, E. F., and Shoemaker, E. M., 1977. Discovery of Asteroid 1976 AA, Icarus, 31, pp. 415-419.

Helin, E. F. and Shoemaker, E. M., 1979. The Palomar Planet-Crossing Asteroid Survey, 1973-1978, Icarus, 40, pp. 321-328.

Hulkower, N. D., 1980. Rendezvous with 1980 PA: Preliminary data, JPL Interoffice Memorandum 312/80.4-641, 2 September 1980.

Hulkower, N. D., and Ross, D. J., 1981. A mission to rendezvous with Anteros, AIAA-81-0314.

Lebofsky, L. A., Lebofsky, M. J. and Rieke, G. H., 1979. Radiometry and surface properties of Apollo, Amor and Aten asteroids, Astron. J., 84, pp. 885-888.

Marsden, B. G., 1979. 1947 XC = 1979 XA, Internat. Astronom. Circ. No. 3432.

Minor Planet Circular Nos. 4823 and 5006, 1979.

Morrison, D., Gradie, J. C., and Rieke, G. H., 1976. Radiometric diameter and albedo of the remarkable asteroid 1976 AA, Nature, 260, pp. 691.

Niehoff, J. C., 1977. Round-trip mission requirements for asteroids 1976 AA and 1973 EC, Icarus, 31, pp. 430-438.

Shoemaker, E. M., and Helin, E. F., 1978a. Earth-approaching asteroids: Populations, origin and compositional types, NASA Conf. Publ. 2053, pp. 161-175.

Shoemaker, E. M. and Helin, E. F., 1978b. Earth-approaching asteroids as targets for exploration, NASA Conf. Publ. 2053, pp. 245-256.

Shoemaker, E. M., William, J. G., Helin, E. F., and Wolfe, R. F., 1979. Earth-crossing asteroids: Orbital classes, collision rates with Earth, and origin, in Asteroids, ed. T. Gehrels, Univ. Ariz. Press, Tucson, Arizona, pp. 253-282.

Stancati, M. L., and Soldner, J. K., 1981. Near Earth asteroids: A survey of ballistic rendezvous and sample return missions, AAS Paper 81-185, pp. 1-16.

Wetherill, G. W., 1976. Where do the meteorites come from: A re-evaluation of the Earth-crossing Apollo objects as sources of chondritic meteorites, Geochim. et. Cosmochemi. Acta, 40, pp. 1297-1317.

Whipple, F. L., 1973. Note on the number and origin of Apollo asteroids, The Moon, 8, pp. 340-345.

SYNTHETIC LIGHTCURVES OF ASTEROIDAL BINARY SYSTEMS

F. Scaltriti, V. Zappalà and E. Anderlucci
Osservatorio Astronomico di Torino
I-10025   Pino Torinese   -   Italia

The idea of satellites of asteroids is tied to observatio=
nal and theoretical reasons. Secondary events in occultation
of stars by minor planets have been interpreted as due to
the presence of small bodies near the primary one. Recen=
tly,a search on photographic plates of possible images of
the suspected satellite of 9 Metis has been made by Chine=
se astronomers(Wang Sichao et al.1981).On the other hand,
lightcurves exist whose trends can hardly be due to chan=
ging projected areas even if the surface of the asteroid
is thought rich of major topographic features.Theoretical=
ly(Farinella et al.1981,Weidenschilling 1981),it has been
shown that quasi catastrophic impacts on medium-size aste=
roids might produce only partial dispersal of the fragments
whose recombination by self-gravitation could result in ra=
pidly spinning equilibrium figures or in fission into bina=
ries.Cook(1971)modelled 624 Hektor with a binary model of
two ellipsoidal components.Wijesinghe and Tedesco(1979)
used a system of spherical bodies revolving in circular
orbits for 171 Ophelia;a similar approach was made by Zap=
palà et al.(1980). The purpose of this research is to in=
vestigate, in a statistical sense,the effect of the pre=
sence of a satellite on a lightcurve.
The model we have adopted consists of two spheres with the
same albedo revolving in circular orbit in a plane contai=
ning the sun and the observer.The radius of the larger sphe=
re(R) was taken equal to 1,2,4,6,8,10, assuming unity the
radius of the smaller sphere; for each R, the distance be=
tween the centers of the two bodies was allowed to vary
from contact components to 8R,via the intermediate cases 2R
and 4R;at last,for given geometrical conditions of the sy=
stem,the phase angle had the values 0,15,30°.On the whole,
69 lightcurves were generated.No particular scattering law
was applied, and the effect of the shadow cast by a body
on the other one was taken into account considering the
intersection on the surface of the smaller (larger) sphere

W. Fricke and G. Teleki (eds.), Sun and Planetary System, 277–278.

of a cylinder(whose axis identifies the sun direction)tan=
gent to the larger (smaller) one.
The following features can be outlined in our lightcurves:
a) the amplitude varies from a few hundredths to about 0.8
mag,i.e. the effect of the presence of the satellite is ma=
sked by observational errors if the ratio R/r is greater or
similar to 10; b) for a given system,varying the phase an=
gle from 0 to 30°,the loss of light at secondary minimum is
constant,whereas the amplitude of the primary minimum in=
creases owing to the more and more pronounced effect of the
shadow; c) secondary minimum has flat bottoms which last
longer as the phase angle increases owing to a later egress
of the companion from the shadow due to the larger sphere;
d)the duration of the primary minimum is larger for greater
phase angles; e) primary minimum developes complex features
as changing slopes and "platforms" of constant luminosity
due to the transit of the satellite on the disk of the com=
panion without appearance of shadowing effects;in extreme
cases, a tertiary extremum is present.
Some features shown by our synthetic lightcurves can easi=
ly seen in some real cases.However,in our opinion,the out=
come of this research deserves some comments;we have seen
that even in the extremely simple case(perhaps not reali=
stic)of two revolving spheres,we can obtain a wide variety
of lightcurves which match observed runs. We can expect
that,growing the complexity of the model(ellipsoids instead
of spheres, plane of the orbit not containing the sun and
the observer),we obtain even more complex lightcurves. Ho=
wever, it is clear that those synthetic lightcurves alone
do not give us a definitive evidence for the existence of
binaries among asteroids;on this respect,for example,Zap=
palà(1980)showed that particular shapes for the asteroids
(parallelepiped,triangular prism,etc.)can reproduce actual
observed lightcurves; we are performing laboratory experi=
ments on isolated bodies(structured in such a way to simu=
late macroscopic roughness)in different conditions of a=
spect,obliquity and phase angles;they might help to separa=
te the effects due to binariety and those attributable to
other causes (mountains, craters, etc.).

Cook,A.F.: 1971, in "Physical Studies of Minor Planets",
        T. Gehrels Ed., NASA SP-267, pp. 155-163.
Farinella,P.,Paolicchi,P.,Tedesco,E.F.,and Zappalà,V.:
        1981, Icarus 46, pp. 114-123.
Wang Sichao, Wu Yuezhen, Bao Mengxian, Deng Liwu, and
        Wu Sufang: 1981, Icarus 46, pp. 285-287.
Weidenschilling,S.J.: 1981, Icarus 46, pp. 124-126.
Wijesinghe,M.P.,and Tedesco,E.F.: 1979, Icarus 40,pp.383-393.
Zappalà,V.: 1980, The Moon and The Planets 23,pp.345-353.
Zappalà,V.,Scaltriti,F.,Farinella,P.,and Paolicchi,P.:
        1980, The Moon and The Planets 22, pp. 153-162.

CCD SCANNING FOR ASTEROIDS AND COMETS

T. Gehrels and R. S. McMillan
The University of Arizona, Tucson, Arizona 85721

ABSTRACT. The sky is to be electronically scanned
for asteroids and comets that come close to the earth.
This is to prevent their collision with the earth and
to provide information on the origins of the solar
system.

A recent report published by the Advisory Council to the United
States National Aeronautics and Space Administration warns that a
large asteroid could someday destroy civilization on Earth similar to
the destruction of about 60% of the animal species during the
Cretaceous-Tertiary Extinction 65 million years ago. The Advisory
Council met during the Summer of 1980 in order to recommend possible
future space programs to NASA. The interest of a part of the report
is focused on asteroids because of studies conducted by Nobel
physicist Luis Alvarez and by others. Measurements were made of the
anomalously high content of the heavy elements iridium, osmium, etc.
in the 1-2 cm thick boundary between the deep sea deposits of the
Cretaceous and the Tertiary Periods. Heavy elements have low
concentrations near the surface of the Earth because they settled
gravitationally towards the center during the early stages of
formation when Earth was mostly molten. In the 1-2 cm layer, on the
other hand, the iridium content is as found in meteorites. Actually
about 7% of the 1-2 cm layer may be due to the meteorite while the
remainder 93% was thrown out from the Earth's mantle when the
meteorite impacted. When one adds together the amount of material of
7% of a 1.5 cm layer all over the earth's surface, assuming that the
meteorite had the same density as that of the layer, about 3 g cm^{-3},
it comes to a mass of 2×10^{15} kilograms and a meteorite diameter
near 10 kilometers.

The Council emphasizes that such an impact could occur again and
that it would have dire consequences. This is seen from an estimate
of the energies that are involved: $E = 1/2 \, mv^2$ where $m = 2 \times 10^{15}$ kg

W. Fricke and G. Teleki (eds.), Sun and Planetary System, 279–284.
Copyright © 1982 by D. Reidel Publishing Company.

and the differential velocity, v, of the Earth-crossing asteroid with
respect to the Earth, is typically 20 kilometers per second. With
the mass in grams and the velocity in centimeters per second, one
obtains the energy in ergs which can be converted into tons of TNT,
tri-nitro-toluene, being the customary unit for expressing explosive
energies; one ton of TNT = 4×10^{16} ergs. It is astounding to
realize that the energy dissipation is on the order of 2.5×10^{13}
tons of TNT, to be compared with 2×10^{4} tons for the A-Bomb
explosion in Hiroshima! The asteroid impact may have an effect on
the ozone layer in the Earth's atmosphere and thereby cause
ultraviolet radiation excesses.

By comparison with the Krakatoa volcanic eruption of 1883 in
Indonesia the dust cloud of the asteroid impact is estimated to be so
massive that sunlight would be extinguished for several years. For
lack of photosynthesis, most green growth would be stopped. The lack
of food would cause unimaginable shortage and chaos among the Earth's
population. Even though the chance may be only 1 in 100 million
years, such an event could occur. Whether or not the impact 65
million years ago actually caused the demise of the dinosaurs among
the extinguished species is debated by biologists who appear to have
evidence of a slow fading out of the dinosaurs, taking at least
10,000 years. There also is not a consensus that the impacting
object at that time was an asteroid or cometary core; the latter has
been proposed because no impact crater of the proper age has been
found. In any case, NASA's Advisory Council urges eloquently: "In
the 130 million years the dinosaurs roamed the earth, they failed to
develop the technology to avoid their own extinction. Homo sapiens
has developed an adequate technology. He can avert any further
extinction by asteroid impact. We think he should."

Ten-kilometer asteroids appear to be the largest that occur
among the Earth-crossers. There is a collisional distribution of the
frequency as a function of size such that there may be as many as
1,000 near the diameter of 1 kilometer. Even the 1 km asteroid could
demolish a large city, the Advisory Council warns. Such a collision
should be avoided by detecting an impending impact in advance and
intercepting the object in order to divert its course. Once the
collisional trajectory is precisely determined, it should be possible
to deflect it by placing an energy source or explosive charge on the
surface by using Shuttle capability that will become available in the
1980s. A small angular deflection may be sufficient if the warning
comes far enough in advance. At least, the Council urges, one should
start Project Spacewatch by searching for the objects that come
through near-Earth space, to determine their orbits precisely, and to
find out if and when a collision might occur. The Council's Report
estimates that at least some of the asteroids larger than 30 meters
in diameter could be found with a moderate size optical telescope.

In the meantime, we had already submitted proposals for building
a Spacewatch Camera. First, in 1977, the proposal was for a "Big

Schmidt", a 1.2-meter camera similar to the one at the Palomar Observatory in California with which we are familiar through previous observing programs of surveying the asteroid belt. On Palomar Mountain there is also a smaller Schmidt camera, of 46 centimeters aperture, that has been used since 1973 by Eleanor Helin and her associates in a continuing survey of near-Earth asteroids (see Helin's paper in this book).

The need was felt to obtain more time on larger Schmidt cameras, but especially the bigger ones are in great demand. The search for an easy solution, of using existing instruments, was not successful. Furthermore, we wanted to reach smaller objects and do it more automatically than is possible with photography. Even though at present only 50 near-Earth asteroids are known, the rate of discovery and size distribution indicate that there are at least 1,000 near-Earth asteroids and comets to be found in a ten-year search with a dedicated camera of 1.5-2.0 meter aperture. We noticed the rapid development in electronic detection devices and fast computers. By 1980 our design concept had grown from a photographic Schmidt camera to an electronically scanning camera combined with computer processing.

A ten-year program of scanning the sky with a dedicated instrument is a large project and it is therefore essential to spell out clearly the motivation and goals. Our own scientific reason is to study the statistics and chemical composition of asteroids and comets. It is to improve the understanding of the primitive material that formed from interstellar dust and gas and of the role that such planetesimals played in the formation of the solar system. We know of the very small dust particles that constitute the interstellar clouds. There are theories of the mechanisms for growth of the grains, for the sticking together of the dust particles in order to form larger planetesimal objects. The asteroidal or cometary planetesimals are an intermediate stage in the formation of the solar system.

We have not as yet seen or spectroscopically recognized an extinct cometary core among the asteroids. It is believed, however, that a piece of comet Encke impacted in 1908 near the Tunguska River in Siberia; there is no crater and very little material has been found, only some glassy beads. This is to be compared to the impact of the solid, more asteroid-like, object in Arizona some 20,000 years ago; the object was of the same 35-meter size as that of the Tunguska Event, but the one in Arizona made the Meteor Crater 1.2 km in diameter and 174 meters deep. The purpose of the Spacewatch Camera is to obtain statistics and physical information on the cometary cores as well as on the asteroids.

The connection between the asteroid belt and the meteorites is one of the study areas for the Spacewatch Camera. We know that in the asteroid belt there is a fair probability of collision. The velocities have a spread of 5 km/sec, on the average. When violent collision and break-up occur near the regions of resonance of their orbit with that of Jupiter then the orbit may be changed to one of greater eccentricity, crossing the orbits of Mars, Earth, Moon and Venus. Ernst Öpik was the first, in 1958, to study the processes of <u>supply</u> of objects from the asteroid belt, and <u>demand</u> by the terrestrial planets sweeping them up. He computed the lifetime of a near-Earth asteroid to be 10 million years on average. Harold Urey, in 1973, mentioned the disastrous effects of impacts on Earth, possibly causing the Cretaceous-Tertiary Extinction. Several other people have worked on the statistics of planet-crossing asteroids and cometary cores: George Wetherill, Edward Anders, Lubor Kresák, Richard Greenberg, to name a few. The statistics and the size-frequency relations are, however, poorly observed and this is a task for the Spacewatch Camera.

The best study of the surface compositions is with laboratory samples, as was done with lunar samples. The observations by the Spacewatch Camera therefore are of interest to NASA in order to consider the need and feasibility of sample return. First, the samples could be collected on the surface of the asteroid and analyzed by proper instrumentation on the spacecraft. The next step could be to return some samples to Earth for detailed laboratory analysis. The ultimate step would be to land an astronaut on an asteroid. Two suitable candidates for such missions are already known, namely asteroids Anteros and Aten. It is estimated that 25 Shuttle launches could put astronauts on Anteros or Aten. Compared to lunar landings, it is easier to land astronauts on these objects that have hardly any gravitation; no energy is needed to take them off the surface. The Spacewatch Camera is to discover even better candidates, the ones having less velocity difference with Earth. The estimated number of 25 Shuttle launches might be reduced to 7. Compared to such a saving, the cost of the Camera and its ten-year survey would be infinitesimal.

As one more motivation for the program, there are some people dreaming about future resources and mining of the asteroids. One would have to find the suitable candidates first and determine their chemical composition as well as their orbit.

A fundamental question for camera design is whether to do it photographically with a Schmidt-type telescope, or in some more modern electronic fashion. The Schmidt is attractive in the large coverage of the sky; with the Palomar "Big Schmidt" one routinely takes photographs with 41 square-degrees coverage. However, the quantum efficiency of photographic plates is low compared to that of charge coupled devices (CCD). Faint fast moving objects can be found only by sensitive detection. A CCD chip consists of rows and columns of pixels that are light-sensitive metal-oxide-semiconductor capacitors.

This program has been helped greatly by the staff of the Steward Observatory of the University of Arizona under the directorship of Peter Strittmatter and financial administration by William Stone. A dome was made available at the Steward Observatory site on Kitt Peak, approximately 80 kilometers west of Tucson, at an altitude of 2091 meters, a location with reportedly excellent image quality.

The RCA CCD 53612 that we presently have as basis for designs, mounted at the prime focus, has 512 pixels in the east-west scanning direction, while there are 320 of these 512-pixel rows. The integration time is 3.3 seconds in which time 163,840 pixels have to be read out, each with a brightness number of at least ten bits, for the various brightnesses of the stars and asteroids. If there were no data compression, the computer would have to process 1.6 million binary digits, "bits", in 3.3 seconds! The computer is to control the camera, analyze the data, find the asteroid, and compute predictions for follow-up observations.

The natural advantage of a CCD is that it operates by transfer of the charges from pixel to pixel, in the case of our CCD 512 times, after which the accumulated charge is read out. The rate of the charge transfer can be electronically controlled. The rate of the camera's scanning on the sky can, of course, also be controlled. Setting these two rates equal allows one to scan continuously. In our design, each star or asteroid image is on the 512 array during a total of 3.3 seconds; the image dwells 6.5 milliseconds on each 30 micrometer pixel. This rate is held for five minutes. The camera is stopped and the scan is exactly repeated. The computer has now stored 2 x 512 x 320 x 10 x 5 x 60/3.3 = 298 million bits (data compression can reduce this by a factor of 10). The two five-minute scans are intercompared by the computer, searching for an image that moved with respect to the stars two pixels or more in the five-minute interval, i.e. more than a quarter of a degree per day.

The sky coverage is to be as extensive as possible so as to minimize the chance that hazardous asteroids escape detection. The practical limits to the scanning are set by the available time and by phase effects. Just like the moon is seen at various phase angles, and thereby at various brightness levels, it follows that the near-Earth asteroids are most efficiently observed near full-phase, that is opposite the sun. We expect to have about nine nights per month available for the scanning, with the other nights used for follow-up of newly discovered objects and scanning of other objects such as more distant comets, stars, galaxies, etc. At the Kitt Peak site, about half of the time is suitable for this type of astronomical observation having clear sky, low wind velocities, and good steadiness of the image ("good seeing"). It then follows that about 17,000 square degrees will be scanned per year, provided that three CCD chips can be used simultaneously. Even this coverage we consider minimal and expect to improve it with better CCD's in the future. The limiting magnitude of the survey is presently set at 19.4 visual magnitude which allows detection of objects larger than about 0.3 km in diameter out to a distance of 0.3 astronomical units from Earth. With better detectors in the future, the sky coverage would be improved, up to a factor of 3; any further improvement would go into detection of smaller objects.

It is expected that about 1,000 near-Earth asteroids and comets will be found in the ten years of scanning, 1985-1995. Approximately half of these will be bright enough for a determination of their reflective properties and thereby of their physical characteristics. At the present phase of our planning it is not clear how much time of the Spacewatch Camera operation could be taken away from the scanning and devoted to such physical studies. Existing programs in Arizona, that use 1.5-2.8-m reflectors, could be asked to observe for the physical studies. In any case, one would like to determine as much as possible whether the objects are carbonaceous, silicaceous, or metallic. Or are they cometary objects?

The present status of the Spacewatch Camera is that of design and detailed preparation for acquisition of major hardware. Funds have already been obtained for a "Phase A" stage of conceptual design of the camera, the CCD detectors, the computer system, and the beginning of the computer programming. The programming will be the most time-consuming part of this project. It is the task of McMillan and Cliff Stoll to develop data compression, recognition of moving objects from consecutive scans, taking "seeing" into account, and computation of the orbits and ephemerides.

The conceptual mechanical drawings have been made under the guidance of Hans Boesgaard of the University of Hawaii. The camera will be directly driven by torque motors directly on the shafts, rather than gear-and-worm drives, and the setting and scanning will be precisely read out with encoders. All of this is, of course, computer controlled, while accelerometers on the camera frame provide refinements due to wind-shaking, etc. The dome has to move fast and be computer controlled in order to follow the rapid scanning. The prime-focus corrector lens system should give near absence of "barrel" or "pincushion" distortions; the design is by Richard Buchroeder. Three CCD detectors will have to be installed together, dry-ice or nitrogen cooled at prime focus. The scanning application of the CCD has to be tested extensively, as it is fairly new. Jack Frecker is in charge of this area. We receive considerable support from the Jet Propulsion Laboratory, where this is a joint project with E. M. Shoemaker and J. Degewij, as well as from the Steward Observatory. Our studies are supported by the Solar System Exploration division of NASA Headquarters. The project relies on the assistance and expertise of many people, more than can be acknowledged by name. We thoroughly appreciate the opportunity to build something new.

COLOR VARIATIONS OF ASTEROIDS DURING ROTATION

H.J. Schober and A. Schroll
Institut für Astronomie
Universität Graz
A-8010 Graz, Austria

A list of asteroids is presented, observed frequently for color variations during rotation as an evidence for spots on the surface.+(-)sign means redder (bluer) at maximum brightness,< means no detection larger than scatter; lightcurve amplitudes are added, polarisation measurements and/ or other color indices are taken from reference 5. Out of a sample of 49 asteroids there is evidence now for ten objects to exhibit spotted surfaces: 3 Juno, 4 Vesta, 6 Hebe, 25 Phocaea, 39 Laetitia, 42 Doris, 71 Niobe, 201 Penelope, 349 Dembowska and 944 Hidalgo.

Asteroid	Type	km	Ampl.	B-V	U-B	$\Delta P/P$	mag.	Base	Ref.
1 Ceres	C	1025	0.04			<0.005			5
3 Juno	S	249	0.05				+0.03	V-I	5
			0.14	-0.025	<0.01				16
4 Vesta	U	555	0.08			0.08	+0.015	U-V	5
			0.11			0.10	+0.046	U-V	5
6 Hebe	S	206	0.15				-0.02	U-V	5
							+0.04	V-I	5
9 Metis	S	168	0.10	<0.007	<0.01				15
15 Eunomia	S	261	0.50				<0.01	B-V	5
16 Psyche	M	249	0.21			<0.10	<0.01:	U-I	5
19 Fortuna	C	226	0.25			<0.03	<0.01	B-V	5
24 Themis	C	249	0.14				<0.02	B-I	5
25 Phocaea	S	73	0.20				+0.02	B-I	5
29 Amphitrite	S	199	0.11/2	<0.01			<0.01	V-I	2,5
36 Atalante	C	124	0.15	<0.01	<0.02				14
37 Fides	S	96	0.16	<0.02	<0.02				13
39 Laetitia	S	158	0.22			<0.08	-0.01	U-V	5
							<0.02	B-R	5
42 Doris	S	104	0.30	+0.02	<0.02				14
44 Nysa	E	68	0.26			<0.17			5
49 Pales	C	175	0.20				<0.02	U-I	5
71 Niobe	S	106	0.11				+0.05	U-V	5
55 Pandora	CMEU	185	0.24	<0.01	<0.02				9
63 Ausonia	S	94	0.95	<0.004	<0.006				5
68 Leto	S	128	0.19	<0.02	<0.03				20

285

W. Fricke and G. Teleki (eds.), Sun and Planetary System, 285–286.

87 Sylvia	CMEU	251	0.40	<0.02	<0.03			15
135 Hertha	M	79	0.17	<0.002	<0.005			6
139 Juewa	C	165	0.20	<0.01	<0.02			4
148 Gallia	S	92	0.32	<0.01	<0.02			19
161 Athor	CMEU	100	0.27	<0.02	<0.03			4
173 Ino	C	169	0.04	<0.01				9
186 Celuta	U	49	0.40	<0.003	<0.006			8
201 Penelope	CMEU	144	0.52	+0.05 ?	+0.06 ?			7,18
218 Bianca	S	59	0.22	<0.002	<0.007			1
247 Eukrate	C	143	0.10	<0.02	<0.03			15
308 Polyxo	U	139	0.20	<0.01	<0.02			3
349 Dembowska	R	145	0.35			<0.12	<0.01 V-I	5
			0.35				+0.04 U-I	5
387 Aquitania	S	113	>0.09	<0.01				10
404 Arsinoe	C	99	0.36	<0.04	<0.06			12
433 Eros	S	20	1.30			<0.025	<0.01 U-I	5
485 Genua	–	–	0.12	<0.003	<0.008			1
563 Suleika	S	59	0.21	<0.02	<0.03			20
628 Christine	U	54	>0.40	<0.03	<0.04			12
679 Pax	U	74	0.07	<0.01	<0.02			11
683 Lanzia	–	–	0.12	<0.004	<0.009			1
736 Harvard	S	14					<0.02 B-I	5
776 Berbericia	C	183	>0.15	<0.01				10
796 Sarita	CMEU	88	0.29	<0.01	<0.02			11
944 Hidalgo	MEU	29	0.35				+0.12 U-V	5
1245 Calvinia	U	38	0.65				<0.02 B-I	5
1580 Betulia	U	6	0.60	<0.02	<0.02			21
1978 CA	S	1	0.80	<0.05	<0.15		<0.03 V-I	5,17
1978 DA	SU	–	0.5:	<0.05	<0.15		<0.03 V-I	5,17

Carlsson,M.,Lagerkvist,C.I.: 1981,Astron.Astrophys.Suppl.44,p.15   (1)
Debehogne,H.,Surdej,A.and J.:1978,Astron.Astrophys.Suppl.32,p.127   (2)
Debehogne,H.,Zappalà,V.: 1980, Astron.Astrophys.Suppl.39,p.163   (3)
Debehogne,H.,Zappalà,V.: 1980, Astron.Astrophys.Suppl.42,p.85   (4)
Degewij,J.,Tedesco,E.F.,Zellner,B.: 1979, Icarus 40,p.364   (5)
Lagerkvist,C.I.: 1981, Astron.Astrophys.Suppl.44,p.345   (6)
Lagerkvist,C.I.,Rickmann,H.: 1981, Moon and Planets, 24, 437.   (7)
Lagerkvist,C.I.,Petterson,B.: 1978, Astron.Astrophys.Suppl.32,p.339   (8)
Schober,H.J.: 1978, Astron.Astrophys.Suppl.34,;p.377   (9)
Schober,H.J.: 1979, Astron.Astrophys.Suppl.38,p.91   (10)
Schober,H.J.: 1981, Astron.Astrophys.99,p.199   (11)
Schober,H.J.: 1981, Astron.Astrophys.100,p.311   (12)
Schober,H.J.: 1981, Astron.Astrophys.,in press   (13)
Schober,H.J.,Schroll,A.:1981, Astron.Astrophys.,submitted   (14)
Schober,H.J.,Surdej,J.: 1979, Astron.Astrophys.Suppl.38,p.269   (15)
Schroll,A.,Schober,H.J.: 1981, Astron.Astrophys.,in press   (16)
Schuster,H.E.,Surdej,A.and J.:1979,Astron.Astrophys.Suppl.37,p.483   (17)
Surdej,J.,Cramer,N.: 1980, IAU Circ. 3527   (18)
Surdej,A.,Surdej,J.: 1979, Astron.Astrophys.Suppl.37,p.471   (19)
Surdej,J.,Schober,H.J.: 1980, Astron.Astrophys.Suppl.41,p.335   (20)
Tedesco,E.F.,Drummond,J.,Birck,P.,Nikoloff,I.,Candy,M.,Zellner,B.:
     1978, Icarus 35,p.344   (21)

RESULTS FROM OCCULTATIONS BY MINOR PLANETS

Gordon E Taylor
Chairman, Working Group on Predictions of Occultations
by Satellites and Minor Planets (IAU Commission 20)
Royal Greenwich Observatory, Herstmonceux Castle,
Hailsham, East Sussex BN27 1RP, UK.

1. INTRODUCTION

Since the minor planets are believed to consist of primordial matte dating from the time of the formation of the solar system there is great interest in determining their composition. It is therefore necessary to calculate their densities, for which we need accurate masses and sizes. On the rare occasions when a minor planets occults a star, timed observations of the event from a number of observing sites enatle an accurate size of the minor planet to be determined.

2. PREDICTIONS

Osculating elements of 184 minor planets, produced by the Institute of Theoretical Astronomy in Leningrad, are used at the Royal Greenwich Observatory to provide daily ephemerides. These ephemerides are then searched against a star catalogue on a disk, containing the position of all SAO and AGK3 stars within 45° of the ecliptic. All appulses of minor planets to stars are noted and a list of possible occultations is produced and printed in various publications. Detailed preliminary predictions are issued to many observatories. It is not possible to improve the predictions until a few days before the event, when a photographic plate showing both the minor planet and the target star can be taken. The positions of these two objects are then determined with reference to the same comparison stars and final predictions are then issued by telex, telegram or telephone.

3. RESULTS

From 1953 until around 1975 predictions of occultations by Ceres, Pallas, Juno and Vesta, only, were issued. An occultation by Juno was observed visually from one place and an occultation by Pallas observed photoelectrically from one observatory in India. Thus only minimum values for the diameters were obtained. A short duration occultation by Eros was also observed in 1975.

W. Fricke and G. Teleki (eds.), Sun and Planetary System, 287–288.
Copyright © 1982 by D. Reidel Publishing Company.

The number of minor planet orbits investigated for occultations
has increased rapidly during the last five years and now stands at
184. As a result the number of occultations observed has also
increased dramatically as Table 1 shows.

Table 1
Occultations observed, 1975-1981

Date	Minor Planet	Area of observation	Derived diameters km
1975 Jan. 23	433 Eros	U.S.A.	20
1977 Mar. 5	6 Hebe	Mexico	195
1978 May 29	2 Pallas	U.S.A.	558 x 526 x 532
1978 June 7	532 Herculina	U.S.A.	217*
1978 Dec. 11	18 Melpomene	U.S.A.	133
1979 Aug. 17	51 Nemausa	U.S.S.R.	153
1979 Oct. 17	65 Cybele	U.S.S.R.	230
1979 Dec. 11	9 Metis	Guyana	(127)
1979 Dec. 11	3 Juno	U.S.A.	290 x 246
1980 Sept. 14	78 Diana	U.S.A.	130*
1980 Oct. 10	216 Kleopatra	U.S.A.	120 x 80*
1980 Nov. 24	134 Sophrosyne	U.S.A.	110*
1981 Aug. 7	18 Melpomene	Australia	(126)
1981 Oct. 5	105 Artemis	S. Africa	(110)

*   Analyses not yet published
    Values in brackets are minimum values (chord lengths)

4.   FUTURE POSSIBILITIES

Predictions for 1982 were issued in 1980 October and those for
1983 will be available shortly. In addition predictions for Ceres,
Pallas, Hygiea and Vesta have been published up to 1989.

A fourth-magnitude star, 1 Vulpeculae, will be occulted by Pallas
on 1983 May 29. The predicted track crosses the southern states of the
U.S.A. from east to west. The event will be easily observable, the
magnitude change at occultation being 4.9 visually and 5.8 photoelectrically. The maximum duration will be 46 seconds.

European astronomers should plan ahead to observe occultations by
two of the largest minor planets in 1988, one by Vesta on March 1 and the
second by Hygiea on March 8. The predicted track of the central line
of the occultation by Vesta, which has a magnitude change of 0.3 visually
and 0.2 photoelectrically and a maximum duration of about 2^m4, is
from northern Libya, across north east Algeria, the Franco-Spanish
border and then just south-west of the British Isles to Iceland. The
track of the Hygiea event, magnitude changes 0.5 visually and 0.7
photoelectrically and maximum duration 0^m6, is from Saudia Arabia,
through Egypt, northern Libya and Algeria, and Iberia to Labrador.

ON THE ROTATIONS OF M ASTEROIDS

C.-I. Lagerkvist   and   H. Rickman
Astronomiska Observatoriet, Uppsala, Sweden

According to the theory of collisional evolution of asteroid rotations by Harris (1979) the densest objects should have the highest average spin rates. Since it is widely believed that M asteroids are of metallic composition, this theory predicts that they rotate more rapidly than the stony C or S objects. Harris and Burns (1979) found indications that this is indeed the case. Their result was confirmed by the present authors (Lagerkvist and Rickman, 1981) in a specialized study. Our conclusion is that both M asteroids and those designated CMEU are characterized by high spin rates and large lightcurve amplitudes.

It is admittedly difficult to draw generalized statistical conclusions about the rotational properties of asteroids. These depend on a number of different parameters and false correlations might easily result. However, we find it significant that the measured M and CMEU asteroids show neither any preference for size ranges in which asteroids often rotate rapidly, nor any clearcut tendency for family membership. Indeed, regarding Hirayama families the converse is true. The M and CMEU asteroids listed by Bowell et al. (1979) clearly appear to avoid membership in these groupings.

In small families, however, such asteroids appear in normal abundance. A particularly interesting case is provided by the M asteroid 338 Budrosa which belongs to the same family as 347 Dembowska and is likely to be a fragment of the same differentiated parent asteroid (Chapman, 1979; Gradie et al., 1979). According to current ideas, Budrosa would be either the metallic core of the parent body or a fragment of it. Evidently information about its shape and rotation is of great interest.

We have observed the M asteroids 76 Freia and 338 Budrosa along with asteroids 125 Liberatrix and 317 Roxane, classified as CMEU and MU, respectively, by Bowell et al. (1979). The observations were performed at the European Southern Observatory in September 1981. The present report gives preliminary results concerning the lightcurves of these four asteroids.

W. Fricke and G. Teleki (eds.), Sun and Planetary System, 289–290.

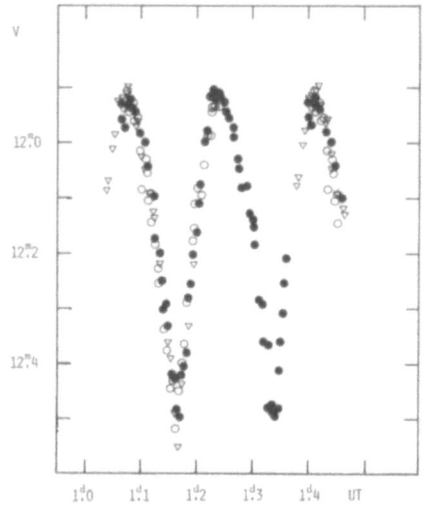

Figure 1. Composite lightcurve of the asteroid 317 Roxane during September 1981. Observations from Sep. 1 (filled circles), Sep. 2 (open circles) and Sep. 4 (triangles).

338 Budrosa was observed during seven nights. There are indications that the period is not very long and that the amplitude is quite small. We consider it likely that the asteroid was seen pole-on, but of course the possibility of a nearly spherical shape can not be excluded. For 76 Freia indications about the rotation period are similar. We have obtained an accurate determination of the synodic period of 317 Roxane, namely, $0\overset{d}{.}340 \pm 0\overset{d}{.}001$. The composite lightcurve is shown in Figure 1. The magnitude scale holds for the Sep. 1 observations. For Sep. 2 and 4 we have adjusted the magnitudes by $\Delta V = 0\overset{m}{.}035$ and $0\overset{m}{.}135$, respectively. These adjustments are due to phase variation, partly caused by the opposition effect. The interval of phase angles is $3°$ to $5°$. For the phase coefficient β_V we derive a value of 0.047 mag/deg. The absolute magnitude is found to be: $V_{max}(1,0) = 9\overset{m}{.}98$. The results for 125 Liberatrix indicate that its rotation period is shorter than 4^h and that the amplitude is $0\overset{m}{.}3$. In conclusion, our new observations tend to confirm the above-mentioned results (Lagerkvist and Rickman, 1981).

REFERENCES

Bowell, E., Gehrels, T., and Zellner, B.: 1979, in *Asteroids*, ed. T. Gehrels, Univ. of Arizona Press, Tucson, pp. 1108-1129.
Chapman, C.R.: 1979, in *Asteroids*, ed. T. Gehrels, Univ. of Arizona Press, Tucson, pp. 25-60.
Gradie, J.C., Chapman, C.R., and Williams, J.G.: 1979, in *Asteroids*, ed. T. Gehrels, Univ. of Arizona Press, Tucson, pp. 359-390.
Harris, A.W.: 1979, Icarus 40, pp. 145-153.
Harris, A.W., and Burns, J.A.: 1979, Icarus 40, pp. 115-144.
Lagerkvist, C.-I., and Rickman, H.: 1981, the Moon and the Planets 24, pp. 437-440.

ASTEROID ROTATION RATES : COMPARISON BETWEEN THEORY AND OBSERVATIONS

P. Paolicchi[+], P. Farinella[+] and V. Zappalà[++]
+ Osservatorio Astronomico di Brera, Merate, Italy
++ Osservatorio Astronomico di Torino, Pino Torinese, Italy

ABSTRACT . Harris' (1979) theory on the collisional evolution of the asteroid spin rates is compared with the observational evidence provided by the statistics of photometric data. The resulting discrepancies can be qualitatively explained as due to different physical processes, occurred in the frame of the asteroid collisional history and connected in particular with the outcomes of catastrophic impact events.

In a recent investigation (Farinella et al.,1981a), we have analyzed the distribution of spin rates for a sample of 280 asteroids whose rotational properties have been determined by photometric techniques. By taking into account the most important selection effects, it has been possible to establish on a good statistical ground some new interesting conclusions, which can be summarized as follows : (a) the spin rate (Ω) distribution of the whole available sample presents a rather broad peak for $1<\Omega<3.5$ rotations/day, with a mean value $\bar{\Omega}$ close to 2.5 rotations/day; the best fit of the distribution by a 3-dimensional Maxwellian curve is not satisfactory from the point of view of the χ^2 test , due to an excess of slow rotators (of period P larger than about 16 hr) and to a corresponding depletion of the Maxwellian's peak (see Fig.1).(b) The observed excess of slow rotators is mainly caused by small (diameter D<100 km) and intermediate objects (100<D<200 km); correspondingly, the mean spin rate increases with the asteroid size (as previously concluded by Tedesco and Zappalà,1980), reaching 3.37 rotations/day for bodies with D>200 km. (c) Two different correlations appear with the taxonomic type : first, non-C/S objects have a very flat spin rate distribution and are on the average faster rotators, and second, the distributions of C and S asteroids, though having close values of $\bar{\Omega}$, show a different shape (the C distribution is broader and more clearly non-Maxwellian).

How can we physically interpret this observational evidence ? Harris (1979) has first proposed a quantitative model for the collisional origin

W. Fricke and G. Teleki (eds.), Sun and Planetary System, 291–294.

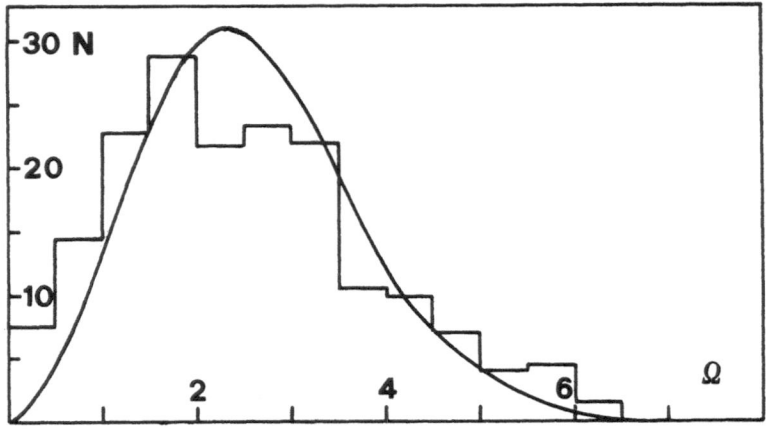

Figure 1. The distribution of Ω (in rots./day) for a sample of about 200 asteroids, chosen in order to minimize observational biases (see Farinella et al.,1981a). The best-fitting Maxwellian curve is also shown for comparison.

of the rotation of asteroids. By using several simplifying assumptions, this model obtains some conclusions of interest for a comparison with observations. In particular, $\bar{\Omega}$ should be independent on the asteroid size for diameters larger than a few kilometers; its value is determined by the equilibrium between a "drag" term caused by the frequent small-scale impacts and the spin-up effect of the rare nearly-catastrophic collisions. This equilibrium spin rate depends on some poorly known parameters, like the mean density of asteroids, the characteristic exponent of their mass distribution and the fraction δ_{KE} of the projectile's kinetic energy delivered to fragments after a catastrophic breakup. However, if δ_{KE} is at most of the order of 0.1, as suggested by laboratory impact experiments (Fujiwara and Tsukamoto,1980),Harris'model yields values of $\bar{\Omega}$ clearly too high, corresponding to periods not larger than 2÷3 hr (Davis et al.,1979). Moreover, if one assumes that the collisional interaction among asteroids is an isotropic "kinetic" process, the spin rate distribution should be approximated by a 3-dimensional Maxwellian (Harris and Burns,1979). These conclusions are clearly contradicted by the available observational evidence (as summarized above), and the problem arises to understand the reasons of these discrepancies.

We believe that the proposal of a more refined and satisfactory theoretical model should be now premature, due to lack of data and physical insight into many relevant processes. However, it is possible to indicate some qualitative explanations for the observed discrepancies, implying at the same time some different areas of future theoretical and observational work needed to understand the evolution of asteroids :

(1) Angular momentum partitioning during catastrophic collisions.
Harris' theory assumes that the spin rate of the fragments is the same of
their parent bodies, but this seems unlikely, since various complex phenomena (see for instance Hartmann,1979) can affect the distribution of
the spin angular momentum among fragments of different size and ejection
velocity. The situation is particularly uncertain for small asteroids,
because their spin rate cannot relax to Harris' equilibrium value during
the interval between two successive breakup events, and so they should
reflect the spin rate distribution arisen from the breakup itself. Indeed,
we recall that small asteroids present the most pronounced deviations from
the Maxwellian distribution, suggesting perhaps that some energy-dissipa-
ting process is effective in causing a slow-down of their rotation just
after the collisional breakup . Laboratory work on this problem appears
feasible, and it could provide very important hints on the behaviour of
real asteroids. Moreover, some observational work on the rotational properties of small family asteroids, which are likely endproducts of some
well-defined breakup events, could be useful to identify some systematic
trend or correlation of more general validity.

(2) Formation of binary and/or multiple objects. If a substantial fraction
of asteroids are or have been during the solar system's lifetime components
of multiple systems (Chapman et al.,1980), the tidal despinning mechanism
could have strongly affected the present spin rate distribution, by trans-
ferring rotational into orbital angular momentum with relatively short
timescales (Zappalà et al.,1980). At the same time, this mechanism seems
the most plausible explanation for the existence of several objects with
spin periods as long as many tens of hours. In this field, theoretical
work is needed to understand better in which cases asteroidal collisions
can give rise to multiple systems (the necessary conditions are probably
high angular momentum transfer from the projectile's motion and strong
self-gravitational binding of the target) and, in these cases, which are
the typical number of components,mass ratio and separation of the systems.
As suggested by Weidenschilling (1981), it is also worthwhile to note that
the possibility of binary fission can change the spin rate distribution
in the high-Ω tail, by implying a rotational stability limit for periods
of about 4 hours (at least for asteroids dominated by self-gravitation).

(3) Ellipsoidal equilibrium shapes. Harris' theory assumes (for the sake
of simplicity) that the shape of asteroids is spherical. This assumption
breaks down in a significant way in the intermediate size range, where a
large fraction of highly elongated bodies is implied by the observed
lightcurve amplitudes (Tedesco and Zappalà,1980). These bodies have momenta of inertia up to two times the spherical value, and therefore, for the
same angular momentum, they rotate slower than spherical objects. Indeed,
in this size range we find that Harris' equilibrium spin rate would be

approached by a population of spherical bodies with a similar distribution
of angular momenta. It is therefore possible that the intermediate size
range (and particularly for 200<D<280 km) is the only one where collisional equilibrium of asteroid rotations can be reached. The large abundance
of highly elongated shapes is correlated with short rotational periods
(just above the limit for binary fission), and this fact is consistent
with the idea that the shape of these asteroids is mainly determined by
gravitational equilibrium, which produces Jacobi triaxial ellipsoids
when the angular momentum is high enough (Farinella et al.,1981b).

(4) Negligible collisional evolution of large asteroids. For D>280 km,
no highly elongated asteroid has been observed. This could be due to the
fact that for these objects the angular momentum transfer caused by collisional events has been low, perhaps smaller than the primordial angular
momentum typical of the accreting proto-asteroids before the present
regime of desruptive collisions was established. In terms of Harris'
theory, this would mean that for large asteroids the timescale of collisional relaxation to the equilibrium spin rate has been longer than the
age of the solar system. Such a conclusion would deserve further analysis,
since it implies a well-defined and quantitative constraint on every
model of the asteroid collisional evolution.

REFERENCES

Chapman,C.R.,Davis,D.R. and Weidenschilling,S.J.: 1980, Bull.Amer.Astron.
    Soc. 12, p.662.
Davis,D.R.,Chapman,C.R.,Greenberg,R.,Weidenschilling,S.J. and Harris,A.W.:
    1979, in Asteroids, ed.T.Gehrels, pp.528-557, University of Arizona
    Press, Tucson.
Farinella,P.,Paolicchi,P. and Zappalà,V.: 1981a, Astron.Astrophys.,in
    press.
Farinella,P.,Paolicchi,P.,Tedesco,E.F. and Zappalà,V.: 1981b, Icarus 46,
    pp.114-123.
Fujiwara,A. and Tsukamoto,A.: 1980, Icarus 44, pp.142-153.
Harris,A.W.: 1979, Icarus 40, pp.145-153.
Harris,A.W. and Burns,J.A.: 1979, Icarus 40, pp.115-144.
Hartmann,W.K.: 1979, in Asteroids, ed.T.Gehrels, pp.466-479, University
    of Arizona Press, Tucson.
Tedesco,E.F. and Zappalà,V.: 1980, Icarus 43, pp.33-50.
Weidenschilling,S.J.: 1981, Icarus 46, pp.124-126.
Zappalà,V.,Scaltriti,F.,Farinella,P. and Paolicchi,P.: 1980, Moon and
    Planets 22, pp.153-162.

ASTEROID COLLISIONAL EVOLUTION : OUTCOMES OF CATASTROPHIC IMPACTS

P. Paolicchi[+], P. Farinella[+] and V. Zappalà[++]
+ Osservatorio Astronomico di Brera, Merate, Italy
++ Osservatorio Astronomico di Torino, Pino Torinese, Italy

ABSTRACT . We discuss how catastrophic collisional events have influenced some outstanding properties of asteroids, like their structure, rotation, shape and family membership. The ultimate outcomes of the collisional evolution are found to be strongly dependent on the asteroid size, because of the different importance of self-gravitational forces and of the different probability of impacts with high projectile-to-target mass ratio.

1. INTRODUCTION

The importance of catastrophic collisions for asteroid evolution is commonly accepted, but only recently its central role against other evolutionary phenomena, such as accretion or orbital evolution, has become apparent. For catastrophic collision we mean an impact event causing a breakup of solid-state forces, sometimes melting phenomena, generally ejection of large fragments. According to Fujiwara et al.(1977) and to Ip (1979), a collision can result into a variety of different outputs, ranging from cratering to spallation to complete disruption of the target asteroid or, at least, of a large part of it. In this latter case we must have a substantial breakup of the solid-state forces (mechanical strength) which keep the asteroid material together. For the disruption of real asteroids we must take into account also self-gravitational forces, which increase the total binding energy of the body in a way dependent on the asteroid size.

2. COLLISIONAL OUTCOMES VS. PROJECTILE-TO-TARGET MASS RATIO

In order to study the outcomes of the major collisions suffered by asteroids of different sizes, we compare the projectile's kinetic energy $\frac{1}{2}mv^2$ with the energy needed to overcome the target's solid-state cohesion (expressed as SM, where M is the target mass and the critical energy density S can be assumed to range from 10^5 to 10^8 erg/g, depending on the material properties) plus the target's self-gravitational binding

295

W. Fricke and G. Teleki (eds.), Sun and Planetary System, 295–298.

energy. This latter (see for instance Davis et al., 1979) is multiplied
by a factor γ, to account for internal energy losses due to dissipative
effects. We expect that catastrophic fragmentation occurs and that a
substantial part of the mass is dispersed to infinity when

$$\tfrac{1}{2}mv^2 > SM + (3\gamma /5)(GM^2/R) , \qquad\qquad (1)$$

where R is the target's radius. A realistic value of γ could be 20, even
if we must keep in mind that this estimate could be wrong by $\pm$ one order
of magnitude for actual asteroidal collisions; moreover, collisions with
widely different values of γ could be commonplace. The response to the
collision is different depending whether the first or the second term in
the right-hand member of (1) dominates. For R smaller than $\sim$ 10 km (a very
indicative value obtained by adopting $\gamma=20$, $S=10^7$ erg/g, density of 2.5
g/cm^3), the collisional process is governed by solid-state effects, and
only two typical outputs are to be expected : (a) cratering, i.e. localized target damage, when the impact energy is less than some critical value,
corresponding, for a typical impact velocity of 5 km/s, to a mass ratio
$\mu \equiv m/M$ lower than a critical value, presumably in the range $10^{-6} \div 10^{-4}$;
(b) if the impact energy is larger than the critical value, the target is
subjected to catastrophic disruption and all the fragments are dispersed
to infinity. On the contrary, for larger asteroids, a strong gravitational
bound is present and they respond to collisions in a more complex way,
depending mainly on the comparison between μ and

$$\tilde{\mu} \equiv (6\gamma/5)(GM/Rv^2) = (3\gamma/5)(v_E/v)^2 , \qquad\qquad (2)$$

where v_E is the target's escape velocity. We can sketch the following
classification : (1) $\mu<\mu_{cr}$ (this latter is the critical value for solid-
state breakup): only cratering phenomena; (2) $\mu_{cr}<\mu<\tilde{\mu}$: extensive fragmentation with deep fractures throughout the asteroid interior, but small
dispersal of fragments, which are mostly reconcentrated by the self-gravitation; the final outcome is a "rubble pile" or a megaregolith (Davis
et al.,1979), shaped roughly as a figure of gravitational equilibrium
(i.e., a Maclaurin spheroid or a Jacobi triaxial ellipsoid, depending
on the spin angular momentum of the body; see Farinella et al., 1981);
(3) $\mu \simeq \tilde{\mu}$: at least some large fragments escape with small relative velocities (comparable with v_E), while some other fragments coalesce again,
as in case (2); this seems a reasonable scenario for the origin of several
dynamical families; (4) $\mu>>\tilde{\mu}$: after the target's disruption all the fragments are dispersed to infinity, achieving widely different heliocentric
orbits. Now, if we can evaluate, for each asteroid size, the mass ratio
of the largest body which collided with it during the solar system's
lifetime, we can assess the different diameter ranges where the previous
collisional outcomes can be preferentially found. Using a collisional
model like that by Dohnanyi (1969), we can deduce that for diameters D
smaller than about 100 km, case (4) must be considered the rule; on the

contrary, for D >200÷250 km, even if super-critical impacts did probably occur, the bodies are megaregolith objects with "regular" shapes and did not loose much of their mass due to collisional evolution (cases (1) and (2)). In the intermediate diameter range, cases (2), (3) and (4) may have occurred, due to different values of various parameters determining the collisional output (e.g., γ, v, etc.).

3. OBSERVATIONAL EVIDENCES

As shown by Farinella et al. (1981), a representative test to investigate the collisional history of asteroids is given by the A vs. ω plot, where A is the lightcurve amplitude (giving roughly the triaxial elongation of the body) and ω is the rotational frequency. Fig. 1(a) presents this plot for family asteroids (from Williams, 1979) larger than 100 km together with all the bodies larger than 200 km. As pointed out by the strong correlation between amplitude and spin rate, the distribution suggests that most objects fit roughly an equilibrium shape (since the Jacobi ellipsoids must be rapid rotators). This fact can be interpreted as follows : (a) for large asteroids (D>200 km) a catastrophic impact caused generally deep fracturing processes, with most ejecta falling down on the remnant "core" (case (2));these objects have probably the appearance of "piles of rubble", and can be considered as non-born families. (b) For smaller diameters (100÷200 km), we may have families formed as in case (3), with a few large fragments escaped in preferential directions ("asymmetrical families") and an equilibrium-shaped largest body. In this range, if the impact angular momentum exceeded a critical value, the splitting into a double or multiple body is also possible (Zappalà et al., 1980).

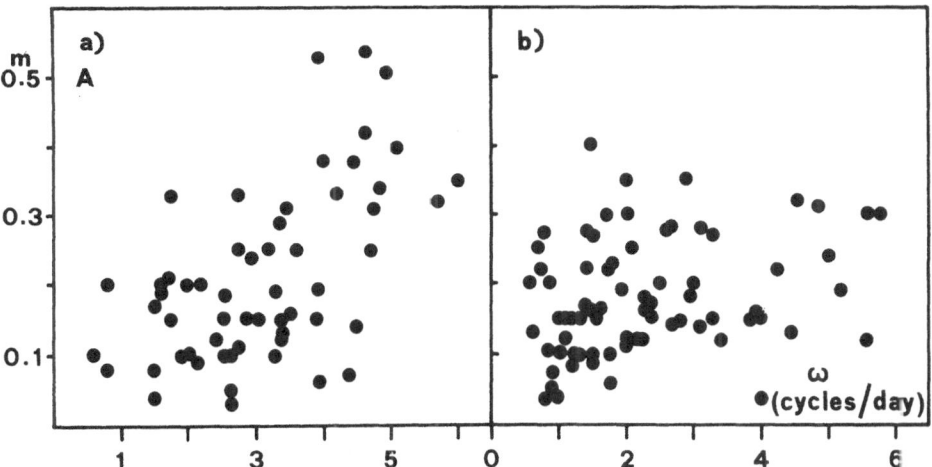

A vs. ω distributions for family (D>100 km) and non-family asteroids (D>200 km) (part (a)), compared with non-family asteroids (100<D<200 km) (part (b)).

The previous conclusions are confirmed when we compare Fig.1(a) with Fig.1(b), where a similar A vs. ω plot is displayed for non-family asteroids in the size range 100÷200 km. The distribution appears much more randomized, with no apparent correlation between the parameters. A similar result is derived from the analysis of objects with D<100 km (see Farinella et al.,1981) : also in this case the situation is conform to that expected for fragments escaped after a complete breakup of the target, with formation of a "dispersed" family or of no family at all.

Finally we have reconstructed the total mass of a large number of families, assuming that it corresponds roughly to the mass of the parent body (PB). Then we have compared the mass of the largest asteroid in the family with that of the PB, obtaining the ratio δ. The result is that δ is close to 1 for large PBs (we recall that for D>200 km most objects could be considered as non-born families), but it decreases significantly for smaller PB size, until below 50 km we have a high dispersion of the fragments with no preferential core. Moreover the "asymmetrical" families (the asymmetry is found in the distribution of proper orbital elements; see Ip, 1979) represent about 70% of those having PB larger than 150 km, 43% for the size range 50÷150 km and only 12% for PBs smaller than 50 km. We note also that the largest asteroid of "asymmetrical" families is very often a triaxial ellipsoid, a spheroid or a possible binary (consistently with the hypothesis of equilibrium shapes),when δ is close to 1, i.e., when the diameter is large enough (>50÷75 km). A particular and different approach seems required for the few largest Hirayama families : they have low values of δ, implying a strong mass dispersion, though their PB mass is fairly large. Several hypotheses are possible : either the breakup event was particularly energetic, or it was not a single event, or these families are composed by different groups with similar proper elements (as it seems the case for the Flora family).

REFERENCES

Davis,D.R.,Chapman,C.R.,Greenberg,R.,Weidenschilling,S.J. and Harris, A.W.:1979, in *Asteroids* (T.Gehrels,Ed.),pp.528-557, University of Arizona Press, Tucson.

Dohnanyi,J.S.:1969, J.Geophys.Res. 74, pp.2531-2554.

Farinella,P.,Paolicchi,P.,Tedesco,E.F. and Zappalà,V.:1981, Icarus 46, pp. 114-123.

Fujiwara,A.,Kamimoto,G. and Tsukamoto,A.:1977, Icarus 31, pp.277-288.

Ip,W.-H.:1979, Icarus 40, pp.418-422.

Williams,J.G.:1979, in *Asteroids* (T.Gehrels,Ed.),pp.1040-1068, University of Arizona Press, Tucson.

Zappalà,V.,Scaltriti,F.,Farinella,P. and Paolicchi,P.:1980, Moon and Planets 22, pp.153-162.

AN IMPROVED REPRESENTATION OF THE AVERAGE
OPPOSITION MAGNITUDES OF ASTEROIDS

Z. Knežević
Astronomical Observatory, Belgrade, Yugoslavia
V. Zappalà
Astronomical Observatory, Pino Torinese, Italy

ABSTRACT: It is well-known that the asteroid mean opposition magnitude - $B(a,0)$ - represent a crude approximation of asteroid average opposition brightness, especially for objects of high eccentricity and/or inclination. In the paper we propose a new simple method of calculating the "quasi-median opposition magnitude", which very closely matches the median apparent brightness of an asteroid in opposition.

The mean opposition magnitude $B(a,0)$ is intended to represent an useful information for observers on the average brightness of an asteroid in opposition. It is frequently used in investigating the effects of observational selection on the statistics of dynamical properties of asteroids (e.g. Kiang, 1966; Kresák, 1967; Knežević, 1981). Currently, it is calculated according to a very simple formula - $B(a,0)=B(1,0)+5\log a(a-1)$ - and it gives the apparent magnitude of an asteroid in a fictitious opposition, at the distance $\underline{a}$ from the Sun and $\underline{a}$ -1 from the Earth, and at zero phase. However, this approximation holds only for nearly-circular, zero-inclination orbits, deviating from the true average opposition magnitude typically by several tenths of a magnitude (Fig. 1). In the extreme cases (Apollos, Amors, 944 Hidalgo) the discrepancy can reach as much as 1^m.

We propose an improved method of representing the average opposition magnitude of asteroids, which makes allowance, at least roughly, for the eccentricity and inclination of the asteroid's orbit. For the sake of simplicity, we neglect the eccentricity of the Earth's orbit, and consider only the Keplerian orbits of asteroids.

The number frequency distribution of the opposition magnitudes of an asteroid will be rather asymmetrical already if its orbital eccentricity exceeds 0.2. Usually, one distribution tail includes few very bright oppositions, while a vast majority of oppositions is concentrated within 1 or 2 relatively faint magnitude intervals (Fig. 2). The true mean magnitude is, therefore, shifted with respect to the corresponding median towards brighter magnitudes, the shift being the larger the more eccentric is the asteroid's orbit. $B(a,0)$ - the calculated mean opposition

299

W. Fricke and G. Teleki (eds.), Sun and Planetary System, 299–302.
Copyright © 1982 by D. Reidel Publishing Company.

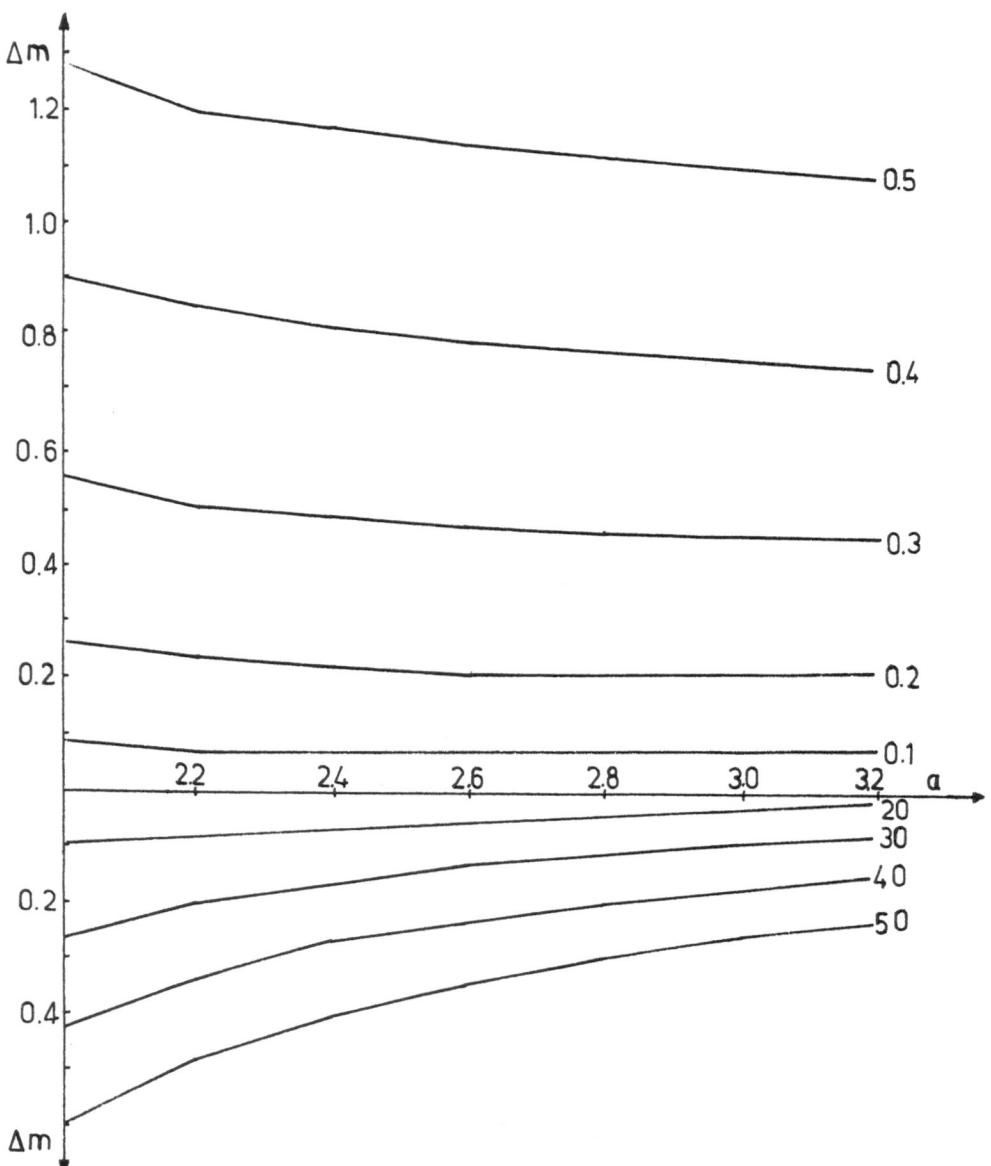

Figure 1. Magnitude difference of the true median magnitude of an asteroid in opposition and $B(a,0)$. Upper diagram: differences for zero-inclination orbits of different eccentricities and semimajor axes, calculated according to $\Delta m=B(1,0)+5\log r_M(r_M-1)-B(a,0)$. Lower diagram: differences for circular orbits of different inclinations and semimajor axes, calculated according to $\Delta m=B(1,0)+5\log a(a^2+1-2\ a\ \cos\ 0.7\ i)^{1/2}-B(a,0)$.

magnitude - is even more shifted and, almost as a rule, it divides the
true oppositions brighter and fainter than this in a approximate rat o
of 1:2. Consequently, it appears more appropriate to use the median opposition magnitude as a measure of the asteroid average opposition
brightness, and we find the following simple way of calculating the
"quasi-median value" - B(QM,0) - which closely approximates the true
median (being, in fact, always between the true mean and median).

It is obvious that an asteroid spends more time in the aphelion than in
the perihelion part of its orbit, and that, therefore, the oppos tion
points tend to cluster around the aphelion. The asteroid spends one half
of the time at distances from the Sun that are greater, and one half at
distances smaller than:

$$r_M = \frac{a(1 - e^2)}{1 + e \cos v_M} \qquad (1)$$

where

$$v_M = \Pi/2 + 2e - 5/6\ e^3 + \ldots \qquad (2)$$

substitutes the exact Kepler's equation; r_M is the median distance of
the asteroid from the Sun. By using it instead of the geometrical mean
distance a in the formula for calculating the average opposition magnitude, one makes allowance for the eccentricity-dependent, non-uniform
distribution of oppositions along the orbit.

The average distance of an asteroid from the Earth depends not only on
the eccentricity of the asteroid's orbit, but also on the orbit inclination and orientation, so that computing the true median distance from
the Earth for each asteroid would be quite complicated. However, we can
easily calculate a pair of distances from the Earth which corresponc to
r_M. The two heliocentric ecliptical latitudes at which the "quasi-median
oppositions" ($r = r_M$) occur, are given by:

$$\sin b_1 = \sin i \sin (v_M + \omega)$$
$$\sin b_2 = \sin i \sin (v_M - \omega + \Pi) \qquad (3)$$

Hence, the distances from the Earth of an asteroid in the quasi-med an
opposition can be expressed as:

$$\Delta_j^2 = r_M^2 + 1 - 2r_M \cos b_j \qquad j = 1,2 \qquad (4)$$

and

$$\Delta_{QM} = (\Delta_1 + \Delta_2)/2 \qquad (5)$$

Finally, by applying (1) and (5) to the standard equation for calculating the apparent magnitude, we find:

$$B(QM,0) = B(1,0) + 5 \log r_M \Delta_{QM} \qquad (6)$$

Clearly, B(QM,0) value gives the apparent magnitude of an asteroid,
again in a fictitious opposition, but at the distance r_M from the Sun
and Δ_{QM} from the Earth, and at zero phase.

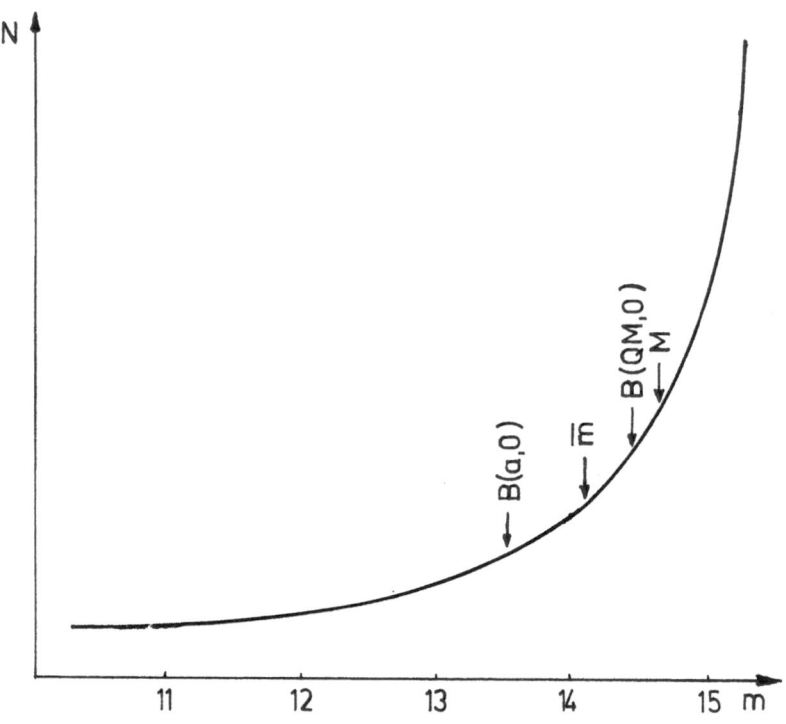

Figure 2. The smoothed distribution of true opposition magnitudes of an asteroid of $B(1,0) = 10^{m}0$, $e = 0.4$, $i = 40^{o}$, $a = 2.8$ a.u. and $\omega = 0^{o}$. Arrows denote $B(a,0)$, mean $(\overline{m})$, $B(QM,0)$ and median (M) magnitudes.

Note that using of Δ_{QM} in (6) instead of the true median produces, even in the most critical cases (e.g. orbit of $a = 2.0$ a.u., $i = 40^{o}$, $e = 0.05$ and $\omega = 90^{o}$), deviation from the true median magnitude not larger than $0^{m}3 - 0^{m}4$.

As an illustration of the goodness of the fit, the distribution of true opposition magnitudes of a fictitious asteroid is shown in Fig. 2, with an indication of the values of distribution mean and median, $B(a,0)$ and $B(QM,0)$. As it can be seen, the value of $B(QM,0)$ is quite close to that of the distribution median, thus making $B(QM,0)$ a better approximation of the asteroid average opposition magnitude than $B(a,0)$.

REFERENCES

Kiang,T.: 1966, Icarus, 5, pp.437-449.
Knežević, Z.: 1981, submitted to Bull.Astron.Inst.Czech.
Kresák, L.: 1967, Bull.Astron.Inst.Czech., 18, pp.27-36.

A COORDINATE PROGRAM FOR POLE DETERMINATION OF ASTEROIDS

V. Zappalà and F. Scaltriti
Osservatorio Astronomico di Torino
I-10025   Pino Torinese   –   Italia

The unresolved problem of the determination of the pole
for the asteroids is of the highest interest for Solar Sy=
stem's studies. The rotational axis direction is a funda=
mental parameter in outlining the collisional history of
the belt; it can allow to select different evolutionary
trends, if they exist, and can help in the exact definition
of impact mechanisms. For this aim what we need is a method
which could create a good statistical set of data, even if
the precision of the results is not very high; we believe
that the method developed by Zappalà(1981) can have the re=
quested attributes.It is essentially based on the hypothe=
sis that the asteroid has a triaxial ellipsoidal shape
(a>b>c)and the rotation is about the shortest axis.The al=
bedo is considered uniform and no particular scattering
law is taken into account; therefore the light variation
should be entirely due to the apparent area seen by the ob=
server.From relations that connect the aspect and the geo=
metric configuration of the body with the observed amplitu=
de as well as with the observed magnitude at maximum light,
it is theoretically possible to deduce the ratio b/c from
each observation.Obviously,a large set of observations(7÷
10),well spread along the asteroid orbit,could refine the
previous result.In the same way it is also possible to de=
duce the aspect angle for these observations and, by combi=
ning them all together,to select the best solution for the
pole which satisfies the data.For a high percentage of the
objects,photometrically known to-date,we can expect an er=
ror in the ecliptic coordinates of the pole not larger than
5÷10°.We believe that this is sufficient for a statistical
research.Now we are obliged to use only observations made
for different purposes and so not homogeneous for our aim;
moreover, only few objects were observed in a sufficient
number of oppositions. However,a check on those asteroids
gave a favourable result.When we compare the pole coordi=
nates obtained by our method with those given by different

303

W. Fricke and G. Teleki (eds.), Sun and Planetary System, 303–304.
Copyright © 1982 by D. Reidel Publishing Company.

procedures and listed by Taylor(1979),the agreement seems
always quite good. The obtained results are the following:

$$
\begin{array}{lll}
7 \text{ Iris}(\lambda_o,\beta_o)=(\ 13,27)\text{or}(192,10) & ; & (\ 11,41) \\
9 \text{ Metis} \quad =(\ \ 0,32)\text{or}(185,50) & ; & (156,15)\text{or}(191,56) \\
22 \text{ Kalliope} \quad =(212,46)\text{or}(\ 11,19) & ; & (215,45) \\
39 \text{ Laetitia} \quad =(116,53)\text{or}(340,62) & ; & (121,37) \\
44 \text{ Nysa} \quad =(101,47)\text{or}(288,53) & ; & (105,30)\text{or}(100,50) \\
433 \text{ Eros} \quad =(\ 18,27)\text{or}(227,10) & ; & (\ 16,12)\text{or}(\ 15,\ 9) \\
511 \text{ Davida} \quad =(\ 92,28)\text{or}(304,31) & ; & (122,10) \\
624 \text{ Hektor} \quad =(146,14)\text{or}(321,\ 0) & ; & (324,10) \\
1566 \text{ Icarus} \quad =(\ 49,11)\text{or}(255,26) & ; & (\ 49,\ 0) \\
1685 \text{ Toro} \quad =(209,34)\text{or}(249,14) & ; & (200,55) \\
\end{array}
$$

where the first two couples represent the results obtained
by the method of Zappalà(1981), whereas the other ones are
previous results.
In order to plan a coordinate international program of ob=
servations, we have to outline some particular procedures
to follow concerning the selection of the objects and the
observations, in order to optimize the data. The most im=
portant of them are the following: a) the asteroid light=
curve should be similar, at a first approximation,to a "si=
nusoid"; b) the observations should be performed at diffe=
rent oppositions(7÷10) differing each other by about 30÷40°
in longitude (in order to cover the amplitude-longitude
diagram sufficiently); c) the observations should be made
at low phase angles (≪10°) and, in different oppositions,
they should be performed at the same phase angle(differing
no more than 5÷6°); this is to minimize the effect of pha=
se and obliquity in the observed amplitude and magnitude;
d) at least in one opposition, it is better to deduce some
magnitude values also at large phase angles, in order to
calculate the phase coefficient of the considered object.
We have selected about 30 asteroids for which the method
seems applicable and we are planning their observational
coverage in the next oppositions by world-wide campaigns.
Moreover, we are also studying (Taylor et al. 1981) a com=
mon subset which can be followed also by the "photometric
astrometry"technique, developed at the Lunar Planetary La=
boratory of Tucson (Taylor 1979). This well known method
starts from completely different parameters and therefore
an agreement in the obtained data should be a decisive test
for the reliability of the results.

REFERENCES
Taylor,R.C.: 1979, in "Asteroids", T. Gehrels Ed.,
          The University of Arizona Press, pp. 480-493.
Taylor,R.C., Tedesco,E.F., Zappalà,V., and Scaltriti,F.:
          1981, in preparation.
Zappalà,V.: 1981, The Moon and The Planets 24, pp. 319-325.

THE WORLDWIDE PHOTOELECTRIC CAMPAIGN FOR THE ASTEROID 51 NEMAUSA

H.J.Schober   and   L.K.Kristensen
Institut für Astronomie   Institut of Physics
Universität Graz   University of Aarhus
A-8010 Graz, Austria   DK-8000 Aarhus C, Denmark

1. EQUATOR POINT AND EQUINOX OF FK4 AND THE ASTEROID 51 NEMAUSA

It is a general problem to determine fundamental declinations by meridian circles with simultaneous solution for latitude, refraction and flexure. Observations of solar system objects therefore have been used to determine the equator point from FK3 and onwards, with an average correction in declination along the entire equator $\Delta\delta_0 = (\delta - \delta_{FK3})$ for $\delta \sim 0^\circ$ near the equator, averaged in right ascension for $0 < \alpha < 24$ h. Especially because of discordances between meridian and vertical circle results B. Strömgren initiated 1939 a program of photographic observations of the asteroid 51 Nemausa with an orbit close to the equator (node $\sim 180^\circ$ and inclination $\sim 10^\circ$). Early work was completed for the FK3 corrections by Naur (1957), investigations were continued and resumed by Møller (1978).

L.K.Kristensen started the work on 51 Nemausa again in 1972, using the collected observational material of the best 2240 photographic positions from 1943-1977 and transit-and refractor measurements from 1858-1902. Residuals O-C were computed and examined by Kristensen (1980); from the Nemausa analysis he obtained the equator point correction for FK4 $\Delta\delta_0 = +0".027 \pm 0".029$ and the equinox for FK4 E $= \Delta\alpha_0 = +0^S.044 \pm 0^S.005$, both epoch 1956. In addition, the orbit of 51 Nemausa might be the best one known today; the mean opposition magnitude $B(a,0)=11.2$ is convenient for all kind of observations, too.

2. PHYSICAL PROPERTIES OF 51 NEMAUSA

Kristensen (1980) also made a linear regression of the residuals in right ascension of 51 Nemausa with phase angle. For a phase angle of 20°, near the stationary points a displacement of $+0".054 \pm 0".018$ toward the illuminated part of the disc of 51 Nemausa was found, to be interpreted as a phase effect due to the diameter of 51 Nemausa ($\sim 0".15$), and the gravitational center being different from the optical center. In order to push forward now the accuracy of the orbit and therefore the accuracy of the FK4 equator and equinox, further account must be taken of the shape and size of the body of 51 Nemausa itself, because:
a) the residuals in right ascension are correlated with the phase angle

W. Fricke and G. Teleki (eds.), Sun and Planetary System, 305–306.
Copyright © 1982 by D. Reidel Publishing Company.

indicating that the observed center of light is displaced toward the illuminated side of the body;

b) the predictions of positions can be made within the accuracy comparable to the radius of the body (occultation of stars);

c) lightcurves are needed to find the true pole and the true flattening of the approximated triaxial ellipsoid.

3. THE PHOTOELECTRIC CAMPAIGN 1980/81/82

Chang and Chang (1963) have obtained the rotation period P= 7^h.785 with a threemaxima lightcurve of 0.14 max.amplitude Δm.Additional short photoelectric measurements were made by Wamsteker and Sather (1974) finding Δm=0.25. 51 Nemausa was classified as type U with a mean diameter of 156 km by Bowell et al.(1979). Due to the accurate position prediction it was possible to observe the occultation of SAO 144417 on Aug.17,1979 at Gissar and Alma-Ata; from the two chords of observations Kristensen (1981) fitted a flattened ellipsoid with Δm = 2.5 log (a/b) = 0.25 and b/a = 0.795 and obtained the mean diameter D=2(ab)$^{1/2}$ with D= 153 ± 7 km. But in order to derive better values for the period (sideral), to find the pole orientation and the shape further observations are needed.

An international photoelectric capaign was organized to observe lightcurves and especially their variations around the two oppositions in Dec.2, 1980 and May 3, 1982. So far the following observers have contributed and have partly observed 51 Nemausa, covering the period from Oct.1980 to Feb.1981 :

P.Birch and I.Nikoloff : Perth Observatory,Australia

K.Pavlovski : Hvar Observatory,Jugoslavia

H.J.Schober : Cerro Tololo Interamerican Observatory and European
              Southern Observatory,Chile,Observatoire de Haute Provence,France

F.Scaltriti and V.Zappalà :Osservatorio Astron.Pino Torinese,Italy.

Full and complete observations near opposition were reported so far only from Bich and Nikoloff;from their preliminary reductions they get again a triple extremum lightcurve as did Chang and Chang (1963),the aspect data being similar. The opposition to be observed next in 1982 will show a different aspect;so we should expect surprising results, might be 51 Nemausa even would turn out to be a binary asteroid ?

REFERENCES

Birch,P.and Nikoloff,I.: 1981, pers.communication

Bowell,E.,Gehrels,T.and Zellner,B.: 1979, in "Asteroids"ed.T.Gehrels,
    Univ.of Arizona Press,Tucson,p.1108

Chang,Y.C. and Chang,C.: 1963, Acta Astron.Sinica 11,p.139

Kristensen,L.K.: 1980, Mitteil.Astron.Ges.48,p.50

Kristensen,L.K.: 1981, Astron.Astrophys.Suppl.44,p.375

Møller,O.: 1978, in "Astronomical Papers dedicated to Bengt Strömgren",
    ed.A.Reiz and T.Andersen,Copenhagen Univ.Obs.,p.355

Naur,P.: 1957, Minor Planet 51 Nemausa and the Fundamental System of
    Declinations,Copenhagen Univ.Obs.,thesis

Wamsteker,W.and Sather,R.E.: 1974, Astron.J.79,p.1465

HALLEY'S COMET AND PLANS FOR ITS OBSERVATION DURING ITS RETURN IN 1986

J. Rahe, Remeis-Observatory Bamberg
        University Erlangen-Nürnberg
J.C. Brandt, Laboratory for Astronomy and Solar Physics
        NASA-Goddard Space Flight Center
L.D. Friedman and R.L. Newburn
        Jet Propulsion Laboratory
        Pasadena, California

1. INTRODUCTION

In historic times, the sudden, unpredictable and often somewhat earie appearance of comets, that violated the elegant order of the heavens, was thought to be a frightening symbol of disaster (which, literally, means a bad star), and especially Comet Halley has been made responsible for more than one tragedy: In 12 BC the comet appeared over Rome and predicted the death of Agrippa. In 66 AD, it was seen over the city of Jerusalem before it was destroyed. It appeared again in 451 during the battle of Chalons when the Roman general Aetilus defeated Attila the Hun, and it appeared in 1066 during the Battle of Hastings and was thus responsible for the defeat of the King Harold's armies by William the Conqueror. When Comet Halley appeared in 1456 at the same time when the Turks were besieging Belgrade, Pope Calixtus III ordered prayers for deliverance from the devil, the Turk, – and the Comet. Occasionally one can also read that he actually excommunicated the comet, but this claim cannot really be substantiated. Also at its latest return in 1910, Comet Halley caused considerable public concern – especially when it was known that the earth was actually to pass through the comet's tail and occasionally special efforts were made to avoid being harmed by the poisoneous cyanogen gas. Today most scientists regard comets as generally harmless.

In the following, we shall only briefly discuss the general characteristics and properties of Halley's comet, as they can be derived from earlier apparitions. For more detailed information, we refer to the excellent review articles on this comet that have recently been written, e.g., Marsden, 1979; Newburn and Yeomans, 1981. The major part of this paper will be devoted to the 1986 apparition.

W. Fricke and G. Teleki (eds.), Sun and Planetary System, 307–320.
Copyright © 1982 by D. Reidel Publishing Company.

2. ORBIT CALCULATION

It was only in 1577, that Tycho Brahe could clearly show that the bright comet of that year was at least 4 times farther away than the moon. More than a century after Brahe's observation, Kepler discovered the laws of planetary motion - but applied them only to planetary movements. He suspected that comets move in straight lines, and actually used Halley's appearance of 1607 to illustrate this claim.

The English astronomer, Edmund Halley, (1656-1742), applied Newton's newly discovered law of gravitation to cometary observations and determined the orbits of 24 well-observed comets. He searched for comets with identical orbital elements. After what he calls "prodigiously" long and troublesome calculations, he makes the first correct prediction for the return of a comet:

"Now many things lead me to believe that the Comet of the year 1531, observed by Apian, is the same as that which, in the year 1607, was described by Kepler and Longomontanus, and which I saw and observed myself at its return in 1682. All the elements agree, except that there is an inequality in the time of revolution; but this is not so great that it cannot be attributed to physical causes....

I may, therefore, with confidence predict its return in the year 1758". He plaintively states:

"Wherefore if it should return according to our prediction about the year 1758 impartial posterity will not refuse to acknowledge that this was first discovered by an Englishman".

As the year 1758 draws closer, the French astronomer Alexis Clairaut, with the help of Joseph Lalande and Hortense Lepaute, undertake the immense effort to calculate the perturbations of the comet's orbit by the influence of Jupiter and Saturn. Working intensively for six months from morning to night, with but little intermission, Clairaut finally predicts on November 14, 1758, that April 13, 1759, is the date of the next perihelion passage. It turns out that his prediction is off by only one month. The actual date is March 13, 1759. The comet was recovered by the farmer and amateur astronomer Johann Palitzsch, living at Prohlis near Dresden, who saw it on the night of Christmas day 1758 with a telescope of 8 feet focus. Independent on Palitzsch's discovery, the first professional astronomer to find the comet, was Charles Messier in Paris, who detected it first on January 21, 1759, although the observatory director, Delisle, would not allow his assistant Messier to disclose his discovery. Messier observed the comet regularly for three weeks before it became lost in the Sun's brightness at the perihelion passage.

For the 1835 return, the best predictions for the time of peri-helion passage by Pontecoulant in France and Rosenberger in Germany, were too early   3.5 and 4.4 days, respectively. The comet was recovered on August 5, 1835, by Etienne Dummouchel in Rome, more than three months before perihelion passage.

For the 1910 return, the best predictions were made by the English astronomers Cowell and Crommelin, but also their predictions were again about three days too early.

The systematic correction of about 4 days for each perihelion passage, suggested the influence of non-gravitational forces as they occur in Whipple's icy-conglomerate model of the cometary nucleus. Including these effects, Michielsen (1968) predicts February 9, 1986, as the date of the next perihelion passage. A more refined prediction was made by Yeomans (1977) who linked the appearance since 837 in one orbit and confirmed the next perihelion date of 9 February 1986, with an uncertainty of about ±0.25 days.

3. PHYSICAL PROPERTIES OF HALLEY'S COMET

Although the data obtained at the time of Comet Halley's last apparition are mostly qualitative, they can still be used to derive a few general properties of this comet. Visual brightness estimates have been compiled by Yeomans (1977) in a light curve, showing the visual apparent magnitude as function of heliocentric distance. As already suggested by observations at earlier appearances, the comet appears to be brighter after perihelion than before.

The linear tail length was also computed by Yeomans (1977) from naked-eye estimates as function of heliocentric distance. The actual tail extension observed, depends naturally on the observing condition and instrument used, but his data seem again to suggest that also the visual tail is after perihelion passage longer than before.

During its last apparition in 1910, Comet Halley came rather close to the earth (0.15 AU) and was relatively bright. In May, the Earth actually passed through parts of the comet's tail. Be-sides visual observations, numerous photographs and spectrograms were obtained at observatories all over the world. The spectra exhibit the usual emission features due to CN, C_2, C_3, etc. The photographs reveal a considerable number of very remarkable features and proces-ses occurring in the coma and the plasma and dust tails. The most comprehensive discussion of the 1910 apparition is given by Bobrovnikoff (1931). Mainly based on his data, a few physical pro-perties of comet Halley's nucleus can be derived. Newburn (1979) estimated a nucleus diameter of 5 km, with 1 and 16 km the extreme limits. With a bulk density of 1 g cm^{-3}, its mass comes out to $6.5 \cdot 10^{16}$g; the actual mass could conceivably be two orders of magnitude higher or lower. The rotation period was determined by Whipple (1980) to $10.^h3$; the rotation is direct, and the axis is

Historical, Physical and Orbital Data of Halley's Comet (Yeomans,1981)

Historical Data

Earliest probable recorded apparition	240 BC
Number of recorded apparitions from 240 BC to 1910 AD; only the 164 BC apparition was not recorded	28
Shortest period between returns	74.42 years (1835-1910)
Longest period between returns	79.25 years (451- 530)
Closest approach to earth	0.04 AU (April 11, 837)
Longest angular tail length recorded	93° (mid-April, 837)
Brightest apparent magnitude recorded	-3.5 (April 11, 837)

Physical Characteristics

Estimated diameter of nucleus	5 km
Estimated density of nucleus	1 g/cm^3
Estimated rotation period	$10^h.3$, direct
Observed spectra in 1910	CN, C_2, C_3, CH, Na, CO^+, N_2^+
Observed tails	plasma and dust
Associated meteor streams	η Aquarid (early May); and Orionid (late October)

Orbital Characteristics

Location of orbit pole	$\lambda = 328^{\circ}.15$; $\beta = -72^{\circ}.24$
Location of perihelion	$\lambda = 305^{\circ}.32$; $\beta = 16^{\circ}.45$
Heliocentric distance of orbit nodes	$r(\Omega)= 1.81$ AU; $r(\mho)=0.85$ AU
Distance of perihelion and aphelion above or below orbit plane	$Z(q)= 0.17$ AU; $Z(Q)= 9.99$ AU
Orbital velocity at perihelion	$V = 54.55$ km/sec
Orbital velocity at aphelion	$V = 0.91$ km/sec

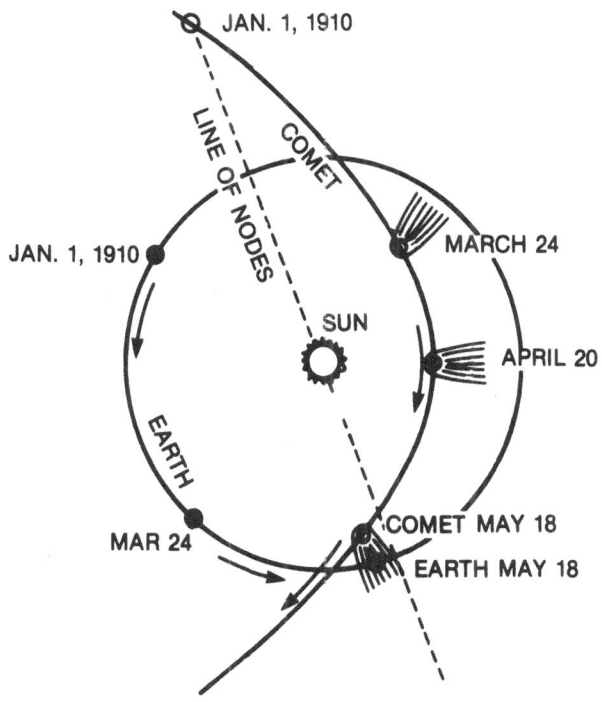

<u>Fig. 1.</u> Schematic drawing of Earth and
Comet Halley's orbit in 1910

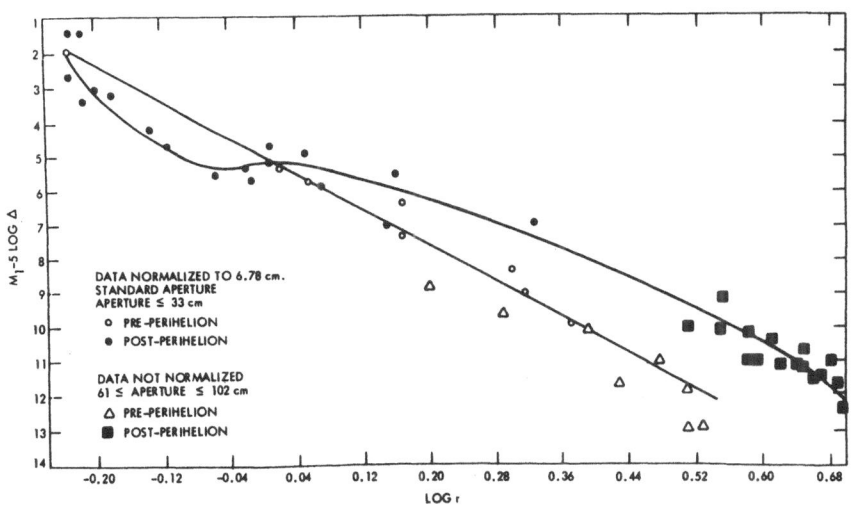

<u>Fig. 2.</u> Comet Halley total magnitude estimates, 1909-1910
(Yeomans, 1981). Only visual estimates are given.

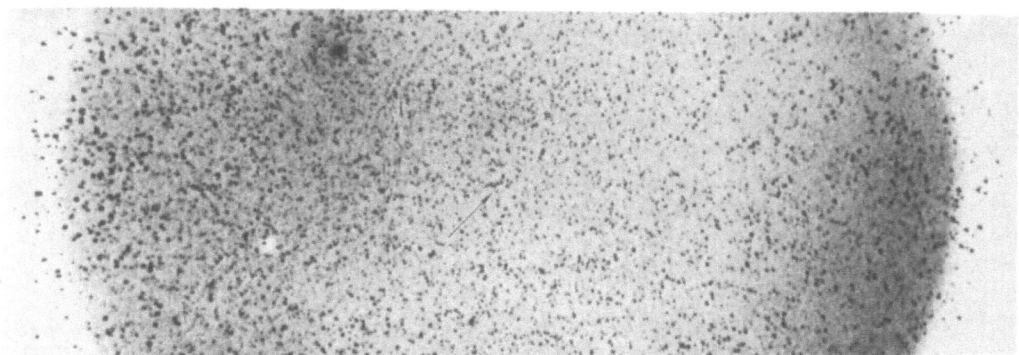

Fig. 3. Comet Halley was discovered by Max Wolf at Heidelberg on Sept. 11, 1909, near Geminorum. The comet appears as a small, fuzzy, 15th mag. image; its position among the stars is marked. It was subsequently also found on plates taken on Aug. 25 in Helwan, Egypt, and on Sept. 9 in Greenwich. Wolf used the 72-cm Heidelberg reflector and exposed his photographic plate for 1 hour. At the time of this observation, the comet was 3.4 AU from the Sun and 3.6AU from the Earth. Halley's Comet was designated 1909c. It reached perihelion on April 20, 1910.

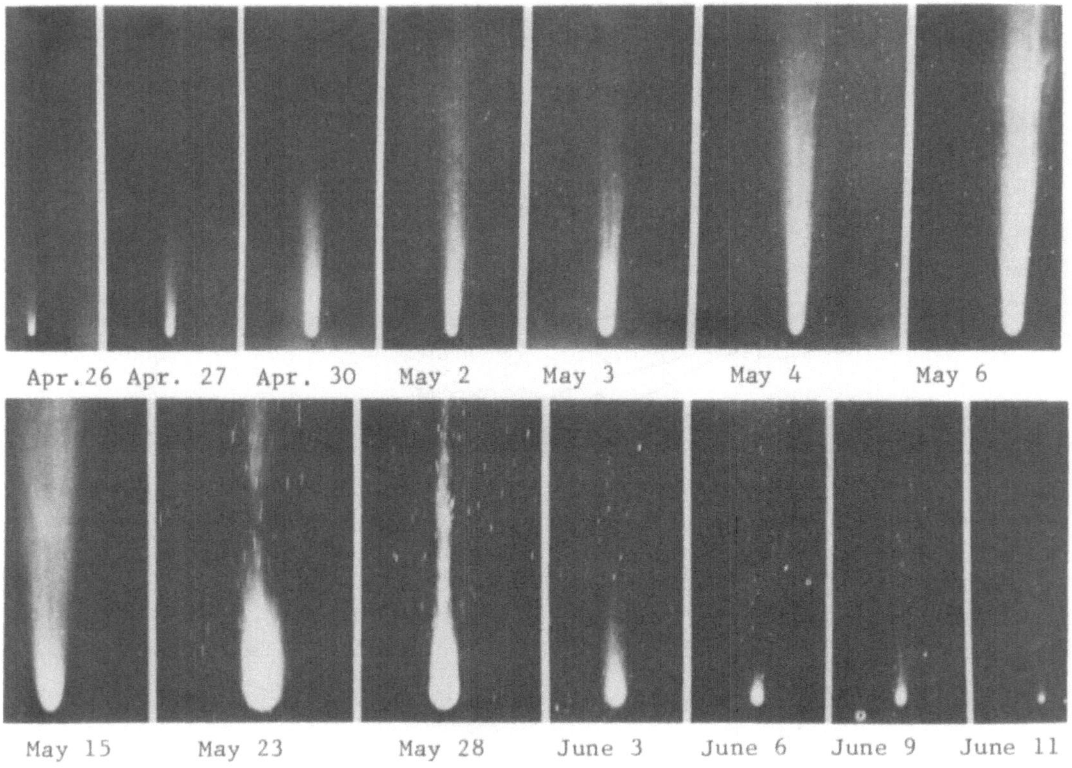

Apr.26 Apr. 27 Apr. 30 May 2 May 3 May 4 May 6

May 15   May 23   May 28   June 3   June 6   June 9   June 11

Fig. 4. Evolution of Halley's Comet from April to June, 1910 (Mt. Wilson Observatory photograph).

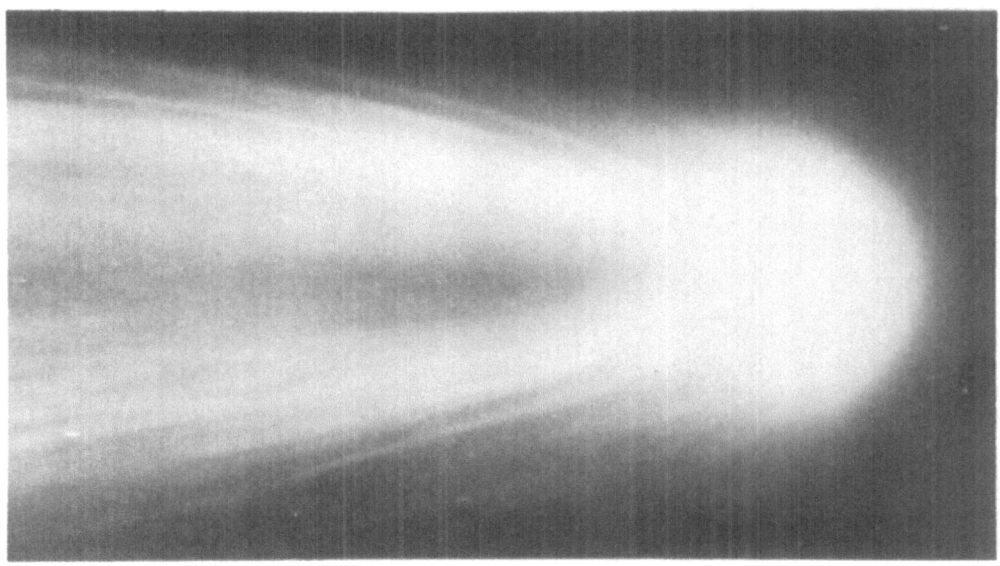

Fig. 5. Head of Halley's Comet as seen on May 8, 1910 (Mt. Wilson Observatory photograph).

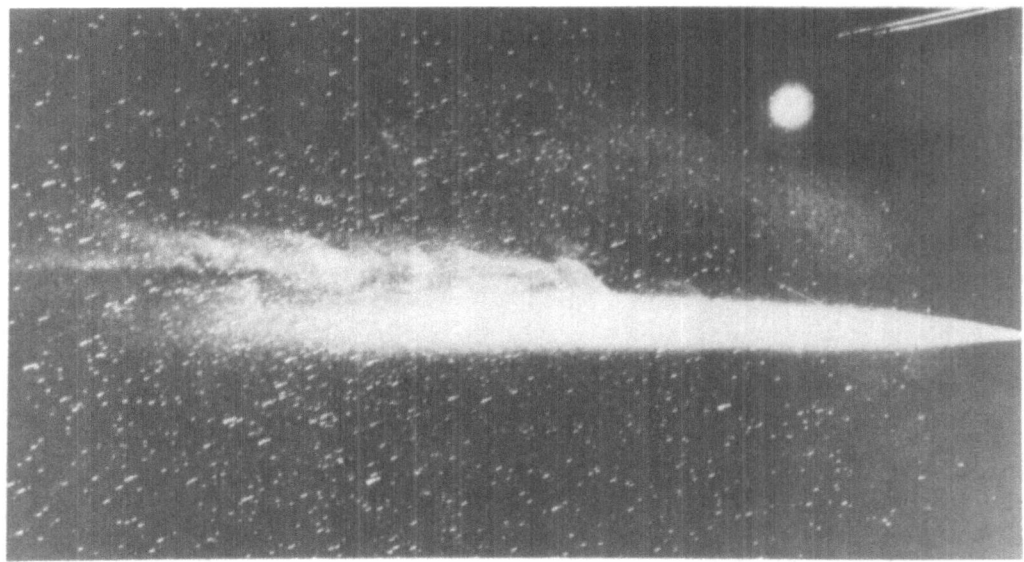

Fig. 6. Photograph of Halley's Comet, taken May 13, 1910, showing a tail some 45° long. Two disconnection events appear in the plasma tail. The strong tail emanating from the head is the dust tail. The bright object near the top is Venus. Near the head, a meteor trail is visible.

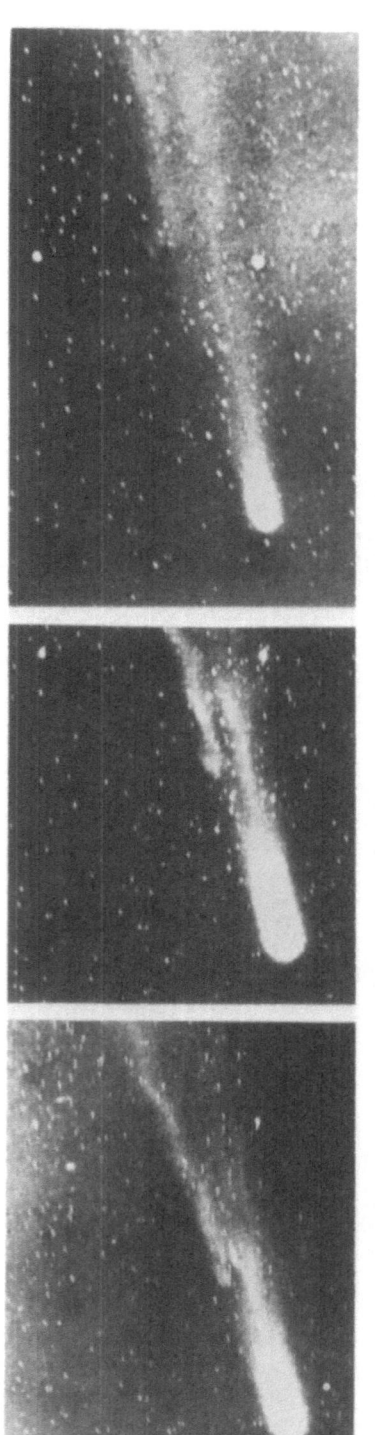
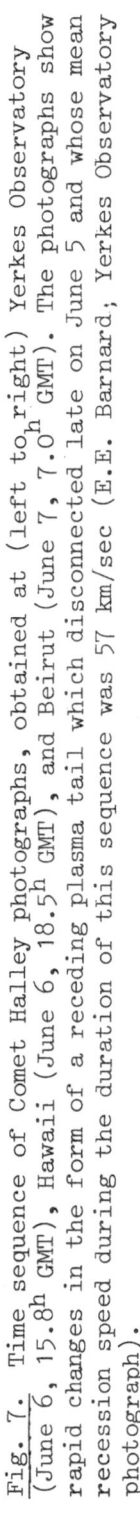

Fig. 7. Time sequence of Comet Halley photographs, obtained at (left to right) Yerkes Observatory (June 6, 15.8^h GMT), Hawaii (June 6, 18.5^h GMT), and Beirut (June 7, 7.0^h GMT). The photographs show rapid changes in the form of a receding plasma tail which disconnected late on June 5 and whose mean recession speed during the duration of this sequence was 57 km/sec (E.E. Barnard; Yerkes Observatory photograph).

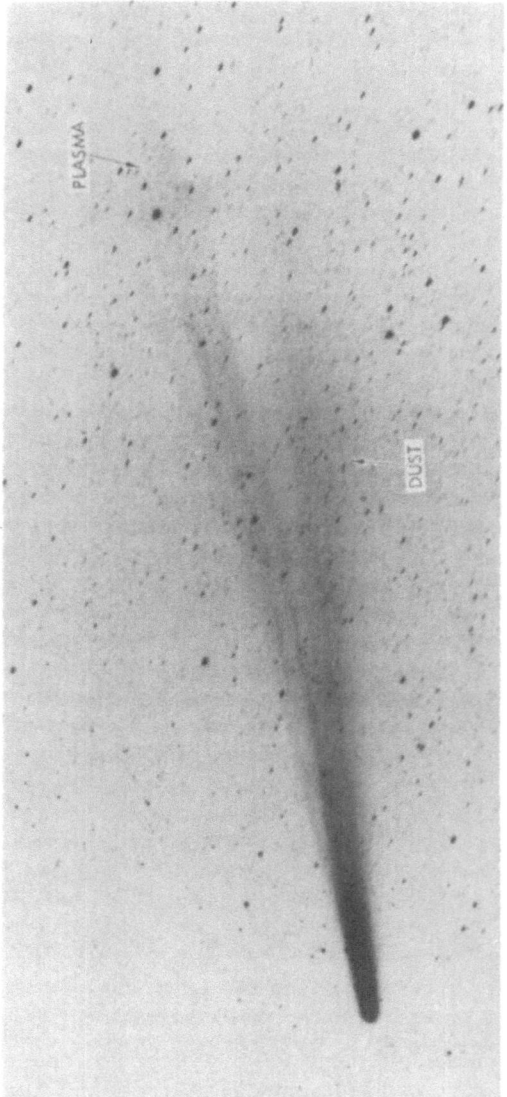

Fig. 8. The two tails of Halley's Comet —the fuzzy dust and the straight plasma tail— can be seen on this photograph, obtained on May 6, 1910, when the comet was 0.7 AU from the Sun and 0.6 AU from the Earth (Lowell Observatory photograph).

nearly normal to the orbital plane.

4. THE 1986 APPARITION

Although Halley's Comet is not nearly as bright or spectacular
as many other comets, it is by far the most famous. As yet, no observations prior to 240 BC have been identified in the ancient records,
but since that time, the comet has faithfully and regularly returned
"once in a lifetime" through all of human history (only the 164 BC
apparition has apparently not been recorded).

With an intermediate period of about 76 years, it has made
relatively few returns to the solar neighborhood since its first
perturbation from the "Öpik-Oort" cloud of comets, far outside and
surrounding our solar system. It is one of the "freshest" comets
with a period of less than 200 years, and it is the only predictable
comet to display the full range of cometary activity. Almost all
known phenomena are present: a large, dense coma; plasma and dust tails;
jets; streamers; outbursts, disconnection events, etc. It is the
only really active comet with a well-established orbit; its gas
production rate is about 100 times greater than that for any other
periodic comet. Finally, it provides a unique opportunity to compare
the properties of a bright active comet at two successive appearances.

Unfortunately, the coming 1985/86 apparition will be very unimpressive to the casual observer, and it is very likely that most
people will not even see the comet. In 1910, the comet passed perihelion on April 20, and the earth was in a relatively favorable
position. In May it went actually through the tail of the comet.
In 1986, the comet will pass perihelion on February 9; but it will not
be observable then, since the earth will be on the other side of the
sun. Those observers, who are away from city light, who own a pair
of binoculars--and who know where to look, should be able to find the
comet in March or early April 1986. Observers in the southern
hemisphere are in a somewhat better position, but also they will probably see only a faint, unimpressive comet. The viewing conditions
have been calculated by Yeomans (1981). Here in--or better: outside
of - Dubrovnik, e.g., one might get a glimpse at the comet in the
morning hours in March and the evening hours in middle and late April.
In March, the comet could conceivably be only a faint 3rd or 4th
magnitude object.

5. SPACE MISSIONS TO COMET HALLEY

Fortunately, we do not depend on visual observations of Comet
Halley through binoculars. We have at our disposal the many observatories with large telescopes around the Earth which will be
motivated to study Halley's comet. In addition, for the first time
we will have platforms in Earth orbit, such as the Space Telescope
and more specialized instrumentation on Spacelab, that can be applied
to observe this Comet. There exists also the opportunity to measure

RELATIVE POSITIONS OF COMET HALLEY AND EARTH 1985-86

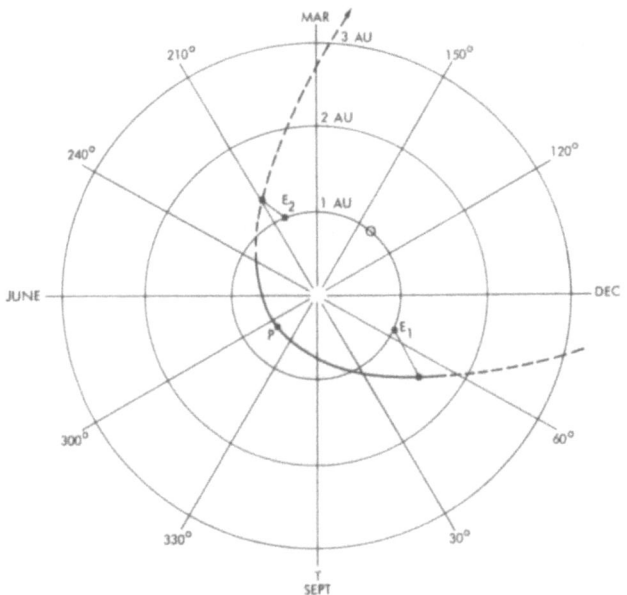

P = PERIHELION OF COMET HALLEY
E₁ = POSITION OF EARTH AT PRE-PERIHELION CLOSE APPROACH OF COMET (NOV. 27, 1985)
E₂ = POSITION OF EARTH AT POST-PERIHELION CLOSE APPROACH OF COMET (APR. 11, 1986)
O = POSITION OF EARTH AT PERIHELION OF COMET HALLEY (FEB. 9, 1986)

Fig. 9. Schematic drawing of Earth and Halley's orbit, 1985-1986.

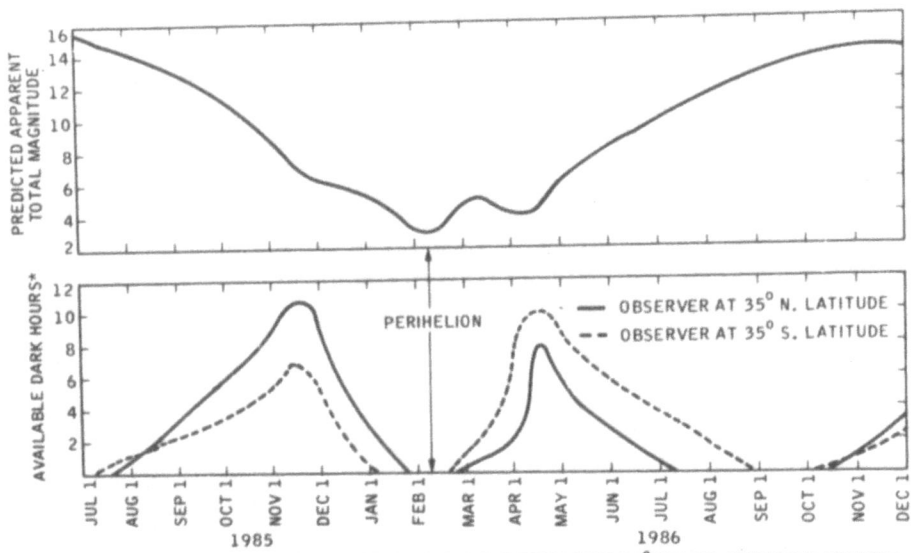

* NUMBER OF HOURS WHEN COMET IS ABOVE, AND SUN IS MORE THAN 18° BELOW, THE LOCAL HORIZON

Fig. 10. Comet Halley 1985-86 ground-based observing conditions (Yeomans, 1981).

cometary properties in situ, through space probes flying through the comet at times of high activity. These plans are still subject to change.

The Space Agencies of Europe, Japan and the Soviet Union have decided to send a special probe to Comet Halley. The European Space Agency ESA will fly a post-perihelion flythrough mission, using a spinning Earth-Orbiting Satellite.

The spacecraft was given the name "Giotto" after the Florentine painter Giotto di Bondone who had seen Halley's Comet in 1301, when the comet streched over more than 70° across the sky, and incorpcrated it very realistically in one of his frescoes in the Arena Chapel of Padova. The painting shows the bright, fiery comet above the marger as the star of Bethlehem. It appears to be one of the first naturalistic representations of a comet.

The spacecraft will carry 10 experiments with a total mass cf 54 kg. It is spin-stabilized and will spin with 15 revolutions/min. A special dual-sheet bumper will protect the spacecraft from destruction by dust particles emitted from the comet's nucleus and impacting the spacecraft with a velocity of about 70 km/sec. The encounter itself lasts only 4 hours.

A prime objective of the ESA mission is to detect and study the comet nucleus. For this purpose, the spacecraft will be equipped with a camera that can obtain color photographs of the nucleus from a distance of 1000 km with a resolution down to 50 m. The visibility of the nucleus, is, however, still a matter of discussion. In the inner coma region, the dust particle density may be so high that the atmosphere is optically thick and the nucleus not visible. Three mass spectrometers will measure the elemental and isotopic composition of the neutral, ionized and dust particles, in order to better understand the physical and chemical processes occurring in the cometary atmosphere. Impact sensors will measure the mass spectrum of the dust particles, and an optical probe will measure the spatial distribution of dust particles and gaseous emissions. The interaction of the solar wind with the cometary material will be studied through plasma experiments and a magnetometer.

Launch of "Giotto" is scheduled for July 13, 1985; encounter date is March 13, 1986; encounter speed is about 68 km/sec. The probe and the comet are then about 1 AU from Earth and Sun. The spacecraft will pass the nucleus in about 500 km.

According to current plans, the Soviet mission will use 2 identical 3-axis stabilized spacecraft. They will be launched in December 1984 and arrive at Venus in June 1985. The encounter distance with Venus is 30.000 km. Each spacecraft will release at least one probe to Venus and will then continue to fly to Halley where they will arrive on March 8, 1986. The spacecraft will have an encounter distance with Halley of about 10.000 km.

Possible Missions to Halley's Comet

	ESA "Giotto"	Japan "Planet A"	USSR Venera
Launch Date	~July 13, 1985	~August 14, 1985	~December 22, 1984
Launch Vehicle	Ariane 2	Mu-3S II	Proton
Spacecraft Mass	750 kg	135 kg	~1000 kg
Spacecraft Type	Spin	Spin/Despin	Three-Axis Stabilized
Communication Rate	53 kbit/s	128 bit/s	>10kbit/s
Encounter Date	March 13, 1986	March 8, 1986	March 8, 1986
Encounter Speed	68 km s^{-1}	70 km s^{-1}	78 km s^{-1}
Targetted Miss Distance	0 km	10^5 km	10^4 km
1 σ Aiming Accuracy	90 km	10^5 km	10^4 km
Nucleus Position Knowledge	500 km	500 km	500 km
Payload Mass	53 kg	10 kg	50 kg
Science Payload	Neutral Mass-Spectrometer	UV Camera	TV Complex
	Ion Mass-Spectrometer	Magnetometer	3-Channel Spectrometer
	Dust Impact Mass-Spectrometer		IR-Spectrometer
	Dust Impact Detector		Dust Mass-Spectrometer
	Electron/Ion Analyser		Dust Particle Counter
	Magnetometer		Ion Mass-Spectrometer
	Narrow-Angle Imaging		Electron/Ion Analyser
	Energetic Particles		Magnetometers
	Optical Probe Experiment		Plasma Waves Analysers

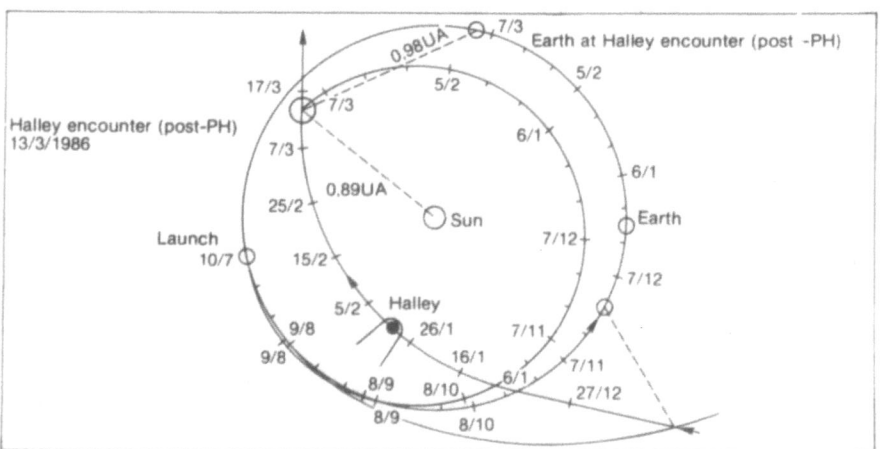

Fig. 11. "Giotto" reference transfer orbit with a launch on 10 July
1985 and a Halley encounter on 13 March 1986. The transfer trajectory lies in the ecliptic plane; it is inside the Earth's orbit
and has a closest distance from the Sun of 0.7 AU. At the time of
the encounter, the distances from Earth and Sun are 0.98 and
0.89 AU, respectively. Encounter duration is 4 hours. The flyby
velocity is 68 km/sec.

Japan is planning to launch its first deep space probe, called
Planet A. It will obtain data on the solar wind and planet Venus,
and approach Halley's Comet in March 1986. It will pass the comet
in a distance of about 100.000 km.

6. INTERNATIONAL HALLEY WATCH

The International Halley Watch (IHW) is a comprehensive program to
prepare for the next apparition of Halley's Comet. A Science Working
Group (Brandt et al.,1980) has studied the IHW concept, and a Lead
Center has been established at JPL. The main goals of the IHW are:

(1) To stimulate, encourage and coordinate scientific
    observations throughout the entire apparition.
(2) To help insure that observing techniques and ins-
    trumentation are standardized whenever possible.
(3) To help insure that the data and results are properly
    documented and archived.
(4) To receive and distribute data to participating
    scientists and to provide information to the public
    and media.
(5) To stimulate relevant instrument development where
    necessary.

The IHW will assume an advocacy role for the study of Halley's
Comet and will provide liaison with facilities (missions, experiments,
observatories) outside the IHW organization itself for an active
program of scientific measurements during the Halley apparition. It
will actively support observations of the comet from deep space and
earth orbit. In situ measurements and near-comet observations from ballistic
ballistic intercept missions will make an important contribution to the
success of the IHW.

A major thrust will be the coordination of the ground-based
observation effort. Observing nets include observers dedicated to the
following subject areas and techniques: large scale phenomena; near-
nucleus studies; spectroscopy; photometry and polarimetry; radio
science; radiometry; and astrometry. A "trial run" of the IHW is
planned for 1984 on the short-period comet Encke or another suitable
comet.

It is important to realize that in 1986, as well as in 2061, 2136,
and 2211, people all over the world will again be anticipating the
return of Halley's Comet. Through a concentrated effort, such as the
IHW, we can certainly learn a great deal about this "Rip-van-Winkel"
of comet that for thousands of years has faithfully displayed some
of its glory and at the same time, terrified our ancestors.

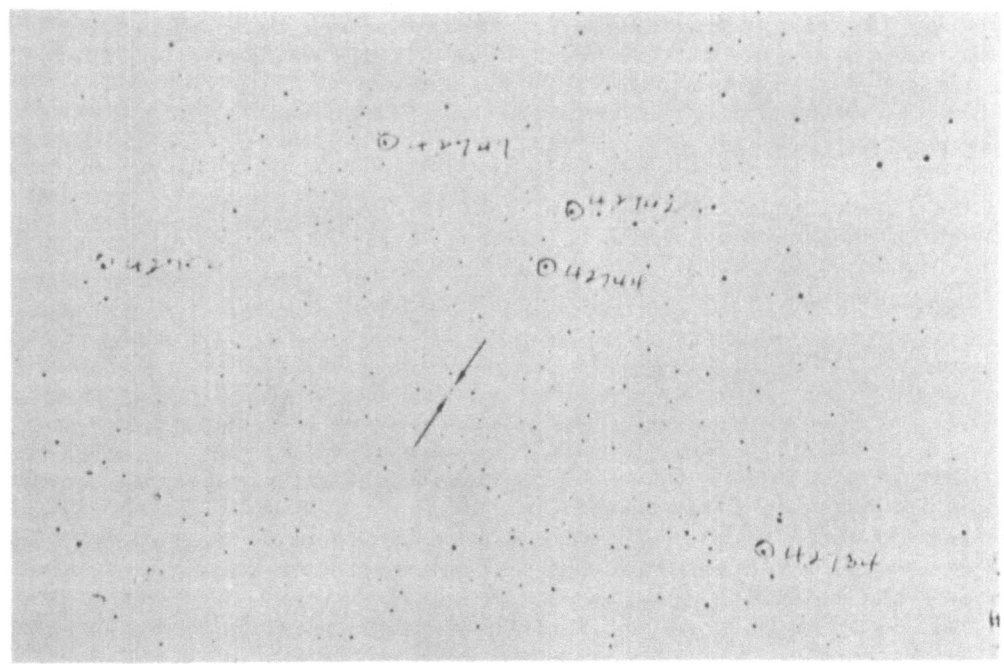

Fig. 12.   Comet Halley on its way back into the deep-freeze of interplanetary space. The figure shows one of the last photographs of this
comet obtained at Lowell Observatory, Flagstaff, Arizona, on May 30,
1911, when the comet was about Jupiter's distance from the Sun. The
position of the comet is marked. The photographic magnitude was about
$18\overset{m}{.}0$. According to Giclas, the comet's brightness had dropped by more
than one magnitude during the previous five days.

REFERENCES

Bobrovnikoff, N.T. 1931. Pub. Lick Obs. 17, 309
Brandt, J.C. et al. 1980. The International Halley Watch. Report of
     the Science Working Group. NASA-TM 82181
Marsden, B.G. 1979. In "Space Missions to Comets". NASA-Conf. Pub. 2089
Michielsen, H.F. 1968. J. Spacecraft and Rockets 5, 328
Newburn, R.L.,Jr. 1979. In "The Comet Halley Micrometeoroid Hazard".
     ESA SP-153, p. 35
Newburn, R.L.,Jr. and Yeomans, D. 1981. Ann.Rev.Astr.Aph., in press
Whipple, F.L. 1980. IAU Circ. 3459
Yeomans, D.K. 1977. Astron. J. 82, 435
Yeomans, D.K. 1981. The Comet Halley Handbook. NASA Doc. JPL 400-91
Yeomans, D.K. and Kiang, T. 1981. MNRAS, in press

PERIHELION ASYMMETRY IN THE PHOTOMETRIC PARAMETERS OF LONG-PERIOD COMETS AT LARGE HELIOCENTRIC DISTANCES

Ján SVOREŇ
Astronomical Institute of the Slovak Academy of Sciences
05960 Tatranská Lomnica, Czechoslovakia

The present statistical analysis is based on a sample of long-period comets selected according to two criteria : (1) availability of photometric observations made at large distances from the Sun and covering an orbital arc long enough for a reliable determination of the photometric parameters, and (2) availability of a well determined orbit making it possible to classify the comet as new or old in Oort's (1950) sense.

The selection was confined to comets with nearly parabolic orbits, as the activity of short-period comets at large heliocentric distances has been investigated in detail by Kresák (1973). 67 objects were found to satisfy our selection criteria. Photometric data referring to heliocentric distances of $r > 2.5$ AU were only used, yielding a total of 2,842 individual estimates and measurements. The lower limit of r approximately coincides with the upper limit of an efficient vaporization of the water ice (Delsemme and Miller, 1971), the radiation of comets beyond $r = 2.5$ AU being controlled by other, poorly known processes.

The conventional photometric parameters M (absolute magnitude) and n (photometric exponent) were determined by the method of least squares, separately for the pre-perihelion and post-perihelion arc of the orbit. The mean distance range was 2.7 to 5.0 AU, with one half of individual observations made at $r > 3.5$ AU. In order to avoid errors in M due to the extrapolation to $r = 1$ AU, the reduced absolute magnitude $H(3.5)$ was introduced; this was defined as the apparent brightness of a comet at $\Delta = 1$ AU, $r = 3.5$ AU. Obviously, this combination of Δ and r cannot occur for an observer on the earth.

One of the results obtained from these data was the evaluation of the perihelion asymmetry of the lightcurves. This was carried out separately for four subsamples of comets, distinguished by the dynamical age (old or new) and by the position of the orbital arc used for the determination of photometric parameters (before or after the perihelion passage). The following table lists the medians of the reduced absolute magnitudes $H(3.5)$ and the medians of the photomet-

W. Fricke and G. Teleki (eds.), Sun and Planetary System, 321–322.

ric exponents n for the four possible combinations of these characteristics :

	H(3.5)	n
old, before perihelion	12.3	0.8
old, after perihelion	12.8	3.7
new, before perihelion	10.2	1.2
new, after perihelion	12.6	3.2

Statistically very significant is the systematic difference between the medians of the exponent n before and after the perihelion
passage. As for the new comets, this effect has already been pointed
out by Whipple (1978) who explained it by the loss of a frosting of
super-volatile materials during the first passage near the Sun. The
suggested process is consistent with the simultaneous appreciable
increase of H(3.5). However, our results show that a post-perihelion
increase of n applies to old comets as well, the change being even
greater than for the new ones; there is also some, statistically insignificant, increase of H(3.5). It is evident that Whipple's explanation does not pertain to this case.

The asymmetry is apparently due to observational selection of
discoveries, which is common to the new and old comets. The discovery probability of a comet depends mainly on its maximum apparent
brightness. In case that the brightness increases slowly during its
approach to the Sun (n small), the chances are greater that the comet will be discovered at a large heliocentric distance, and that it
will enter our statistics. This effect is intensified by the extended period of observability which increases the probability that the
comet will attain a favourable position with respect to the Earth,
not far from the opposition, as a detectable object. A similar selection does not apply to the post-perihelion arc of the orbit, because every comet already discovered remains under observation until
it gets out of the reach of large telescopes. Therefore, the post-
perihelion observations generally provide more objective statistical
information as to the photometric parameters of long-period comets.

A similar behaviour of old and new comets casts doubts on the
explanation of the perihelion asymmetry of new comets in terms of a
loss of their primordial surface frosting.

REFERENCES

Delsemme, A.H., Miller, D.C.: 1971, Planet. Space Sci. 19, pp. 1229-
      1257.
Kresák, L.: 1973, Bull. Astron. Inst. Czechosl. 24, pp. 264-283.
Oort, J.H.: 1950, Bull. Astron. Inst. Netherl. 11, pp. 91-110.
Whipple, F.L.: 1978, Moon Planets 18, pp. 343-359.

ULTRAVIOLET SPECTROSCOPY OF COMETS

J. Rahe
Remeis-Observatory Bamberg
Astronomical Institute, University Erlangen-Nürnberg

1. INTRODUCTION

Ultraviolet observations are for cometary studies especially important since the four basic atomic elements, H, O, C, and N have their strongest resonance transitions between 1200 and 1700 Å. The most prominent features in all cometary spectra are the Lyman-alpha emission of atomic hydrogen at 1216 Å, and the (0,0) band of OH near 3090 Å. The intensity of these transitions is due to the large abundances of H and OH in cometary comae, (about 2 orders of magnitude larger than those of CN or C_2), and their large g-factors of about 10^{-3} sec^{-1} at 1 AU.

In the following, recent spectroscopic ultraviolet observations of comets are presented, and the results derived from them, are discussed.

2. OBSERVATIONS AND RESULTS

The first ultraviolet observations of comets were made in 1970 when two bright comets, Tago-Sato-Kosaka (1969 IX) and Bennett (1970 II), were observed by Orbiting Observatories OAO-2 and OGO-5. These measurements revealed (Code et al., 1972) that both objects have a strong Lyman-a radiation which can be explained by resonance scattering of solar Ly-a radiation on the neutral hydrogen atoms surrounding the cometary nucleus. The hydrogen cloud extends to about 10^7 Km and is somewhat elongated into the anti-solar direction due to the effect of solar radiation pressure.

OAO-2 also provided data on the (0, 0) bands of the hydroxyl radical OH at 3090 Å, indicating an optical depth of $\tau > 1$ out to about 10^5 km. H and OH are several hundred times as abundant as CN or C_2, which show the most intense bands in the visible region. OH and H very probably result mainly from photodissociation of H_2O molecules vaporized from the nucleus. They are produced in the ground state and subsequently excited by fluorescence.

W. Fricke and G. Teleki (eds.), Sun and Planetary System, 323–330.

Two factors contribute to the very large extent of the H-coma: the relatively low ionization probability of H, and a large outflow velocity. The H_2O velocity is given by the thermal radial ouflow velocity of about 1 km/sec. A detailed analysis of the Ly-a isophotes from several comets shows that the two velocity components of 8 and 20 km/sec are needed to fit the data. These two velocities very probably correspond to a two-step dissociation of water, the most likely source of H and OH. The H_2O molecules are dissociated in the inner coma region into H and OH by solar photons. Most of the excess energy is converted into kinetic energy of the H-atoms, and the resulting velocity is about 20 km/sec if there is no thermalization by collision. If we look farther away from the center, we observe the H-atoms that come from the dissociation of OH; but this dissociation of OH into O and H is unfortunately hardly known.

The H-production rates amount to about 10^{30} molecules/sec; they are comparable to the OH production rates, and about 2 orders of magnitude larger than the gas production rates derived from the visible spectrum.

Theoretical models for the interpretation especially of the H and OH observations have been constructed and discussed in the literature (Keller, 1976). Especially the determination of the OH production rate is model-dependent and depends on the exact value of the OH lifetime.

The models dealing with the H emission usually assume an optically thin emission (which is not correct out to about 10^6 km from the nucleus) and take the orbital motion of the comet as well as gravitation and radiation pressure into account. With such a model, the hydrogen production rate and its variation with heliocentric distance, the mean hydrogen outflow velocity and lifetime can be determined.

There are strong indications that H_2O is indeed an important, probably even the dominant, parent molecule of the observed hydrogen: the ratio of the H and OH production rates remains constant with changing heliocentric distance, which suggests that H and OH have a mutual parent, and the two species are produced in a ratio of about two to one, which one would indeed expect, if water is this common parent.

For a few comets, the H_2O-production could be determined. For Comet Kohoutek, e.g., we find at 1 AU, about 3×10^{29} molecules/sec, or a total mass loss of H_2O of 10^{15} g, which is about 1% of the total mass. Short period comets, on the other hand, have spent a much longer time in the inner solar system. They seem to have lost most of their volatile material and have a much smaller evaporation surface. E.g., Comet Encke's (P = 3.3 years) or Comet Bradfield's 1979 X (P = 250 years) H_2O-production rates are more than one order of magnitude smaller than those for bright "new" comets, such as e.g.,

Comet Seargent 1978 XV.

Before the IUE satellite was launched, the most comprehensive
vacuum ultraviolet cometary spectra were obtained for Comet West
(1976 VI) from rocket observations (Feldman and Brune, 1976; Smith
et al., 1981). In addition to known emission features, previously
unobserved features due to C, O, CO, CS, C^+, CO^+, CN^+, and CO_2^+ were
identified.

Until today, two relatively brighter comets have been observed
with the IUE satellite: Comet Seargent (Jackson et al., 1978) and
Comet Bradfield (Feldman et al., 1980). Comet Seargent 1978 XV had
a perihelion distance of its parabolic orbit of 0.37 AU; the orbital
inclination was $66.^o5$. Comet Bradfield 1979 X had a retrograde orbit
with a period of 250 years. This comet was the first for which
ultraviolet observations were made over a wide range of heliocentric
distances (0.7 to 1.6 AU).

In the UV spectra of Comet Bradfield which are reproduced here,
one notices in the short-wavelength region, the highly overexposed
hydrogen Ly-a, and also OI (1302), CI (1561, 1657, 1931), and
SI (1813). In the long-wavelength part of the spectrum, the dominant
features are CO , CS, CO_2 , and the OH bands. Shortward of 2500 $\overset{o}{A}$,
the Mullikan bands of C_2 (Δ v=0) and Ly-a in 2. order appear. Several
weak emission features can be noticed which are as yet unidentified.

The continuum is very weak and indicates a very small dust-to-gas
ratio for this comet, which is consistent with ground-based filter
photometry observations in the optical region.

In addition to the IUE observations of the two brighter comets
Seargent and Bradfield, several faint periodic comets, such as
P/Encke, P/Tuttle, and P/Stephan-Oterma, were observed by IUE
(Feldman, 1981).

The IUE observations of Comet Bradfield allowed the study of
spatial variations of different emissions (Feldman et al., 1980).
For the OH (0, 0) band e.g., several different exposures were made
with the spectrograph slit offset by different amounts from the
nucleus, i.e., the peak of visual brightness of the coma. The
observed intensity profiles were then compared with model calculations.
With an assumed H_2O outflow velocity of 0.5-1 km s^{-1}, a water
production rate at 1 AU of (0.5-1) x 10^{29} s^{-1} was derived. This
value is considerably lower than the value found for long-period
"new" comets. It agrees with the fact that Comet Bradfield is a
short-period "old" comet (P$\approx$250 years).

The spectra of Comet Bradfield show that of all cometary emissions observed in the UV, only those from CS and S appear strongly
concentrated near the nucleus. This suggests (Jackson et al., 1981)
that CS and S have both a common, short-lived parent, such as

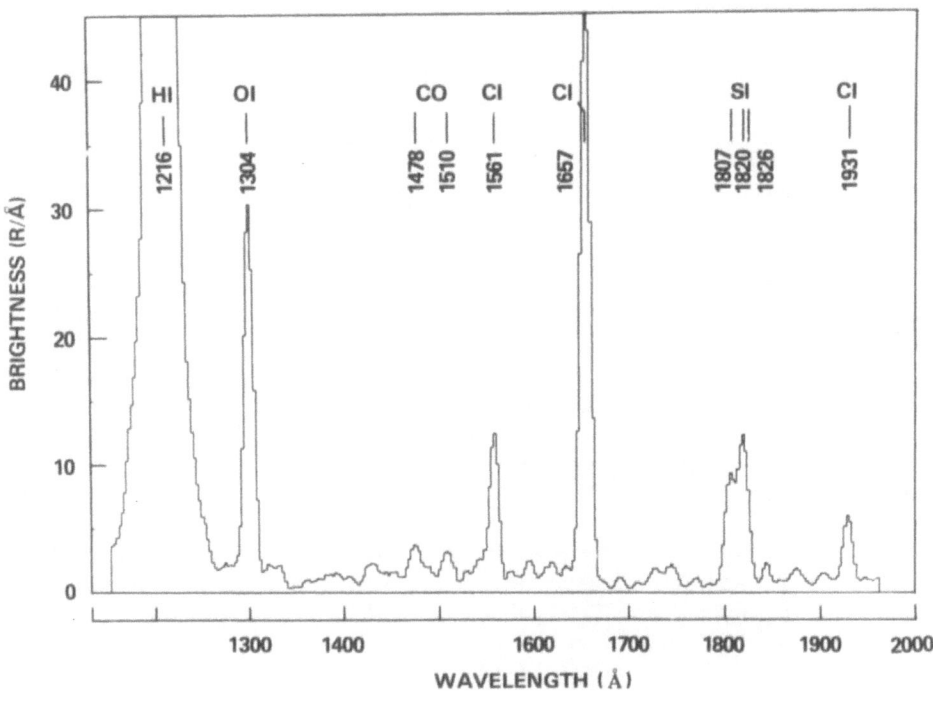

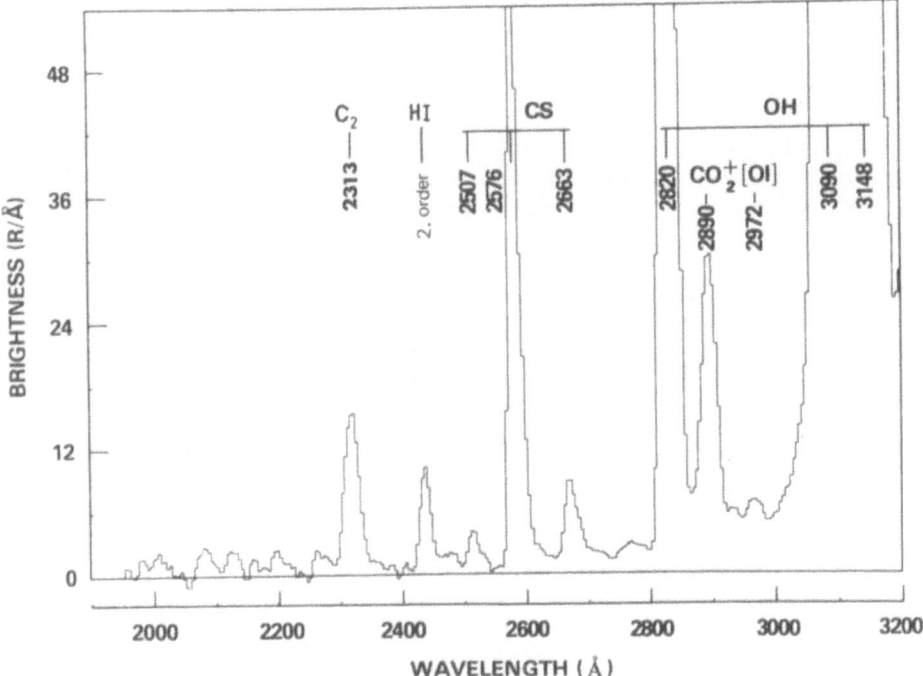

Fig. 1. IUE spectrograms of Comet Bradfield (1979 X).

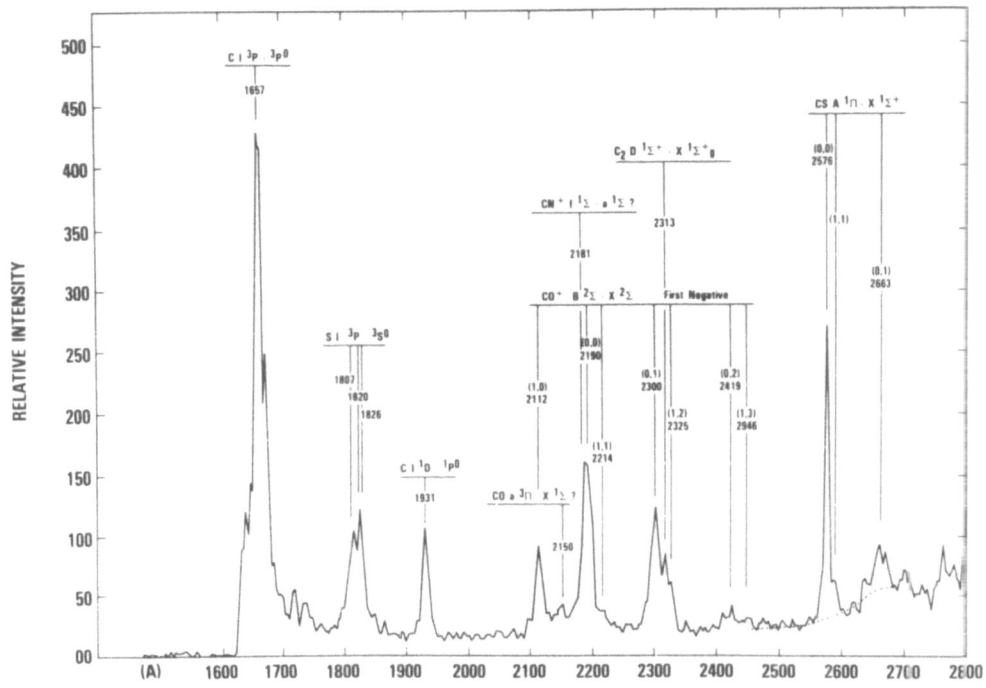

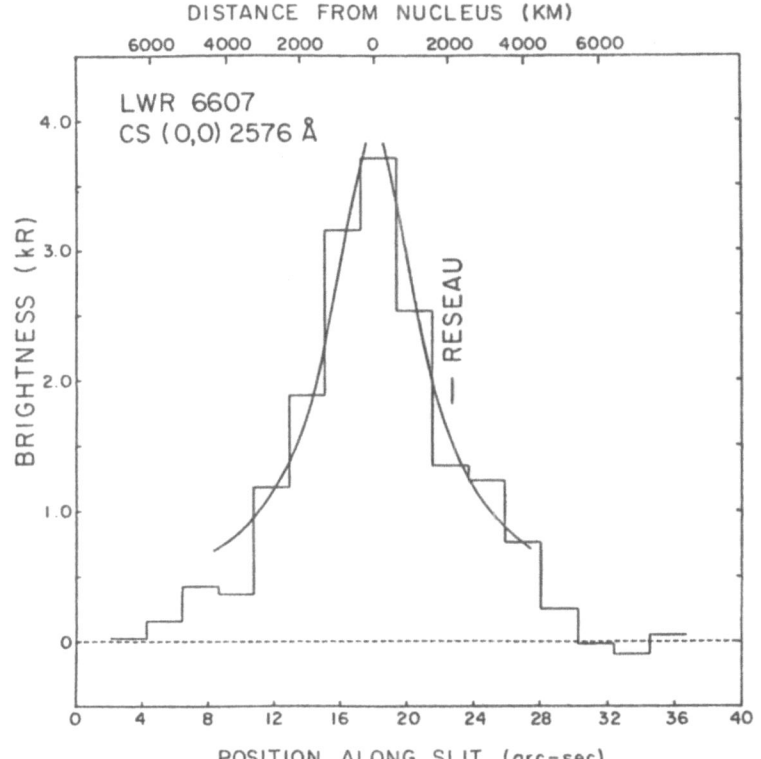

Fig. 2 (top).
Rocket-borne spec-
trum of Comet West
(Smith et al,1981).

Fig. 3 (left).
Variation of the
CS (0,0) band in
the large aperture
of IUE (Jackson et
al, 1981). The
solid line illu-
strates the bright-
ness variation of
CS, if CS was pro-
duced from CS_2 via
photodissociation.

e.g., CS_2. The lifetime of CS_2 at 1 AU is only 100 sec, i.e., shorter than that of any other known cometary radical. With an outflow velocity of 1 km/sec, it has a scale length of 100 km. Fig. 3 shows the variation of the CS (0,0) band brightness in the large (20") aperture of IUE. The smooth lines indicate the expected response if the CS was indeed produced by photodissociation of a parent molecule, such as CS_2. The CS production rate is considerably smaller than the water production rate.

The UV observations of Comet Bradfield over a large range of heliocentric distances (0.7 to 1.6 AU), allowed the determination of the variation of the production rates of various species, especially H and OH, with heliocentric distance. While it is usually assumed that the OH (and presumably also the H_2O) production rate varies with r^{-2}, these observations showed a decrease as $r^{-3.7}$ (Weaver et al., 1981). A satisfactory explanation for this surprisingly strong decrease has not been found.

For comets Seargent and Bradfield, the rotational structure of the (0,0) and (1,0) bands of OH could be resolved in IUE high dispersion spectra. It could be shown (A'Hearn et al., 1981) that the radicals are excited by resonance pumping of the OH by solar radiation. Differences in the observed relative intensities of individual lines are essentially due to the differences in radial heliocentric velocities of the comets at the time of observation.

Table 1 - Observed species in cometary spectra

(species observed in the UV between 1216 and 3085 $\overset{\circ}{A}$, are underlined)

Organic	:	C	C_2	C_3	CH	CN	CO	CS	HCN	CH_3CN
Inorganic	:	H	NH	NH_2	O	OH	H_2O	S		
Metals	:	Na	K	Ca	V	Mn	Fe	Co	Ni	Cu
Ions	:	C^+	CO^+	CO_2^+	CH^+	H_2O^+	OH^+	Ca^+	N_2^+	CN^+
Dust	:	silicates (infrared reflection spectrum)								

3. CONCLUSIONS

What can we learn from these UV observations of comets?

First, several species that were previously not observed, could be discovered in UV spectrograms: they are C, C^+, S, CO, and CS.

A list of known species is given in Table 1. Strong indications were found for the presence of CS_2 as a new parent molecule.

Second, already the first UV observations of a comet proved the existence of an enormous H-cloud surrounding the cometary nucleus. The study of the H and OH emissions in a number of comets, strongly support Whipple's "icy-conglomerate" model of the nucleus. H_2O has only once been detected in one comet in the radio region. In the UV, its emission cannot directly be observed, only its three dissociation products, H, OH, and O. If one assumes that all, or at least most, of the observed HI Lyman-alpha, the OH (0,0) band, and OI (1312 Å) come from the photodissociation of H_2O, i.e., water, one can almost immediately conclude that normal, frozen water is a major constituent of a cometary nucleus.

For "new" comets, the typically found H_2O production rates amount to about 10^{29}-10^{30} molecules/sec; they are of the same order of magnitude as those predicted by Whipple (1950, 1951) already in 1950, 1951 from his analysis of non-gravitational forces acting on cometary nuclei. For short-period "old" comets, the H_2O production rates are about one order of magnitude smaller.

The third, very interesting result of these UV observations is the following: especially with the IUE-satellite, a number of different comets have been observed. These comets were observed at various heliocentric distances; they looked very different; they showed a different gas-to-dust ratio; and some of them were quite "old", others rather "new". Only the dust, and the CO^+ abundances seemed to vary from one comet to the other. The UV spectrum of the "new" Comet Seargent, looked practically identical to spectrum of the "old" Comet Encke. The UV spectra of all comets observed so far, are surprisingly practically identical. From this, one can perhaps conclude that the comets observed so far, have a very similar, homogenous chemical composition, and that they have perhaps also a common origin (Donn and Rahe, 1981).

In summary, we can say that the UV observations of comets have not drastically changed the generally accepted ideas about the nature of these fascinating and often still puzzling objects. But they have confirmed many until then unproven assumptions and considerably expanded our understanding of the physical and chemical processes occurring in their head and tail.

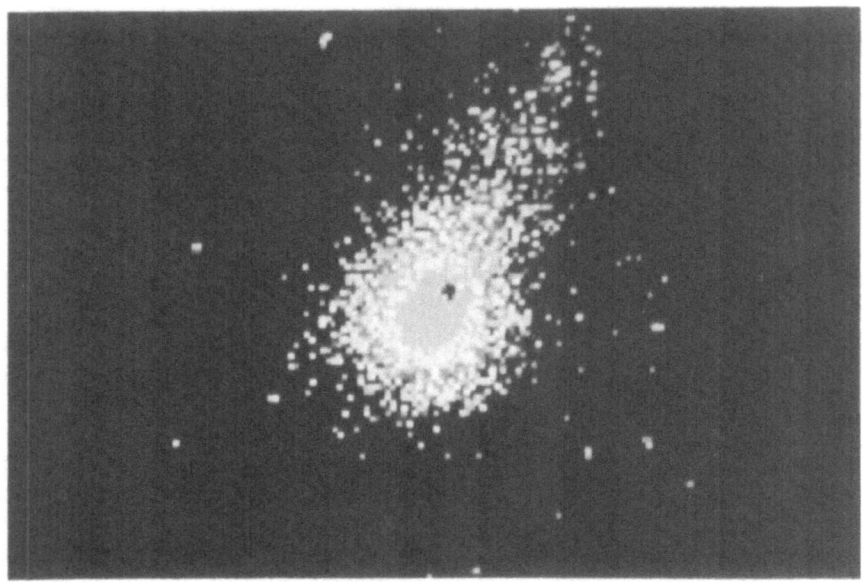

Fig. 4. Comet Bradfield 1979 X, as observed on January 10, 1980, with
the finder telescope of the IUE satellite.

REFERENCES

A'Hearn, M.F., Schleicher, D.G., Donn, B., Jackson, W.M. 1981. In "The
    Universe at Ultraviolet Wavelengths". NASA Conf. Pub. 2171, p. 73.
Code, A.D., Houck, T.E., Lillie, C.F. 1972. In "Scientific Results from
    Orbiting Astronomical Observatory OAO-2". NASA SP-310, p. 109.
Donn, B., Rahe, J. 1981. In Proc. IAU-Coll. No. 61.
Feldman, P.D., Brune, W.H. 1976. Ap.J. Letters 209, L145.
Feldman, P.D., Weaver, H.A., Festou, M.C., A'Hearn, M.F., Jackson, W.M.,
    Donn, B., Rahe, J., Smith, A.M., Benvenuti, P. 1980. Nature 286, 132.
Jackson, W.M., Rahe, J., Donn, B., Smith, A.M., Keller, U.H., Benvenuti,
    P., Delsemme, A.H., Owen, T. 1978. Astr. Astroph. 73, L7.
Jackson, W.M., Halpern, J., Feldman, P.D., Rahe, J. 1981. Astr. Astroph.
    In press.
Keller, H.U. 1976. Space Sci. Rev. 18, 641.
Meier, R.R., Opal, C.B., Keller, H.U., Page, T.L., Carruthers, G.R. 1976.
    Astr. Astroph. 52, 283.
Opal, C.B., Carruthers, G.R., Prinz, D.K., Meier, R.R. 1974. Science 185.
Smith, A.M., Stecher, T.P., Casswell, L. 1980. Ap.J. 242, 402.
Weaver, H.A., Feldman, P.D., Festou, M.C. 1981. In "The Universe at
    Ultraviolet Wavelengths". NASA Conf. Pub. 2171, p. 65.
Whipple, F.L. 1950. Ap.J. 111, 375; Ap.J. 113, 464.

PLASMA – DUST INTERACTIONS IN THE SOLAR AND COMETARY ENVIRONMENT

H.J. Fahr, H.W. Ripken and G. Lay
Institut für Astrophysik der Universität Bonn
Auf dem Hügel 71, D-5300 Bonn 1 (FRG)

By zodiacal light observations, direct dust particle detections, white light F-corona observations, and infrared corona observations it has become evident that interplanetary dust particles are present even very close to the sun, i.e. up to solar distances of about 4 solar radii. These particles, being either of a siliceous or a carbonaceous type, are known to have an amorphous rather than a highly organized cristalline structure. Within the solar wind domain they give rise to a specific form of a plasma – solid body interaction that will be described here in detail.

The main processes, occurring when 1 KeV solar ions mpinge on the surfaces of interplanetary dust particles, essentially consist of an ion absorption by the mineral surface layer, a subsequent deionization in this layer, an occasional desorption from the surface in the form of a neutral molecule, and finally a reionization and an antisolar convection by the ambient solar wind. Thus, whenever saturation with trapped solar wind particles is reached in the outer mineral layer, dust surfaces enable a prompt conversion of ionic into neutral gas species and give rise to the existence of neutrals even close to the sun. Among these neutral desorption products by far the most interesting components are H and He atoms since they can uniquely be identified by their resonance luminescence in the UV or EUV.

Fahr et al. (1981) have discussed the details of the desorption. In brief the following picture arises: Solar wind α-particles are deionized and desorb as neutral helium atoms. For solar wind protons the chain of chemical processes is not unique. For the cristalline surface and a sputtering rate sufficiently high to supply enough oxygen at the dust surface, most of the incoming protons desorb as H_2O or OH molecules, yielding atomic H in subsequent dissociations. For insufficient oxygen supply, protons exclusively desorb as H_2 molecules, of which only a small percentage later dissociates into H atoms. For amorphous surface structures the desorption of deionized protons in the form of SiH and CH molecules also becomes likely. These hybrids are then dissociated with a high probability to yield H atoms. To cover the whole

331

W. Fricke and G. Teleki (eds.), Sun and Planetary System, 331–333.

range of possible H atoms yields, calculations were carried out for efficiency values of 0.9 (best case) and 0.05 (worst case) for the conversion of protons into H atoms.

Due to the high ionization rates for H and He in the solar vicinity, convection and diffusion of these atoms need not be considered. The local equilibrium densities of these atoms are thus obtained by equating corresponding desorption and reionization rates. Consistent numerical calculations on the basis of the current-ly most plausible parameters yield H and He densities shown in Fig. 1.

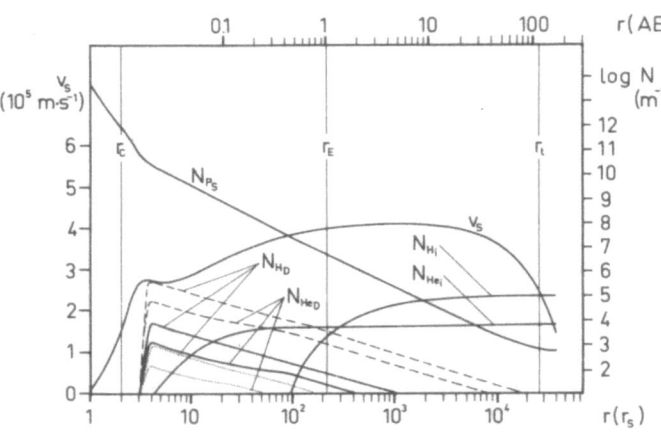

Fig. 1: Model solutions for solar wind - dust interaction. Shown are solar wind velocity v_s and density $N_{P,s}$, inter-stellar H and He densi-ties $N_{H,i}$ and $N_{He,i}$, and dust-generated H and He densities $N_{H,D}$ and $N_{He,D}$, each for three specific dust cross sec-tions Γ_E: 8.0E-18 1/cm (dashed), 2.0E-19 1/cm (solid), and 1.0E-19 1/cm (dotted line).

As pointed out by Fahr et al. (1981), the best method of detecting dust-generated hydrogen and helium atoms is to observe the spectral profiles of the resonance lines at 121.6 nm and 58.4 nm, respectively. To calculate these spectral intensities a theoreti-cal description of the velocity distribution functions of dust generated H and He atoms is required. Compared to the orbital velocities of the dust particles, thermal velocities of the desorbing atoms can be neglected. In a sufficient approximation the neutral gas distribution functions are identical with the velocity distribu-tion $f_D(\underline{r},\underline{v})$ of the dust particles. The latter function is deter-mined using the following assumptions:
1) Dust particles are moving in quasi-circular Keplerian orbits.
2) Orbital motions are prograde.
3) The distribution over the orbital inclinations i is such that the first moment of $f_D(\underline{r},\underline{v})$ fits the dust density model of Leinert et al. (1978): $N_D(\underline{r}) = \bar{N}_D(r) N_D(\phi)$, ϕ being the ecliptic latitude.

Under these assumptions the function $f_D(\underline{r},\underline{v})$ is reduced to an expression $f_D(i)$ that can be explicitly determined as the solution of an Abel-type integral equation:

$$N_D(\phi) = 2 \int_\phi^{\pi/2} f_D(i) \, (tgi / \cos\phi) \, (tg^2 i - tg^2\phi)^{-1/2} \, di$$

With the help of $f_D(i)$ the spectral intensities of the H-121.6 nm and He-58.4 nm resonance luminescence can be calculated. For a specific observation from the earth (into a direction in the ecliptic plane with a solar offset of 5°) the results shown in Figures 2 and 3 are obtained. With high resolution observations of these spectra, the dust-typical intensity features could be uniquely identified.

As a result of simultaneous observations of both hydrogen and helium resonance spectra relevant information can be obtained concerning solar wind parameters, physical and dynamical properties of the zodiacal dust, and desorption and reaction rates of deionized neutrals in the solar vicinity.

The interaction between the solar wind and cometary dust tails can be treated analogously to the plasma – dust processes described in this paper. Depending on the saturation level of cometary dust surface layers, one expects to find free neutral H and He atoms in the vicinity of dust tails, providing a new access to cometary research.

References: Fahr, H.J., Ripken, H.W., and Lay, G.:1981, Astron. Astrophys. (in print). Leinert, C., Hanner, M., and Pitz, E.: 1978, Astron. Astrophys. 63, 183.

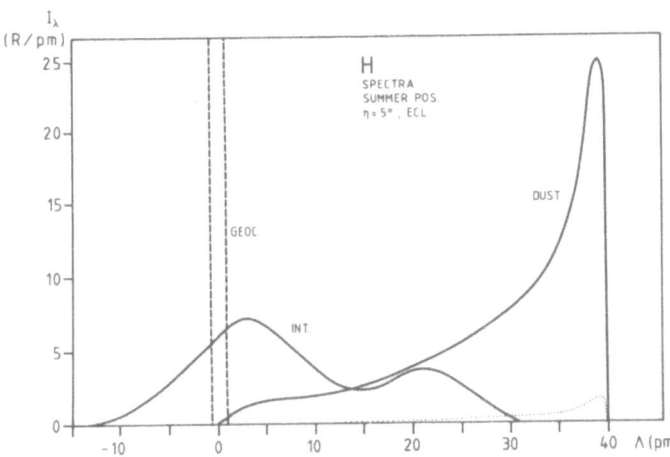

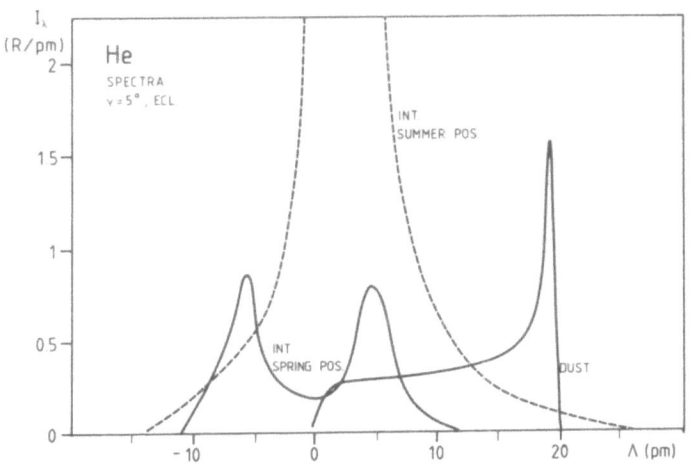

Figs. 2 and 3: Spectral intensity distributions I_λ of hydrogen Ly-α and He I-58.4 nm resonance radiation versus line center distance Λ. The geocoronal Ly-α peak is marked by vertical dashed lines. The observation-time independent dust-generated intensities are given for efficiencies of 0.9 (solid line) and 0.05 (dotted line). Interstellar radiation intensities are calculated for June 21 (H: solid line, He: dashed line) and March 21 (He: solid line).

THE TOTAL MASS AND STRUCTURE OF THE METEOR STREAM ASSOCIATED WITH
COMET HALLEY

Anton HAJDUK
Astronomical Institute of the Slovak Academy of Sciences
84228 Bratislava, Czechoslovakia

The meteor stream associated with Comet Halley has been studied by the author on the basis of a long series of visual and radar observations during the Eta Aquarid and Orionid meteor shower periods (Hajduk, 1970, 1973, 1980). It appeared that the stream exhibited inhomogeneities in both directions, across it and along the orbit. A stable zone of higher density and variable filaments have been detected. This variability makes the determination of the average particle density and of the mass of the stream a rather complicated task. The problem was solved by analyzing two extensive sets of data, the Springhill observations from 1958-1967 kindly supplied by Dr. B. A. McIntosh ($\sim$ 500,000 radar meteor echoes) and the Ondrejov observations from 1961-1978 ($\sim$ 100,000 echoes), taking also in account other results.

The mean hourly rates of meteors of different categories of echo duration have been determined from both sets of data within the solar longitudes of 39° to 55° and 203° to 218°, corresponding to the Eta Aquarid and Orionid shower periods, respectively. During these periods the Earth crosses the stream in two different sections, approaching the centre of the stream to $0.065 - 0.11$ AU and $0.154 - 0.18$ AU, respectively. The echo rates reduced to the zenith, and to the collecting area corresponding to the sensitivity for the particular echo duration category, yield the mean reduced particle flux $f = 2.5 \times 10^{-11}$ $m^{-2}s^{-1}$ at the distance of 0.065 AU from the comet's orbit, for all echoes. This value corresponds to a limiting visual magnitude of 7.0 to 7.5; the flux was also determined separately for the 1s and 8s echo duration levels (a detailed analysis will be published elsewhere). The crucial point of the determination of the total mass of the stream is the construction of the integrated mass distribution of shower meteors along the scale of mass. The shower-to-background ratios of the mean hourly rates at the particular echo duration levels were applied to the sporadic background mass distribution according to Millman (1975). The magnitude indices of the Eta Aquarids, Orionids and the background obtained from the Skalnaté Pleso data (Kresáková, 1966) and the Orionid-to-sporadic ratios obtained from the Budrio data (Hajduk and Cevolani, 1981) were also taken into account.

335

W. Fricke and G. Teleki (eds.), Sun and Planetary System, 335–336.
Copyright © 1982 by D. Reidel Publishing Company.

The relative contribution to the total mass of the stream reaches a maximum for particle masses of $\sim 10^{-5}$ kg. Both distributions were extrapolated over the interval between the limits set by the radiation pressure cut-off, 10^{-12} kg (Kresák, 1976), and the ejection cut-off, 10^2 kg (Whipple, 1951). Combining the mass distribution with the particle fluxes, a total flux of 3.1×10^{-16} kg $m^{-2}s^{-1}$ was obtained for the distance D = 0.065 AU from the comet's orbit at the heliocentric distance of 1 AU. Corresponding values at other distances are 2.0×10^{-16} kg $m^{-2}s^{-1}$ at D = 0.11, 1.4×10^{-16} kg $m^{-2}s^{-1}$ at D = 0.16, and 1.1×10^{-16} kg $m^{-2}s^{-1}$ at D = 0.18. An extrapolated value of 5×10^{-16} kg $m^{-2}s^{-1}$ was adopted for D = 0.01. The mass flux in the stream filaments and zones producing local shower maxima may reach up to twice these mean values. The spatial density between D = 0.18 and 0.065 ranges from 10^{-21} kg m^{-3} to 3×10^{-21} kg m^{-3}, with an extrapolated value of 5×10^{-21} kg m^{-3} at D = 0.01 AU from the comet's orbit.

The total mass of the stream was determined at 5×10^{14} kg. The present mass of the nucleus of Comet Halley, according to Newburn and Yeomans (1982), is 6.5×10^{13} kg, or 8 times smaller than that obtained for its meteor stream. Combining the estimates of the distribution of the mass loss into different constituents with the present H_2O production rate of the comet according to Delsemme (1976), our figure is found to be equivalent to the production of solid particles over 5000 revolutions - even when the losses by dynamical ejection and disintegration are disregarded. Thus the required lifetime turns out to be more than one order of magnitude longer than the average lifetime of comets of this type according to Kresák (1981). There are two possible explanations of this discrepancy : either the conversion factors between meteor magnitudes and masses, as generally adopted, are in error, or Comet Halley was originally an exceptionally large object, with a much higher dust production than at present. The latter explanation appears more probable.

REFERENCES

Delsemme, A.H.: 1976, Lecture Notes in Phys. 48, pp. 314-318.
Hajduk, A.: 1970, Bull. Astron. Inst. Czechosl. 21, pp. 37-45.
Hajduk, A.: 1973, Bull. Astron. Inst. Czechosl. 24, pp. 9-13.
Hajduk, A.: 1980, IAU Symp. 90, pp. 149-152.
Hajduk, A., Cevolani, G.: 1981, Bull. Astron. Inst. Czechosl. 32, pp. 304-310.
Kresák, L.: 1976, Bull. Astron. Inst. Czechosl. 27, pp. 35-46.
Kresák, L.: 1981, Bull. Astron. Inst. Czechosl. 32, pp. 321-340.
Kresáková, M.: 1966, Contr. Astron. Obs. Skalnaté Pleso 3, pp. 75-112.
Millman, P.M.: 1975, in B. Field and A.G.W. Cameron (eds) "The Dusty Universe", McGraw-Hill, New York, pp. 185-209.
Newburn, R.L., Yeomans, D.K.: 1982, Ann. Rev. Earth Planet. Sci. 10, in press.
Whipple, F.L.: 1951, Astrophys. J. 113, pp. 464-474.

OUTBURSTS OF COMETS

S. Gąska
Institute of Astronomy, N. Copernicus University,
Toruń, Poland.

P. Gronkowski
Institute of Physics, High College of Educaticn,
Rzeszów, Poland

The aim of this report is an attempt to explain the sudden brightness variations of the following comets: Schwassmann – Wachmann (1925 II), Kritzinger (1914a), and Ikeya – Seki (1965f).
, First of them is elliptical with eccentricity e=o.135 and radius vector $5.3 < r < 7.3$ A.U., so that the comet remains all the time between the orbits of Jupiter and Saturn. Between discovery in 1927 and 1973 the minimal distances comet – Jupiter and comet – Saturn were equal to 2.5 A.U. and 1.8 A.U., respectively.
Within this period 52 outbursts were observed (Grudzińska, 1979). These outbursts are not uniformly distributed along the orbit. For example: in 1933, 3 outbursts were observed in 1941-8 events, and in years 1928, 1936 and 1950 – none.
Various reasons have been considered to explain the outbursts. Richter, Whitney, Donn, Urey and others have suggested the correlation between cometary and Solar activities. However, there are some evidences that this relation may be accidental (Dreyer, 1972; Grudzińska, 1976; Weber and Barnes, 1972). Grudzińska has shown that there is not contradiction with the supposition that the outbursts are caused by the collisions of the comet with the meteor streams.
Our detailed investigations show that the outbursts occur at the fixed points of the cometary orbit (Gąska, 1982). These points are determined by the commesurabilities of the true orbital motions of the comet with those of Jupiter and Saturn.
Let n_1, n_2, n_3, denote orbital motions (in arc seconds) of Jupiter, Saturn and comet respectively and k_1, k_2, k_3, small integers. We search for the moments, when the resonance condition is satisfied:

$$f = k_1 n_1 + k_2 n_2 + k_3 n_3 = 0$$

These moments have been compared with observations collected by Grudzińska (1979) in the accompanying table.

W. Fricke and G. Teleki (eds.), Sun and Planetary System, 337–338.

Table.

T_o-T_c (days)	Percentage of outbursts
0	54
$\pm$ 20	84
$\pm$ 50	96
$\pm$100	100

T_o and T_c denote the range of observed and calculated moments of outbursts. In second column the percentage of outbursts which occurred in time intervals $(T_o$-$T_c)$ is given. It shoud be noted that for all the outbursts with $|T_o$-$T_c| > 20^d$, f $< 4"$, and $|k_1 + k_2 + k_3| \leq 10$.

The same method was adopted to parabolic comets. The comet Kritzinger (1914a) had the orbital elements: q=1.1984 A.U., e=0.999665. It changed brightness for about 2^m on June 3, 1914 (Chofardet and van Biesbrock, 1914). For the system; Earth-Venus-comet: k_1:k_2:k_3=1:2:(-4) and f=0 on this day. For the system Earth-Mars-comet: k_1:k_2:k_3=(-1):7:(-2) f=0 on June 1, and June 3, 1914.

The comet Ikeya-Seki had q=0.0077527 and e=0.9999123. It exploded on Oct.21, 1965 and divided itself into two separate parts. In this cass we have taken into account the interactions of the following systems:

1. Earth-Venus-Mercury-comet - k_1:k_2:k_3:k_4=1:1:1:(-1)

2.   Venus-Mercury-comet - k_2:k_3:k_4=3:4:(-1)

3.   Mercury-comet - k_3:k_4=2:(-1)

The function f reaches 0 on the days: Oct. 17 and Oct. 21,1965 for all the above systems (Gąska and Gronkowski, 1982).

These results strongly suggest that the outbursts of comets occur at fixed points closely determined by the commensurabilities of the motions of the investigated comets and planets.

References
Chofardet 1914 A.N. 197
Dreyer M. 1972, Circ. Bur. centr. Telegr. astr. No 2432
Gąska S. 1982, Acta Astr. in press
Gąska S. and Gronkowski P. 1982, Acta Astr. in press
Grudzińska S. 1976, Acta Astr. 26, 117
                    1979, Studia Soc. Sci. Toruń, vol. 6 No. 2
                    1980, Acta Astr. vol. 30 No. 3
Weber E. and Barnes R. 1972, Circ. Bur. centr. Telegr. Astr.
          No. 2439

ON THE ELECTRIC CHARGE OF INTERPLANETARY GRAINS

Jean-Pierre J. LAFON
Observatoire de Paris-Meudon   (DESPA)
92190 - MEUDON (FRANCE)

In this communication we describe a work by J.-P.J.Lafon, J.M.Millet and Ph. Lamy which emphasizes the importance of surface effects in what concerns the mechanism of grain charging in the interplanetary plasma. Numerical computations performed under various conditions of interplanetary plasma show that the photoemission and the thermoemission of electrons can govern the charge and even reverse it at radial distances of a few solar radii.

Indeed any body embedded in a ionized medium carries an electric charge Q such that the total electric current generated by the charged particles striking it exactly cancels the current generated by those which leave its surface. Then

$$\frac{dQ}{dt} = \sum_i j_i = 0 \qquad (1)$$

where j_i denotes the ith contribution to the total current. Since each j_i depends on Q through Coulomb attraction or repulsion. A lot of papers have been published concerning this problem (Lafon, 1976 ; Lafon et al, 1981 including references), because of its importance in grain dynamics, grain evolution and even chemistry. Although the physics of solid body surfaces is still not well known, recent fairly reliable data concerning photoemission and thermoemission of electrons allow not so bad determinations of emitted electric currents. At the same time a better knowledge of the interplanetary plasma (Millet et al, 1980) enables to find fairly reliable values for the grain charge.

Hereafter we only illustrate some properties woth thinking about. Fig. 1 shows quite different temperature profiles of solar wind in high speed streams for which computations have been made ; they are derived from experimental data (thinlines) or from a model (thicklines); in what we call case 2, there is a sharp discontinuity in proton temperature whereas the other profiles are smooth. For details see the works of Millet et al (1980). Photoemission by the enlightened side has been taken into account on the basis of Grard's (1973) work. A distribution function of the thermoelectrons was de-

W. Fricke and G. Teleki (eds.), Sun and Planetary System, 339–341.
Copyright © 1982 by D. Reidel Publishing Company.

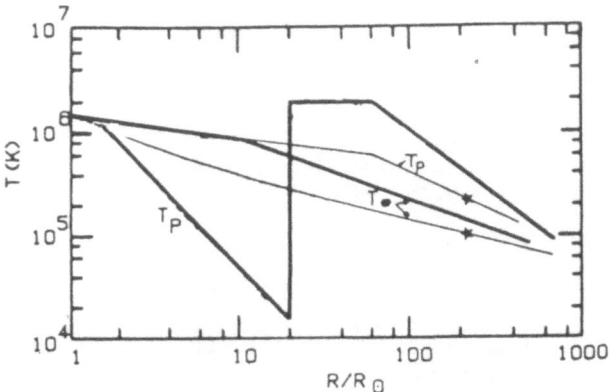

Fig.1 - Case 1 : Thin lines; Case 2 : Thick lines.

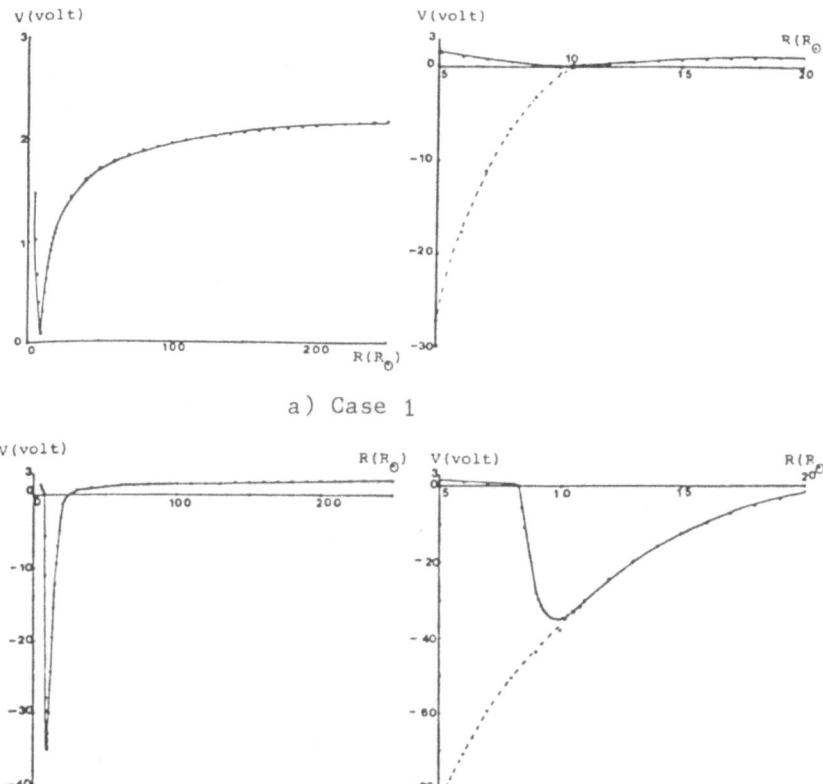

a) Case 1

b) Case 2

Fig. 2

rived theoretically and included in the numerical code.

On fig.2 we have plotted the electric potential reached by the grains versus radial distance between 5 and 200 R_θ, on the left hand side, and the same between 5 and 20 R_θ (enlarged scale) on the right hand side; fig.2 a,b respectively correspond to cases 1 and 2 of fig 1. Since all collected currents are proportional to the square p^2 of the grain radius p when expressed in terms of the potential, the equilibrium ("floating") potential is characteristic of the grain material, independent of p,whereas the electric charge is proportional to p (Lafon et al., 1981)

The main features are as follows :

- The potential reaches saturationlike values at large distances of the sun.
- Electric photoemission weakens the negative potential at small radial distances in a non negligible way, and renders the potential positive at large distances (it would be negative without photoemission).

- Electron thermoemission is dominant for radial distances lower than 11 R_θ (the dotted curve represents the potential if thermoemission is neglected).

- Thermoemission prevents grain disruption (disruption potential can no longer be reached.

- The characteristics of the plasma are important parameters : whereas in case 1 the potential is positive everywhere (fig 2a), it reaches high negative values in case 2 (fig 2b). Thus fluctuations in plasma properties should induce charge fluctuations,

- Under such conditions the floating potential can have a unique value (Lafon et al., 1981)

This theoretical work on grain charge is still in progress. This requires also new laboratory measurements and theoretical models concerning surface effects described above. We may also introduce other surface effects. One must also not forget that the relative drift velocities between the components of the plasma and the grains can be important (Lafon et al, 1981). Finally the accuracy of the determination of the charge of interplanetary grains is coupled with that with which the structure of the solar wind is known.

Grard R.J.L., 1973, J. Geophys.Res., 78, pp. 2885-2906.

Lafon 1976, Thèse de Doct.d'Etat, Univ. Paris 7,Rapport DESPA 159bis
    Observatoire de Paris-Meudon, 92190-Meudon (France)

Lafon, Lamy, Millet, 1981, Astron.Astrophys. 95, pp. 295-303

Millet, Lafon, Lamy, 1980, Astron.Astrophys., 92, pp. 6-12

OBSERVATIONS OF ZODIACAL LIGHT AT ABU-SIMBEL, EGYPT, DURING THE PERIOD 1975-1979

A.S. Asaad
Institute of Astronomy and Geophysics, Egypt

The curves published by Weinberg (1964) indicated the existance of large discrepancies between measured brightness and polarization of Zodical Light (ZL) published by different observers. This could be partially explained by the solar activity, in such a way, that the brightness increased with decreasing solar activity while the polarization behaved in the opposite sence. Such relations were noted by Asaad (1967) and by Dufay (1966) and Weill (1966). Besides Asaad et al. (1979 a) found that the brightness of the ZL was higher if observed from low geographical latitudes by about 30% for a latitude difference of 10°. They also noted that the brightness was higher by 50% for an increase in the elevation of the observing site above sea level by about 8000 feet.

To investigate the problem further it was decided in Egypt to carry out observations of ZL for one complete solar cycle at Abu-Simbel in the southern part of the country where the transparency and observing conditions were much favourable. The station was built and observations of ZL were carried out by A.S. Asaad, J.S. Mikhail and S. Nawar since 1975 for two months each year usually April and November. Analysis of the results for half of a solar cycle showed the following:

a) The brightness of the ZL increases with decreasing solar activity.
b) The colour indices CI(B - V) and CI(B - R) are not different from that of the sun by no more than 0.2 magnitude. Colour indices become redder with increasing elongation from the sun during quite solar activity. While during higher solar activity the CI(B - V) becomes bluer with increasing elongation from the sun. In between quiet and active sun no systematic variations in both colour indices of the ZL.
c) Observations at Abu-Simbel show that the brightness of the ZL change with the lunar phase, Asaad et al. (1979 b). It has a peak at new moon, decreases and tends to increase at full moon.

CONCLUSIONS

The results obtained so far at Abu-Simbel for the ZL brightness showed short period variations with lunar phase as well as long period

W. Fricke and G. Teleki (eds.), Sun and Planetary System, 343–344.

variations with the solar cycle. The lunar phase variations were detected from ground based observations at Abu-Simbel by Asaad et al. (1979 b) as well as by Divari (1963) but have not been detected from outher space, Renger and Vande Noord (1967), Vande Noord (1970) and Burnett (1972). As such is the case, the lunar variations of the ZL maybe due to tides of the moon on the earth's atmosphere rather than on the dust clouds surrounding the earth.

As for the solar cycle changes, the process suggested by Asaad (1967) seems to hold. The solar wind during maximum solar activity may destroy, evaporate or push away some of the fine particles in the interplanetary medium. The variations in the intensity of cosmic rays with the 11 years solar cycle found by Kudo and Wada (1968) is in favour of solar activity explanation of ZL variations. During minimum solar activity the brightness of ZL increase by florescence of dust particles by collision. This may also explain the colour indices variations with elongation and with solar cycle.

REFERENCES

Assad,A.S.: 1967, Observatory 87, 83.
Asaad,A.S.: 1967, Nature, 214, 259.
Assad,A.S., Mikhael,J.S., Nawar,S.: 1979a, Helwan Obs. Bull. No. 204;
                                     1979b, Helwan Obs. Bull. No. 203.
Burnett,G.B.: 1976, Interplanetary dust and ZL, Lecture notes in Phys.
              No. 48, ed. by Elsasser, H. and Fechig,H., 53.
Divari,N.B.: 1964, Sov. Ast. 7, 547.
Dufay,J.: 1966, Comptes Rendus 263, 947.
Kudo,S. and Wada,M.: 1968, Eleven years variations of cosmic ray inten-
                     sity, Report of Ionosphere and Space Res., Japan,
                     22, No. 3, 137.
Regner,V.H. and Vande Noord,E.L.: 1967, Nasa SP-150 ed. by Weinberg,
                                  J.L., 45.
Vande Noord,E.L.: 1970, Astrophys, J., 161, 309.
Weill,M.G.: 1966, Comptes Rendus, 263, 943.
Weinberg,J.L.: 1964, Summary Report on ZL, Hawaii Inst. of Geophys. July.

THE BRIGHTNESS INTEGRAL EQUATION. A DIFFERENT APPROACH TO THE STUDY OF THE ZODIACAL LIGHT.

J. Buitrago, R.Gómez and F.Sánchez
Instituto de Astrofísica de Canarias; Canary Islands. Spain.

In this short report, we present the preliminary results of a new method developed for obtaining the three dimensional distribution of interplanetary dust from observations performed in the Ecliptic (Earthbound or Satellite). This new approach is essentially based in the analytical trasformation of the Zodiacal Light Brightness Integral into a Volterra integral equation of the second kind and its subsequent numerical resolution. The implementation of this method requires a careful set of observations confined to a plane perpendicular to the ecliptic, passing through the Sun, and ranging from an elongation angle $\varepsilon = \beta = \pi$ (Gegenschein) to $\varepsilon = \beta = \pi/2$ (Ecliptic Pole).
The following assumptions are made :
The medium is homogeneous.
The spatial distribution can be represented in the form $n \sim r^{-\nu} \cdot f(\beta_\odot)$
Under these conditions, and for the case $\nu = 1$, the unknown function $f(\beta_\odot)$ can be related to the intensity $I(\beta, R)$, and the scattering function $\sigma(\theta)$ through the integral equation :

$$\frac{\partial^2 G}{\partial \psi^2} + G = \int_0^\psi \left[\sin(\alpha + \pi - \psi) \frac{\partial^2 \sigma}{\partial \psi^2} - 2 \cos(\alpha + \pi - \psi) \frac{\partial \sigma}{\partial \psi} \right] \cdot$$

$$\cdot f(\alpha) \, d\alpha + \sigma(\pi) \cdot f(\psi) \tag{1}$$

where

$$G(\psi, R) = (F_0 R_0^3 n_0)^{-1} \cdot R^2 \operatorname{sen}^2 \psi \cdot I(\psi, R) \tag{2}$$

Fo: solar flux at the distance Ro, no : particle number density,
$I(\psi, R)$: measured intensity, at distance R, for an elongation $\varepsilon = \beta = \pi - \psi$
For the numerical resolution of equation (1), we used experimental data from the 1964-1975 all-sky survey of Zodiacal Light at the Teide Observatory (Dumont and Sánchez, 1976).

W. Fricke and G. Teleki (eds.), Sun and Planetary System, 345–347.

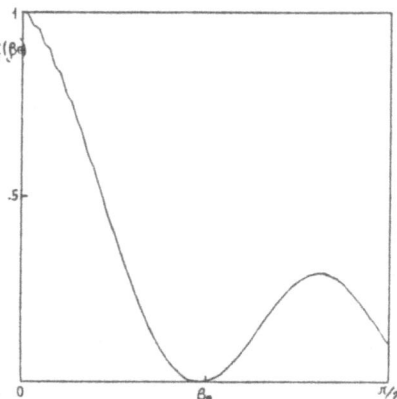

Fig.1.- Angular part of the spatial distribution.

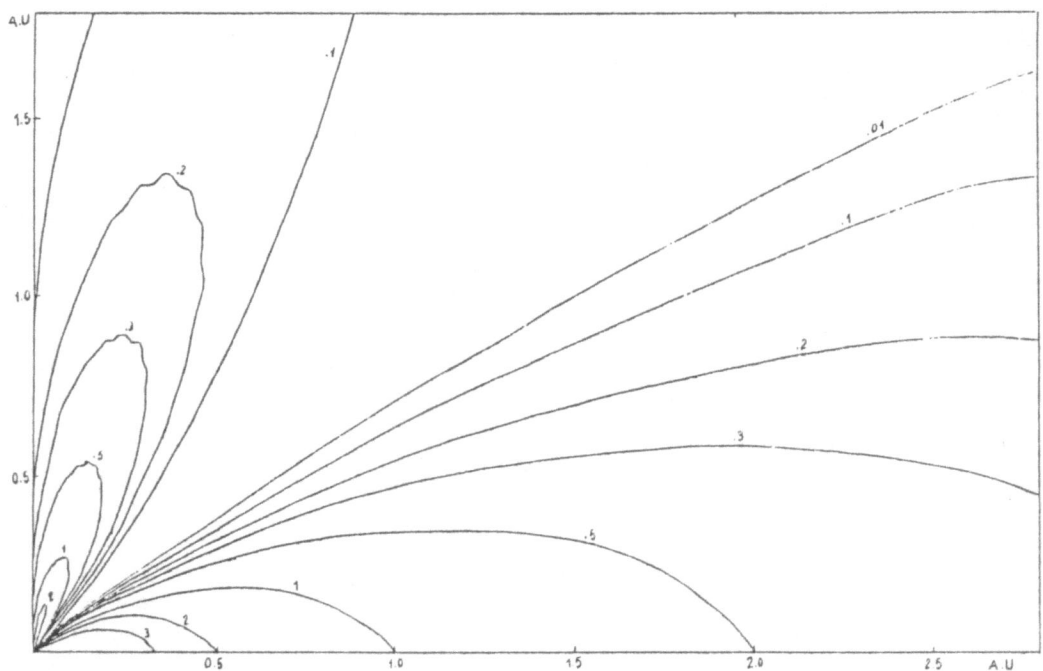

Fig.2 .- Equidensity lines of Interplanetary Dust in a plane
perpendicular to the ecliptic passing through the Sun.

The angular part of the spatial distribution and the resulting equidensi
ty lines, without smoothing, are shown in Fig. 1 and Fig.2 respectively.
The secondary maximum appearing in the angular distribution give raise
to the relatively small lobes in the set of equidensity curves. This
could be interpreted as an enhancement of orbits of high inclination
(about 70º) near the Sun.
As already indicated, we consider these results as provisional. It seems
necessary to apply the method to other set of observations for comparison. Measurements from a spacecraft would be specially interesting.
A full paper with a formal derivation of the method and discussion of
the results is in preparation.

REFERENCES :

Buitrago, J: 1979, Planet Space Sci. 27, pp. 1043-1044
Buitrago, J: Alvarez, P; López, G; Mujica, A and Sánchez F.; 1981,
Planet. Space Sci. 29, pp. 137-138.
Dumont, R. and Sánchez F.; 1976, Astron. and Astrophys. 51, pp. 393-399
Mujica, A.; López, G. and Sánchez, F.: 1980, Planet Space Sci. 28, pp.
657-660.
Mujica, A.: 1981, Thesis, Univ. La Laguna. Canary Islands.
Schuerman, D.W.: 1979, Planet Space Sci. 27, pp. 551-556.

SARAJEVO OBSERVATORY FIREBALL PATROL

Vladimir Miličević[1] and Muhamed Muminović[2]
1) Faculty of Science and Mathematics - Sarajevo
2) Astronomical Observatory - Sarajevo

1. INTRODUCTION

The Czechoslovak Ondrejov Observatory has a long tradition in the study of meteors, especially fireballs. As a result of co-operation of the said Observatory and the Yugoslav observatories at Sarajevo and Hvar an idea was launched to establish fireball photographing stations. In 1980 special Czech cameras were mounted, first in Sarajevo and later in Hvar. The stations are an embryo for possible future Yugo - slav network for registering and studying fireballs.

The interest has increased of late in watching and re- covery of fireballs which have possibly landed. Pioneering work by Ceplecha et al. of Ondrejov resulted in photogra - phing the Pribram meteorite's fall and its subsequent reco- very (1959). In 1970. the Lost City meteorite was photogra- phed by the American Prairie Network stations and recovered. The Canadian network MORP managed to recover the photogra - phed fireball (later meteorite) named Innisfree (1977).

2. OBSERVATIONS

Both Yugoslav stations are equipped with cameras made at the Ondrejov Observatory. Lenses are fish-eye Zeiss Distagon of 30 mm focal length at f/3.5. There is a rotating shutter in front of plate holder. Fireball trail break is at 0.08 sec. intervals. Kodak Tri X Pan glass plates are used. The came- ras have inbuilt thermostats to prevent misting of the len- ses during exposures.

Experience so far has shown that in clear nights of average quality one plate can be exposed for the whole dura- tion of the night. When the moon is older than 4 days one exposure is made with the moon and one without it. When the moon is full two plates are exposed for equal periods each.

W. Fricke and G. Teleki (eds.), Sun and Planetary System, 349–350.

3. RESULTS

By September 25th, 1981, the Sarajevo Observatory made a
total of 102 protographs. The Hvar Observatory, which began
photographing later, made about two dozen. Sarajevo Obs.
managed to photograph four of the brighter fireballs: on
5th Sep.80, 9/10th Dec.'80, 13th Dec.80, and 25/25th Sep.
1981. The magnitudes were respectively:-8; -12; -9; -9;.

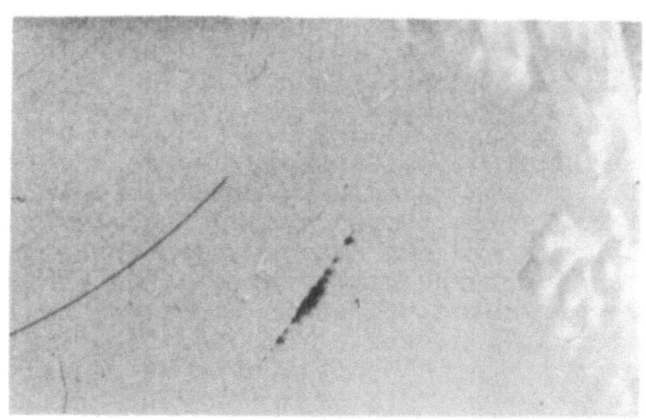

Fig.1. Bright Fireball of 9/10th December, 1980.

Brightest recorded fireball, of 9/10th Dec.'80, be -
longs to the Geminid Stream; it travelled through the atmosphere approx. 1.33 secs. The sky distance covered is cca
10°. Only preliminary analysis of the fireballs has been
made and the detailed one will be done at Ondrejov.

4. ACKNOWLEDGEMENTS

Our special thanks to Dr.Z.Ceplecha for initiating this
programme and advice. Thanks to K.Pavlovski of Hvar Observatory and the staff of the Sarajevo Observatory, notably
N.Grubić, M.Stupar and J. Mulaomerović.

REFERENCES

Ceplecha,Z., Boček,J., Ježkova,M., Porubčan,V. and
Polnitzky,G.: 1980, Bull.Astron.Inst.Czechosl.31.
Halliday,I.: 1973 "Evolutionary and Physical Properties of
Meteorids", ed.Hemenway, IAU Colloquium 13 (NASA SP-319),1.

SECTION IV

MOTIONS IN THE PLANETARY SYSTEM

REVIEW OF THE DYNAMICS OF SATELLITES AND PLANETARY RINGS

P. J. Message
Department of Applied Mathematics and Theoretical Physics,
Liverpool University, Liverpool, England

Abstract

Some of the new discoveries and theoretical work of the most recent few years are reviewed.   Many new satellites have been discovered in the systems of Jupiter and Saturn, and the planets Jupiter, Saturn and Uranus have been found to possess a number of very narrow rings.

The recent few years have seen a great number of exciting new discoveries in the satellite and ring systems of the outer planets, from Earth-based observations as well as from the observations from the Pioneer and Voyager spacecraft.   The present review will be largely devoted to developments since the author's earlier review of theories of the motion of natural satellites in the Solar System (Message 1979).

The mass of Mars' satellite Phobos has been determined from a Viking flyby to be $(9.9 \pm 1.2) \times 10^{15}$ kg. (Christiansen et al., 1977), Lambeck (1979) has studied possible evolutionary histories of the orbit of Phobos under the influence of tidal friction.   The slow change of the inclination argues against the theory that Phobos was captured by Mars.

Considerable interest has been aroused by observations of the occultation on 1978 June 7 of the star AGK3 + 4° 1839 by the minor planet 532 Herculina, which provided evidence of a satellite of diameter 150 km of this minor planet, whose diameter is 217 ± 3 km, their distance apart being indicated to be 975 ± 1 km.   The stability of such a weakly bound system has been considered by Donnison (1979).

The visit in 1979 of the Voyager spacecraft to Jupiter led to the discovery of a narrow ring at a distance of 1.8 Jupiter radii (128,000 km) from the planet's centre, the width being no more than about 30 km.   (Smith, et al., (1979), and Owen et al., (1979)), and of two small satellites, whose temporary designations, in the system

353

W. Fricke and G. Teleki (eds.), Sun and Planetary System, 353- 359.
Copyright © 1982 by D. Reidel Publishing Company.

proposed by Aksnes and Franklin (1978), are 1979 J1 and 1979 J2, the
former being found by Voyager 2, and having a major semiasis of
1.80 ± 0.01 Jupiter radii, an orbital period of 7 hours 8 minutes,
no detectable eccentricity or inclination (Jewitt et al., 1979), and
the latter having been found by Voyager 1, having a major semi-axis of
3.1054 Jupiter radii, period of 16 hours 11 minutes 21.25 ± 0.50 seconds,
and an inclination of approximately 10.25 (Synnott 1980).   Detection
of the ring of Jupiter in infra-red observations at Mauna Kea, Hawai
has been reported by Becklin and Wynn-Williams (1979).

    Work has continued on the theory of the motion of the great
satellites of Jupiter. Ferraz-Mello (1976) found that consistency could
not be obtained,on the basis of Sampson's theory, between the values of
the masses of the satellites found from Pioneer 10 observations and the
values found by Sampson and de Sitter for the eccentricity of Callisto.
Lieske (1980), continuing to develop the methods employed by Sampson,
and values obtained from Pioneer 10 and 11 for the satellite masses
and parameters of Jupiter's gravitational field, has produced an improved ephemeris of the great satellites, which was used for the Voyager
mission, and found to be in error by less than 200 km.   Brown (1977)
has produced a new theory of the galilean satellites to second order,
and Vu (1981), following the reformulation by Sagnier (1975) of Sampson's
theory, has constructed a theory of the great inequalities and librations
of the satellites Io, Europa, and Ganymede.   An adaptation by Duriez
(1982) of his global theory to incorporate resonant terms has been
applied to the Galilean satellites.

    Tidal friction in Io, enhanced by the forced eccentricity due to
the resonances of orbital period with Europa and Ganymede, has been
shown to be a heat source of great significance for the physical state
of the satellite, as well as important in the long-term evolution of
the orbits of these satellites (Peale, et al. (1979), Yoder (1979),
Greenberg (1981)), though it is not yet clear whether the evolution of
the system is towards, or away from, deeper 2:1 resonances.
Hadjidemetriou and Michalodimitrakis (1981) have found, by numerical
integration, periodic solutions of the general four body problem
related to the motion of these satellites.

    There has been an arresting wealth of new discovery in Saturn's
system in the past few years, both from spacecraft visits and from
Earth-based observations.   The coincidence of the Earth with the plane
of the rings during the 1966 and 1979/80 oppositions of Saturn gave a
brief period of visibility from the Earth of faint inner satellites
usually lost in the glare of light from the rings, and although
observations made in 1966 by Dollfus et al. at the Pic-du-Midi and other
observatories had not provided unambiguous orbit determinations and
hence confirmation of the discovery of any satellite (see Bobrov (1972)
for a review, and Fountain and Larson (1978), and Aksnes and Franklin
(1978), for discussions of the observations), the situation has become
clearer with the analysis of observations from Pioneer 11 in November
1979 (Gehrels et al., (1980), Van Allen et al. (1980)), and from

observations from the Earth in early 1980 (see Marsden (1980), Larsen
and Fountain (1980), and Larsen et al. (1981), for more detailed
reviews), and observations from Voyager 1 in November 1980.   It has
become apparent that there are two satellites which featured in the
1966 observations, which were given the temporary designations 1980 S1
and 1980 S3, and now, after confirmation of their identities, have the
designations SX and SXI respectively, moving in very nearly the same
orbit about Saturn, with major semi-axes of 151,472 km. and 151,422 km.,
respectively, and eccentricities of 0.007 and 0.009 respectively.
The former has largest dimension 70 km., and shortest 35 km., and was
imaged well during a close approach by Voyager 1, and the latter is of
size about 100 km.   The inner one will catch up with the outer, when
their mutual attractions will cause them to exchange their order of
distance from Saturn, until the new inner one catches up the new outer
one next, this process being repeated indefinitely.   In a frame of
reference rotating at an appropriate uniform angular speed, one satellite
moves in a horseshoe-type orbit, and the other in an elongated oval
situated within the open end of the horseshoe.

Another satellite newly discovered through Earth-based observations
in 1980 is "Dione B", temporarily designated 1980 S6, and, after
confirmation, SXII, which has radius 40 ± 20 km., and an orbit of major
semi-axis 378,060 km., of very small eccentricity, and probably
librates about the Lagrangian position (L4) making an equilateral
triangle with Saturn and Dione, preceding Dione (Lecacheux et al.,(1980),
Reitsema et al., (1980), Lamy and Mauron (1981), and see Marsden (1980)).
Suggestions of a satellite (1980 S13)[1] librating about the Lagrangian
point (L5) following Dione have not been confirmed.   Further satellites
were found through the visit of Voyager 1 to Saturn (Beatty 1981),
designated SXIII (temporarily 1980 S26), of radius 125 ± 50 km., with
major semi-axis 141,700 km., SXIV   (1980 S27), of radius 100 ± 50 km.,
and major semi-axis 139,400 km, and SXV (1980 S28) of radius about
50 km., and major semi-axis 138,200 km.   The orbits of the first two
are on either side of the newly discovered F ring, and SXV moves
closely outside the A ring (whose outer edge is at a distance 137,400 km.
from the centre of Saturn).   These three satellites are believed to
play an important part in the confining of the rings, as discussed
later.   There are observations suggesting that there may be satellites
in the Lagrangian configurations with Tethys, and, possibly, one
which is sharing an orbit with Enceladus.

Developments in the theory of the motions of the previously-known
satellites of Saturn include Rapaport's (1978) improvement of the
agreement between theory and observations of Iapetus, by inclusion of
further terms of the disturbing function than in previous theories,
and the treatment by Rose (1979) of the motion of Phoebe by numerical
integration, and a new set of orbital elements for Hyperion by Hatanaka
(1979) from recent observations made at the Tokyo Observatory.   Provisional determinations of the mass of Titan, from a second-order theory
of the motion of Hyperion by the author, using observations between
1887 and 1922 are (see Seidelmann 1978), from the free libration in

longitude of period 21 months, 1: (4208 ± 40) times that of Saturn, from
the secular motion of the apse, 1: (4256 ± 20), from the 18¾-year
term in the eccentricity, 1: (4245 ± 89), from the term of the same
period in the apse longitude, 1: (4098 ± 148), and, from the 21-monthly
term in the eccentricity, 1: (3966 ± 353), with an overall mean of
1: (4239 ± 16).

Ward (1981), studying possible evolutionary histories of the orbit
plane of Iapetus, and of its Laplacian plane, concludes that it could
well have originated, with the other satellites, in one accreting disc.

The rings of Saturn have provided one of the most spectacular
sights of the heavens since Huygens in 1657 first correctly interpreted
the appendages on either side of the planet, and G.D. Cassini in 1675
found the gap between the A and B rings which bears his name.
Maxwell (1859) concluded that the stresses within a solid ring of the
dimensions of Saturn's would inevitably cause it to break up, and
studied the conditions for stability of a ring composed of small
particles. (See the rediscussion of this study by Cook and Franklin
(1964).) In 1895 Keeler's early use of the spectroscope to detect
relative velocity by use of the Doppler effect showed that the rotational speed of the rings decreases with distance from Saturn, as it must
if the rings are made of a large number of individual particles moving
under Saturn's gravitational attraction. Jeffreys (1947) concluded
that, as a result of collisions between the particles, the rings would
very quickly thin out to single particle thickness. Indications that
the ring thickness was more than 1 km seemed difficult to reconcile
with this, but Voyager 2 observations suggest a maximum thickness of no
more than 150 metres. Attempts to understand the Cassini division
have generally involved study of the disturbing effect of Mimas, since
the distance from Saturn, at which the orbital period of a ring particle
would be exactly half that of Mimas, lies within the gap. (Goldsbrough
1921 & 1941). However the orbit of exact commensurability does not
fall symmetrically within the gap, but rather just close to its inner
boundary. Franklin, Colombo and Cook (1971) offer the explanation
that the mass of the B ring itself displaces the position of exact
resonance, while Goldreich and Tremaine (1978), also noting the unexpect-
edly large width of the ring, suggest that spiral density waves
(similar to those proposed in some theories of galactic structure)
would be generated at the distance of exact resonance, and would clear
particles out on the outer side of the resonance only. Observations
from the Voyager spacecraft show the rings to be composed of a large
number of separated individual narrow rings, and that the Cassini
division also contains such rings, though less densely populated with
particles. Also present in the B ring are darker "spoke-like"
features, which rotate as would solid spokes of a wheel. Reaching
theoretical understanding of all of these newly-discovered structures
poses a more difficult task. Goldreich and Tremaine (1979) proposed
that particles could be confined within a narrow ring by the "shepherding" or "herding" action of two satellites, in orbits one on either
side of the ring, since the effect of a satellite on a smaller ring

particle which came into close configuration with it would be to change
its angular momentum in such a way as to move it into an orbit of major
axis slightly more different from that of the satellite.   This theory
has received support from the discovery, from Pioneer 11 observations,
of the narrow F ring, just outside the A ring, and, from Voyager 1
observations, of the satellites SXIII and SXIV, in orbits on either
side of the F ring, and also of the satellite SXV, just outside the A
ring, and believed to confine ring particles within the edge of that
ring.   The "braided" appearance of the F ring, seen in Voyager 1
pictures, presented a considerable challenge to theoretical understanding.
Dermott (1981) shows how such an appearance could arise from the effect
of a small-integer commensurability with the shepherding satellite.
Voyager 2 pictures show that the spoke-like features in the B ring lie
slightly off the plane of the ring, perhaps offering some support to
the idea that they result from the interaction of changed particles
with the magnetic field of Saturn, and also that the F ring is itself
composed of a number of even finer narrow rings.

Turning now to work on the system of Uranus, Greenberg (1976)
found an upper bound can be put on the product of the masses of Ariel
and Umbriel, from a study of the three-satellite commensurability
between the mean motions of these satellites and Miranda.   Uranus
became the second planet to be known to be encircled by rings, with the
observation by Elliott et al. (1977), Millis et al (1977) and Hubbard
and Zeller (1980) of the occultation on 10th March 1977 of the star
SAO 158687.   Nine narrow rings (designated α, β, γ, δ, ε, η, 4 (or θ),
5 (or ι), and 6 (or κ) were indicated, and confirmed at a number of
later occultations of other stars (see, e.g., Nicholson et al. (1978),
Elliott et al. (1981),).   Some of these rings have ellipticities
large enough to be determined from the data, especially ε, which pre-
cesses as if rigid.   Goldreich and Tremaine (1979) suggest this to
be due to the mutual gravitational attraction of the particles of the
ring, and Dermott, Gold, and Sinclair (1979) that it is due to the
presence within the ring of an unseen satellite, keeping the ring
particles locked into rotating horseshoe type orbits, precessing with
the satellite.   Studies have been made to relate the positions of the
ring to resonances each involving two of the known satellites of Uranus
(Dermott and Gold, 1977, Aksnes, 1977, Goldreich and Nicholson, 1977,
Steigmann, 1978).

If Neptune also possesses rings, Dobrovobskis (1980) points out
that they may not lie in the plane of the planet's equator due to the
variation of the orientation of the Laplacian plane with distance from
Neptune which results from the considerable inclination of Triton's
orbit to Neptune's equator.

Pluto has been added to the roll of planets suspected of possessing
satellites following the noticed elongation of photographic images of
the planet and the resolving of images by speckle interferometry
(Christy and Harrington, 1978, Harrington and Christy, 1980 & 1981,
Bonneau and Foy, 1980).

Notes

[1] More recent observations suggest that this was a sighting of a
satellite near the Lagrangian configuration L_4 with Tethys.

REFERENCES

Aksnes, K.: 1977, Nature 269, 783.
Aksnes, K. and Franklin, F. A.: 1978, Icarus 36, 107.
Beatty, J. K.: 1981, Sky and Telescope 61, 7.
Becklin E. E. and Wynn-Williams C. G.: 1979, Nature 279, 400.
Bobrov, M. S.: 1972, Sov. Astron. J. 16, 348.
Bonneau, D. and Foy, R.: 1980, Astronomy and Astrophysics, 92, L1.
Brown, B.C.: 1977, Celestial Mechanics, 16, 229.
Christy, J.W. and Harrington, R. S.: 1978, Astron. J. 83, 1005.
Christiansen, E. J. et al.: 1977, Geophysics Research Letters, 4, 555.
Cook, A. F. and Franklin, F. A.: 1964, Astron. J. 69, 173.
Dermott, S. F.: 1981, Nature 290, 454.
Dermott, S. F. and Gold, T.: 1977, Nature 267, 590.
Dermott, S. F., Gold, T. and Sinclair, A. T.: 1979, Astron. J., 84, 1225.
Dermott, S. F., Murray, C. D.: 1979, Icarus, 43, 338.
Dermott, S. F., Murray, C. D. and Sinclair, A. T.: 1980, Nature 284, 309.
Dobrovolskis, A. R.: 1980, Icarus, 43, 222.
Donnison, J. R.: 1979, Monthly Notices of the Royal Astron. Soc., 186,35P.
Duriez, L.: 1982, Celestial Mechanics 26, 231.
Elliott, J. L. et al.: 1977, Nature 267, 328.
Elliott, J. L. et al.: 1981, Astron. J. 86, 444.
Ferraz-Mello, S.: 1976, Science, 192, 1127.
Fountain, J. W. and Larson, S. M.: 1978, Icarus, 36, 92.
Franklin, F. A., Colombo, G. and Cook, A. F.: 1971, Icarus, 15, 80
Gehrels, T. et al.: 1980, Science, 207, 434.
Goldreich, P. and Nicholson, P.: 1977, Nature, 269, 783.
Goldreich, P. and Tremaine, S.: 1978, Icarus, 34, 240.
Goldreich, P. and Tremaine, S.: 1979, Astron. J., 84, 1638.
Goldreich, P. and Tremaine, S.: 1979, Nature, 277, 97.
Goldsbrough, G. R.: 1921, Phil. Trans., A 222, 101
Goldsbrough, G. R.: 1941, Phil. Trans., A 239, 183.
Greenberg, R.: 1976, Icarus, 29, 427.
Greenberg, R.: 1981, Icarus, 46, 415.
Hadjidemetriou, J. D. and Michalodimitrakis, M.: 1981, Astronomy and
       Astrophysics, 93, 204.
Harrington, R. S. and Christy, J. W.: 1980, Astron. J., 85, 168.
Harrington, R. S. and Christy, J. W.: 1981, Astron. J., 86, 442.
Hatanaka, Y.: 1979, in "Dynamics of the Solar System" (ed. R. L.
       Duncombe: Proceedings of I.A.U. Symposium No.81).
Hubbard, W. B. and Zellner, B. M.: 1980, Astron. J., 85, 1663.
Jeffreys, H.: 1947, Monthly Notices of the Royal Astron. Soc., 107, 263.
Jewitt, D. C., Danielson, G. E. and Synott, S. P.: 1979, Science, 206,
       951.
Lambeck, K.: 1979, Journal of Geophysics Res., 84, 5651.
Larson, S. M. et al.: 1981, Icarus 46, 175.

Lassell, W.: 1847, Monthly Notices of the Royal Astron. Soc., 7, 167.
Marsden, B. G.: 1980, Icarus, 44, 29.
Marsden, B. G.: 1980, Journal of Geophys. Res., 85, 5957.
Maxwell, J. C.: 1859, "On the stability of the motion of Saturn's
    rings" (London).
Message, P. J.: 1979, in "Dynamics of the Solar System" (ed R. L.
    Duncombe: Proceedings of I.A.U. Symposium No.81).
Millis, R. L. et al.: 1977, Nature, 267, 330
Nicholson, P. D. et al.: 1978, Astron. J., 83, 1240.
Owen, T. et al.: 1979, Nature 281, 442.
Peale, S., Cassen, P. and Reynolds, R.: 1979, Science, 203, 892.
Rapaport, M.: 1978, Astronomy and Astrophys., 62, 235.
Rose, L. E.: 1979, Astron. J., 84, 1067.
Sagnier, J. L.: 1975, Celestial Mechanics, 12, 19.
Seidelmann, K.: 1978, Report on a meeting on 21 September 1978 at U.S.
    Naval Observatory, on "Observations and Ephemerides of Planetary
    Satellites", and related correspondence.
Smith, B. A. et al.: 1979, Science 204, 951.
Steigmann, G. A.: 1978, Nature 274, 454.
Synnott, S. P.: 1980, Science, 210, 786.
Van Allen, J. M. F. et al.: 1980, Science, 207, 415.
Vu, D. T.: 1981, Astron. and Astrophys., 94, 140.
Ward, W. R.: 1981, Icarus, 46, 97.
Yoder, C. F.: 1979, Celestial Mechanics, 193 and Nature, 279, 769.

DYNAMICAL EVOLUTION AND DISINTEGRATION OF COMETS

Ľubor KRESÁK
Astronomical Institute of the Slovak Academy of Sciences
84228 Bratislava, Czechoslovakia

ABSTRACT. Current concepts of the origin and evolution of comets are reviewed. The place of their formation from which they have been delivered into the Oort reservoir is still an open problem, but the region of the outermost planets appears most probable. The interplay of stellar and planetary perturbations can be traced by model computations which reveal both the general trends and the variety of individual evolutionary paths. The present structure of the system of comets is controlled by the dynamical evolution of its individual members limited by their physical aging by disintegration. Where the lifetimes are short, as in the Jupiter family of short-period comets, an equilibrium between elimination and replenishment is established. The role of different destructive processes and the resulting survival times are discussed.

Our planetary system is imbedded in a huge, nearly spherical cloud of comets, extending to solar distances of more than 1000 times that of the outermost planet, and occupied by a number of individual objects which possibly exceeds the number of stars in the Galaxy. From this point of view it may appear anachronous to call the comets "interplanetary objects". Yet it is only in interplanetary space, mostly within the orbit of Jupiter, where the comets are observable. Since it takes some comets millions of years to enter this region once, and since many of them do not come there at all, we cannot pretend to know more than a minute fraction of their whole population.

Even more conspicuous than the abundance of comets and the vast space occupied by them, is their role in the evolutionary history of the Solar System. The "new" comets are samples of the most primitive state of matter left essentially intact by the processes which governed the evolution of the planets, satellites, and asteroids. In spite of this antiquity and a slow general evolution, individual comets are subject to a variety of high-speed processes going on before our eyes : drastic transformations of orbits at encounters with the planets, formation and disappearance of large comae and tails, splitting of their nuclei, sudden brightness bursts and sudden extinction.

W. Fricke and G. Teleki (eds.), Sun and Planetary System, 361–370.

The evolutionary history of comets can be divided into two main phases : formation of the comet cloud and its subsequent evolution. In his classical paper published thirty years ago, Oort (1950) has initiated a major progress in this research area by making a clear distinction between the two questions, where did the comets originate and where are they coming from today. Thanks to Oort's concept of interaction of stellar and planetary perturbations and to the application of modern computing techniques, we are witnessing a rapid progress in understanding the second phase of evolution. The first phase still remains a matter of controversy.

Different authors put the origin of comets practically anywhere between the asteroid belt and the interstellar medium interacting with the outskirts of the Solar System. A significant continuing replenishment from the outside is ruled out by the evidence that not a single comet observed so far was moving in an hyperbolic orbit compatible with the relative motions in the environment of the Solar System. The main objection against effective accretion processes taking place far beyond the orbits of the outer planets is an excessive density of matter required for the production of kilometer-sized cometary nuclei. As we go closer to the Sun and the density becomes plausible, serious difficulties arise with the overshooting effect. An excessive original population and total mass would be necessary to leave a sufficient number of objects in nearly parabolic orbits. It appears, however, that this problem is easier to overcome than that of the low density (Safronov, 1972, Öpik, 1973, Cameron, 1973, Dermott and Gold, 1978, Fernández, 1980). It has also been suggested that the comets are remnants of the disruption of a former major planet 5 million years ago (Van Flandern, 1979), or that they are being produced even nowadays by volcanic eruptions on planets and satellites (Vsekhsvyatskij, 1972), or by agglomeration of meteor particles within meteor streams (Alfvén and Arrhenius, 1976); but neither of these hypotheses is compatible with observational evidence. A tabular summary and discussion of the various theories of comet origin can be found in Delsemme (1977).

However strange it may appear at first glance, the observed lack of flattening of the comet cloud to some particular plane (e.g., to the ecliptic) lends support to their origin inside, and with, the Solar System. Model computations show that stellar perturbations must have randomized the orientations of orbital planes of comets within the age of the Solar System. On the other hand, recent capture of an interstellar cloud without preserving a preferred plane of motion would require an unrealistic encounter velocity of about 1/100 of the velocity of neighbouring stars (Fernández, 1981). It appears most likely that comets have accreted in the Uranus-Neptune region, with Jupiter and Saturn participating strongly in their transport into the Oort cloud. The boundaries of the cloud were set by the orbital stability with respect to the galactic perturbations (Chebotarev, 1966, Antonov and Latyshev, 1972), and the cloud was gradually shrinking with subsequent star passages through it, stripping away its outer shells (Weissman, 1980a).

Turning to the ensuing dynamical evolution of individual comets it must be emphasized that their arrival from distances of 30,000 to 100,000 AU is unambiguously borne out by their observed motion near the Sun. This is illustrated by Figure 1 showing the distribution of binding energies E (= reciprocal semimajor axes 1/a in AU^{-1}) of 111 best determined long-period cometary orbits (Marsden et al., 1978). The upper histogram refers to the original orbits before entering the planetary zone, the lower one to those of the same comets after leaving it. The sharp peak at $0 < E < 10^{-4}$, formed by the new comets, is smeared out completely just by the first passage between the planets. The dotted curve indicates the expected distribution of energies immediately after the passage (Everhart, 1969), and the full curve the same values corrected for the probability of the next appearance falling on a fixed time span. The latter curve would tend to level off with the increasing number of revolutions. For a better discrimination, the curves are scaled by a factor of 10 with respect to the histograms.

While the lower histogram corresponds fairly well (within the limits of random fluctuations in a small sample) with the dotted curve, there is a striking discrepancy between the upper histogram and the full curve. This discrepancy demonstrates that almost all of the new comets must have been observed at their first passage near the Sun, and that they will not return as observable objects anymore. Since they do not display destructive changes on the observable arc of the orbit (often exceeding 180° in true anomaly, where one half of the total insolation per revolution is received) it seems that their

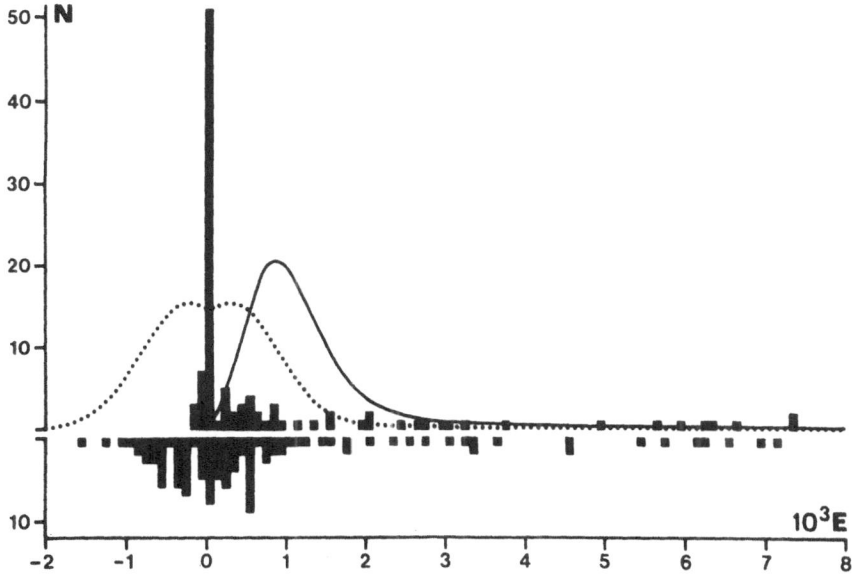

Figure 1. Distribution of the binding energies E = 1/a of long-period comets before entering the planetary region (up) and after leaving it (down). For explanation of the curves see text.

fading away is due to the removal of a thin primordial surface layer (Whipple, 1977a) rather than to a total destruction. A puzzling point is that a simultaneous loss of the physical and dynamical signatures of new comets is required. In fact, changes of $\pm 2 \times 10^{-5}$ in E, which would be detectable in the statistics of comet orbits, are typical for perihelion passages at $q \backsim 15$ AU (Fernández, 1981), while strong outgassing only appears at $q < 3$ AU, with the vaporization of water ice. The lack of comets dynamically old but physically new would require substantial physical aging at solar distances greater than 10 AU, which appears improbable, or some rejuvenation process going on at the outskirts of the Oort cloud (Kresák, 1977a).

After the first passage through the planetary zone, nearly one half of the new comets are thrown out on hyperbolic orbits. These are no longer gravitationally bound to the Solar System, but continue to follow its motion through the Galaxy. Their outward drift is very slow because their residual velocities with respect to the Sun are low : typically 0.3 to 1 km/s (or pc/Myr) for the comets which passed inside the orbit of Jupiter. The probability of escape drops to less than 1/4 after the second passage and continues to decrease further. At the same time the comets become essentially immune to changes of the perihelion distance by stellar perturbations, unless planetary perturbations happen to return their aphelia to the outskirts of the cloud. This is because in the original orbit of $E \backsim 5 \times 10^{-5}$ there was, on the average, one star passing the Sun within the distance of the comet's aphelion per one revolution of the comet (Weissman, 1981), while after an average increase to $E = 7 \times 10^{-4}$ there is but one such case per 10,000 revolutions. The statistical effects of stellar and planetary perturbations have been thoroughly investigated by Everhart (1973), Rickman (1976), Fernández (1980), Weissman (1981) and others, and consistent results have been obtained by using widely different approaches and techniques. As shown by Everhart (1977), the average revolution period of the remaining comets tends to decrease with the number of returns N in such a way that E (=1/a) is proportional to $N^{1/2}$ and the period P to $N^{-3/4}$. For original perihelion distances q < 4 AU it takes $\backsim 1000$ revolutions to reduce the period to 200 years.

At this stage, however, a definite selection of objects for an accelerated evolution takes place. These are the low-inclination comets ($i < 10^\circ$) with perihelia near the orbit of Jupiter (4 AU < q < 6 AU), for which the average number of returns necessary for reaching the same effect is reduced by a factor of $\backsim 20$. Everhart calls this the capture zone; similar zones pertain to the other outer planets but their capturing efficiency is much smaller (Everhart, 1977). Once a short period, comparable with that of Jupiter, is attained, the applicability of the conventional random walk or Monte Carlo procedures fails, as period-to-period resonances can store the comets for long time spans in librating orbits.

Figure 2 shows a few interesting examples of orbital patterns described by short-period comets in the reference frame rotating with the heliocentric motion of Jupiter. The two comets on the left are

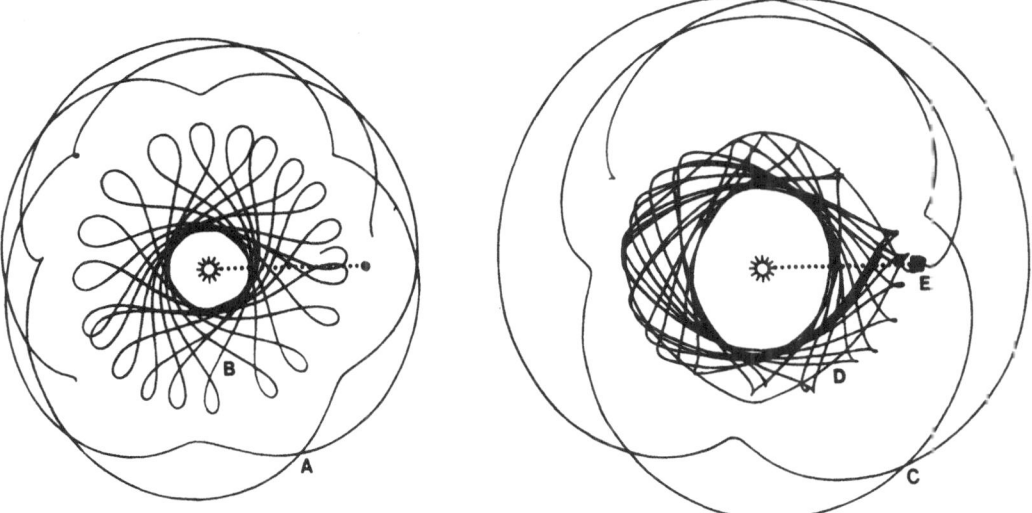

Figure 2. Examples of orbital patterns of short-period comets in the reference frame rotating with the heliocentric motion of Jupiter : P/Schwassmann-Wachmann 1, 1858 - 1974 (A); P/Tempel 2, 1858 - 1967 (B); P/Oterma, 1858 - 1937 (C); model objects preceding P/Oterma by 1.0°, 1937 - 2192 (D) and by 1.5°, 1937 - 2037 (E).

moving in fairly stable orbits, one outside and the other inside the orbit of Jupiter (Carusi and Valsecchi, 1981). A period closer to the resonance of 2:1 would even make the loops on the latter trajectory avoid the vicinity of Jupiter for centuries, arranging them into a horseshoe-shaped pattern. As the sample of about 100 known short-period comets is too small to display all interesting possibilities, the other part of the figure is composed of a real object and two related model objects (Carusi et al., 1981). The outer trajectory shows the approach of P/Oterma to Jupiter. After the encounter, this comet made three revolutions in the inner orbit; due to a 3:2 resonance it passed again near Jupiter, and was thrown out into an orbit similar to that pursued before the capture. The two model objects were supposed to approach the planet in exactly the same orbit as P/Oterma, but preceding it by 1.0° and 1.5° in mean anomaly. One of them stays in the inner orbit for more than 250 years, while the other becomes a satellite of Jupiter for at least a century. Under exceptional circumstances (nongravitational deceleration, assistance of the Galilean satellites) their motions might get stabilized completely.

The main problem in explaining the origin of short-period comets has been for long the theoretical expectation that 1/4 of the captures by Jupiter from long-period orbits should result in retrograde motion, which is clearly at variance with observation. This problem has been solved, in excellent agreement, by backward integrations of the orbits of real comets (Kazimirchak-Polonskaya, 1972) and by forward model integrations (Everhart, 1973, 1977). It turned out that multi-stage captures at repeated close planetary encounters, reducing

the eccentricities and inclinations, were involved. A clear insight
can be obtained by plotting the Tisserand invariant T with respect to
Jupiter (i.e., the energy integral in the reference frame rotating
with the planet's revolution) against the binding energy E. In this
plot the regions in which encounters with different planets are possible and the resulting directions of displacement can be specified,
as shown in Figure 3. The longest, most efficient interactions take
place along the boundaries of these regions and at low inclinations.
Curvature of these boundaries prevents single-stage captures or ejections in the strips between them and the dotted chords. Therefore, an
object of high Tisserand invariant (T > 0.544) and low unperturbed
jovicentric velocity (< 5.4 km/s) cannot be ejected by Jupiter on an
hyperbolic orbit, and cannot have been captured by Jupiter alone from
the Oort cloud. A moderate overlapping is possible due to the simplifications inherent in the definition of the invariant and due to the
size of the sphere of action of Jupiter. Nevertheless, it is worth
noting that one half of the known short-period comets occupy the region where assistance of the other planets was necessary - in spite
of their lower detection probability due to larger perihelion distances.

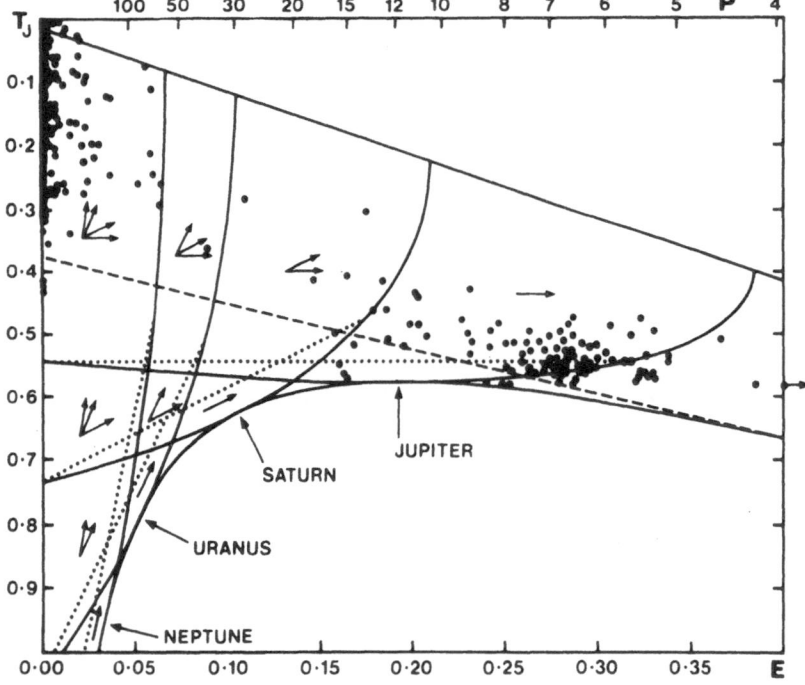

Figure 3. Distribution of known comets (dots) in binding energy E and
Tisserand invariant with respect to Jupiter T. Curves, boundaries of
the regions where encounters with different planets are possible; arrows, displacement directions for decelerating encounters (for acceleration reverse); dashed line, q = 2.5 AU for i = 0°.

Figure 3 also explains why only few comets are ultimately captured into short-period orbits. Only 20% of the known long-period comets fall within the margin of the diagram (in its upper left), and not a single one has the Tisserand invariant high enough to represent a would-be member of the Jupiter family - even if subject to close planetary encounters. It is also explained why observational evidence cannot confirm the presence of another reservoir of comets in the region of the outer planets (Whipple, 1964, 1972). Whether a comet arrives from this region (lower left corner, $E > 0.02$) or from the Oort cloud (left edge, $E < 10^{-4}$), the evolutionary paths converge before the perihelion distance is reduced enough to render the comet detectable. This ambiguity, combined with insufficient information about the size distribution of the largest cometary nuclei (Öpik, 1973), makes it very difficult to estimate the present and original mass of the whole comet system.

With the knowledge of the original state it would be possible, in principle, to recount the long-term statistical evolution of the comet system between the open outer boundary set by unbinding from the Sun (escape on hyperbolic orbits), and the reflecting inner boundary set by temporary binding of the aphelia to Jupiter. However, the dynamical evolution is accompanied with the decay of comets limiting their survival times which, for some types of orbits, are very much shorter than the age of the system. Hence, a purely dynamical solution cannot reproduce the real situation. While the continuing loss of comets rules out a steady state of the system as a whole, there must exist some quasi-steady state in its innermost part where short evolutionary time scales require an equilibrium between the depletion by ejection and destruction, and the delivery of new objects from other regions.

The first signature of aging is the change of a new comet into and old one, probably due to the removal of a primordial, very active surface layer (Whipple, 1977a) after the very first passage near the Sun. As already mentioned, this incident eliminates a great majority of new objects, at least by making them much fainter for the subsequent returns (Delsemme, 1979). During each passage, the comet loses some of its volatile constituents by vaporization, and the outgoing gases also take away some solids. The gases and finer dust escape from the Solar System on hyperbolic orbits (with the size limit of these dust grains depending strongly on the binding energy E and on their optical properties), while larger particles remain in elliptic orbits, forming meteor streams. Thermal inertia of the surface displaces the area of most violent mass loss from the subsolar point in the direction of the rotation of the nucleus. The resulting momentum is responsible for nongravitational effects in the comet's motion. These are easily detectable at the reappearances of short-period comets, resulting in semi-regular changes of the revolution period, typically by 0.001% (at $q = 2$ AU) to 0.005% (at $q = 1$ AU) per revolution (Marsden, 1972). Together with the limited accuracy of positional measurements, they prevent a reliable integration of the motion of short-period comets over more than a few centuries (strictly speaking: over more than two close approaches to the planets), both backward and forward in time. The progressive ablation of the outer shells tends to reduce the size of the nucleus and, at the same time, to cover it by a spotty solid crust shielding the deeper supplies of volatiles against direct insolation. Irrespectively of the relative contribution of these two processes, which may be different for different objects and evolutionary phases (Shulman, 1972) the inherent aging must result in a progressive brightness decrease. Such a decrease has in fact been detected by the comparison of photometric observations of short-period comets at different returns (Vsekhsvyatskij, 1958), but quantitative extrapolations were at variance with continuing observations of comets for which early death dates had been predicted. This failure was apparently due to a number of biases (Kresák, 1974), in particular to an increasing proportion of observations with large telescopes which only record the innermost condensation in the coma. With a few exceptions, irregular brightness variations of comets seem to mask the general trends completely.

In addition to the general aging, there are discrete events which can affect the rate of evolution quite appreciably. Collisions with the planets are possible but too rare to have statistical significance. No such case has been observed so far, the probability of collision being about 1.3×10^{-7} per passage through the whole planetary region in a long-period orbit (Weissman, 1980b). This figure already includes passages through the Roche limits of the planets which result in tidal splitting of the nucleus, as observed on P/Brooks 2 in 1889 after its close approach to Jupiter in 1886. Tidal splitting by the Sun is common to the Kreutz group of sungrazing comets which in itself is a product of branching disruptions of a single, exceptionally large comet (Marsden, 1967). There were about 18 observed cases of splitting without any evident reason. This occurs, on the average, once per 12 revolutions of long-period comets and once per 90 revolutions of short-period comets (Kresák, 1981). It was argued that splitting of new comets is significantly more frequent than splitting of old comets (Weissman, 1980b), but this seems to be an effect of observational selection (Kresák, 1981). Some model computations identify splitting with a complete elimination of the comet, which is incorrect; observations imply rather quiet separation of minor fragments, which would not reduce appreciably the lifetime of the main component. On the other hand, splitting of large parent objects could produce a temporary increase of the number of active objects, as in the case of the Kreutz group. It was even suggested by Öpik (1971) that the distribution of comet orbits reflects a common origin of tens of comet groups and pairs, and that such families may constitute a majority of the total comet population. Later analyses have shown, however, that these orbital similarities can be explained by observational selection and chance coincidence (Whipple, 1977b, Kresák, 1982).

Combination of all the aging processes controls the lifetimes of individual comets as active objects, depending on their original size, composition and strength, and changing perihelion distance and revolution period. Observational evidence suggests average lifetimes of about 400 revolutions for the Jupiter family of comets and 200 revolutions for periodic comets of Halley type, at q = 1 (Kresák, 1981). These figures are in agreement both with the estimate of 200 to 500 revolutions by Fernández (1981), based on fitting the observed energy distribution to his dynamical model, and with the estimate of 300 revolutions by Dobrovolskij (1972), based on the theoretical sublimation rates. Weissman (1980b) finds considerably longer lifetimes for sublimation and suggests that the decay is controlled by random disruption. Possibly the composition of individual comets is not uniform and only some of them, 15% according to Weissman, are resistive to rapid destruction. Surprisingly enough, there is no observational evidence of significant physical differences between new and old, or long-period and short-period comets (Donn, 1977, Kresák, 1977b). It must also be remembered that there may be more than one active period of physical aging for each comet. Transition into an orbit of large perihelion distance can stop the decay process until perturbations bring the comet again closer to the Sun. It can also happen that volatiles hidden under a surface crust restore the activity of a dormant object, being exposed by a collisional or rotational breakup.

A puzzling feature of the distribution of long-period comets in absolute magnitude is a cutoff near H = 12, indicating that fainter objects do not exist at all. This is borne out by the independence of the absolute magnitude distribution on the minimum geocentric distance, and by the fact that almost all long-period comets observed close to the Earth were bright naked-eye objects (Kresák, 1978). What happens at the end of the comet's active lifetime is still obscure. A larger inactive object moving in a comet-like orbit has never been observed. Among the nearly 2,500 numbered asteroids there is only 944 Hidalgo, 2060 Chiron and four or five Apollo objects for which cometary origin appears dynamically plausible (Kresák, 1979), and identification of the Apollo objects as parent bodies of meteorites appears incompatible with their cometary origin (Levin and Simonenko, 1981). According to Froeschlé and Rickman (1980), less than 1% of the Apollo and Amor objects may be derived from dead comets. Linking up of the size distribution of comets with that of low-density fireball-producing meteoroids suggests that most cometary nuclei break apart into sizeable boulders when their supply of volatiles becomes exhausted (Kresák, 1978).

REFERENCES

Alfvén, H., Arrhenius, G.: 1976, "Evolution of the Solar System", NASA SP-345, Washington, pp. 231-238.
Antonov, V.A., Latyshev, I.N.: 1972, IAU Symp. 45, pp. 341-345.
Cameron, A.G.W.: 1973, Icarus 18, pp. 407-450.
Carusi, A., Valsecchi, G.B.: 1981, private communication.
Carusi, A., Kresák, L., Valsecchi, G.B.: 1981, Astron. Astrophys. 99, pp. 262-269.
Chebotarev, G.A.: 1966, Astron. Zh. 43, pp. 435-440.
Delsemme, A.H.: 1977, in A.H. Delsemme (ed.), "Comets, Asteroids, Meteorites", Univ. Toledo, Ohio, pp. 453-457.

Delsemme, A.H.: 1979, IAU Symp. 81, pp. 265-271.
Dermott, S.F., Gold, T.: 1978, Astron. J. 83, pp. 449-450.
Dobrovolskij, O.V.: 1972, IAU Symp. 45, pp. 352-355.
Donn, B.: 1977, in A.H. Delsemme (ed.), "Comets, Asteroids, Meteo-
    rites", Univ. Toledo, Ohio, pp. 15-23.
Everhart, E.: 1969, Astron. J. 74, pp. 735-750.
Everhart, E.: 1973, Astron. J. 78, pp. 329-337.
Everhart, E.: 1977, in A.H. Delsemme (ed.), "Comets, Asteroids, Me-
    teorites", Univ. Toledo, Ohio, pp. 99-104.
Fernández, J.A.: 1980, Icarus 42, pp. 406-421.
Fernández, J.A.: 1981, Astron. Astrophys. 96, pp. 26-35.
Froeschlé, C., Rickman, H.: 1980, Astron. Astrophys. 82, pp. 183-194.
Kazimirchak-Polonskaya, E.I.: 1972, IAU Symp. 45, pp. 373-397.
Kresák, L.: 1974, Bull. Astron. Inst. Czechosl. 25, pp. 87-112.
Kresák, L.: 1977a, in A.H. Delsemme (ed.), "Comets, Asteroids, Meteo-
    rites", Univ. Toledo, Ohio, pp. 93-97.
Kresák, L.: 1977b, Bull. Astron. Inst. Czechosl. 28, pp. 346-355.
Kresák, L.: 1978, Bull. Astron. Inst. Czechosl. 29, pp. 103-125.
Kresák, L.: 1979, in T. Gehrels (ed.), "Asteroids", Univ. Arizona,
    Tucson, pp. 289-309.
Kresák, L.: 1981, Bull. Astron. Inst. Czechosl. 32, pp. 19-40 and
    321-339.
Kresák, L.: 1982, Bull. Astron. Inst. Czechosl. 33, in press.
Levin, B.J., Simonenko, A.N.: 1981, Icarus, in press.
Marsden, B.G.: 1967, Astron. J. 72, pp. 1170-1183.
Marsden, B.G.: 1972, IAU Symp. 45, pp. 135-143.
Marsden, B.G., Sekanina, Z., Everhart, E.: 1978, Astron. J. 83, pp.
    64-71.
Oort, J.H.: 1950, Bull. Astron. Inst. Netherl. 11, pp. 91-110.
Öpik, E.J.: 1971, Irish Astron. J. 10, pp. 35-91.
Öpik, E.J.: 1973, Astrophys. Space Sci. 21, pp. 307-398.
Rickman, H.: 1976, Bull. Astron. Inst. Czechosl. 27, pp. 92-105.
Safronov, V.S.: 1972, IAU Symp. 45, pp. 329-334.
Shulman, L.M.: 1972, IAU Symp. 45, pp. 271-282.
Van Flandern, T.C.: 1978, Icarus 36, pp. 51-74.
Vsekhsvyatskij, S.K.: 1958, "Fizicheskie Kharakteristiki Komet", Gos.
    Izdatelstvo Fiz. Mat. Literatury, Moskva, pp. 17-23.
Vsekhsvyatskij, S.K.: 1972, IAU Symp. 45, pp. 413-418.
Weissman, P.R.: 1980a, Nature 288, pp. 242-243.
Weissman, P.R.: 1980b, Astron. Astrophys. 85, pp. 191-196.
Weissman, P.R.: 1981, in L.L. Wilkening (ed.), "Comets", Univ. Ari-
    zona, Tucson, in press.
Whipple, F.L.: 1964, Proc. Nat. Acad. Sci. 51, pp. 711-718.
Whipple, F.L.: 1972, IAU Symp. 45, pp. 401-408.
Whipple, F.L.: 1977a, in A.H. Delsemme (ed.), "Comets, Asteroids, Me-
    teorites", Univ. Toledo, Ohio, pp. 401-408.
Whipple, F.L.: 1977b, Icarus 30, pp. 736-746.

A DYNAMICAL STUDY OF POSSIBLE BIRTHPLACES OF COMETS

Julio A. Fernández
Max-Planck-Institut für Aeronomie,
D-3411 Katlenburg-Lindau 3, FRG

ABSTRACT. The orbital evolution of comets with different assumed birth-places, namely the Uranus-Neptune region, the Jupiter region and 'in-situ' from where 'new' comets seem to come, is followed numerically throughout the solar system lifetime. Perturbations of the four giant planets and random passing stars are considered. Several evolutionary paths are thus defined for the comet orbits. The efficiency of each one of them in bringing comets into the planetary region is discussed in connection with the origin of comets.

1. INTRODUCTION

The discussion of the origin of comets has so far been closely related to the study of their dynamical properties. Oort (1950) noted that a significant fraction of the observed long-period comets has reciprocal semimajor axes (1/a) concentrated in the very narrow range $0 < 1/a < 10^{-4}$ AU^{-1}. From this property Oort suggested that comets are stored at very large heliocentric distances forming a structure known as the 'cometary cloud'. According to Oort's view, stellar perturbations are responsible for bringing comets into the planetary region, which he de-fined as 'new' comets. Studies with larger samples of long-period com-ets have further confirmed such concentration at values of 1/a close to zero (Marsden and Sekanina, 1973; Marsden et al., 1978).

Accepting as a starting point the existence of a cometary reser-voir at large heliocentric distances, the question is how it was formed. Two distinct views have been discussed: 1) <u>Planetary origin</u>. Comets were formed in the planetary region and scattered to the cometary re-servoir by planetary perturbations (e.g. Oort, 1950; Safronov, 1972; Öpik, 1973). 2) <u>In-situ origin</u>. Comets were directly formed in the com-etary reservoir (e.g. Cameron, 1973; Biermann and Michel, 1978).

W. Fricke and G. Teleki (eds.), Sun and Planetary System, 371–374.

2. THE NUMERICAL MODEL

The likelihood of an assumed comet origin may be assessed by the efficiency of the evolutionary path so defined in bringing comets into the planetary region at present. To this purpose, five models of comet origin were considered: three 'planetary' and two 'in-situ' origins (see Table I).

Table I: Models of comet origin

Origin	Initial conditions	
Uranus-Neptune region (UN1)	$20 < q_i < 30$ AU $0 < 1/a_i < 10^{-4}$ AU^{-1}	$0 < t_j < 4.5$ billion yr
Uranus-Neptune region (UN2)	$20 < q_i < 30$ AU $0 < 1/a_i < 10^{-4}$ AU^{-1}	$0.5 < t_j < 1.5$ billion yr
Jupiter region (J)	$4 < q_i < 6$ AU $0 < 1/a_i < 10^{-4}$ AU^{-1}	$0 < t_j < 1.0$ billion yr
In-situ (IS1)	$q_i = a_i = 2.5 \times 10^4$ AU	$t_0 = 0$
In-situ (IS2)	$q_i = a_i = 10^4$ AU	$t_0 = 0$
	(For all the cases: $0 < i_i < 10°$)	

The initial orbital elements: perihelion distances (q_i), inclinations (i_i) and semimajor axes (a_i), were taken at random within the adopted intervals. For a planetary origin, t_j represents the 'injection' times of the comets scattered by planetary perturbations into the cometary reservoir. Later injection times were adopted for comets coming from the Uranus-Neptune region, following the suggested long time scale for bodies under the gravitational control of these two planets to reach near-parabolic orbits (Fernández and Ip, 1981). For an 'in-situ' origin, t_0 is the formation time. The size of the samples ranged between 10^3 and 10^4 comets.

The evolution of comet orbits was followed up to the present. When a test comet passed through the planetary region, a random energy change caused by the perturbing action of the four giant planets was applied as devised by Fernández (1981). At any moment, perturbations of random stars passing at distances smaller than 2.5×10^5 AU were considered following the procedure described by Fernández (1980). The computations of a test comet were terminated if: 1) it was ejected in a hyperbolic orbit (either by stellar or planetary perturbations), 2) its semimajor axis became smaller than 10^3 AU.

3. RESULTS

The larger the perihelion distance, the smaller the probability that
comets will be driven into the planetary region by stellar perturbations. Let us thus focus our attention on the surviving comets whose
perihelia are smaller than a certain upper limit (400 AU). The largest
fraction of surviving comets with q < 400 AU was obtained at any time
for an origin in the Uranus-Neptune region (UN1 and UN2). This fraction
reaches a value ~ 0.1 for the model UN1, that is considering injection
times distributed throughout the solar system lifetime (Fig. 1). For an
assumed origin in the Jupiter region the diffusion of comet orbits
proceeds much faster, so only about 10^{-3} of the initial population remains at present with q < 400 AU.

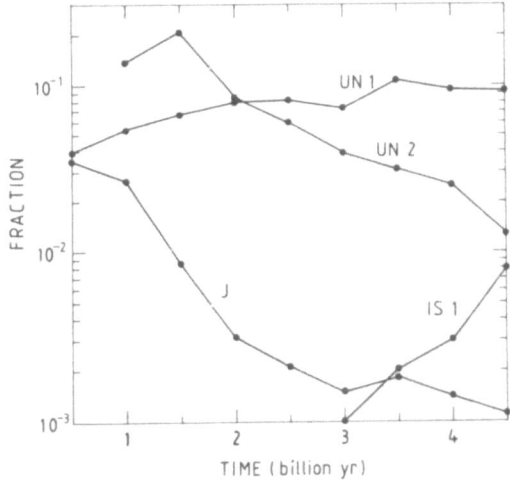

Figure 1. Fraction of surviving comets with q < 400 AU as a
function of time. The plots correspond to the models of comet
origin defined in Table I.

Comets formed in-situ should diffuse inwards under the perturbing
action of passing stars in order to reach the planetary region. This
effect can be noticed in the progressive filling of the region of small
perihelion distances for an origin at 2.5×10^4 AU. Nevertheless, the
current fraction of comets with q < 400 AU is still smaller than that
obtained for the model UN2. On the other hand, as the diffusion proceeds,
comet losses to hyperbolic orbits increase reaching roughly 2/3 of the
initial comet population. The more tightly bound comets at 10^4 AU (not
shown in Fig. 1) are less affected by stellar perturbations, so the
diffusion inwards of comets proceeds much slower.

The results regarding the small-q population of the cometary reservoir can be compared with the computed current rates of incoming 'new'

comets ($a > 10^4$ AU) shown in Table II. In agreement with the larger
fraction of surviving small-q comets, greater rates of comet passages
are found for the models of comet origin in the Uranus-Neptune region,
in particular for UN1. The rate greater than one obtained in the latter case shows that most comets pass through the planetary region as
'new' ones more than once.

Table II: Current rate of comet passages with $q < 30$ AU
            (per 10^8 yr and per 'injected' comet)

Model	Rate
UN1	3.80
UN2	0.11
J	0.011
IS1	0.017
IS2	0.002

In summary, a greater rate of comet passages is found for evolutionary paths of comet orbits starting in the Uranus-Neptune region.
Comets starting in the Jupiter region evolved rapidly due to the much
stronger perturbations of Jupiter, so only a small fraction of them are
expected to remain bound to the solar system at present. For an 'in-
situ' origin, let us say at 10^4 AU, the problem seems to be the long
time scales required for comets to reach the planetary region. At several 10^4 AU the diffusion speed increases but the comet losses to
hyperbolic orbits become more and more significant.

A more detailed account of the above work will be presented elsewhere.

4.   REFERENCES

Biermann, L., and Michel, K.W.: 1978, The Moon and the Planets, 18,
    pp. 447-464.
Cameron, A.G.W.: 1973, Icarus 18, pp. 407-450.
Fernández, J.A.: 1980, Icarus 42, pp. 406-421.
Fernández, J.A.: 1981, Astron. Astrophys. 96, pp. 26-35.
Fernández, J.A., and Ip, W.-H.: 1981, Icarus (in press).
Marsden, B.G., and Sekanina, Z.: 1973, Astron. J. 78, pp. 1118-1124.
Marsden, B.G., Sekanina, Z., and Everhart, E.: 1978, Astron. J. 83,
    pp. 64-71.
Oort, J.J.: 1950, Bull. Astron. Inst. Neth. 11, pp. 91-110.
Öpik, E.J.: 1973, Astrophys. Space Sci. 21, pp. 307-398.
Safronov, V.S.: 1972, The Motion, Evolution of Orbits, and Origin of
    Comets (G.A. Chebotarev, and E.I. Kazimirchak-Polonskaya, Eds.),
    pp. 329-334, IAU Symp. No. 45.

AN INTERSTELLAR ORIGIN FOR COMETS

W.M. Napier
Royal Observatory
Edinburgh
Scotland, UK

1. INTRODUCTION

It is usually supposed that the outer regions of the Oort cloud have been thermalised by the cumulative effect of stellar encounters and that the long-period comets are perturbed into the inner solar system by passing stars. This heating, or relaxation, is a gentle process, Weissman (1980) for example estimating that only $\sim$ 9% of the primordial comet has been lost through ejection. However it is a well known result in stellar dynamics (e.g. Ogorodnikov, 1965) that the relaxation of stellar motions in the Galaxy is almost entirely due to the presence of large masses, such as nebulae or star groups, individual stellar encounters taking $\sim 10^4$-10^5 times as long to produce the same degree of relaxation. It is therefore necessary to consider the effect of large perturbing masses on a primordial cloud of comets, and it will turn out that the outer regions of such a cloud (say beyond 2a $\sim$ 20,000 AU) would probably have boiled off by now.

2. THE STRUCTURE OF THE INTERSTELLAR MEDIUM

The existence of individual molecular cloud complexes, with masses in the range 10^5-10^6 $M_\odot$, has been known for about a decade. It is only recently, however, with large scale surveys of the millimetre wave emission from carbon monoxide, that the importance of these clouds has been recognised (see Gordon and Burton, 1980; Solomon and Sanders, 1980). Modelling of the CO emission by Gordon and Burton (1980) gives $\sim$ 2-20 pc radius for most large molecular clouds, with masses $\gtrsim$ a few 10^4 $M_\odot$ and a striking tendency for these large structures to occur in clusters. According to Solomon and Sanders molecular clouds are gathered into complexes, or giant molecular clouds (GMCs). These have mean mass M = 5 x 10^5 $M_\odot$ and mean radii R = 20 pc; there are $\sim$ 4000 in the Galaxy, comprising a very flat system ($Z_{\frac{1}{2}} \sim$ 50-60 pc) with number density at the solar distance $\sim$ 1/5th that at 6 kpc, where it peaks. The derived masses are uncertain by at least a factor of 2 and quite possibly more. We follow Talbot and Newman (1976) in assuming that the Galaxy has been richer in gas in the past, yielding a mean encounter

W. Fricke and G. Teleki (eds.), Sun and Planetary System, 375–378.
Copyright © 1982 by D. Reidel Publishing Company.

rate $\sim$ 50% higher than at present. With this factor, we find that the solar system has experienced N past penetrations into GMCs, given by

$$N \sim 9(\overline{V}_{\odot}/20) \ \overline{f}(\overline{V}_{\odot})$$

$\overline{V}_{\odot}$ being the mean peculiar speed of the sun in km/s, averaged over its history, $\overline{f}$ being a mean gravitational focussing factor. Typically $1.3 \leq \overline{f} \leq 4$, and a reasonable range for the number of past encounters is $12 \leq N \leq 20$, corresponding to a mean interval $200 \leq \Delta t \leq 400$ Myr. It is assumed that the GMCs are randomly distributed. If, as now seems more probable (Cohen et al., 1980), they are concentrated in the spiral arms, the most plausible value is $N \sim 20$.

Mass inhomogeneities on all scales seem to be present in GMCs. From the point of view of the tidal disruption it is the larger mass inhomogeneities, usually comprising clusters of smaller structures, which are important. For example in a GMC towards M17 studied by Elmegreen and Lada (1976), about a third of the mass is in fragments of masses $(5, 10, 10, 20) \times 10^4$ $M_{\odot}$, with radii $(3, 3, 4, 6)$ pc respectively; the HI in a GMC studied by Hasegawa et al. (1980) comprises about half a dozen fragments with virial masses ~ 1–4×10^4 $M_{\odot}$; and so on. A reasonable model of a GMC of mass 5×10^5 $M_{\odot}$ might be to suppose it comprises ~ 25 fragments of masses $\sim 2 \times 10^4$ $M_{\odot}$; this yields a filling factor ~ 0.05, and neglecting internal focussing effects one expects the solar system to pass within ≤ 5 pc of such substructure during each random encounter with a GMC. There may therefore have been ~ 20 such encounters in the past, with the number unlikely to have been less than ~ 12 on the present assumptions.

3.   THE EFFECT OF CLOSE ENCOUNTERS

The effects of encounters with such masses were studied by 3-body integrations using a standard numerical package. Orbits of given size and shape were randomly oriented, with typically 50 or 100 orientations in bins typically 2500 AU wide in a and 0.2 wide in e. Nebulae of various masses M, impact parameters P, asymptotic approach speeds V were flown past the solar system and the number of comets escaping from each bin recorded. It was found that even a single flyby may have a drastic effect on the Oort cloud, confirming that stellar encounters are more or less irrelevant to its long-term dynamics. For example a flyby of mass $M = 2 \times 10^4$ $M_{\odot}$ at $P = 5$ pc leads to ~ 20–60% loss of long-period comets. If there have been $(2, 3, 4, 5)$ past encounters at $(2, 3, 4, 5)$ pc respectively – a total of 14 encounters – with nebulae of $M = 2 \times 10^4$ $M_{\odot}$ and $V = 20$ km/s, one finds a survival probability $\sim 5 \times 10^{-4}$ for long-period comets, with the number unlikely to be outside the range 5×10^{-6}– 5×10^{-3}. Thus at least the outer regions of a primordial cloud would by now have been swept away.

The effect of a close encounter is not only to cause comets to escape: the surviving cloud is energised, making subsequent escape easier. This was studied numerically by following 3350 individual

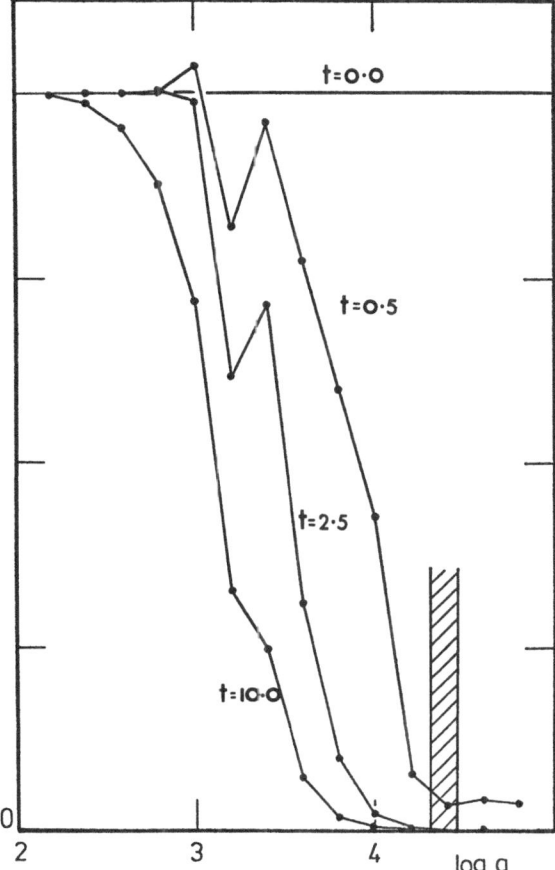

Fig. 1. Depletion of a primordial comet cloud initially uniformly
populated in (log a,e), semi-major axis a in AU. The hatched region
represents the zone from which long-period comets arrive. If N = 28
time t is in units of 1000 Myr. Thus for 14 past encounters the curve
t = 2.5 represents the structure of the primordial cloud after 5000 Myr.

orbits for successive random encounters. The results are shown in Fig.
1. There is a rapid clearing of the outer regions of a primordial
cloud and a slower eating into the core. Hypothetically there could be
a remnant primordial core of $\leq 10^4$ AU radius.

The radius of this core can be approximately calculated on the
impulse approximation. For encounters with nebulae of M = 2 x 10^4 $M_\odot$,
P = 5 pc, V = 20 km/s, one finds $r_5 \sim 0.9/N^{1/3}$, the radius r_5 in units
of 50,000 AU. Hence the initial depletion of the primordial cloud is
very rapid, much of the damage being done within the first few encounters, and a precise knowledge of the encounter rate is not therefore
critical. If the potential of a GMC as a whole is considered, then
grazing encounters with nebular complexes of mass 2 x 10^5 $M_\odot$, radius
20 pc, yield $r_5 \sim 1.2/N^{1/3}$. Thus the indications are that the effect

of a GMC as a whole is comparable with that of its substructure. This
has been neglected in the numerical study, suggesting that the derived
results are conservative.

4.   AN INTERSTELLAR SCENARIO

There are two Jeans' masses characteristic of cold, dense molecular
clouds. That of the gas is of order of a stellar mass, that of the
dust alone is of order 10^{16}–10^{18}g. Thus if effective dust/gas separation mechanisms operate within a dense cloud one may expect the formation of icy planetesimals of a few km diameter (Napier and Humphries,
in preparation). If $\sim$ 5% of the mass of a GMC is in planetesimal form,
consistent with a long-standing discrepancy between the expectations of
nucleosynthesis theory and the observed metal abundances of young stars
(e.g. Greenberg, 1974; Tinsley and Cameron, 1974) then there may be
$\sim 10^{-2}$ planetesimals AU^{-3}. Now the sun, encountering a GMC or the substructure examined, will have a sphere of influence of radius $\sim$ 50,000
AU, and so $\sim 10^9$ planetesimals/yr will enter its sphere of influence.
Of these $\sim 10^5$/yr will enter at less than the local escape velocity,
for moderate velocity dispersions, and be temporarily captured. In a
static potential field such a comet would escape again in a few Myr
when it again leaves the solar sphere of influence. However in a few
Myr the sun may have moved away from the molecular cloud and its sphere
of influence will have extended to $\sim$ 200,000 AU trapping the comet.
Thus something like 10^5 planetesimals/yr may be captured for something
like the Myr or so during which the sun is in the environment of, but
receding from, a massive nebula. Thus one expects a quasi-equilibrium
cloud of order 10^{11} comets around the sun, with a sharp concentration
of aphelia around 50,000 AU. These numbers are more or less ad hoc
features of the 'primordial Oort cloud' hypothesis but follow naturally
from the properties of the ISM on the interstellar one.

REFERENCES

Cohen, R.S., Cong, H., Dame, T.M., and Thaddeus, P.: 1980, Ap. J.
          Lett. 239, L53.
Elmegreen, B.G., and Lada, C.J.: 1976, Astron. J. 81, p.1089.
Gordon, M.A., and Burton, W.B.: 1980, Proc. Third Gregynog Astrophysics
          Workshop on Giant Molecular Clouds (eds. P.M. Solomon and
          M.G. Edmunds), Pergamon Press, Oxford, p.25.
Greenberg, J.M.: 1974, Astrophys. J. 189, L8.
Hasegawa, T., Sato, F., and Fukui, Y.: 1980, IAU Symposium 87, "Inter-
          stellar Molecules" (ed. B. Andrew), Reidel, Dordrecht, p.159.
Ogorodnikov, K.F.: 1965, "Dynamics of Stellar Systems", Pergamon,
          Oxford, London.
Talbot, R.J., and Newman, M.J.: 1977, Astrophys. J. Suppl. 34, p.295.
Tinsley, B.M., and Cameron, A.G.W.: 1974, Astrophys. Sp. Sci. 31, p.31.
Solomon, P.M., and Sanders, D.B.:1980, Proc. Third Gregynog Astrophysics
          Workshop on Giant Molecular Clouds (eds. P.M. Solomon and
          M.G. Edmunds), Pergamon Press, Oxford, p.41.
Weissman, P.R.: 1980, Nature 288, p.242.

STRONG PERTURBATIONS AT CLOSE ENCOUNTERS WITH JUPITER

A. Carusi and G.B. Valsecchi
IAS - CNR, Reparto di Planetologia, Roma (Italy)

ABSTRACT.   The phenomena occurring at a close encounter of a minor body with Jupiter are reviewed and the relations between the characteristics of the pre-encounter orbit and the possible outcomes of the interaction are discussed.

The orbits of those minor bodies that pass close to the outer planets can be greatly perturbed by them (Everhart, 1973; Kazimirchak-Polonskaya, 1967); in particular Jupiter is the most efficient transformer of cometary orbits (Carusi and Valsecchi, this volume). The perturbations at a close encounter can be expected to be stronger either if the minimum approach distance D, computed simply from the initial keplerian orbit of the minor body, is low or if the encounter velocity is low and the encounter geometry is favourable. These last conditions imply that the Tisserand invariant $T = 1/a + 2\sqrt{a(1-e^2)}\cos i$ (a in units of Jupiter semiaxis) must be high - about 3 or more, see Kresák (1979) - and that the initial orbit of the minor body must be nearly tangent to that of the planet (Carusi et al., 1979; Froeschlé and Rickman, 1981). When they are satisfied, the actual minimum approach distance d can be much smaller than D (Carusi et al., 1981a,b; Kresák et al., 1982).

A good way to look at the problem of strong perturbations is to analyze the behaviour of the heliocentric and jovicentric energy of the minor body for an extended period of time including the encounter. To do so we can plot the two quantities $-1/a_{\odot}$ vs. $-m/(m_{\odot}a_{\jmath})$, where $a_{\odot}$ and $a_{\jmath}$ are respectively the osculating heliocentric and jovicentric semimajor axes of the minor body, and $m_{\odot}$, m are the masses of the Sun and Jupiter. In Figs. 1 and 2 eight examples of such a plot are given: the evolutions shown in Fig. 1 refer to the encounters with Jupiter of P/Smirnova-Chernykh (1949-1975), of P/Shajn-Schaldach (1931-1971) and of P/Gehrels 3 (1962-1977); for comparison, a couple of jovicentric orbits of the jovian satellite J XII (Ananke) are reported in the fourth column. The first two evolutions shown in Fig. 2 refer to the encounters with Jupiter of P/Oterma (1931-1969) and of a varied version of it (Oterma 14, see Carusi et al., 1981a,b); the other two evolutions are taken from the samples of fictitious bodies RANDUN and ASTRID (Carusi and Pozzi, 1978b). Also plotted in the two figures are the

379

W. Fricke and G. Teleki (eds.), Sun and Planetary System, 379-384.

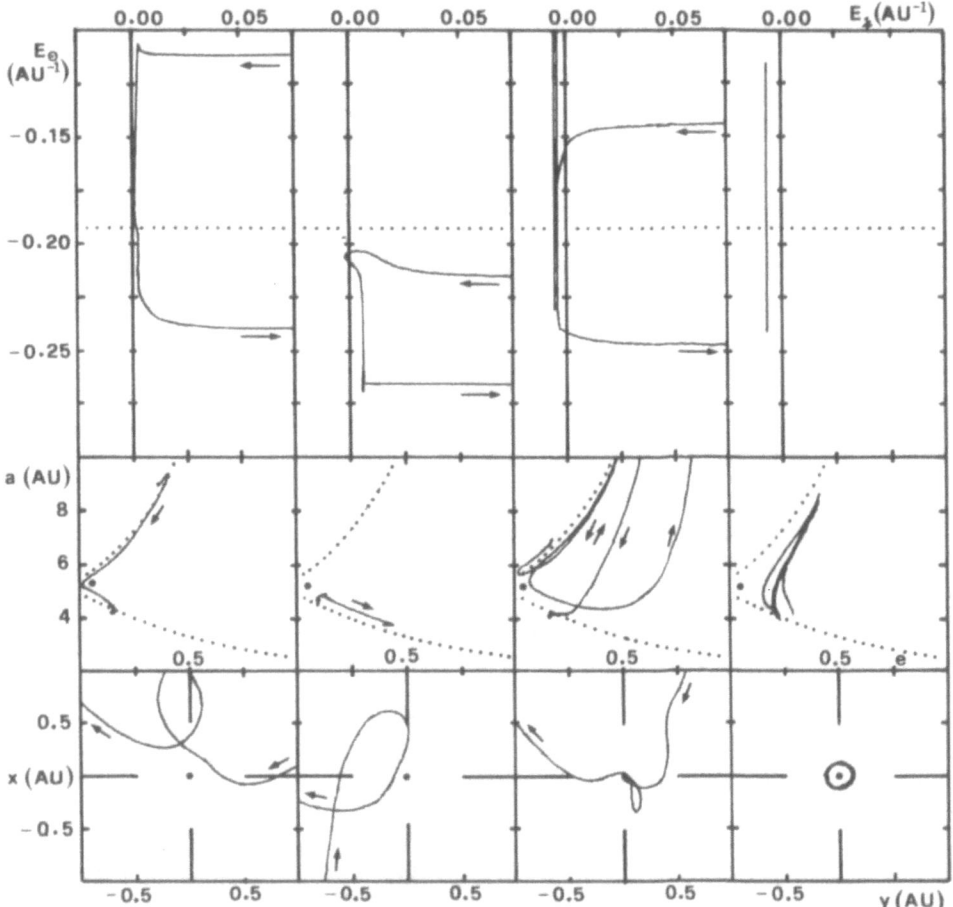

Fig. 1 – First column. Top: energy plot of the evolution of P/Smirnova-Chernykh between 1949 and 1975; $E_\odot = -1/a_\odot$, $E_4 = -m /(m_\odot a_4)$ ($m_\odot$, m : Sun and Jupiter masses; $a_\odot$, a_4: heliocentric and jovicentric semiaxes of the comet). The dotted line refers to Jupiter's heliocentric semiaxis. Centre: heliocentric semiaxis-eccentricity diagram of P/Smirnova-Chernykh; dotted lines refer to the orbits tangent in aphelion (lower curve) or perihelion (upper curve) to that of Jupiter. Bottom: ecliptic projection of the jovicentric path followed by P/Smirnova-Chernykh; the rotating frame is centred on Jupiter and the Sun is on the negative x-axis. The arrows indicate the direction of the motion.
     – Second column: same for P/Shajn-Schaldach between 1931 and 1971.
     – Third column: same for P/Gehrels 3 between 1962 and 1977.
     – Fourth column: same for two revolutions of J XII (Ananke) around Jupiter.

evolutions of the same objects on the (heliocentric) semiaxis-eccentricity diagram, and the paths followed by them in a rotating ecliptical frame centred on Jupiter and with the Sun always on the negative x-axis. All the evolutions were computed in a simple Sun–Jupiter–minor body model, u-sing Greenspan's Discrete Mechanics (Carusi and Pozzi, 1978a); the starting points were taken from Marsden (1979) for P/Smirnova-Chernykh and P/Shajn-Schaldach; from Rickman (private communication) for P/Gehrels 3; from Herget (1968) for J XII and from Carusi et al. (1981b) for P/Oterma.

    These evolutions have been selected because they summarize the major phenomena that can occur in an effective close encounter. P/Smirnova-Chernykh underwent a "transition", i.e. an exchange of perihelion with aphelion, shifting from an orbit nearly tangent to that of Jupiter in perihelion to one nearly tangent in aphelion. This type of process was observed by, among others, Kazimirchak-Polonskaya (1967), Sitarski (1968) and Rickman (1979); its importance was stressed by Carusi et al. (1979), Carusi and Valsecchi (1979, 1980) and, recently, by Froeschlé and Rickman (1981). The evolution of P/Smirnova-Chernykh in the energy plot, like those of all the other objects plotted here, except J XII, is characterized by two regimes: the horizontal one is typical of the heliocentric motion with Jupiter as perturber, while the vertical one represents a jovicentric motion perturbed by the Sun. In this case the passage from one regime to the other is not sharp; the jovicentric energy never becomes negative and, even to the scale of the plot, it is not constant. It therefore appears that a matched conic section approximation would not satisfactorily represent the actual motion of this comet.

    The second column of Fig. 1 shows the evolution of P/Shajn-Schaldach, which underwent a strong deceleration by Jupiter during the encounter of 1940–1947. Between 1942 and 1944 the jovicentric energy of the comet became negative (Carusi and Valsecchi, 1981, 1982), i.e. the comet underwent a temporary satellite capture (TSC) by Jupiter. The capture corresponds to the upper part of the loop recognizable in the jovicentric pattern; during it the comet was at more than 1 AU from Jupiter (note that the jovicentric plot is a projection on the ecliptic plane!) and in the energy plot the jovicentric energy varied continuously. Therefore, this TSC does not correspond to a motion mainly controlled by Jupiter and is rather different from those that we will discuss later.

    Comet P/Gehrels 3, shown in the third column of Fig. 1, underwent both the mentioned processes, since it was transferred from an orbit well outside that of Jupiter to one well inside it, and was captured as temporary satellite for an extended period of time (Carusi and Valsecchi, 1979, 1981; Rickman, 1979; Rickman and Malmort, 1981). The TSC of P/Gehrels 3 lasted from 1967 to 1974 and, during it, the jovicentric energy was rather stable, whereas the heliocentric one became positive in correspondence to a passage of the comet extremely close to the planet that occurred in 1970.

    It is interesting to compare the three evolutions just discussed with that of J XII reported in the fourth column of Fig. 1. Its jovicentric energy looks more stable than those of the comets, as would be expected. It has been often proposed that J XII and its companions J VIII, J IX and J XI have been captured from heliocentric orbits with the aid of some dissipative mechanism (see, for instance, Pollack et al., 1979): the comparison of its energy plot with that of P/Gehrels 3, the comet that has had the longest and deepest TSC known so far, suggests that the proposed capture scenario for J XII is not implausible.

    The evolution of P/Oterma between 1931 and 1968 (Fig. 2) shows a

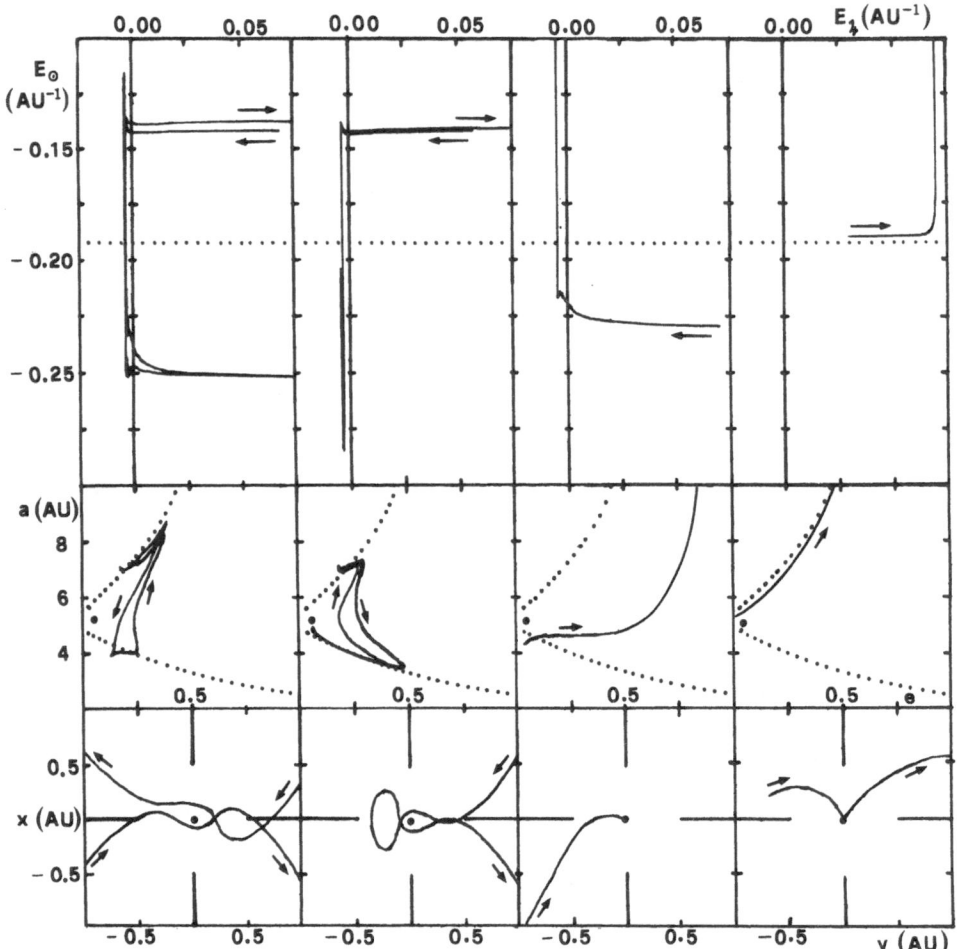

Fig. 2 - First column: same as Fig. 1 for P/Oterma between 1931 and 1969.
  - Second column: same for the fictitious object Oterma 14.
  - Third column: same for the fictitious object RANDUN327.
  - Fourth column: same for the fictitious object ASTRID 891.

variety of interesting phenomena connected with its two successive encounters. It became a satellite of Jupiter between 1935 and 1939 (Chebotarev,
1967; Carusi and Valsecchi, 1981, 1982) and was transferred to an inner
"quasi-Hilda" type orbit (Kresák, 1979). Also P/Gehrels 3 and P/Smirnova-
Chernykh have been injected into orbits of this type, but only for P/Oter-
ma was the ratio of the mean motion to that of Jupiter very close to 3/2.
Therefore in 1950, at the time of the comet's second perihelion passage in
the inner orbit, the Sun, Jupiter and P/Oterma were on a straight line,
with the velocity vector of the comet orthogonal to the conjunction line,
so that the perturbations suffered thereafter compensated the previous o-
nes (Carusi et al., 1979) and the comet came to aphelion in 1961, exactly
in time to encounter Jupiter again. The second encounter was almost the
mirror image of the first, as it is recognizable in Fig. 2, and P/Oterma

was ejected back to a heliocentric orbit resembling the initial one. Also during the second encounter P/Oterma suffered a TSC, lasting from 1960 to 1965. Carusi et al. (1981a,b) have studied the evolution of a chain of objects moving on the same orbit that P/Oterma had in 1931. A wide variety of orbital patterns at the close encounter emerged as a result of this study (Kresá k et al., 1982), some of them closely resembling that of P/ Gehrels 3. One of those Otermas, the object n° 14, is shown in the second column of Fig. 2. It has been chosen because there is a nice similarity of its final orbit to that of the true comet, with the interesting difference that Oterma 14 was not injected into an inner heliocentric orbit meanwhile. It instead was in conjunction with the Sun and Jupiter while it was at the aphelion of a very perturbed heliocentric orbit of semiaxis and eccentricity very similar to those of Jupiter (the leftmost point in its a-e plot). After that, again because of the orthogonality of the comet velocity vector to the conjunction line, the mechanism of the compensation of the perturbations acted, and the rest of the evolution was the mirror image of the first part. The TSC of Oterma 14 was rather longer than that of the true comet, as it lasted more than eight years.

There are two other possible outcomes of a close encounter that can be relevant for the evolutionary history of the Solar System, i.e. the collision of the minor body with the planet and the ejection from the Solar System on a hyperbolic orbit. The third column of Fig. 2 shows the collision of the fictitious object RANDUN 327 (Carusi and Pozzi, 1978b) with initial parameters a = 4.354 AU, e = 0.0343, i = 0.61°; the last column shows the ejection by Jupiter of the fictitious object ASTRID 891, with initial parameters a = 5.273 AU, e = 0.0134, i = 33.53°. It is noticeable that the final orbit of this object resulted markedly hyperbolic, its eccentricity being 1.23. Kresák et al. (1982) have shown that it is very improbable that objects with values of the Tisserand invariant T greater than 2.83 be ejected from the Solar System by a single encounter. Therefore, if this object had a smaller initial inclination (less than about 25°) it would have probably not been ejected.

A common feature of all the objects shown in Figs. 1 and 2 is the high pre-encounter value of their Tisserand invariant T (P/Smirnova-Chernykh: 3.01; P/Shajn-Schaldach: 2.96; P/Gehrels 3: 3.02; P/Oterma: 3.03; RANDUN 327: 3.02; ASTRID 891: 2.66), with the exception of ASTRID 891. In fact, the role of T in determining the possible outcomes of a close encounter is very important. We have already pointed out that T must be less than 2.83 to allow an ejection from the Solar System. On the other hand, TSC's tend to occur if T is greater than about 2.9 and if T is greater than 3 the TSC's may be as long enduring and deep as those of P/Oterma and P/Gehrels 3. This does not necessarily imply, as it is shown by the case of P/Smirnova-Chernykh, that a TSC will occur at each close encounter with Jupiter of a comet with T greater than 2.9.

The authors wish to thank Dr. L. Kresák for useful discussions on the subject of this paper.

REFERENCES

Carusi, A., Kresák, L., and Valsecchi, G.B.: 1981a, IAS-Reparto di Planetologia, Internal Report n°2, February 1981.
Carusi, A., Kresák, L., and Valsecchi, G.B.: 1981b, Astron. Astrophys., 99, pp. 262-269.
Carusi, A., and Pozzi, F.: 1978a, Moon and Planets, 19, pp. 65-70.
Carusi, A., and Pozzi, F.: 1978b, Moon and Planets, 19, pp. 71-87.
Carusi, A., Pozzi, F., and Valsecchi, G.B.: 1979, in "Dynamics of the Solar System" (R.L. Duncombe ed.), IAU Symp. 81, Dordrecht - Holland, pp. 185-189.
Carusi, A., and Valsecchi, G.B.: 1979, in "Asteroids" (T. Gehrels ed.), Tucson - USA, pp. 391-416.
Carusi, A., and Valsecchi, G.B.: 1980, Moon and Planets, 22, pp. 113-124.
Carusi, A., and Valsecchi, G.B.: 1981, Astron. Astrophys., 94, pp. 226-228.
Carusi, A., and Valsecchi, G.B.: 1982, in "Comparative Study of the Planets" (A. Coradini and M. Fulchignoni eds.), Dordrecht - Holland, pp. 131-138.
Chebotarev, G.A.: 1967, "Analytical and Numerical Methods of Celestial Mechanics", New York - USA, p. 239.
Everhart, E.: 1973, Astron. J., 78, pp. 316-328.
Froeschlé, C., and Rickman, H.: 1981, Icarus, 46, pp. 400-416.
Herget, P.: 1968, Astron. J., 73, pp. 737-742.
Kazimirchak-Polonskaya, E.I.: 1967, Sov. Astron. - A.J., 11, pp. 349-365.
Kresák, L.: 1979, in "Asteroids" (T. Gehrels ed.), Tucson - USA, pp. 289-309.
Kresák, L., Valsecchi, G.B., and Carusi, A.: 1982, Bull. Astron. Inst. Czechosl., in press.
Marsden, B.G.: 1979, "Catalogue of Cometary Orbits", Smithson. Astrophys. Obs., Cambridge - USA.
Pollack, J.B., Burns, J.A., and Tauber, M.E.: 1979, Icarus, 37, pp.587-611.
Rickman, H.: 1979, in "Dynamics of the Solar System" (R.L. Duncombe ed.), IAU Symp. 81, Dordrecht - Holland, pp. 293-298.
Rickman, H., and Malmort, A.M.: 1981, Astron. Astrophys., in press.
Sitarski, G.: 1968, Acta Astron., 18, pp. 171-195.

STATISTICS OF CLOSE ENCOUNTERS OF MINOR BODIES WITH THE OUTER PLANETS

A. Carusi and G.B. Valsecchi
IAS - CNR, Reparto di Planetologia, Roma (Italy)

ABSTRACT. For each of the four outer planets a sample of 1000 fictitious minor bodies in heliocentric orbits has been generated, and a single close encounter of each minor body with the corresponding planet has been computed. In this paper we present the distributions of the initial and final orbital elements, together with the statistics of various types of orbital evolutions.

The aim of the present work is to improve and extend the statistical research on close encounters with Jupiter of Carusi and Pozzi (1978b) in order to get a deeper understanding of the evolution of objects moving on cometary orbits in the outer Solar System. The method followed here is very similar to that used in Carusi and Pozzi (1978a,b), so that we will only describe the differences from the previous research. They are:

a) the research has been repeated for Jupiter and extended to Saturn, Uranus and Neptune, in order to allow a comparison of the effects of the four outer planets on similar populations of 1000 fictitious minor bodies each; hereafter we will refer to them using the names JUP, SAT, URA, NEP;

b) we have used an improved version of the program computing the close encounter, with a new procedure to compute the time step; this version was described in Carusi and Valsecchi (1979);

c) the initial distributions of the elements of the four populations of fictitious minor bodies differ from those of the previous research.

The initial distributions used in Carusi and Pozzi (1978b) were a little unrealistic, since they were flat in eccentricity and too rich in low inclination classes. The form of the new distributions is identical for the four populations having encounters with the four outer planets, except for a scaling factor multiplying the 1/a distributions. As was suggested to us by Dr. L. Kresák, the initial distribution functions of e and i were:

$$P(e)=\sin(\pi e) \quad \text{for} \quad 0<e<1$$

$$P(i)=\sin(6i) \quad \text{for} \quad 0°<i<30°.$$

385

W. Fricke and G. Teleki (eds.), Sun and Planetary System, 385–388.
Copyright © 1982 by D. Reidel Publishing Company.

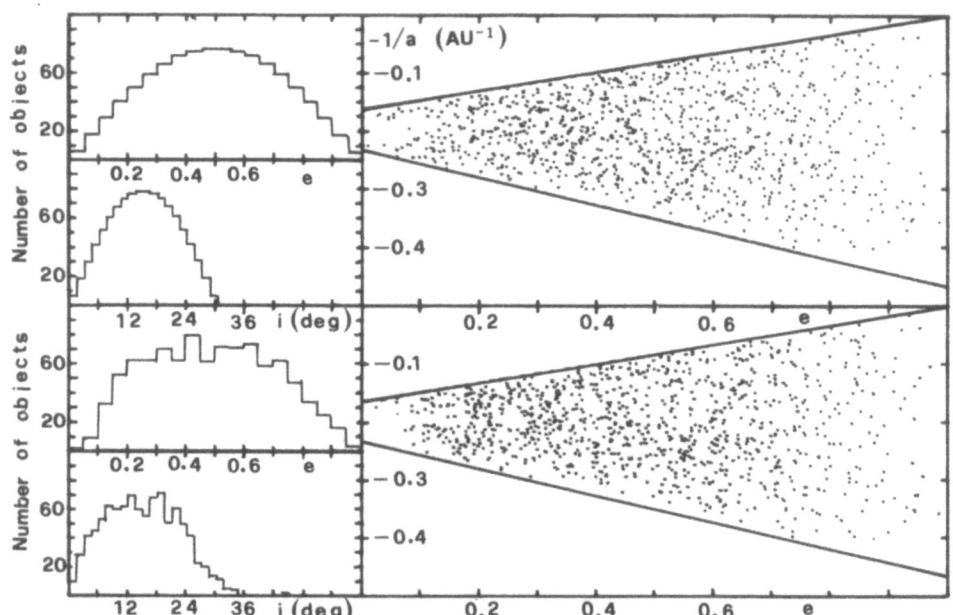

Fig. 1 - Upper part. <u>Left</u>: initial distributions of e (top) and i (bottom)
         for the sample JUP. <u>Right</u>: -1/a vs. e diagram of the initial or-
         bits of the sample JUP.
       - Lower part. Same as in the upper part for the final distributions
         of the sample JUP.

The whole intervals of variation of e and i were divided in 40 classes,
and for each class of eccentricity a flat distribution in 1/a between the
limits:

$$\frac{1+e}{q-R} \geq \frac{1}{a} \geq \frac{1-e}{Q+R}$$

was generated, where q and Q are the planet's perihelion and aphelion
distances, and a and e refer to the minor body; R is 2/3 AU for Jupiter
(as in the previous research), Saturn, Uranus, and 4/3 AU for Neptune.
In order to take into account the lack of low eccentricity short period
comets of high inclination, an additional constraint was added, namely:

$$i < 2.4° + 80.8° \ e$$

This condition gives a line that roughly divides, in the e - i plane, the
region populated by known short period comets from the empty one. The
resulting distributions for the sample JUP are shown in the upper part of
Fig. 1. We do not report the same distributions for the other samples since they are identical for e, i and 1/a, provided that a is expressed in u-
nits of the planet's semiaxis.
        The values of ω and Ω were chosen at random between 0° and 360° and
it was then checked if the resulting orbits had a minimum distance from

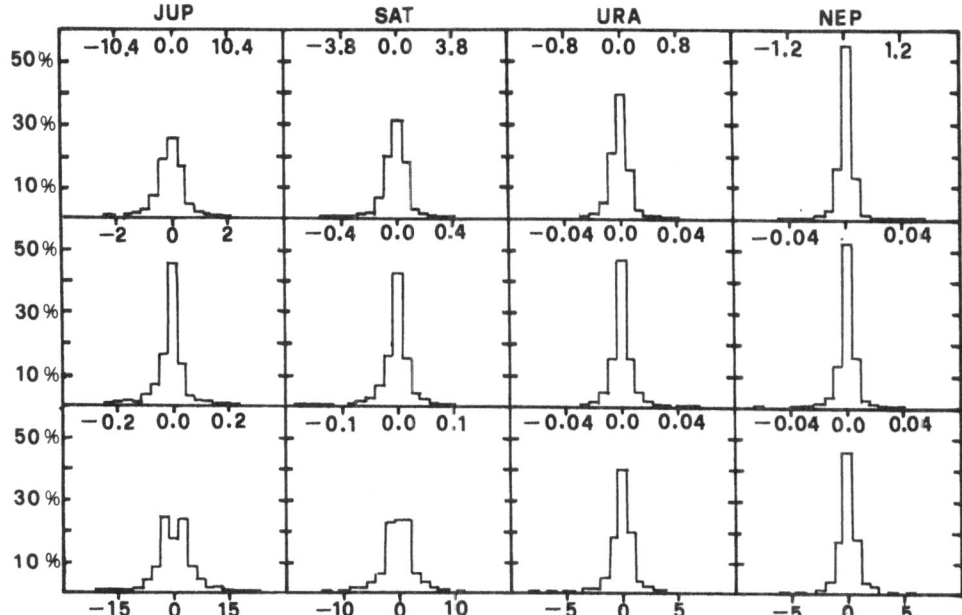

Fig. 2 – Histograms of the central parts of the distributions of $\Delta(1/a)$
(first row), Δe (second row) and Δi (third row) for the four
samples. The $\Delta(1/a)$ scale is 100 AU ; on the top, also the
scale for 100 $\Delta(ap/a)$ is given; the Δi scale is in degrees.

the planet's orbit of less than R. If not, they were discarded and replaced by new orbits also conforming to the given constraints. The initial
values of the true anomalies were chosen with the same procedure used in
Carusi and Pozzi (1978a), moving the planet and the minor body backwards,
along their unperturbed orbits, from the point of minimum distance to a
relative distance of 4R.

The final distributions for the sample JUP are shown in the lower
part of Fig. 1. It is possible to see that the distributions of e and i
are not very different from the initial ones. Since this is true, to a
greater extent, also for the final distributions of SAT, URA and NEP, we
do not report them here.

In Table 1 the frequency of collisions and of temporary satellite
captures (TSC) found in the present research are given (for TSC we mean
that the planetocentric orbital elements of the minor body become ellipti-

Table 1 – Frequency of collisions with the planet (C) and of temporary
satellite captures (TSC)

Sample	C	TSC
JUP	8	46
SAT	6	16
URA	2	4
NEP	–	4

Table 2 - Statistics of the variations of orbital elements

Sample	$\Delta 1/a$ (AU^{-1})		$(\Delta(ap/a))$		Δe		Δi (deg)	
	mean			SD	mean	SD	mean	SD
JUP	−0.0034	(−0.018)	0.029	(0.15)	−0.010	0.073	0.27	5.9
SAT	−0.0012	(−0.012)	0.011	(0.11)	−0.0056	0.049	−0.11	4.5
URA	−0.00014	(−0.0027)	0.0018	(0.035)	−0.0027	0.027	0.11	2.9
NEP	−0.000056	(−0.0017)	0.0015	(0.045)	−0.0015	0.021	0.020	2.2

cal for some time during the encounter). At variance from Carusi and Pozzi (1978b), where some ejections from the Solar System were found, we have had no final hyperbolic orbits in all our samples, probably because of
the choice of the initial distributions. Few collisions were found in
JUP, SAT, URA, and none in NEP. The number of TSC's decreases by a factor
3 passing from JUP to SAT, by a factor 4 passing from SAT to URA, and remains constant passing from URA to NEP.

In Table 2 the general statistics of the variations of the orbital
elements of the four samples are given, i.e. the means and the standard
deviations (SD) of $\Delta 1/a$, Δe and Δi. Also given in parentheses are the
values of $\Delta(ap/a)$, where ap is the semiaxis of the planet. In Fig. 2 the
histograms of the central parts of the distributions corresponding to these values are shown. As already noted by Carusi and Pozzi (1978b) and by
Carusi and Valsecchi (1979), these distributions are non-Gaussian, since
they are too peaked and with too high tails to be fitted by a normal distribution. Moreover, the exclusion of the wing objects does not improve
the situation, as has been checked using the χ^2 test. It appears that these distributions cannot be explained by a Gaussian with the addition of a
few very large variations.

A comparison of the distributions shows the different effects of the
various planets on similar populations of minor bodies that can encounter
them. Uranus and Neptune seem to have comparable efficiencies, although
the minimum distances between the pre-encounter orbits of the minor bodies
of NEP and Neptune were, on the average, about the double than in the case
of all the other planets. The SD's of the perturbations are slightly greater in URA than in NEP although, if the energy perturbations are computed
in units of the planet's semiaxis, Neptune is more efficient. Passing to
SAT the SD's increase by a factor of almost 2 for e and i, and of almost 8
for 1/a (of almost 3 if computed in planetary semiaxis units). Jupiter is
the most efficient: its SD's are always greater than those of Saturn, e-
specially in the case of the energy perturbations.

The allowed space limited us to only a few comments. More detailed
analyses of the data presented here, and discussions and comparisons with
previous researches will be published elsewhere.

REFERENCES

Carusi, A., and Pozzi, F.: 1978a, Moon and Planets, 19, pp. 65-70.
Carusi, A., and Pozzi, F.: 1978b, Moon and Planets, 19, pp. 71-87.
Carusi, A., and Valsecchi, G.B.: 1979, in "Asteroids" (T. Gehrels ed.),
Tucson - USA, pp. 391-416.

KEPLERIAN ESTIMATES OF PRE-DISCOVERY ENCOUNTERS WITH JUPITER FOR SHORT-PERIOD COMETS

H. Rickman   and   J. Karm
Astronomiska Observatoriet, Uppsala, Sweden

Orbital perturbations at close encounters with Jupiter are known to have a predominant influence on the population of short-period comets. Pre-discovery encounters are of particular interest concerning the origin, evolution and discovery circumstances of the comets. For the purpose of identifying such encounters and roughly estimating the approach dates and distances one can often use unperturbed, Keplerian orbits. We have made such estimates for all the 118 short-period comets known at the end of 1980, using as a rule the orbital elements in Marsden's (1979) catalogue. The treated time interval was chosen as the maximum of 100 years or 10 orbital revolutions. However, each calculation was stopped after the first approach to within 0.5 AU of Jupiter, and such encounters occurred for nearly half the comets.

Fig. 1 shows bivariate distributions of the minimum approach distances (Δ) and numbers of revolutions (R) from encounter to discovery, as found in this survey. The distributions are formed for the Jupiter family (aphelion distances Q < 6.5 AU) and the remaining comets separately. They appear very different from one another. The Jupiter family comets show a strong (60%) concentration to $\Delta < 0.5$ AU, while the others are rather uniformly scattered over the range $\Delta < 4$ AU. To a large part this difference is due to differring orbital geometries, one particular reason being that the Jupiter family comets in general have much lower inclinations than the other ones. However, Fig. 1 also shows that comets of the Jupiter family, in contrast to the others, have a tendency to be discovered shortly after close encounters with Jupiter. In the domain of (a,e,i) space occupied by the Jupiter family the random probability per orbital revolution to encounter Jupiter at $\Delta < 0.5$ AU may be estimated at about 5%, using the results by Froeschlé and Rickman (1981). Among the real Jupiter family comets, however, about 20% experience such encounters during the first revolution backward from discovery (R < 1). With increasing values of R the corresponding percentage decreases rapidly.

Of course a reason for this behaviour is that a large fraction of the Jupiter family comets have been "captured" at close encounters with

389

W. Fricke and G. Teleki (eds.), Sun and Planetary System, 389–390.

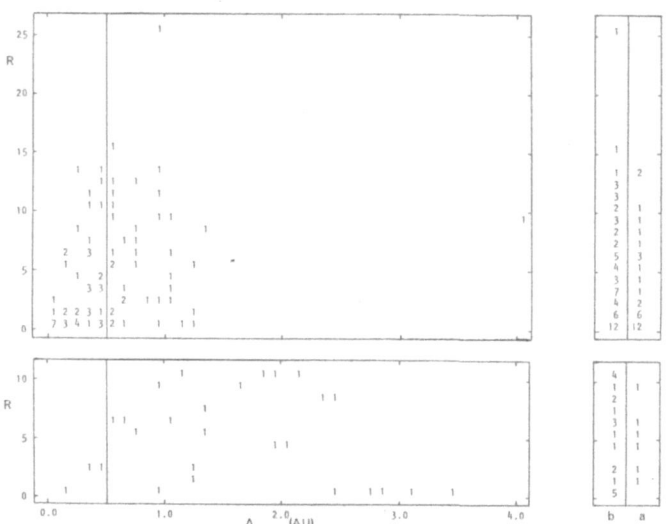

Figure 1. Distributions of R vs. Δ for the Jupiter family
(above) and remaining short-period comets (below).

Jupiter in the sense that their perihelion distances were considerably
reduced, thereby drastically increasing the likelihood of discovery.
This post-capture discovery likelihood appears to have been quite high
at least in recent time. An indication about this is provided by the
marginal distribution of R shown at the right of Fig. 1, separately for
comets discovered before (b) and after (a) 1950. While in the Jupiter
family as a whole 26% of the comets have R < 1, this fraction was only
20% before 1950 but has increased to 38% afterward. A probable reason
is the more efficient sky coverage provided in recent time by the use
of Schmidt telescopes.

    It must be kept in mind that the Keplerian orbits may deviate con-
siderably from the perturbed ones, and that these deviations statistically increase with the computation interval. From comparisons of our
Keplerian results with those of published orbital integrations for 52
pre-discovery encounters, we find that up to a limit of 50 years of computation interval the Keplerian estimates are almost always correct to
within a couple of months in encounter date and 0.1 AU in approach distance. Consequently, the important lower parts of Fig. 1 are reasonably
reliable, while the upper parts are not.

REFERENCES

Froeschlé, C., and Rickman, H.: 1981, Icarus 46, pp. 400-414.
Marsden, B.G.: 1979, *Catalogue of Cometary Orbits*, 3rd ed., IAU Central
    Bureau for Astronomical Telegrams, Cambridge, Massachusetts.

THE MOTIONS OF COMETS NEAR THE 2/1 RESONANCE WITH JUPITER

S. Vaghi
Mission Analysis Office, ESOC, Darmstadt, FRG

H. Rickman
Astronomiska Observatoriet, Uppsala, Sweden

ABSTRACT.   We present the results of orbital integrations in a 9-body model for comets P/Lexell, P/Tempel-Swift, P/Wirtanen, P/Kohoutek, P/West-Kohoutek-Ikemura, P/Haneda-Campos and P/Wild 2. All these comets have spent some time near the 2/1 resonance with Jupiter. We study in particular the time evolution of the orbital energy in order to evaluate the role of a repeated encounter mechanism (capture into the resonance followed by ejection) for producing the observed depletion of comets near the resonance.

As shown by Kresák (1965, 1966, 1974) and Marsden (1967), the distribution of cometary mean motions exhibits a prominent gap at the 2/1 resonance with Jupiter. The gap extends approximately from 0.158 yr^{-1} to 0.178 yr^{-1} of inverse orbital period. In Marsden's (1979) catalogue 33 cometary apparitions are found in this range: 13 by P/Pons-Winnecke, 4 by P/Kopff, 3 by P/Tempel 1, two each by P/d'Arrest and P/de Vico-Swift, and one each by P/du Toit-Neujmin-Delporte, P/Kohoutek, P/Kojima, P/West-Kohoutek-Ikemura, P/Wild 2, P/Haneda-Campos, P/Pigott, P/Wirtanen and P/Tempel-Swift.

Several mechanisms for depleting the cometary population near resonances with Jupiter have been described. In broad outline one may distinguish between two principal ideas. If a comet enters an orbit near a simple mean motion commensurability at a close encounter with Jupiter, then after a small number of revolutions there will be another encounter at which the comet is likely to be expelled from the neighbourhood of the resonance (Marsden, 1967, 1970b; Kresák, 1974). On the other hand, a comet might also enter into a temporary libration around a resonance, characterized by a large amplitude of the critical argument and the mean motion, possibly involving moderately close encounters with Jupiter. During such a libration the comet would stay mainly near the borders of the gap, passing quickly through the central part (Kresák, 1974).

At the 2/1 resonance both types of motion are known. A famous example of the first one is comet P/Lexell (Kazimirchak-Polonskaya, 1972,

W. Fricke and G. Teleki (eds.), Sun and Planetary System, 391–394.

1976), which was however remarkably far from the resonance at both perihelion passages between the encounters in 1767 and 1779. Indeed, with an inverse orbital period of 0.1786 yr^{-1} in 1770 it was slightly outside the resonance gap as defined above. Comet P/Pons-Winnecke was mentioned by Kresák (1974) as an example of the second type of motion, the libration having been discovered by Marsden (1970b). However, in the apparitions observed so far the comet has not contributed in creating the gap but instead filled it rather uniformly.

In order to establish more clearly the occurrence of the different types of motion near the 2/1 resonance, we are investigating all the above-mentioned comets. In this preliminary report we present results of orbital integrations for P/Lexell, P/Tempel-Swift, P/Wirtanen, P/Kohoutek, P/West-Kohoutek-Ikemura, P/Haneda-Campos and P/Wild 2. The latter four comets had not been discovered at the time of the investigations quoted above.

For the orbital integrations we have used a gravitational 9-body model (Mercury added to the Sun, Earth-Moon system represented by a point-mass at the barycenter, Pluto neglected). Starting conditions for the planets were normally taken from the JPL Development Ephemeris Nr. 96 (Standish Jr. et al., 1976). Cometary coordinates and velocities were calculated from the osculating orbital elements in Marsden's (1979) catalogue. The integration of the equations of motion of the N-body problem (Danby, 1962) was performed with the aid of Hamming's predictor-modifier-corrector method (Ralston, 1964) as implemented in the DHPCG program of the IBM Scientific Subroutine Package. Each integration was started with the comet situated in the 2/1 resonance gap, and it proceeded backward or forward in time until one of the following criteria was fulfilled. Either the comet had left the gap, or it had entered an extended period of orbital stability with no encounters to within 1.5 AU of Jupiter, or a further integration was judged unreliable because of the neglect of nongravitational forces and the uncertainty of the starting elements.

Figure 1 shows the time evolution of the inverse orbital period (P^{-1}) for the seven comets of the present report along with P/Pons-Winnecke. Data for three unobserved apparitions of the latter comet have been taken from the catalogue by Hasegawa (1967). For comet P/Tempel-Swift our integration proceeds forward from apparition nr. 8, and data for the unobserved apparitions nr. 4, 6 and 7 have been taken from Hasegawa's catalogue. Pre-discovery values of P^{-1} for comets P/West-Kohoutek-Ikemura and P/Kohoutek have been taken from an investigation by Rickman and Carlborg (Rickman, 1979).

Clear evidence for the mechanism of repeated encounters is seen for comets P/Lexell, P/Wirtanen, P/Kohoutek, P/West-Kohoutek-Ikemura and P/Wild 2. Two apparitions near the 2/1 resonance are surrounded on both sides by encounters with Jupiter implying considerable perturbations of P^{-1}. Encounter distances before and after the resonant motion are 0.020 AU and 0.0015 AU for P/Lexell, 0.28 AU and 0.46 AU for P/Wirtanen, 0.14

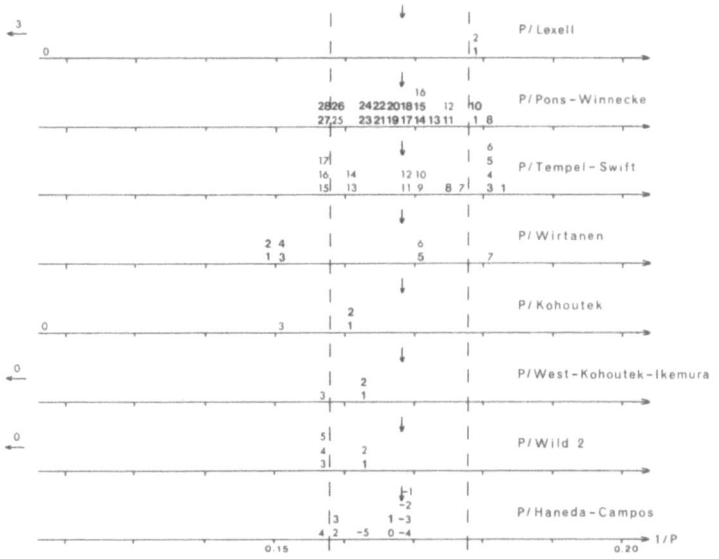

Figure 1. Evolutions of inverse orbital periods (unit: yr^{-1}, for eight comets. Osculating values at perihelion passages are marked by apparition numbers (1 = discovery). Thin numbers denote unobserved apparitions.

AU and 0.67 AU for P/Kohoutek, 0.011 AU and 0.58 AU for P/West-Kohoutek-Ikemura and 0.0057 AU and 1.01 AU for P/Wild 2.

The evolution shown for P/Pons-Winnecke corresponds to libration around the resonance, and the one for P/Tempel-Swift covers a time interval shortly before and during the start of a libration (Marsden, 1970b). These evolutions do not appear to imply an efficient clearing of the resonance gap. P/Pons-Winnecke is indeed the main contributor of observed cometary apparitions over the whole gap. P/Tempel-Swift does participate in forming the gap as far as observed apparitions are concerned, but this is due only to the fact that the faintness of the comet precludes observations except at the short-period edge of the gap (Marsden, 1970b).

Comet P/Haneda-Campos shows a remarkable evolution. Due to the uncertainty of the one-apparition orbital elements and the ignorance of nongravitational force parameters, we have only integrated the motion back to 1941. This integration shows the comet to remain within the resonance gap in spite of three encounters with Jupiter at minimum distances of 0.29 AU in 1945 and 1957, and 0.31 AU in 1969. Again the comet would have acted in filling the gap, had it only been discovered at an earlier apparition. The forward integration shows the comet to leave the central part of the gap already in 1981 at an encounter to within 0.32 AU of Jupiter. The result is interesting, since it shows the possibility of a certain orbital stability in the presence of repeated encounters with Jupiter. However, the uncertainty of the starting orbit makes it unclear whether the actual comet follows this particular evolution.

Our results for comet P/Lexell are in good agreement with those by Kazimirchak-Polonskaya (1972). For P/Tempel-Swift we find a good agreement with the data given by Marsden (1963), and for P/Wirtanen we confirm the results by Marsden (1967, 1970a). For P/Wild 2 our backward integration shows an orbit with a period of 42 yrs and a perihelion distance of 5.0 AU before the very close pre-discovery encounter in 1974. This is in good agreement with the results by Marsden (1978).

The above list of temporary visitors in the 2/1 resonance gap, spending only two apparitions there, adds to the examples given by Marsden (1967), namely: P/Lexell, P/Wolf-Harrington, P/Harrington and P/d'Arrest. Indeed, the repeated encounter mechanism appears to account very well for the formation of the gap. One must not oversimplify the model, however. A very close encounter with Jupiter leading to an orbit relatively far from the resonance may be followed after two revolutions by another very close encounter (example: P/Lexell), while an intermediate orbit very close to the resonance may lead to a relatively distant second encounter (example: P/Wild 2). This fact is of course due to the perturbations of the cometary orbits during the 12 years between the encounters, and its effect is to smear out the gap somewhat.

REFERENCES

Danby, J.M.A.: 1962, *Fundamentals of Celestial Mechanics*, McMillan Co., New York.
Hasegawa, I.: 1967, Mem. Coll. Sci. Kyoto Univ. 32, pp. 37-83.
Kazimirchak-Polonskaya, E.I.: 1972, IAU Symp. 45, pp. 373-397.
Kazimirchak-Polonskaya, E.I.: 1976, in *the Study of Comets*, NASA SP-393, pp. 490-536.
Kresák, L.: 1965, Bull. Astron. Inst. Czech. 16, pp. 292-296.
Kresák, L.: 1966, Mém. Soc. Roy. Sci. Liège, Sér. 5, 12, pp. 459-467.
Kresák, L.: 1974, IAU Coll. 22, pp. 193-203.
Marsden, B.G.: 1963, Astron. J. 68, pp. 795-801.
Marsden, B.G.: 1967, Science 155, pp. 1207-1213.
Marsden, B.G.: 1970a, Astron. J. 75, pp. 75-84.
Marsden, B.G.: 1970b, Astron. J. 75, pp. 206-217.
Marsden, B.G.: 1978, private communication.
Marsden, B.G.: 1979, *Catalogue of Cometary Orbits*, 3rd ed., IAU Central Bureau for Astronomical Telegrams, Cambridge, Massachusetts.
Ralston, A.: 1964, in *Mathematical Methods for Digital Computers*, eds. A. Ralston and H.S. Wilf, Wiley and Sons, New York, pp. 95-109.
Rickman, H.: 1979, IAU Symp. 81, pp. 293-298.
Standish Jr., E.M., Keesey, M.S.W., and Newhall, X.X.: 1976, NASA TR-32-1603.

TEMPORARY SATELLITE CAPTURES BY JUPITER FOR ORBITS RESEMBLING THE ONE OF COMET P/GEHRELS 3

H. Rickman
Astronomiska Observatoriet, Uppsala, Sweden

A.M. Malmort
Stockholms Observatorium, Saltsjöbaden, Sweden

Recent investigations (Rickman, 1979; Carusi and Valsecchi, 1979, 1981) have indicated that comets P/Gehrels 3 and P/Oterma experienced temporary captures as Jovian satellites shortly before discovery. In order to verify that these results are not jeopardized by the uncertainties of the starting orbits, extended "clouds" of such orbits should be investigated. We have done this for comet P/Gehrels 3 (Rickman and Malmort, 1981), attempting also a general examination of the conditions for satellite captures in orbits similar to the one of this comet. We require of a temporary satellite capture that the interval of gravitational binding to Jupiter lasts for more than 1000 days. Furthermore, the captured object should perform at least one revolution around Jupiter during the gravitational binding. The number of revolutions (N_r) is given by the number of minima of the distance from Jupiter ($N_m = N_r + 1$). Our dynamical model is the elliptic restricted three-body problem (Sun-Jupiter-comet), and we integrate regularized equations of motion. We proceed backward in time from a starting date in June 1974, close to the end of the gravitational binding as found by Rickman (1979) with a ten-body model. One of the four closely spaced orbits from that work is taken as the "central orbit" in the present investigation. Its evolution through the encounters with Jupiter in 1973 and 1970 is practically identical in the three- and ten-body models.

The other orbits were formed by varying the osculating orbital elements of the comet at the starting epoch. Fig. 1 shows the results of simultaneous variation of the semimajor axis (a) and eccentricity (e), and Fig. 2 shows corresponding results for the perihelion longitude ($\tilde{\omega}$) and starting mean anomaly (M_0). The number plotted for each orbit is N_m, and the symbols immediately following N_m denote: m and s – the distance from Jupiter, d(t), has a shallow minimum not counted in N_m, or the curve d(t) exhibits a saddle-point; r – one of the counted minima is comparatively remote; c and n – a collision, or near-collision, with Jupiter is found; + – further distance minima are expected after the end of our integration in December 1962.

Both diagrams exhibit very complicated patterns. Each one shows an

W. Fricke and G. Teleki (eds.), Sun and Planetary System, 395–396.

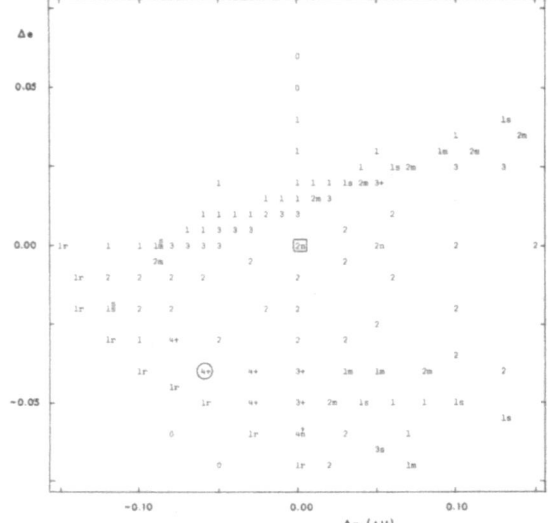

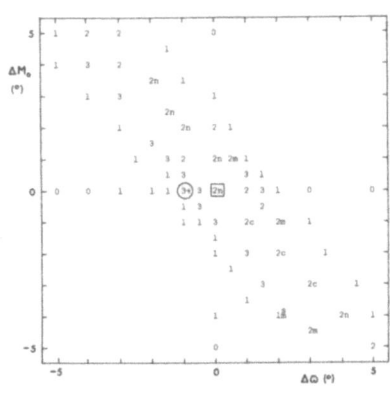

Figure 1. Results of the variations of a and e.

Figure 2. Results of the variations of $\tilde{\omega}$ and M_o.

extended region enclosing the origin (= central orbit) where at least
one revolution is performed in the satellite motion. There is an obvious
tendency for saddle-points, shallow minima and relatively stable satellite captures to occur near the borders of these regions. The elongated
structure (upper left to lower right) observed in Fig. 2 is caused by
the low eccentricity of the starting orbit. Different orbits with the
same value of $\tilde{\omega} + M_o$ are hence nearly equivalent and in general exhibit
the same number of distance minima. Two of the orbits marked by plus
signs (those encircled in Figs. 1 and 2) were integrated further backward, and the expectation of additional distance minima was thus confirmed. The orbit from Fig. 1 shows nine minima, while the one from Fig. 2
exhibits at least 22 minima over 60 years, back to August 1913 when our
extended integration was stopped.

We have found that the observational uncertainties of the cometary
orbit are far smaller than the typical variations considered in this
work. Hence it is established at a high level of confidence that comet
P/Gehrels 3 was really a temporary Jovian satellite shortly prior to
discovery, even according to our restrictive criteria.

REFERENCES

Carusi, A., and Valsecchi, G.B.: 1979, in *Asteroids*, ed. T. Gehrels,
    Univ. of Arizona Press, Tucson, pp. 391-416.
Carusi, A., and Valsecchi, G.B.: 1981, Astron. Astrophys 94, pp. 226-228.
Rickman, H.: 1979, IAU Symp. 81, pp. 293-298.
Rickman, H., and Malmort, A.M.: 1981, Astron. Astrophys. (in press).

ON THE LONG-TERM ORBITAL EVOLUTION OF COMET P/BOETHIN

D. Benest
Observatoire de Nice, France

R. Bien
Astronomisches Rechen-Institut, Heidelberg, FRG

H. Rickman
Astronomiska Observatoriet, Uppsala, Sweden

ABSTRACT

The possible effect of nongravitational forces on the motion of comet P/Boethin is investigated. Besides, the orbital period is varied. The authors find in all cases that the comet librates temporarily around the 1/1 resonance with Jupiter as a remote Jovian satellite. A time interval of 2000 years backward and forward is treated. It is found that the long-term orbital evolution of comet P/Boethin is likely to be governed by major perturbations at close encounters with Jupiter.

Comet P/Boethin (1975 I) has been found to librate around the 1/1 resonance with Jupiter during at least a few centuries, if purely gravitational motion is assumed (Benest et al., 1980). This libration has the characteristics of a remote Jovian satellite motion as described by Benest (1971, 1978). We are extending our investigation of comet P/Boethin in two ways. First, we include the action of nongravitational forces in order to determine, at least in broad outline, their influence on the librational motion. Besides, we attempt to trace the orbital evolution over a long time interval (2000 years backward and forward).

The present report gives results obtained in a four-body model (Sun-Jupiter-Saturn-comet). For the integration we used the Bulirsch and Stoer (1966) method. Masses and starting conditions for Jupiter and Saturn were taken from Schubart and Stumpff (1966). Evidently, the origin and ultimate fate of the comet can not be specified with any confidence because of the uncertainty of the cometary starting elements and the ignorance of the nongravitational force parameters. Indeed, varying these conditions we find a remarkable diversity of possible evolutionary histories.

We have integrated nine orbits for comet P/Boethin. One of these

397

(hereafter referred to as the "central orbit") is based on the elements
in Marsden's (1979) catalogue and corresponds to purely gravitational
motion. Besides, we have treated four different gravitational orbits varying the orbital period (P) at the starting epoch (1974 Dec. 19) by ΔP
= $\pm 8^d$ and $\pm 16^d$. These variations are comparable to the observational uncertainty of the osculating orbital period in 1975, stated by Marsden
(1979) to be about one week. We also treated four nongravitational orbits
with the same starting period as the central orbit. For lack of better
information about the nongravitational force on comet P/Boethin, we adopted the standard expression (Marsden, 1974) with constant values of the
parameters A_1, A_2 and A_3. For simplicity, non-zero values were used only
for the transverse component in the orbital plane (i.e. $A_2 \neq 0$). Our set
of values, A_2 = ± 0.15 and ± 0.30 (in units of 10^{-8} AU/day^2), spans the
range observed for short-period comets. Among the 89 sets of nongravita-
tional parameters listed by Marsden (1979), $|A_2|$ < 0.15 in 82 cases and
$|A_2|$ < 0.30 in 87 cases.

All our nine orbits show temporary librations around the 1/1 resonance with Jupiter enclosing the starting date. These librations have
the properties of remote Jovian satellite motion, and the minimum duration is about 180 years, slightly less than one complete libration period. Figures 1 and 2 show the evolutions of the semimajor axis (a) of
the cometary orbit in a few typical cases.

In all our nongravitational orbits as well as the gravitational
ones with $\Delta P \leq 0$ the libration started around 1830 in connection with
encounters with Jupiter to within about 0.6 AU around December 1829 and
with Saturn to within 0.7 AU in 1836-37. In the two orbits with ΔP > 0
the approach distances at these encounters are considerably larger, and
the libration continues further backward in time. It should be noticed,
however, that comet P/Boethin may sometimes pass moderately close to Jupiter without the libration being terminated. Such encounters occur regularly at the extreme values of $\lambda - \lambda_J$ (λ = the cometary mean longitude; λ_J = Jupiter's mean longitude), when a passes the value of Jupiter's
semimajor axis. The most recent example occurred in 1909 with an approach distance of about 0.5 AU.

Comet P/Boethin is at present near the short-period border of the
stability band for remote Jovian satellite motion (Benest, 1971). Decreasing the orbital period (ΔP < 0) one withdraws from the band, and
increasing the period (ΔP > 0) one enters into it. Indeed we find the
shortest interval of libration for the orbits with ΔP < 0. The end of the
present libration interval comes already in 2007 for ΔP = -16^d and in
2019 for ΔP = -8^d in connection with encounters with Jupiter to within
0.5 AU and 0.6 AU, respectively. The orbits with ΔP > 0 show very long-
lasting librations intermitted by short periods of circulation (a few
centuries). More definitive departures from the 1/1 commensurability are
found before 560 for ΔP = 16^d and after 2530 for ΔP = 8^d. In particular,
the orbit with ΔP = 16^d appears to librate around the 1/2 resonance before 280 and then around the 2/3 resonance for slightly less than one
period before in 562 it enters the 1/1 commensurability. Close encounters

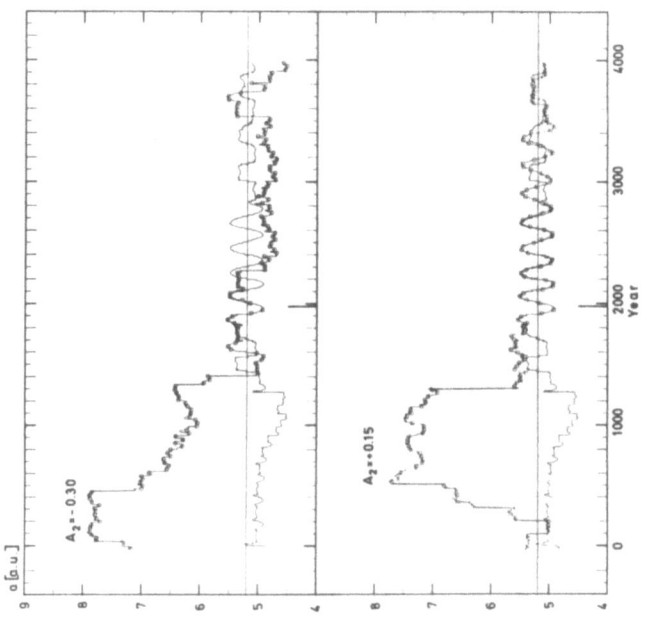

Figure 2. Evolutions of the semimajor axis are shown for two values of A_2, each together with the central orbit (full-drawn curve without symbols). The major mark on the time axis indicates the starting epoch.

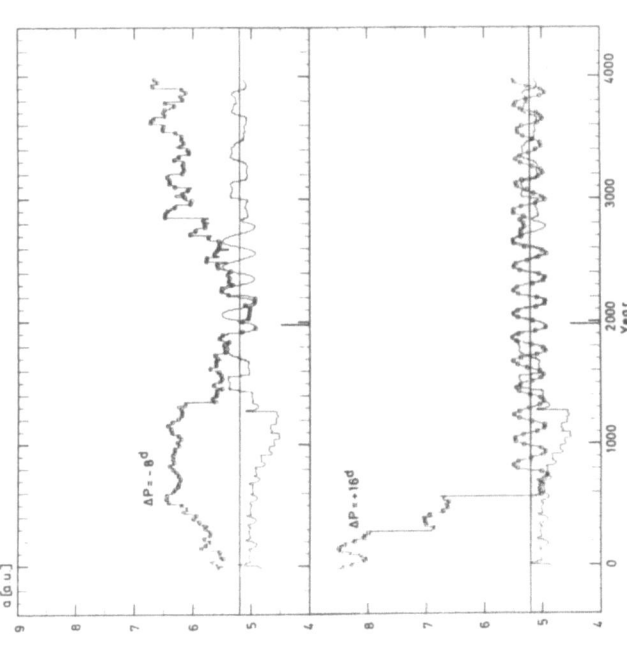

Figure 1. Evolutions of the semimajor axis are shown for two values of ΔP, each together with the central orbit (full-drawn curve without symbols). The major mark on the time axis indicates the starting epoch.

with Jupiter are found in 278 and 562.

Generally speaking, the long-term orbital evolution of comet P/Boe-
thin appears to be governed by close encounters with Jupiter. None of
our investigated orbits shows any major perturbation by Saturn in spite
of the fact that they all start (in 1974) as Saturn-tangent orbits and
are hence relatively sensitive to Saturnian perturbations (see Carusi
and Valsecchi, 1979; Froeschlé and Rickman, 1981). Likewise, the nongravitational forces seem to influence the orbital evolution only indirect-
ly by changing the encounter geometries at approaches to Jupiter. In general our forward integrations show the orbits to remain relatively close to the 1/1 resonance during 2000 years, while the backward integrations reveal a more drastic divergence. No evolutions to long-period orbits have been found, however. The largest deviation from the 1/1 resonance is the one shown in Fig. 1 for $\Delta P = 16^d$.

It must be emphasized that the rapid divergence of the varied orbits invalidates any detailed conclusions about the long-term evolution
of comet P/Boethin on the basis of present knowledge. Our results only
indicate in a statistical sense the most likely types of evolution.

REFERENCES

Benest, D.: 1971, Astron. Astrophys. 13, pp. 157-160.
Benest, D.: 1978, Thesis, Université de Nice.
Benest, D., Bien, R., and Rickman, H.: 1980, Astron. Astrophys. 84,
    pp. L11-L12.
Bulirsch, R., and Stoer, J.: 1966, Numer. Math. 8, pp. 1-13.
Carusi, A., and Valsecchi, G.B.: 1979, the Moon and the Planets 22,
    pp. 113-124.
Froeschlé, C., and Rickman, H.: 1981, Icarus 46, pp. 400-414.
Marsden, B.G.: 1974, Ann. Rev. Astron. Astrophys. 12, pp. 1-21.
Marsden, B.G.: 1979, *Catalogue of Cometary Orbits*, 3rd ed., IAU Central
    Bureau for Astronomical Telegrams, Cambridge, Massachusetts.
Schubart, J., and Stumpff, P.: 1966, Veröff. Astron. Rechen-Inst.
    Heidelberg No. 18.

ON THE DISPLACEMENT OF METEOR SHOWERS ACTIVITY

P.B. Babadzhanov, Yu.V. Obrubov
Astrophysical Institute, Dushanbe, USSR

The influence of planetary perturbations on the orbits of meteor streams causes the appearance of new meteor showers and disappearance of the known ones (Babadzhanov, Obrubov, 1980).

Now we consider another consequence of planetary perturbations: the regular displacement of dates of the maximum activity of meteor showers, which is clearly seen for some of them during comparatively short intervals of time, and negligible for others, and contrary to the accepted opinion (Adams, 1867; Plavec, 1950) does not always correspond to the secular variation of the longitude of ascending node Ω of the stream orbit.

The tables 1 and 2 present the elements of meteor streams orbits (Cook, 1973) and their secular variations for the last 100 yrs (for Leonids, Abelman's (1898) results are given); ΔT_{obs}-the observed displacement of maximum activity.

Table 1. The orbital elements of meteor streams

Shower	a	e	q	i	Ω	ω
Leonids	10.3	0.905	0.977	162.6	234.5	170.9
Quadrantids	3.08	0.683	0.977	72.5	282.7	170.0
Bielids	3.52	0.756	0.861	12.6	247.3	223.2
Geminids	1.36	0.896	0.142	23.6	261.0	324.3
η-Aquarids	13.0	0.958	0.560	163.5	42.4	95.2
Orionids	15.1	0.962	0.571	163.9	28.0	82.5

Table 2. The secular variations of orbital elements and dates of maximum activities

Shower	Δe	Δi	$\Delta \omega$	$\Delta \Omega$	ΔT_{obs}	ΔT_{cal}
Leonids	0.0002	-0.036	1.57	1.39	1.45	1.39
Quadrantids	-0.024	0.53	1.80	- 0.37	- 0.36	- 0.37
Bielids	0.008	-2.89	16.6	-17.9	-17.9	-17.9
Geminids	-0.001	0.77	1.58	- 1.62	0	- 0.04
η-Aquarids	0.00008	-0.05	1.64	1.94	0.16	0.18
Orionids	0.00009	0.07	1.29	1.54	0.59	0.41

W. Fricke and G. Teleki (eds.), Sun and Planetary System, 401–402.
Copyright © 1982 by D. Reidel Publishing Company.

402

P. B. BABADZHANOV AND YU. V. OBRUBOV

It is evident that $\Delta T_{obs}=\Delta\Omega$ only in the case when the value of the radius-vector at the node in which the stream intersects the Earth's orbit is constant:

$$d/dt \left[a\,(1-e^2)/(1\pm e\cdot\cos\omega)\right]= 0 \tag{1}$$

Here and later the top sign corresponds to the ascending node, the bottom sign (-) to the descending one.
Taking into account only the secular perturbations of the first order we have from (1):

$$(2ae\pm\cos\omega)\,de = \pm(1-e^2)e\cdot\sin\omega d\omega. \tag{2}$$

In the given case the following correlations are satisfied (Lidov, 1961):

$$(1-e^2)\cos^2 i=C_1=\text{const.}, \quad (0.4-\sin^2\omega\sin^2 i)=C_2=\text{const.} \tag{3}$$

It follows, that if $\omega=0^o$ (or 180°), then differential $de=0$, the condition (2) is satisfied and the displacement of the maximum activity is equal to $\Delta\Omega$. The Leonid and Quadrantid showers satisfy the condition (2). The Bielids have the greatest $\Delta T_{obs}=18^o/100$ yrs which is in good accordance with $\Delta\Omega$ of Biela comet's orbit. But here $|\sin\omega|\gg 0$ and it follows that the condition (2) is not only satisfied in the cases $\omega=0^o$ (or 180°). The changed form of (2) can be written by using the differential form of (3):

$$(2ae\pm\cos\omega)e\cdot\cos\omega\,\sin^2 i = \pm(C_2-e^4(0.4-\sin^2\omega)). \tag{4}$$

The condition (4) is satisfied for Bielids and $\Delta T_{obs}=\Delta\Omega$.
Now let us consider the case when the orbit is slightly inclined to the ecliptic plane and the conditions (2) and (4) are not satisfied. In the case $i=0^o$ (or 180°) the orientation of the orbit is determined only by the ecliptic longitude of orbit perihelion $L=\Omega+\text{arctg}\,(tg\omega\cos i)$. The variation of the orbit's form also leads to the displacement of the intersection point of orbits of the stream and the Earth and displacement of the dates of maxima are determined by the variation of the perihelion longitude and eccentricity:

$$\Delta T_{cal}=\Delta\Omega+\Delta\omega\,\text{sign}(\cos i) \pm (2ae\pm\cos\omega)\Delta e/e\cdot\sin\omega. \tag{5}$$

The calculated values of ΔT_{cal} for Geminids, η-Aquarids and Orionids (Table 2) are in satisfactory accordance with the observations.

Abelman, I.S.: 1898, Izv. Russk. Astr. Ob-va, 7, 1-3, p. 35.
Adams, J.C.: 1867, Month. Not. Roy. Astron. Soc., 27, 247.
Babadzhanov, P.B., Obrubov Yu.V.: 1980, "Solid Particles in the Solar System", eds. I. Halliday, B.A. McIntosh,D. Reidel Publ. Comp. Dordrecht-Holland, p. 157.
Cook, A.F.: 1973, NASA SP-319, p. 183.
Lidov, M.L.: 1961, Iskusst. Sputniki Zemli, 8, p. 5.
Plavec, M.: 1950, Nature, 165, 4192, p. 362.

SOME CHARACTERISTICS OF THE ASTEROID BELT STRUCTURE

Mike Kuzmanoski
Institute of Astronomy, Belgrade, Yugoslavia

Proceeding from the vectors normal at the orbit planes of the numbered minor planets, listed in "EMP for 1979" we determined the mean plane of the asteroid orbits (Kuzmanoski, 1981). Its position relative to the ecliptic plane is defined by the node longitude $\Omega_s = 78°.141$ and the inclination $i_s = 0°.594$. All further investigations have been carried out with reference to the mean plane.

By computing the angles between the projections of the perihelior directions of the minor planets on the mean plane and the cross-section straight line of the mean plane and the ecliptic plane, as well as by grouping them into classes by 10^0, we obtained a distribution analogous to that of the perihelion longitudes. Due to the small inclination of the mean plane relative to the ecliptic plane the distribution, found in this way, did not bring out any novel, more essential, characteristics in comparison with these established earlier by similar investigations (Kresak, 1967, Popović, 1973, Chebotarev, 1976). In order to provide a more complete insight into the distribution of perihelion points them-selves we included in our study also the inclinations of the perihelion directions to the mean plane and the perihelion distances. By grouping the inclinations of the perihelion directions into classes by 1^0 we obtained the distribution illustrated in Figure 1, which looks, at first glance, like a normal one. However, it proved that the expectation of a normal distribution can be rejected with high probability. Of the total of 2042, the perihelia of 965 asteroids are below the mean plane and those of another 1077 are above it. The perihelion directions of almost half the asteroids make angles of between -4^0 to $+4^0$, distributed in such a way that there is an equal number of them above and below the mean plane. By taking into consideration the previous distributions and including in our inquiry the perihelion distances themselves, one is enabled to find out the spatial distribution of the perihelion points of the minor planets. By plotting the perihelion points for different inclinations one realizes that the perihelia are distributed within a ring whose inner boundary (the one towards the Sun) is more sharply defined than the outer one. The inner boundary of this ring is characterized by the clearly noticeable asymmetry. The inner boundary in broader

W. Fricke and G. Teleki (eds.), Sun and Planetary System, 403–404.

terms, arround the longitude 270 , is at a distance of about 1.7 AU from
the Sun, while the one at the diametrally opposite side is about 2.0 AU.

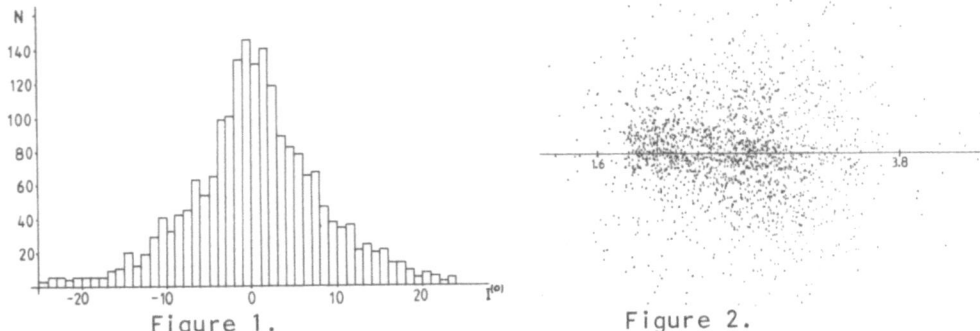

Figure 1.   Figure 2.

In continuation of investigations made by Kresák (1967, 1978) we searched
out the ring's shape by means of the points of intersection of the
asteroid orbits and the planes perpendicular to the mean plane. The intersection points of the orbits and the perpendicular plane containing the
Jovian perihelion direction, are illustrated in Figure 2. By rotating
the normal plane by 60^0, we obtained the cross-sections of the belt in
its different zones. Some global characteristic of the belt can thereby
be studied, alike concerning its density and grouping of these intersection points and the shape in different directions. By means of the
coordinates of the points of intersection we determined the radius-vectors
of the middle of the belt (whose coordinates are, in fact, arithmetical
means of the coordinates of these points intersection) for different
positions of the perpendicular plane. In rotating the perpendicular plane
by 10^0, starting from the cross-section straight line of the mean and
ecliptical planes, we obtained 36 points. It proved that all the points
obtained in this way are below the mean plane, yet their distances from
it are very small (0.001 to 0.004 AU). In view of such small deviations
the middle of the belt was assumed to lie in the mean plane. Accordingly,
the middle line of the belt, resulting from these 36 points, is
represented by an ellipse, whose dimensions and the orientation are as
follows:

$$a = 2.709 \qquad e = 0.050 \qquad \omega = 287°012$$

REFERENCES:

Chebotarev,A. and Shor,A.: 1976, Trudy ITA AN SSSR 15, pp. 60-90.
Kresák,L.: 1967, Bull. Astron. Inst. Chechoslovakia, 18, 27-36.
Kresák,L.: 1978, Three-dimensional distributions of the potential sources
          of interplanetary dust, COSPAR III.C.2.4.
Kuzmanoski,M.: 1981, Publ. Dept. Astron. Univ. Belgrade, 10, in press.
Popović,B and Elizar,S.: 1973, Matematički vesnik, Belgrade, 10 (25)
          pp. 249-254.

RESONANCES IN THE MOTIONS OF MINOR PLANETS AND THEIR USE FOR
THE DETERMINATION OF MASSES

H. Scholl
Astronomisches Rechen-Institut, Heidelberg

ABSTRACT

   The secular behavior of resonant orbits is described on the basis
of Greenberg's and Schubart's theories. We review the results of numerical investigations obtained by Froeschlé and Scholl and the results obtained recently by Wisdom, who found a mapping for resonant motion. In
addition we review the possibilities for mass determinations of planets
and of minor planets using resonant motion.

INTRODUCTION

   The minor planets span a belt between the orbits of Mars and
Jupiter. The main feature of their dynamics can be demonstrated by plotting the frequency distribution of orbital energies or equivalently of
their mean motions (Fig.1). The narrow gaps in the main part, the
Kirkwood gaps, and the isolated small groups in the outer part of the
belt are related to orbital resonances with Jupiter. An orbit is in resonance with Jupiter if a conjunction Sun - minor planet - Jupiter re-
peats at the same location in a fixed reference system within a few orbital periods. Jupiter's and the minor planet's mean motions, n_J and n,
yield a ratio of small integer numbers: $n/n_J = (p+q)/p$.

   In such a case, Jupiter for instance can always add orbital angular
momentum over 10 to 100 orbital periods before Jupiter starts to take
angular momentum out of the minor planet's orbit. This long lasting
transfer in one direction implies a perturbation enhancement.

   It is therefore tempting to explain the Kirkwood gaps only by especially strong Jovian perturbations. The difficulty, however, is to
explain with the same theory the isolated groups in the outer part of the
belt which cluster around resonances. The recent attempts in order to
understand the dynamics of resonant motion are reviewed in the following
two chapters. Besides the theoretical challenge of understanding resonant motion, which is related to the cosmogony of the solar system,

405

W. Fricke and G. Teleki (eds.), Sun and Planetary System, 405–410.
Copyright © 1982 by D. Reidel Publishing Company.

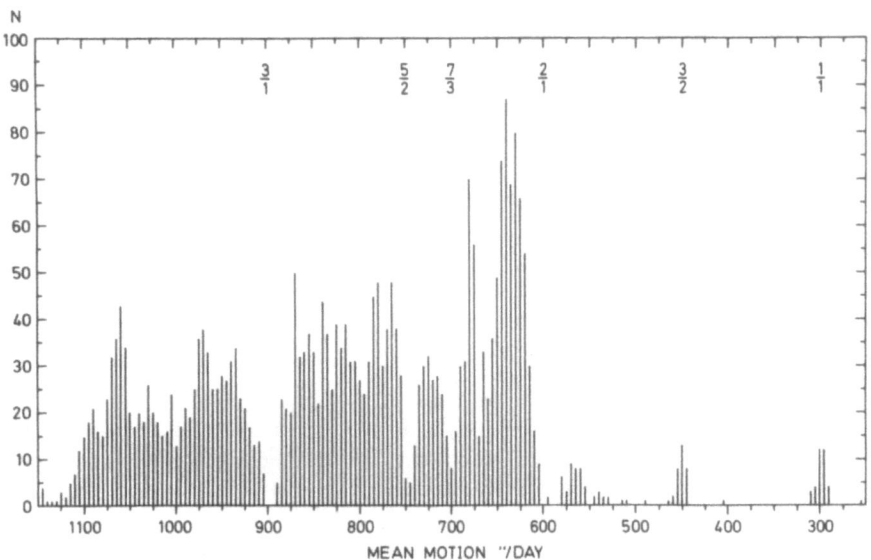

Fig.1 Histogram of minor planet distribution with mean
       motion, 1979.

resonant orbits offer a very practical astronomical application, namely
the determination of planetary and asteroidal masses. The efforts and
prospects of mass determinations are reviewed in the last chapter.

Greenberg (1) developed a secular perturbation theory for low eccentric, planar orbits at the 2/1 resonance. He used a perturbation
function R which we generalized to the form:

$$\overline{R} = G\, m_J\ [F_0(\alpha) + F_1(\alpha)\ e^2 + F_2(\alpha)\ ee_J \cos(\tilde{\omega} - \tilde{\omega}_J) + F_3(\alpha)\ e^q \cos\sigma]/a_J$$

where G is the gravitational constant, m_J is Jupiter's mass; $\alpha = a/a_J$;
$q\cdot\sigma = -q \cdot \tilde{\omega} - \ell + \ell_J(p+q)$; $n/n_J = (p+q)/p$. The functions F are of order of
unity.

For the case q = 1 and any p we find the simple solutions:

$$e \sin \sigma = C \sin(At + \delta)$$
$$e \cos \sigma = C \cos(At + \delta) + m_J\, n\, \alpha\, F_3/A$$

(1)

where amplitude C and phase δ are constants of integration.
$A = (p+1)n_J - pn - d\ell_0/dt$.

Using this solution we can describe the behavior of small eccentric
orbits in an e-σ space (Fig.2). The resulting orbit is formally similar
to a non-resonant orbit resulting from the classical secular theory
which is described in an e - $\tilde{\omega}$ space. In the classical theory, the eccentricity is decomposed in a proper and forced term by vector addition.

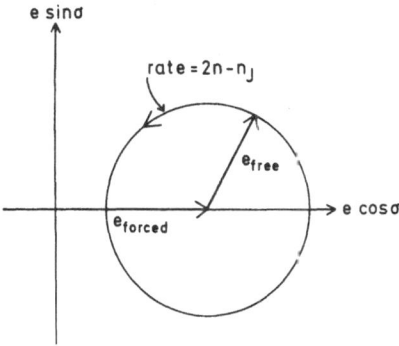

Fig.2   Trajectory in e - σ space
corresponding to the
solution in Equation (1).

In the resonant case, the proper eccentricity can be defined as in
the non-resonant case by the integration constant C. From Fig.2 and
equations 1 one can easily derive the properties for low eccentric
resonant motion (1).

A more general theory without restrictions with respect to eccentricities and order of resonances was developed by Schubart (2) who followed an idea of Poincaré. Schubart modified Poincaré's canonical set
of differential equations with a time dependent Hamiltonian H.
Schubart's idea then was to average the Hamiltonian H over a resonance
period. He replaced the averaged, time independent Hamiltonian $\bar{H}$ in the
canonical equations which became thus for the planar circular case completely integrable. Fig.3 shows the topology for resonant motion in a
S - σ space, where $S = \sqrt{a/a_J}\,(1 - \sqrt{1 - e^2})$. S and σ are conjugate variables in the canonical system. Schubart's topology contains fully
Greenberg's topology. The advantage of Greenberg's theory is to have an
analytical solution for the equations. In Schubart's theory, a solution
is given in form of an implicit equation.

An orbit in Schubart's topology (Fig.3) appears as a closed curve,
circular - like or banana shaped. Using Schubart's topology one can
easily predict the behavior of orbits even in the three-dimensional elliptic problem. In the planar circular problem, an orbit is either a
σ-librator, an inner or outer circulator. The regions are separated by
the dark 'separatrix'. In the more general case, an orbit is no longer
represented by a closed curve. It can even cross the 'separatrix' and
change its behavior. Librators can became circulators and vice versa.

NUMERICAL STUDIES OF RESONANT MOTION

In order to investigate the secular behavior of resonant motion
quantitatively, Schubart's set of differential equations was solved
numerically over time intervals up to 10^5 years at the 3/1, 7/3, 5/2,
2/1, 3/2, and 1/1 resonances (1). Unstable orbits were only found when

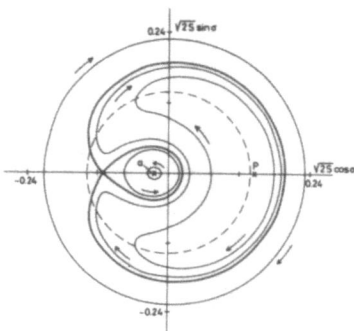

Fig.3   Trajectories in σ versus $\sqrt{2S}$ space from Schubart's
        averaged circular model at the 2/1 commensurability
        for K = 0.802. The arrows indicate directions of motion
        in this space. The darker lines correspond to critical
        bifurcation trajectories. Paths immediately around
        point a are apocentric librators; those about p are
        pericentric librators. The dashed circle corresponds
        to the exact center of the resonance.

the starting values yielded an orbit which implied an immediate close
approach to Jupiter. None of the initially stable orbits became unstable in the sense that they left the resonance zone. In addition,
applying the surface-of-section method and calculating Liapunov's maximal
characteristic number, Froeschlé and Scholl (3) demonstrated the stability of resonant motion in the planar case.

     Recently, Wisdom (4) investigated resonant motion at the 3/1 re-
sonance by a new method. Wisdom succeeded to find a mapping for resonant motion, which enabled him to simulate orbits over some 10^6
years. A mapping consists of an algebraic formula which uses as input
data the orbital elements of a minor planet at time T_1. The formula
yields orbital elements after one siderial revolution of the minor
planet at time T_2, for instance. The advantage of a mapping versus
numerical integration is obvious. In a numerical integration one has to
calculate the orbital elements between T_1 and T_2 at several steps by
time consuming arithmetic, while a mapping needs only one step. According to Wisdom's results, orbits which start as librators and which
remain librators over 10^5 years may cross the separatrix of Schubart's
topology after 10^6 years.

     None of his orbits became unstable in the sense of leaving the re-
sonance zone. His results, however, indicate the possibility that any
orbit enters a region which causes such a strong increase in eccentri-
ty that the orbit becomes a Mars crosser. Then, Mars might eliminate
the minor planet. Further investigations and applications of Wisdom's
mapping might help to solve the riddle of the asteroidal energy

distribution (Fig.1).

THE USE OF RESONANT MOTION FOR MASS DETERMINATIONS

According to the preceding chapters, resonant orbits are especially strongly perturbed. It is therefore natural to investigate the possibility to determine the mass of the perturbing body by an analysis of
the perturbed resonant orbit. Perturbations in the orbital elements of
resonant and non-resonant orbits depend linearly on the mass of the perturbing body. However for a resonant orbit, perturbations act in the
same direction over much longer time intervals as compared to non-resonant orbits. Consequently, we can expect more accurate mass values
from resonant minor planets than from non-resonant ones which have about
the same distance to the perturbing body during conjunction.

In the past, the masses of Jupiter and Saturn were determined by an
analysis of resonant motion at the 5/2, 2/1, 3/2 and 4/3 resonances in
the case of Jupiter and at the 5/2 resonance (Trojans) in the case of
Saturn. Jupiter's mass derived from the Pioneer and Voyager spacecrafts
did not yield a significantly differing mass as was already obtained by
Newcomb in the last century from the motion of 33 Polyhymnia. Polyhymnia is close to the 5/2 resonance with Jupiter. Resonant minor planets
at the 2/1, 3/2 and 4/3 resonance which are much closer to Jupiter's
orbit than the 5/2 resonance did not yield a value superior to Newcomb's
value. This is not too surprising when we look at the results of our
numerical investigations. High eccentric orbits at the 5/2 resonance
like Polyhymnia's (e = 0.34) orbit increase for instance their semimajor
axes stronger and over much longer periods than comparable orbits at the
2/1 and 3/2 resonance.

Though there is no immediate interest at present for the determination of planetary masses from resonant minor planets, there is a high
interest in new masses of minor planets. The masses are needed in order
to derive densities for the different types of minor planets. The first
mass of a minor planet, 4 Vesta, was determined by Hertz (5). He used
197 Arete as the perturbed body which is in a 4 : 5 resonance with Vesta.
These two minor planets approach each other at least within 0.03 AU
every 18 years at a relative speed of 2.1 km/sec. on the average. Later,
Schubart improved Vesta's mass and determined new masses of Ceres and
Pallas which are close to a 1 : 1 resonance. Between 1802 and 1848, Ceres
and Pallas approached each other within 0.3 AU, however at a high relative velocity of more than 5 km/sec. Since this period their mutual
minimum distance and relative velocity increased to 1.2 AU and 16 km/sec.
respectively. The known minor planet masses are (6) in 10^{-10} solar units:

1	Ceres	5.9 ± 0.3
2	Pallas	1.08 ± 0.22
4	Vesta	1.38 ± 0.12

In a systematic search for close encounters between the largest minor planets with diameters $\geq$ 200 km and the other numbered minor planets, I found more resonant pairs which approached each other closely since their discovery. The pairs 1-1054, 4-1063, 24-656, 45-20, 423-1364 are some examples. The complete list will be published elsewhere. In the future, it appears that the pair 20 and 44 which is in a 1 : 1 resonance is observable by the astrometric satellite HIPPARCOS between 1987 and 1990. During this whole period, the two minor planets stay within 0.2 AU and have a minimum distance of 0.04 AU with a relative velocity of only 1.7 km/sec.

REFERENCES

(1) Greenberg, R. and Scholl, H.: 1979, Resonances in the Asteroidal
     Belt; in 'Asteroids', Ed. T. Gehrels, The University of Arizona
     Press, Tucson, Arizona, pp. 310-333.
(2) Schubart, J.: 1978, New Results on the commensurability cases of
     the Problem Sun-Jupiter-Asteroid; in 'Dynamics of Planets of
     Satellites and Theories of their Motion', Ed. V. Szebehely,
     Dordrecht, Holland, pp. 137-143.
(3) Froeschlé, C. and Scholl, H.: 1981, The Stochasticity of Peculiar
     Orbits in the 2/1 Kirkwood Gap; Astron. Astrophys. 93, 62-66.
(4) Wisdom, J.L.: 1981, The Origin of the Kirkwood Gaps: A Mapping for
     Asteroidal Motion near the 3/1 Commensurability; Thesis, California
     Institute of Technology, Pasadena.
(5) Hertz, H.G.: 1968, Mass of Vesta; Science 160, 299-300.
(6) Schubart, J. and Matson, L.: 1979, Mass and Density; in 'Asteroids',
     see Ref.(1), pp. 84-97.

NOTE ADDED IN PROOF

    After my presentation of this paper at the ERMA meeting in Dubrovnik, Dr. Kiang conjectured that my reason for not mentioning his work on resonant motion was motivated by the fact that his results contradict my results. I explained him that one of the reasons for not mentioning his work was the following: His treatment of Hill's equations for resonant motion, which he calls the Enabling Principle, is mathematically incorrect since he drops important terms. He also cannot give any physical justification for dropping these terms. Dr. Kiang justifies the application of the Enabling Principle, which implies wrong mathematics and which has no relation to physics, only by the statement: 'I (Dr.Kiang) take the stand that the Enabling Principle is an unproven assumption and must be justified by the results deduced from it'.

    Such a principle can, however, yield only meaningless results.

STABILITY OF REAL HECUBA AND HILDA ASTEROIDS

T. Kiang
Dunsink Observatory, Castleknock, Ireland

In a recent paper (*Vistas in Astronomy* 24 (1980) 17-37, 'Paper I' here-after), I presented a method of determining the orbital stability of asteroids in near-resonance with Jupiter, together with some results of applying the method to two sets of fictitious asteroids around the 2/1 (Hecuba) and 3/2 (Hilda) resonances. Here I shall report on the results on some real asteroids in the same two regions and some practical experience gained in applying the method.

It is convenient to express all lengths in units of a'=5.202803 AU, all times in T_0=1.888834 yr. Then the mean motion of Jupiter is n' = $\sqrt{(1+m')}$, and the mass of Jupiter in solar units was taken to be 1/1047.39. The computation consists of 3 parts. 1) From the listed elements of the asteroid, a, e, M, ω, Ω, the mean longitude of Jupiter at the same epoch λ', and the appropriate value of the resonance parameter α (1 for Hecuba, 2 for Hilda), we calculate in turn, $L=\sqrt{a}$, $G=L\sqrt{(1-e^2)}$, $U=(1+\alpha)L-\alpha G$, $S=L-G$, $\sigma=(1+\alpha)(\lambda'-\omega-\Omega)-\alpha M$. In the well-known theory, U is a constant of motion, and S and σ are a pair of canonical variables governed by the Hamiltonian $F = F(S,\sigma;U) = 1/2 L^2 + n'G + \bar{R}$, where $\bar{R}$ is the 'averaged perturbing function' averaged over one synodic period and containing the small quantity m' as a factor. The solutions are closed curves F = const. in the (S,σ)- or (e,σ)-plane. Depending on the behaviour of σ, we may distinguish two types of solutions, the *librators* and the *circulators*, the former keep on crossing and re-crossing the resonance, while the latter remain always on one side. 2) We integrate numerically the canonical equations $\dot{S}=F_\sigma$, $\dot{\sigma}=-F_S$. Since the solution curves are symmetrical about $\sigma=0°$, we need only do this for half the 'orbit', beginning say at $\sigma=0°$ (or 180°) and $e=e(\min)$ and ending at $e(\max)$. The idea of Paper I is that, at the end of each integration step, we evaluate the Hessian of F, $K = K(S,\sigma) = F_{SS}F_{\sigma\sigma}-F_{S\sigma}F_{\sigma S}$. It was found that the second partial derivative with respect to S had to be evaluated 'numerically' after calculating F at 5 adjacent values of S and this meant a big increase in computer time. 3) Let the solution curve have period P_0. After completing the integration, we calculate $f=\Sigma f_n(P_0)$, where

$$f_n(t) = \int_0^t ds \int_0^s f_{n-1}(\tau).K\{S(\tau),\sigma(\tau)\} \, d\tau \, , \, (f_0(t) = 1).$$ We keep on adding

W. Fricke and G. Teleki (eds.), Sun and Planetary System 411–412.

α	CLASS	ASTEROID(NUMBER)	U^2(AU)	e_{min}	P_0(yr)	step-size	f	P_*
2	Lib	Wild(1941)	4.644	.2572	254.	$T_0/4$	-0.000 11	1.1×10^5
2	Lib	Marsden(1877)	4.315	.1998	266.	$T_0/8$	-0.000 035:	2.0×10^5:
2	Lib	Bononia(361)	4.311	.1970	262.	$T_0/4$	-0.000 12:	1.1×10^5:
2	Lib	Schubart(1911)	4.203	.1618	265.	$T_0/8$	-0.000 111	1.12×10^5
2	Lib	Hilda(153)	4.145	.1410	268.	$T_0/2$	-0.000 066:	1.5×10^5:
2	Lib	Vogtia(1439)	4.095	.1130	268.	$T_0/2$	-0.000 034:	2.0×10^5:
2	i.Cir	Rollandia(1269)	3.989	.0900	111.	$T_0/2$	-1.34	6.19×10^2
2	i.Cir	Normania(1256)	3.953	.0076	199.	$T_0/2$	+1.77	—
2	i.Cir	Chicago(334)	3.901	.0176	121.	$T_0/2$	-1.17	5.43×10^2
1	o.Cir	Henrietta(225)	3.645	.2561	139.	$T_0/2$	+3.90	—
1	o.Cir	Freia(76)	3.507	.1643	114.	$T_0/2$	+0.559	—
1	Lib	Zulu(1922)	4.082	.4495	359.	$T_0/8$	-0.000 21:	3.5×10^5:
1	Lib	Pala(1921)	3.836	.3625	303.	$T_0/8$	+0.001 3	—
1	Lib	Griqua(1362)	3.693	.3162	357.	$T_0/4$	+0.000 93:	—
1	i.Cir	Moravia(1901)	3.264	.0588	272.	$T_0/2$	-1.982	8.995×10^3
1	i.Cir	Hecuba(108)	3.262	.0539	234.	$T_0/2$	-1.30	1.16×10^3

more terms until f ceases to change by more than the pre-assigned amount of 0.1%. For the librators, this often required 16 or 17 terms as initial fluctuations of about 10^2 settled down to final values of 10^{-4}. If the final f is between 0 and -2, then the asteroid is stable and has a hyper-period $P_* = 2\pi P_0 / \arctan\{\sqrt{[1 - (1+f)^2]}/|1+f|\}$, otherwise it is unstable.

The results on 9 Hilda librators and inner circulators and 7 Hecuba librators, inner and outer circulators are given in the Table above. For the librators, f (always small) was found to depend critically on how finely the integration was done. Take Schubart (1911) as example. For successively halved step-sizes of T_0, $T_0/2$, $T_0/4$, $T_0/8$, we got respectively +153, -54, -101, -111 in units of the 6th decimal, thus changing from instability to stability. In the Table, the last (smallest) step size used in each case and the resulting f are given, the latter to 4 figures if it did not differ from its predecessor by more than 1%; to 3, if the difference was between 1 and 10%; to 2, if between 10 and 20%, and followed by a colon, if more than 20%. It may be seen that even with a stepsize as small as $T_0/8$ (requiring a whole night run on a Nova 2/10 Computer), a convergent value of f cannot be said to have been attained in many cases. Nevertheless, the indications are that the Hilda librators are all stable with hyper-periods on the order of 10^5 yr. More computation is being planned and further discussion has to be given elsewhere.

Addendum : The method employed in this paper is based on the use of an unconventional principle about which there is controversy .

FIRST-ORDER THEORY OF CERES, PALLAS, JUNO, AND VESTA.

Wolfgang Höppner
German Hydrographic Institute,
Bernhard-Nocht-Str. 78, D-2000 Hamburg 4

In the last years considerable efforts have been made to construct analytical theories of the major planets and the moon. But I do not know of any new attempt to construct a theory of an asteroid since Duncombe's numerical integration (1969) of the orbits of Ceres, Pallas, Juno, and Vesta.

Therefore I thought it worth-while to compute new first-order theories of these minor planets. For computing the periodical and secular perturbations I have used Hansen's method (1857) in the form used by Hill (1890) and Clemence (1949; see also Brouwer and Clemence, 1961), with further modifications. The starting orbital elements for these theories (at epoch 1950.0) have been taken from polynomial fits to osculating elements derived from Duncombe's integration (1969); this has been extended to cover the time span from 1900 to 2000. The elements of the perturbing planets have been taken from Simon and Bretagnon (1975; Mercury to Neptune) and Seidelmann & al. (1974; Pluto).

The results of the first-order theories cannot be presented here in detail (listings of the perturbation terms may be obtained upon request). They may be summarised as follows: The amplitude of the largest perturbations in the mean anomaly of Ceres and Vesta is of the order of magnitude of 500"; the perturbations of Juno reach 900". The theories of these asteroids show no prominent small divisors.

The case of Pallas is quite different. The theory not only shows the famous Jupiter term with argument $18 \, g(\text{Jup.}) - 7 \, g(\text{Pallas})$, with a coefficient of nearly 25°, but also a Venus term with argument $15 \, g(\text{Pallas}) - 2 \, g(\text{Venus})$ and a coefficient of 6".

The orbital elements at epoch of the theories of Ceres, Juno, and Vesta have been adjusted to Duncombe's integration

W. Fricke and G. Teleki (eds.), Sun and Planetary System, 413–414.

(1969), the rectangular coordinates being taken as "observed positions". The adjusted theories represent the integrated positions within 0.0007 a.u., or 50" in heliocentric spherical coordinates, during the 20th century. The adjusting process has not yet converged in the case of Pallas. But it seems that, after an adjustment, the elements of Pallas will be such that the Jupiter term cited in the last paragraph will be reduced to about 5° (which means that it will still be considerably larger than the perturbations of the other three asteroids), while the Venus term will be enlarged to about 120".

I plan to extend these theories, at least those of Ceres, Juno, and Vesta, to the second order using a series manipulator. Under consideration are the manipulators SAP by Richardson (1980) and NOMAN by Van Flandern and Pulkkinen (1981). Both manipulators being developed on IBM machines I am testing at the moment which of them will be better suited for the CDC system I am using now.

When checking the formulae for computing the second-order perturbations I have detected the following error in the texts on Hansen's method since Hill (1890): The expression of the quantity E" (cf. e.g. Brouwer and Clemence, 1961, p. 463) contains the factor $a/r = 1/(1 - e \cos E) = dE/dg$. This factor should be replaced by unity. This error can be traced to Hill's (1890) and Clemence's (1949, 1961) using the mean anomaly g as independent variable while Hansen uses the excentric anomaly E in his original paper (1857). Admittedly the effect of this correction will be very minute in the case of small excentricities and inclinations.

References

Brouwer,D. and Clemence,G.M.: 1961, "Methods of Celestial Mechanics", Academic Press, pp. 416-464.
Clemence,G.M.: 1949, Astron.Pap.Am.Eph. 11, pp. 225-500.
Clemence,G.M.: 1961, Astron.Pap.Am.Eph. 16, pp. 261-333.
Duncombe,R.: 1969, Astron.Pap.Am.Eph. 20, pp. 135-308.
Van Flandern,T.C. and Pulkkinen,K.F.: 1981, private communication (NOMAN, Release 1).
Hansen,P.A.: 1857, Abhandl.Kön.Sächs.Ges.Wiss. 5, pp. 41-218.
Hill,G.W.: 1890, Astron.Pap.Am.Eph. 4.
Richardson,D.L.: 1980, private communication (SAP User's Guide).
Seidelmann,P.K., Doggett,L.E., and De Luccia,M.R.: 1974, Astron.J. 79, pp. 57-60.
Simon,J.L. and Bretagnon,P.: 1975, Astron.Astrophys. 42, pp. 259-263.

METHODS OF COMPUTATION OF THE PERTURBED MOTION OF SMALL BODIES
IN THE SOLAR SYSTEM

Yu.V. Batrakov
Institute for Theoretical Astronomy, Leningrad, USSR

ABSTRACT

   In this review the following topics are considered: the equations
of motion and regularization, numerical and Taylor integration methods,
Encke's method and intermediate orbits.

   Orbits of minor planets and comets, in general, have larger eccentricities and inclinations than planets. Since in addition, the number
of minor planets and comets exceeds by far that of the major planets,
it is difficult to develop analytical theories for the minor bodies.
Therefore, numerical methods are almost the only methods to investigate
the motions of small bodies.

1. EQUATIONS OF PERTURBED MOTION

   Equations of motion in rectangular coordinates are most widely
used for orbit computations. The successful computation of Halley's
apparition in 1910 by Cowell, who used a numerical integration of the
equations in rectangular coordinates, made Cowell's method very popular.
Equations for osculating elements and for general variables (Herrick
1978) are used less intensively because of their rather complicated
form. Recently, the regularized form of equations attracted attention
of those who use perturbed orbit computations because regularization is
convenient for treating close encounters. Encke's method for deviations
of the real motion from the intermediate one is often used for the same
purpose.

2. NUMERICAL METHODS FOR SOLVING THE EQUATIONS OF MOTION

   Numerical integration is the general method for obtaining the
solution of equations of motion, which gives the perturbed trajectory
with sufficient accuracy for the time interval needed. Historically,
the first methods really used for computing the perturbed trajectories

W. Fricke and G. Teleki (eds.), Sun and Planetary System, 415–419.

of minor planets and comets were the multi-step methods among which the
Gauss method of quadrature and its modification known as the Cowell
method are widespread. These methods, especially the second one, are
intensively used even nowadays when orbits are improved from observations or when the ephemerides of celestial bodies are computed. Cowell's
method is widely used in ITA for solving many practical problems connected with the motion of comets. There is no need to describe these
methods because these are well known (Subbotin 1937). The methods of
such a kind are based on approximating the solution by polynomials.
Their powers determine the order of the methods. The specific feature
of multi-step methods is to obtain the numerical solution at some point.
One must use the information on the solution at preceding points. That
is why these methods were called the multi-step ones. The multi-step
methods are very convenient for integrating the equations with constant
step, when the perturbations are small and do not vary strongly. When
perturbations within the integration interval are large and change
quickly as it is the case at encounters of minor bodies with the major
planets, changing the step-size becomes necessary. Changing the step-
size in multi-step methods is a rather tedious procedure.

Recently, the so-called one-step methods were improved or newly
developed for the purposes of orbit computations. For obtaining the
solution at some point with such methods, one must know only the solution at one preceding point. One-step methods have advantages as
compared to the multi-step methods when integrating with a variable
step; changing the step does not require additional cumbersome computations and the step-size is adapting itself to the local peculiarities
of the trajectory. The one-step fourth order Runge-Kutta method (Hall
and Watt 1979) was intensively used to compute trajectories of interplanetary and lunar probes. The solution of equations in this method
was approximated by a second order parabola. Later on, different modifications appeared, based upon an approximation of the solution by more
high-order parabolas. Among them the most known are the Runge-Kutta-
Fehlberg and the Runge-Kutta-Nyström-Fehlberg methods (Fehlberg 1969,
Fehlberg 1972).

Everhart (1974) has developed a method which gives the solution at
a point by a segment of series in powers of time. The coefficients are
empirically obtained. For computing the series' coefficients, the integration step is divided into a number of smaller parts using the criterion of maximum accuracy of the result (Gauss distribution of the
points). This is one of the most effective modern methods. Basically,
it is close to Runge-Kutta methods of high order.

In the extrapolation method (Hall and Watt 1979), developed by
Bulirsch and Stoer (1966) and Stoer (1974), the solution is approximated by a rational function of special form (the order of the polynomials in the denominator is equal to that of the polynomials in the
numerator, or exceeds it by unit). The integration step is divided into
a number of smaller parts. The solution is obtained by the 'modified
mean rule'. Using several such solutions, which correspond to different

numbers of the points which divide the step, a rational interpolating function is found. Their value, which corresponds to an infinitely great number of smaller steps, is called the extrapolated solution and is used as an initial one for the next step. In order to obtain the extrapolated solution, reccurence formulae are used. The extrapolation method was compared with other methods many times and was found to be of high efficiency (Moore 1974, Makarova and Nikol'skaja 1976, Kuzelev 1981). The advantages of the extrapolation method are mainly due to the possibility of a choice of relatively large integration steps and to the slow growing of the accumulated integration errors.

For the computation of the perturbed motion of small bodies, computer programmes based on the extrapolation method are also regularly used at ITA. The perturbations of all 9 major planets are taken into account. To ensure more rapid computations, positions of major planets are stored in the computer memory in tabular form. Equations of motion are used in rectangular coordinates (helio- or barycentric).

3. ENCKE'S METHOD AND INTERMEDIATE ORBITS

For orbits with small perturbations, trajectory computations can be made less time consuming by the use of Encke's method (Subbotin 1937), according to which the real motion is a sum of the unperturbed motion (the reference orbit) and of small deviations from the latter. The deviations are considered as perturbations. Equations of motion for these deviations have small and slowly changing right-hand parts and can be solved by numerical integration using larger integration steps as compared with that for the original equations of motion. As a rule, it allows to save computer time and to decrease the integration errors. The initial osculating orbit is mostly used in Encke's method as a reference orbit which ensures the coincidence of the initial position and velocity vectors of the real orbit and that of the intermediate one. The reference orbit is close in first order to the real trajectory. The deviations are small in the vicinity of the starting time and grow with time in square.

At ITA, non-perturbed reference orbits were constructed which are much closer to the real trajectory at initial time than it can be ensured by the initial osculating orbit (Batrakov and Makarova 1979, Batrakov 1981). This reference orbit consists of two motions: the non-perturbed two-body motion around some fictitious attracting centre and the motion of the centre itself. Parameters of the two-body motion and of the motion and the mass of the centre are chosen in order to ensure the required order of the initial vicinity of the real and the reference orbits. For the reference orbit we can ensure the coincidence at initial time not only for the reference and the real position and velocity vectors, but also for the second and even third time derivatives of the position vectors depending on the type of reference orbit. Deviations of the real motion from the reference grow in this case with third and even fourth order of time. In the vicinity of the initial

time, these deviations grow much slower than that ones for the initially
osculating reference orbit. When constructing reference orbits of this
kind, it is also possible to minimize the velocity of the growth of the
deviations by a proper choice of the reference orbit's parameters. The
model computations showed that these reference orbits save computing
time as compared to the cases of no reference orbits and even of no
osculating reference orbit. Using these reference orbits seems to be
promising in cometary dynamics.

4. ON THE COMPARISON OF THE EFFICIENCY BETWEEN NUMERICAL METHODS FOR PERTURBED MOTION

A large variety of papers deals with this subject. The paper of
Moore (1974) is of special interest. Moore compared Runge-Kutta methods
of high orders, methods of rational extrapolation, Cowell's method and
some other methods. The results showed that Cowell's method of 12-th
order exceeds all others with respect to efficiency. The rational extrapolation method is inferior in this respect because it requires more
computations of the right-hand sides of equations which results in an
increase of computer time. Integration errors for Everhart's method
grow slower than those for the rational extrapolation method (Kuzelev
1981). Numerical experiments by E.N. Makarova at ITA showed that
Everhart's method is less time consuming than the rational extrapolation
method when integrating the equations for osculating elements or for
deviations from a reference orbit (Encke's method). When the equations
for rectangular coordinates are integrated, the rational extrapolation
method requires less computing time. Thus, the efficiency depends on the
type of equations.

5. REGULARIZATION

Regularization of the equations of motion according to the
Kustaanheimo-Stiefel method (KS-method) is obtained by transforming the
coordinates and time to new variables. As a result, the equations of
motion in the non-perturbed case take a linear form and correspond to
harmonic oscillations. In regularized equations the integration step
can be taken to be almost constant which allows to use very efficient
multi-step methods. The regularization may show an impressive effect
for elongated cometary and asteroidal orbits. The essentials of the KS-
method are given in the book by Scheifele and Stiefel (1975). The efficiency of numerical integration of regularized equations was considered
by Bordovicyna and Suhopljueva (1980).

6. USE OF TAYLOR SERIES FOR NUMERICAL INTEGRATION

The idea of using Taylor series for numerical integration have
been developed by Steffensen (1956, 1957) who offered reccurent formulae
which allow to compute the coefficients of these series one after an-

other. The ideas of Steffensen have been used by Sitarski (1979a, b) who has compiled a programm for numerical integration by Taylor expansions for comets and who has used this programm for improving the orbit of the asteroid Adonis. At ITA the ideas of Steffensen are intensively developed by V.F. Mjacin and his associates. In 1970 they managed to compile a programm (Mjacin and Sizova 1970) for integrating the non-regularized equations of motion by Taylor series which have been used for solving some practical problems (Sizova 1976). Later, a method was developed which uses Taylor integration for regularized equations of motion (Mjacin and Sizova 1980). This method has been especially adopted for cometary orbit computations when one may expect enccunters of the comet with major planets.

REFERENCES

Batrakov,Yu.V., Makarova,E.N.: 1979, Bull.ITA, 14, N7 (160), p.397

Batrakov,Yu.V.: 1981, in "Opredelenie koordinat nebesnyh tel", Latv.Gos. Universität, Riga, p.3

Bordovicyna,T.V., Suhopljueva,L.E.: 1980, Bull.ITA, 14, N10 (163), p.591

Bulirsh,R., Stoer,J.: 1966, Numer.math., Bd. 8, Heft 1

Everhart,E.: 1974, Celest.Mech., 10

Fehlberg,E.: 1969, Computer, 4

Fehlberg,E.: 1972, Computing, 10

Herrick,S.: 1978, Astrodinamika, 3, Mir, Moscow, p.360 (in russian)

Hall,J., Watt,J. (eds.): 1979, Sovremennye cislennye metody resenija obyknovennyh differencial'nyh uravnenij, Mir, Moscow, p.312 (in russian)

Kuzelev,S.V.: 1981, in "Opredelenie i modelirovanie dvizenija ISZ i gravitacionnogo polja", NIIGAIK, Novosibirsk, p.57

Makarova,E.N., Nikolskaja,T.K.: 1976, Bull.ITA, 14, N4 (157), p.210

Makarova,E.N.: 1979, Bull,ITA, 14, N8 (161), p.486

Mjacin,V.F., Sizova,O.A.: 1970, Bull.ITA, 12, N5 (138), p.389

Mjacin,V.F., Sizova,O.A.: 1980, Bull.ITA, 14, N10 (163), p.597

Moore,H.: 1974, in "Lecture notes in Math.", 362, Springer-Verlag, Berlin-Heidelberg-New York, p.149

Sitarski,G.: 1979a, Acta Astron., 29, N3, p.401

Sitarski,G.: 1979b, Acta Astron., 29, N3, p.413

Sizova,O.A.: 1976, Bull.ITA, 14, N4 (157), p.238

Steffensen,J.F.: 1956, Kong.Danske Videnskab.Selskab.Math.-Fys.Medd., 31, 3

Steffensen,J.F.: 1957, Kong.Danske Videnskab.Selskab.Math.-Fys.Medd., 31, 3

Stiefel,E., Scheifele,G.: 1975, Linejnaja i reguljarnaja nebesnaya mehanika, Nauka, Moscow, p.303 (in russian)

Stoer,J.: 1974, in "Lecture notes in Math.", 362, Springer-Verlag, Berlin-Heidelberg-New York, p.3

Subbotin,M.F.: 1937, Kurs nebesnoj mehaniki, 2, ONTI, Moscow-Leningrad, p.404 (in russian)

A NEW METHOD FOR EXPRESSION OF THE PERTURBATION FUNCTION

V.G. Shkodrov
Section of Astronomy
Bulgarian Academy of Sciences, Sofia, Bulgaria

The perturbation function R is usually expressed in a spherical reference system S'. To this end the addition theorem of spherical functions in S' is applied. Consequently, $R = R(\theta, \lambda)$. To solve the differential equations of the movement, the transformation

$$R(\theta, \lambda) \rightarrow R(E_k) \qquad (1)$$

is needed, where E_k is a set of selected orbital elements. Nearly always the transformation (1) is made by means of identities well-known in spherical trigonometry. The papers (Kaula, 1961; Brumberg, 1967) are typical examples of this.

Recently, when examining the rotating-orbital movement of the real celestial bodies, a preference to Delaunay-Andoyer variables is observed. This is due to the circumstance that there is a wide class of analytic methods in the celestial mechanics, based on the use of the equations of motion in the osculating elements. As a rule, authors introduce Andoyer's variables following (Deprit, 1967). They define the rotating movement of a celestial body in the phase space $(\alpha, \beta, \gamma; p_\alpha, p_\beta, p_\gamma)$ of the three coordinates α, β, γ (Euler andles) and the three moments. After that by means of the canonical transformation:

$$(\alpha, \beta, \gamma; p_\alpha, p_\beta, p_\gamma) \rightarrow (l_1, g_1, h_1; L_1, G_1, H_1) \qquad (2)$$

they pass into a new six-dimensional space, consisting of three coordinates l_1, g_1, h_1 and three moments L_1, G_1, H_1, known as Andoyer variables.
In spite of the convenience, which results from the introduction of Andoyer variables, this method is accompanied by considerable difficulties. They appear when we transform the perturbation function expressed in variables from one phase space into variables from another space (Kinoshita, 1972; Barkin, 1977).
These difficulties are by and large avoided by the approach given here. The essence of the method is the generalization of the addition theorem of spherical functions, which are defined in reference system with common origin, but rotated one to the other (Shkodrov, 1981).
Let S' be the source reference system, S is the rotated one, and let the transformation of S' to S be described by α, β, γ - three Euler angles, i.e.

$$S' \xrightarrow{(\alpha, \beta, \gamma)} S \qquad (3)$$

W. Fricke and G. Teleki (eds.), Sun and Planetary System, 421–422.
Copyright © 1982 by D. Reidel Publishing Company.

If r_1 is the position vector of the body causing a gravitational field,
r is that of the moving body, $r_1 \in S'$, $r \in S$, the addition theorem has the
form:

$$P_n(\cos\psi)=(2n+1)^{-1} \sum_{mm'} D^{n*}_{m'm}(\alpha,\beta,\gamma) Y^*_{nm'}(\theta_1',\lambda_1') Y_{nm}(\theta,\lambda), \qquad (4)$$

where θ,λ are the spherical coordinates of r in S, and θ_1,λ_1 those of r_1
in S'; ψ is the angle between the two vectors r and r_1; $Y_{nm}(\theta,\lambda)=\bar{P}^m_n(\cos\theta)$
$e^{-im\lambda}$, $\bar{P}^m_n(\cos\theta)$ is the normalized Legendre function; $D^n_{m'm}(\alpha,\beta,\gamma)$ is the
Wigner function; with (*) denoting complex conjugate. If S' transforms it-
self to S by means of an intermediate system $\tilde{S}$, i.e. according to scheme

$$S' \xrightarrow[(\alpha_1,\beta_1,\gamma_1)]{} \tilde{S} \xrightarrow[(\alpha_2,\beta_2,\gamma_2)]{} S \qquad (5)$$

instead of (4), we shall have:

$$P_n(\cos\psi)=\frac{1}{2n+1}\sum_{\tilde{m}mm'} D^{n*}_{\tilde{m}'\tilde{m}}(\alpha_1,\beta_1,\gamma_1) D^{n*}_{\tilde{m}m}(\alpha_2,\beta_2,\gamma_2) Y^*_{nm'}(\theta_1',\lambda_1') Y_{nm}(\theta,\lambda) (6)$$

The following relation exists between the Wigner functions in (4) and (6)

$$D^{n*}_{m'm}(\alpha,\beta,\gamma)=\sum_{\tilde{m}=-n}^{n} D^n_{m'\tilde{m}}(\alpha_1,\beta_1,\gamma_1) D^{n*}_{\tilde{m}m}(\alpha_2,\beta_2,\gamma_2) \qquad (7)$$

The use of (4) and (6) gives an opportunity to obtain elegantly many of
the known transformations (1) of the perturbation function. For insrance,
if the motion of the body is described in S, S' is connected with the
orbit and in (4) we assume that $\alpha=\Omega$, $\beta=I,\gamma=\omega$, $\theta=0$, $\lambda=v$, then the perturbation function is expressed in Kepler elements. We can reach the same
results from (6) assuming that S' is rigidly fixed in the planet, in S
the motion is described and S is connected with the orbit. In case, that
in the right-hand side of (1) Andoyer variables should take part in (5),
we assume (Shkodrov, 1981a)

$$\alpha_1=h_1-\frac{\pi}{2}, \beta_1=I_1, \gamma_1=\frac{\pi}{2}; \alpha_2=g_1-\frac{\pi}{2}, \beta_2=J_2, \gamma_2=1_1+\frac{\pi}{2}, (\cos J_1=L_1 H_1^{-1}, \cos I_1=H_1 G_1^{-1}) \quad (8)$$

and by means of (6) we transform $R(\theta,\lambda)$ in a perturbation function, containing Andoyer variables. In case, that R is described in the phase
space $(\alpha,\beta,\gamma;p_\alpha,p_\beta,p_\gamma)$ it is needed to use (7), which is a generalized
analogy of the addition theorem of spherical functions.
The method summarized here gives an opportunity for a simple and compact
expression of the perturbation function and its transformation from
some variables into other variables. It admit very interesting generalizations.

REFERENCES

Barkin, Y.: 1977, Astron. J. (Moscow), 52, p. 1076 (in Russian).
Brumberg, V.A.: 1967, Bull. ITA, 11, 2(125), p.73, (in Russian).
Deprit, A.: 1967, Am. J. Phys., 55, p. 424.
Kaula, W.: 1961, Geophys. J. Roy. Astron. Soc., 5, p. 104.
Kinoshita, H.: 1972, Publ. Astron. Soc. Japan, 24, p. 423.
Shkodrov, V.: 1981, Compt.rend. Acad. bulg. Sci., 34, 5, p. 605.
Shkodrov, V.: 1981a, Compt. rend. Acad. bulg. Sci. 34, 6, p. 755.

RELATIVISTIC DYNAMICS OF THE EARTH-MOON SYSTEM

V.A.Brumberg and T.V.Ivanova
Institute of Theoretical Astronomy
191187 Leningrad
U.S.S.R.

ABSTRACT. Equations of motion of the Earth and Moon convenient for numerical integration have been derived within the framework of the mass-point PPN formalism taking into account parameters β, γ and coordinate parameter α. The Lagrangian of the geocentric lunar motion is expanded in powers of the eccentricity of the orbit of the Sun and the parallax parameter. The first degree terms in lunar motion have been found by analytical iterations with the aid of the Poissonian processor up to order m^7. Variational terms and mean motions of the perigee and node have been obtained up to order m^8. Relativistic reduction of observations of the Moon is discussed with relation to the problem of determining the lunar orbital constants.

The internal accuracy of the modern analytical theories of the Newtonian motion of the Moon is characterized by coefficients of order $1''\cdot10^{-5}$ in longitude or latitude and $5\cdot10^{-11}$ in radius vector (Henrard,1979; Chapront-Touzé, 1980; Schmidt,1980). Within this precision it is necessary to include the relativity effects more accurately than it was done before (Brumberg,1958; Baierlein,1967; Krogh and Baierlein,1968; Finkelstein and Kreinovich,1976). This is the aim of the present paper.

Taking into account parameters β, γ of the PPN formalism (Will and Nordtvedt,1972; Anderson,1974) and coordinate parameter α (Brumberg,1972) the post-Newtonian equations of motion of point masses M_i may be written

$$\ddot{\underline{r}}_i = -\sum_{j \neq i} M_j \underline{r}_{ij} r_{ij}^{-3} + \sum_{j \neq i} m_j (A_{ij}\underline{r}_{ij} + B_{ij}\dot{\underline{r}}_{ij}) \tag{1}$$

with

$$A_{ij} = \dot{\underline{r}}_i^2 r_{ij}^{-3} - (\alpha+\gamma+1)\dot{\underline{r}}_{ij}^2 r_{ij}^{-3} + (3/2)(\underline{r}_{ij}\dot{\underline{r}}_j)^2 r_{ij}^{-5}$$

$$+ 3\alpha(\underline{r}_{ij}\dot{\underline{r}}_{ij})^2 r_{ij}^{-5} + 2[(-\alpha+\beta+\gamma+\tfrac{1}{2})M_i + (-\alpha+\beta+\gamma)M_j]r_{ij}^{-4}$$

423

W. Fricke and G. Teleki (eds.), Sun and Planetary System, 423–428.
Copyright © 1982 by D. Reidel Publishing Company.

$$+\sum_{k\neq i,j} M_k\{(-\alpha+2\beta+2\gamma)r_{ij}^{-3}r_{ik}^{-1}+(-2\alpha+2\beta-1)r_{ij}^{-3}r_{jk}^{-1}+\alpha r_{ij}^{-1}r_{ik}^{-3}$$

$$+2(\gamma+1)r_{ij}^{-1}r_{jk}^{-3}+(\alpha-2\gamma-3/2)r_{ik}^{-1}r_{jk}^{-3}-\alpha r_{ik}^{-3}r_{jk}^{-1}+[\alpha r_{ik}^{-3}$$

$$-(\alpha+\tfrac{1}{2})r_{jk}^{-3}-3\alpha r_{ij}^{-2}r_{ik}^{-1}+3\alpha r_{ij}^{-2}r_{jk}^{-1}](\underline{r}_{ij}\underline{r}_{ik})r_{ij}^{-3}\},\qquad (2)$$

$$B_{ij}=[2(-\alpha+\gamma+1)(\underline{r}_{ij}\underline{\dot{r}}_{ij})+\underline{r}_{ij}\underline{\ddot{r}}_{j}]r_{ij}^{-3},\qquad (3)$$

where $\underline{r}_{ij}=\underline{r}_i-\underline{r}_j$, $m_i=M_i/c^2$ and for barycentric coordinates

$$\sum_i M_i(1+\tfrac{1}{2}\underline{\dot{r}}_i^2/c^2-\tfrac{1}{2}\sum_{j\neq i}m_j/r_{ij})\underline{r}_i=0.\qquad (4)$$

Equations of motion of massive bodies in the model of a perfect fluid include the Nordtvedt effect (Will,1971) proportional to factor $4\beta-\gamma-3$. Discussion of lunar laser measurements did not reveal such effect (Mulholland,1980) leading to the conclusion $4\beta-\gamma-3=0$. But this conclusion implicitly employs the assumptions of the perfect fluid model. It seems to us that the point mass model did not loose its meaning for the Earth-Moon dynamics.

From (1) it is easy to derive the equations of relative motion. Introducing geocentric vector of the Moon ζ and heliocentric vector of the Earth-Moon barycenter $\underline{R}$ yields

$$\underline{\ddot{\zeta}}=-M_c\underline{\zeta}\zeta^{-3}+M_2(\underline{r}_{12}r_{12}^{-3}-\underline{r}_{32}r_{32}^{-3})+m_2(A\underline{R}+B\underline{\dot{R}}+C\underline{\zeta}+D\underline{\dot{\zeta}}),\quad (5)$$

$$\underline{\ddot{R}}=-M(M_1\underline{r}_{12}r_{12}^{-3}+M_3\underline{r}_{32}r_{32}^{-3})/M_c+m_2(A'\underline{R}+B'\underline{\dot{R}}+C'\underline{\zeta}+D'\underline{\dot{\zeta}}),\quad (6)$$

where indices 1,2,3 are referred to the Earth, Sun and Moon, respectively. Here $M=M_2+M_c$, $M_c=M_1+M_3$ and

$$A=A_{32}-A_{12},\quad C=(M_1A_{32}+M_3A_{12})/M_c+(M_1A_{31}+M_3A_{13})/M_2,\quad (7)$$

$$A'=(M_1A_{12}+M_3A_{32})/M_c+(M_1A_{21}+M_3A_{23})/M_2,$$

$$C'=M_1M_3(A_{32}-A_{12})/M_c^2+M_1M_3(A_{23}-A_{21}+A_{31}-A_{13})/(M_2M_c).\qquad (8)$$

Changing A_{ij} for B_{ij} gives B,D,B',D' in place of A,C,A',C'.

Within the framework of the post-Newtonian approximation Equations (5),(6) are rigorous. For modern actual applications one may use in the right-hand member of (6) $C'=D'=0$ and Schwarzschild's expressions for A', B'.

For analytical treatment it is reasonable to use the Lagrangian. The first step is to expand it in powers of ζ/R. Then we use the Schwarzschild's solution for $\underline{R}=X',Y',Z'$:

$$(X'+iY')/A=\exp i\lambda'+\tfrac{1}{2}e'\{[-3+\sigma(\alpha-\beta+\gamma+1)]\exp i\pi'$$

$$+[1+\sigma(\alpha-\beta+\gamma+1)]\exp i(2\lambda'-\pi')\}+\dots(9)$$

and $Z'=0$. The mean longitude of the Earth λ' and the longitude of the perihelion π' are linear functions of time with frequencies N and $\sigma(-\beta+2\gamma+2)$N respectively, mean motion N and major semi-axis A being related by the relativistic generalization of the third Kepler law

$$N^2A^3[1+\sigma(-3\alpha+2\beta+\gamma)]=M \qquad (10)$$

where $\sigma=N^2A^2/c^2$ is relativistic small parameter $\approx10^{-8}$.

Our analytical method is similar to that of Musen (1971) being still better adapted to computer calculations. We have found the variational inequalities and the first degree terms with respect to eccentricity and inclination of the lunar orbit, eccentricity of the orbit of the Sun and the ratio of the major semi-axes of the lunar and solar orbits. Coefficients are expanded in powers of Hill parameter m=N/(n-N), n being mean motion of the Moon. Due to presence of the small analytical divisors the convergence of the obtained series is rather slow. Final results are presented as series for ρ/a_0 (a_0 is defined by $n^2a_0^3=M_c$), difference $v-\lambda$ between true and mean longitudes and latitude φ. Trigonometric arguments are usual Delaunay arguments D, ℓ, ℓ', F with relativity contributions in the frequencies of ℓ, ℓ', F leading to the indirect effects in Newtonian terms. Numerical expansions are shown in Tables 1-3. Each coefficient of the final series is represented as the sum of five components: Newtonian part marked by N, relativity free-of-parameters part marked by R and three relativity parts proportional to α, β, γ. Right columns contain the general relativity terms (marked by E) in harmonic coordinates ($\alpha=0$). Symbolically, E=R+β+γ and E+α is general relativity term in standard coordinates ($\alpha=1$). The mean motions of the perigee and node are

$$\dot{\pi}=14849682''.424(N)+0''.8328(R)-0''.2568\beta+1''.1520\gamma\ , \qquad (11)$$

$$\dot{\Omega}=-6928790''.0675(N)+0''.5902(R)+0''.0435\beta+1''.2673\gamma\ . \qquad (12)$$

General relativity contributions are 1''.7280 and 1''.9010. One should add to $\dot{\pi}$ the Schwarzschild advance 0''.06.

Table 1. Radius vector of the Moon ρ/a_0

D	ℓ	ℓ'	$N\cdot10^1$	$R\cdot10^8$	$\beta\cdot10^8$	$\gamma\cdot10^8$	$\alpha\cdot10^8$	$E\cdot10^8$
0	0	0	9.9912	-0.2273	-1.3132	-0.6086	0.4999	-2.1491
2	0	0	-0.0717	0.2486	0.0141	0.0145	-0.4966	0.2772
4	0	0	-0.0002	0.0016	0.0001	0.0001	-0.0033	0.0018
0	1	0	-0.5438	0.0060	0.0707	0.0327	-0.0132	0.1094
2	-1	0	-0.0968	-0.0378	0.0206	0.0093	0.0628	-0.0079
2	1	0	-0.0041	0.0204	0.0008	0.0008	-0.0408	0.0221
4	-1	0	-0.0008	0.0040	0.0002	0.0002	-0.0082	0.0045
0	0	1	0.0012	-0.0583	0.0665	-0.0333	0.0088	-0.0251
2	0	-1	-0.0051	0.0167	-0.0133	0.0056	-0.0225	0.0089
2	0	1	0.0007	-0.0062	0.0114	-0.0030	0.0139	0.0021
1	0	0	0.0029	0.0841	-0.1239	0.0246	0.0007	-0.0152

Table 2. Longitude of the Moon $v - \lambda$

D	ℓ	ℓ'	$N \cdot 10^{-4}$	$R \cdot 10^3$	$\beta \cdot 10^3$	$\gamma \cdot 10^3$	$\alpha \cdot 10^3$	$E \cdot 10^3$
2	0	0	0″.2106	−0″.5104	−0″.0138	−0.0305	1″.0179	−0″.5547
4	0	0	0.0008	−0.0052	−0.0001	−0.0003	0.0104	−0.0056
0	1	0	2.2641	0.0121	0.0001	0.0001	−0.0242	0.0123
2	−1	0	0.4606	0.0814	−0.0368	−0.0229	−0.1120	0.0217
2	1	0	0.0175	−0.0559	−0.0011	−0.0025	0.1115	−0.0595
4	−1	0	0.0035	−0.0106	−0.0005	−0.0007	0.0216	−0.0118
0	0	1	−0.0655	1.8731	−3.6707	1.0139	−0.0001	−0.7838
2	0	−1	0.0152	−0.0402	0.0406	−0.0153	0.0457	−0.0150
2	0	1	−0.0022	0.0182	−0.0345	0.0091	−0.0288	−0.0071
1	0	0	−0.0129	−0.2943	0.5379	−0.1155	−0.0000	0.1281

Table 3. Latitude of the Moon ψ

D	F	$N \cdot 10^{-4}$	$R \cdot 10^5$	$\beta \cdot 10^5$	$\gamma \cdot 10^5$	$\alpha \cdot 10^5$	$E \cdot 10^5$
0	1	1″.8516	4″.6799	0″.0036	0″.0059	8″.9159	4″.6894
2	−1	0.0618	2.6523	−0.3835	−0.3391	−4.2662	1.9297
2	1	0.0094	−2.2669	−0.0616	−0.1367	4.6136	−2.4652

Coordinates ζ, v, ψ are not measurable quantities. As such quantities we may consider (Brumberg, 1981a,b) time interval T in lunar laser ranging, geocentric angular distance Υ between the Sun and the Moon and geocentric angular distance θ between the Moon and a distant celestial object (quasar). Coordinate parameter α disappears in T, Υ, θ. The coefficients of $\sin \ell$ and $\sin F$ in Υ and θ (provided that the chosen quasar lies near the pole of ecliptic) may be adopted for the definition of lunar eccentricity and inclination constants respectively.

REFERENCES

Anderson, J.D., 1974. In: Experimental Gravitation
    (ed. B.Bertotti), p.163, Academic Press.
Baierlein, R., 1967. Phys. Rev. 162, p.1275.
Brumberg, V.A., 1958. Bull. ITA 6, p.733.
Brumberg, V.A., 1972. Relativistic Celestial Mechanics,
    Nauka, Moscow (in Russian).
Brumberg, V.A., 1981a. Astron. J. (USSR) 58, p.181.
Brumberg, V.A., 1981b. In: Reference Coordinate Systems
    for Earth Dynamics (eds. E.M.Gaposchkin and B.Kola-
    czek), p.283, Reidel Publ. Comp.
Chapront-Touzé, M., 1980. Astron. Astrophys. 83, p.86.
Finkelstein, A.M. and Kreinovich, V.Ja., 1976. Celes.
    Mech. 13, p.151.
Henrard, J., 1979. Celes. Mech. 19, p.337.
Krogh, C. and Baierlein, R., 1968. Phys. Rev. 175, p.1576.
Mulholland,J.D.,1980. Rev. Geophys. Space Phys. 18, p.549.
Musen, P., 1971. Celes. Mech. 3, p.289.
Schmidt, D.S., 1980. Moon and Planets 23, p.135.
Will, C.M., 1971. Astrophys. J. 163, p.611.
Will,C.M. and Nordtvedt,K.,1972. Astrophys. J. 177, p.757.

APPENDIX

It may be of interest to make two supplements. The first one is to give some details concerning the relativistic reduction of lunar observations.

Ignoring effects due to the Earth-Moon gravitational field and using the relations (16),(65),(56) and (59) of (Brumberg, 1981a) we have within our approximation

$$2cT = \rho\{2+(m_2/R)[(\gamma+1)(2-\rho d/R)-\alpha(2-3\rho d/R)(1-d^2)]\},$$

$$\cos\psi = -d+(m_2/R)[-\alpha d-\tfrac{1}{2}(\alpha+\gamma+1)\rho/R+(3/2)\alpha\rho d^2/R](1-d^2),$$

$$\cos\theta = -\underline{s}\,\underline{\rho}/\rho +(m_2/R)\{-(\gamma+1)(\underline{R}\times\underline{s})/(R-\underline{s}\,\underline{R}) + [\alpha d$$

$$+\tfrac{1}{2}(\alpha+\gamma+1)\rho/R-(3/2)\alpha\rho d^2/R](\underline{\rho}\times\underline{R})/(\rho R)\}(\underline{s}\times\underline{\rho})/\rho,$$

where $d=(R\rho)/(R\rho)$ and $\underline{s}$ denotes unit vector characterizing the direction of light from a quasar in the remote past. Substituting the coordinates of the Moon and the Sun we obtain the expansion of $2cT$ shown in Table 4. Coefficients are expressed in metres. Newtonian values are written in form of floating-point numbers with decimal order on the right. As everywhere in this paper our Newtonian values are very approximate and are given only for comparison.

Table 4. Round-trip time $2cT$ in lunar laser ranging

D	ℓ	ℓ'	N		R	β	γ	E
0	0	0	7.68807	+8	5.840	-10.105	2.906	-1.359
2	0	0	-5.51983	+6	1.858	0.109	0.057	2.024
4	0	0	-1.53955	+4	0.013	0.000	0.000	0.013
0	1	0	-4.18449	+7	-0.367	0.544	-0.161	0.016
2	-1	0	-7.44601	+6	-0.365	0.159	-0.002	-0.208
2	1	0	-3.17355	+5	0.154	0.006	0.003	0.163
4	-1	0	-6.52331	+4	0.030	0.002	0.001	0.033
4	1	0	-2.11285	+3	0.002	0.000	0.000	0.002
0	0	1	8.98179	+4	-0.321	0.511	-0.128	0.062
2	0	-1	-3.90411	+5	0.124	-0.103	0.039	0.060
2	0	1	5.68863	+4	-0.048	0.088	-0.023	0.017
4	0	-1	-2.20332	+3	0.002	-0.001	0.000	0.001
1	0	0	2.24142	+5	0.659	-0.953	0.201	-0.093
3	0	0	-3.48977	+3	0.002	-0.007	0.002	-0.003

Considering that in function of the true longitudes of the Sun and Moon $d \approx -\cos(v-v_\odot)$ we may simplify the expression for ψ as follows

$$\psi = v - v_\odot + \sigma\{-\alpha(A/R)\cos(v-v_\odot) +\tfrac{1}{2}(\rho/A)[\alpha+\gamma+1$$
$$-3\alpha\cos^2(v-v_\odot)]\}\sin(v-v_\odot) \quad .$$

It is easy to find the expansion of the reduction $\psi - \lambda + \lambda_\odot$. In particular, the coefficients of the families of variation, elliptic and evection inequalities coinside with corresponding coefficients in Table 2 for $\alpha = 0$.

The angular distance χ between the Sun and a quasar considered above may be expressed by an approximate formula

$$\cos\chi = \underline{R}\;\underline{\varsigma}/R - (m_2/R)(\gamma+1)(1+\underline{R}\;\underline{\varsigma}/R) \quad .$$

Let this quasar be situated practically in the pole of ecliptic in such a way that $\underline{\varsigma} = (0,0,-1)$, $\cos\chi = -(\gamma+1)m_2/R$. The angular distance θ for such a quasar will be

$$\theta = 90° - \psi + \varsigma[(\gamma+1)A/R + \alpha\psi\cos(v-v_\odot)]\cos(v-v_\odot) \quad .$$

The inclination inequalities in the expansion of $90° - \theta$ are the same as those in Table 3 for $\alpha = 0$ and may be used for determining the lunar inclination constant.

The disappearance of α in T, ψ, θ may be also demonstrated without actual substituting lunar and solar coordinates. In fact, if harmonic coordinates ($\alpha = 0$) are designated by a tilde, then we have

$$\tilde{\underline{\rho}} = \underline{\rho} + \alpha(m_2/R)[-\underline{\rho} + \underline{R}(\underline{R}\;\underline{\rho})/R^2] + \alpha(m_2/R^3)\{(\underline{R}\;\underline{\rho})\underline{\rho}$$
$$+ \tfrac{1}{2}[\rho^2 - 3(\underline{R}\;\underline{\rho})^2/R^2]\underline{R}\}$$

or in spherical coordinates

$$\tilde{\rho} = \rho\{1 - \alpha(m_2/R)[1 - (3/2)\rho\, d/R](1-d^2)\} \quad ,$$

$$\tilde{v} - v = \alpha(m_2/R)[d + \tfrac{1}{2}(\rho/R)(1-3d^2)]\sin(v-v_\odot) \quad ,$$

$$\tilde{\psi} = \psi[1 - \alpha(m_2/R)d^2] \quad .$$

These relations demonstrate the disappearance of in T, ψ, θ.

The second remark refers to the recent paper presented by J.-F. Lestrade, J. Chapront and M. Chapront-Touzé at the IAU Colloquium No. 63 and kindly delivered to us by the authors. This paper deals with separate determination of the indirect relativistic effects in the lunar theory caused by the relativity terms in the heliocentric motion of the Earth-Moon barycenter. The problem is treated in osculating elements. However, to have consistent theory one has to solve the "main" relativistic lunar problem also in osculating elements but that was never done. So far the direct comparison of these results with ours seems to be very cumbersome.

In great detail our results are intended to be published in the Transactions of the Institute of Theoretical Astronomy.

ON THE SECULAR EFFECTS IN THE MOTION OF A PLANETARY SATELLITE

V.G. Ivanova
Section of Astronomy
Bulgarian Academy of Sciences
Sofia, Bulgaria

The perturbation function development obtained in (Shkodrov, 1980) gives an opportunity for considerable generalizations of the theory of motion of natural and artificial bodies (Shkodrov, 1980a) and introduce greater clarity into the qualitative and quantitative analyses of the solution. A qualitative analysis has been made in this paper and secular disturbances of the first order in the motion of a trial body in a zonal gravitation field have been obtained.

For the purpose use was made of the general representation of the perturbation function R obtained in (Shkodrov, 1980), i.e.

$$R = \sum_{nm}\sum R_{nm}, \tag{1}$$

where

$$R_{nm} = GM(-1)^m i^m A^*_{nm} \frac{a_e^n}{r^{n+1}} i^{m'} \beta_{nm'} D^{*n}_{mm'}(\Omega, I, \omega+v), \quad (i=\sqrt{-1}). \tag{2}$$

In (2): G=gravitational constant; M=planetary mass; a_e=its equatorial radius; r=planetocentric radius-vector of the trial body; A^*_{nm}=complex dimensionless constants characterizing the distribution of the planet's matter; $D^{*n}_{mm'}(\Omega, I, \omega+v)$=Wigner's function with arguments: Ω=the length of the orbit ascending node; I=its inclination; ω=argument of the pericentre and v=true anomaly.

$$\beta_{nm'} = \begin{cases} (-1)\frac{n+m'}{2} \left[\dfrac{(2n+1)(n+m'-1)\ !!\ (n-m'-1)\ !!}{(n+m')!!(n-m')!!} \right]^{1/2}, & (n+m') - \text{even} \\ 0, & (n+m') - \text{odd}. \end{cases} \tag{3}$$

The coefficient β_{nm} is connected with the zeros of the Legendre polynomials. To avoid the difficulty in (2) connected with the oddnes of the sum (n+m'), we set down m'=n-2k, m'∈[-n,n], k∈[n,0], and change the summing-up index in (2). Further we express β_{nm} by Euler gamma-functions. To separate the zonal part of the perturbation function in (2) we set m=0 and take into account that

$$A^*_{no} = -\overline{J}_n = -(2n+1)^{-1/2} J_n, \tag{4}$$

W. Fricke and G. Teleki (eds.), Sun and Planetary System, 429–430.

$$D\overset{n^*}{\text{o}}_n \cdot 2k(\Omega, I, \omega+v) = (-1)^n (2n+1)^{-1/2} Y_{nn-2k}(I, \omega+v) \qquad (5)$$

we obtain

$$R_n = \frac{(-1)^{n+1} GM}{r^{n+1}} \frac{J_n}{2n+1} a_e^n \Sigma i^{n-2k} \beta_{nn-2k} Y_{nn-2k}(I, \omega+v). \qquad (6)$$

The first order secular disrurbances in the motion are obtained using the Lagrange equations for the averaged problem of two bodies. This is equivalent to a separation of the constant part of the perturbation function not depending on the angular variables v and ω. Solving the averaged Lagrange equations we obtain the secular effects in the slowly changing Kepler elements caused by the zonal part of the perturbation function with index 2n. These effects we write in the following form:

$$[\omega]_{2n} = (-1)^{n+1} \frac{\Gamma(n+1/2)}{2\Gamma(n+1)} \left(\frac{a_e}{p}\right)^{2n} J_{2n} \{ [2(2n+1)P_{2n}(\cos I) - \text{ctg} I \cdot P_{2n}^1(\cos I)] \cdot$$
$$M_{2n-1}^o + (2n+1) \frac{1-\tilde{e}^2}{2\tilde{e}} P_{2n}^1(\cos I)(M_{2n-2}^1 + M_{2n-2}^{-1}) \}, \qquad (7)$$

$$[\Omega]_{2n} = (-1)^{n+1} \frac{\Gamma(n+1/2)}{\Gamma(n+1)} \left(\frac{a_e}{p}\right)^{2n} J_{2n} \text{cosec} I \cdot P_{2n}^1(\cos I) M_{2n-1}^o, \qquad (8)$$

$$[M_o]_{2n} = (-1)^n \frac{\Gamma(n+3/2)}{\Gamma(n+1)} \frac{(1-\tilde{e}^2)^{\frac{3}{2}}}{\tilde{e}} \left(\frac{a_e}{p}\right)^{2n} J_{2n} P_{2n}(\cos I)(M_{2n-2}^1 + M_{2n-1}^{-1}), \qquad (9)$$

where $[\ ]_{2n}$ stands for the secular changes in the Kepler elements, $p = a(1-\tilde{e}^2)$; $M_n^m = M_n^m(\tilde{e})$ are the coefficients defined analogously in (Aksenov, 1977). The total secular effect in the orbital elements can be determined by the formula:

$$[E_k] = \sum_{n=0}^{\infty} [E_k]_{2n} \quad (k=1,2,3). \qquad (10)$$

The secular variations in ω, Ω and M_o depending on the inclination are depended only by the properties of $P_{2n}(\cos I)$ and $P_{2n}^1(\cos I)\text{cosec} I$. The influence of the eccentricity ẽ on the secular changes is determined by $(M_{2n-2}^1 + M_{2n-1}^{-1})(1-\tilde{e}^2)/\tilde{e}$ for ω and M_o, and by M_{2n-1}^o for Ω. The formulas obtained for the secular effects are general and with their help the effects of an arbitrary zonal harmonic can be calculed.

REFERENCES

Shkodrov, V.G.: 1980, Compt. rend.Acad.bulg.Sci. 33,8, p. 1021.
Shkodrov, V.G.: 1981, Compt. rend. Acad.bulg. Sci. 34, 5, p. 605.
Aksenov, E.P.: 1977, Motion theory of the artificial satellite of the
          Earth.

L´INVARIABILITÉ SÉCULAIRE DES GRANDS DEMIAXES DES ORBITES PLANÉTAIRES

B. Popović
Ognjena Price 80, Belgrade, Yougoslavie

Proceeding from the perturbing force (2) by which a planet of mass m_j affects the motion of a given planet, the equations of perturbation (6) assume the form (7) for $d(\mu a)/dt$ if use is made of (8). Hence the secular term (7). Due to the periodicity of the function R_j with respect to M, as a consequence of (4), (2) and (3), we obtain, after performing the integration, the equation (10). Thus, the semi-axes of the planetary orbits are seculary invariable.

On a trouvé un membre séculaire dans les perturbations du troisième ordre (aucun tel membre se trouve dans les perturbations du premier et du second ordre). Mais cela peut être une conséquence seulement du procédé du développement en série, comme - par example - $m^3 t$ peut être le premier membre de la série pour $\sin(m^3 t)$. Il existe, peut-être, une autre voie du développement n´ayant aucun membre séculaire. Pour montrer que cela est vrai, je prouve (avec un peu plus de détails, à cause des gens qui ne peuvent pas facilement accepter cela) que le membre séculaire du demiaxe est toujours nul.

Quand nous avons une planète de la masse m, avec le vecteur de position (par rapport au Soleil) $\vec{r}$ et avec la vitesse héliocentrique $\vec{v}$, attirée par Soleil et par les planètes des messes m_j (avec les vecteurs de position $\vec{r}_j$ et les vitesses héliocentriques $\vec{v}_j$), le mouvement héliocentrique de cette planète est déterminé par les équations:

$$\frac{d\vec{v}}{dt} = f(1 + m)r^{-3}\vec{r} + f\sum m_j \vec{F}_j \tag{1}$$

$$\vec{F}_j = \frac{\partial R_j}{\partial \vec{r}}, \qquad R_j = r_j^{-3}\vec{r}_j \cdot \vec{r} - |\vec{r}_j - \vec{r}|^{-1} \tag{2}$$

L´orbite presque-elliptique de la planète est donné par les expressions:

W. Fricke and G. Teleki (eds.), Sun and Planetary System, 431–433.

$$\vec{r} = a\left[(\cos u - e)\vec{e}^0 + \sqrt{1-e^2} \sin(\vec{c} \cdot \vec{e})^0\right], \quad e = |\vec{e}|$$

$$\vec{v} = \left[-\sin u \ \vec{e}^0 + \sqrt{1-e^2} \cos u \ (\vec{c} \cdot \vec{e})^0 \cdot \frac{1}{r}\sqrt{\mu a}\right. \tag{3}$$

$$u - e\sin u = M, \qquad M = \frac{1}{a}\sqrt{\frac{\mu}{a}} \cdot (t - T) \tag{4}$$

où $\vec{c}$, $\vec{e}$, T sont des quantités presque-constantes et le grand demiaxe a (une quantité presque-constate aussi) est exprimable par quelconque des deux expressions:

$$\frac{\mu}{a} = 2\frac{\mu}{r} - \vec{v}^2, \qquad a = c^2\mu/(1 - e^2), \tag{5}$$

Les perturbations dans le mouvement de la planète sont généralement définies par les relations

$$\left(\frac{d\vec{r}}{dt} = \frac{\partial\vec{r}}{\partial t}\right) = \frac{\delta\vec{r}}{\delta t} = 0, \quad \left(\frac{d\vec{v}}{dt} = \frac{\partial\vec{v}}{\partial t}\right) = \frac{\delta\vec{v}}{\delta t} = f \cdot \sum m_i\vec{F}_i \tag{6}$$

Par cela, de la première égalité (5), on a

$$\mu a^{-2}\frac{da}{dt} = 2f \cdot \vec{v} \cdot \sum m_i\vec{F}_i = 2f \cdot \frac{\partial\vec{r}}{\partial t} \cdot \sum m_i\vec{F}_i$$

Dans ces équations le temps t n'entre directement - d'après (3) - que par u, et d'après (4) seulement par M. Par cela on a

$$\frac{\partial\vec{r}}{\partial t} = \frac{\partial\vec{r}}{\partial M} \cdot \frac{\partial M}{\partial t} = \frac{1}{a}\sqrt{\frac{\mu}{a}} \cdot \frac{\partial\vec{r}}{\partial M}$$

et ensuite

$$\frac{d}{dt}(\sqrt{\mu a}) = f\sum m_i\vec{F}_i \cdot \frac{\partial\vec{r}}{\partial M} = f\sum m_i\frac{\partial R_i}{\partial M} \tag{7}$$

En considérant que la même fonction (3) entre aussi dans les expressions pour $\vec{r}_i$, donc

$$\vec{r} = \vec{h}(\vec{c}, \vec{e}, M), \qquad \vec{r}_i = \vec{h}(\vec{c}_i, \vec{e}_i, M_i)$$

où $(\vec{c}, \vec{e})$, resp. $(\vec{c}_i, \vec{e}_i)$ ne présentent que 5 quantitiés indépendantes (à cause de $\vec{c} \cdot \vec{e} = 0$, $\vec{c}_i \cdot \vec{e}_i = 0$), Dans chacune de ces fonctions on n'y a qu'une quantité changeante rapidement (M ou M_i) et 5 quantitiés presque-constantes - comme les variables indépendantes. En conséquence

$$R_i = R_i \ (\vec{c}, \vec{e}, M; \vec{c}_i, \vec{e}_i, M_i)$$

est une fonction de 12 variables indépendantes (à travers $\vec{r}$, $\vec{r}_i$). On a aussi

$$\vec{F}_i \cdot \frac{\partial\vec{r}}{\partial M} = \left(\frac{\partial R_i}{\partial x}\vec{i} + \frac{\partial R_i}{\partial y}\vec{j} + \frac{\partial R_i}{\partial z}\vec{k}\right) \cdot \left(\frac{\partial x}{\partial M}\vec{i} + \frac{\partial y}{\partial M}\vec{j} + \frac{\partial z}{\partial M}\vec{k}\right),$$

donc

$$\vec{F}_i \cdot \frac{\partial \vec{r}}{\partial M} = \frac{\partial R_i}{\partial x} \frac{\partial x}{\partial M} + \frac{\partial R_i}{\partial y} \frac{\partial y}{\partial M} + \frac{\partial R_i}{\partial z} \frac{\partial z}{\partial M} = \frac{\partial R_i}{\partial M} \qquad (8)$$

où tous les autres variables indépendants restent sans aucun change.

Nous devons trouver les variations séculaires en partant de l'équation (7), en y considérant R_i comme une fonction périodique de M, resp. de M_i. Nous voyons de (2) que - pour un M_i fixe - R_i est une fonction périodique de M (et pour un M fixe, R_i est une fonction périodique de M_i). En prenant strictement, la périodicité concerne la variable u (resp. u_i), mais (4) montre que la périodicité provient de M (resp. M_i). En vérité, pendant qu'on change M de 0 jusque 2π (ou également de $-\pi$ a $+\pi$), on change aussi u de 0 à 2π, donc

$$\vec{r}(2\pi) = \vec{r}(0), \quad resp. \quad \vec{r}(\pi) = \vec{r}(-\pi),$$

sans regard au traitement de $\vec{r}$ - comme une fonction de u ou fonction de M.

Après tous cela, nous pouvons passer au membre séculaire à droite de (7), c'est à dire à l'intégral

$$\left[\frac{d}{dt}(\sqrt{\mu a})\right]_{sec} = \int_{-\pi}^{\pi} \int_{-\pi}^{\pi} \frac{\partial R_i}{\partial M} \, dM_i \, dM = \int_{-\pi}^{\pi} \int_{-\pi}^{\pi} \left[\frac{\partial R_i}{\partial M} \, dM\right] dM_i \qquad (9)$$

En prenant un M_i fixe, l'expression en $\left[\ \right]$ serra - par (2) et (8) -

$$\int_{-\pi}^{\pi} \frac{\partial R_i}{\partial M} \, dM = R_i \, (M, M_i, \vec{c}, \vec{e}, \vec{c}_i, \vec{e}_i)\Big|_{M=-\pi}^{M=+\pi} =$$

$$= \left[r_i^{-3} \vec{r}_i \cdot \vec{r}(\pi) - |\vec{r}_i - \vec{r}(\pi)|^{-1}\right] -$$

$$- \left[r_i^{3} \vec{r}_i \cdot \vec{r}(-\pi) - |\vec{r}_i - \vec{r}(-\pi)^{-1}\right]$$

Et finalement, l'équation (9) devient

$$\left[\frac{d}{dt}(\sqrt{\mu a})\right]_{sec} = 0, \quad resp. \quad (\sqrt{\mu a})_{sec} = const. \qquad (10)$$

ON LINEAR STABILITY OF TRIANGULAR LIBRATION POINTS OF
THE PHOTOGRAVITATIONAL RESTRICTED THREE-BODY PROBLEM
WHEN THE MORE MASSIVE PRIMARY IS AN OBLATE SPHEROID

Ram Krishan Sharma
Applied Mathematics Section
Vikram Sarabhai Space Centre
Trivandrum, INDIA

1. INTRODUCTION

It is well known that the triangular points of the
restricted three-body problem are linearly stable for the mass ratio
$0 < \mu < \mu_0 = \frac{1}{2} - \sqrt{69}/18 = 0.03852089\ldots$, the critical mass value. The
range of the mass parameter giving rise to stable triangular solutions
decreases when the more massive primary (Subba Rao - Sharma, 1975) or
both the primaries (Bhatnagar - Hallan, 1979) are oblate spheroids with
their equatorial planes coincident with the plane of motion. If the
more massive primary is a source of radiation, the value of critical
mass decreases with the increase in the radiation force (Chernikov,1970).

This paper considers the case when the more massive primary
is an oblate spheroid with its equatorial plane coincident with the
plane of motion and also is a source of radiation having its radiation
force Fp exactly opposite to the gravitational attraction force Fg and
that it changes with the distance by the same law. Then it is possible
to consider the result of action of this force to reduce the effective
mass of the radiating primary or a particle. It is acceptable to speak
about a reduced mass of a particle as the effect of reducing its mass
depends on the properties of the particle itself. The resultant force
acting on the particle is $F = Fg-Fp = Fg (1-Fp/Fg) = q\ Fg$, where q is
the mass reduction factor constant for a given particle.

The critical mass μ_c is found to decrease with the increase
in oblateness and radiation pressure force and becomes zero for
$q = (2+3A)(3A/2)^{3/2}/4$, where A is the oblateness coefficient. The
eccentricity of the conditional retrograde elliptic periodic orbits
around the triangular points at the critical mass increases with the
increase in A and Fp and becomes unity when μ_c is zero.

2. EQUATIONS OF MOTION AND LOCATION OF THE TRIANGULAR POINTS

Following the terminology and notations of Szebehely (1967),
the equations of motion are

435

W. Fricke and G. Teleki (eds.), Sun and Planetary System, 435–436.

$$\ddot{x} - 2 n \dot{y} = \Omega_x, \qquad \ddot{y} + 2 n \dot{x} = \Omega_y, \tag{1}$$

with

$$\Omega = n^2\left[(1 - \mu) r_1^2 + \mu r_2^2\right] + q (1 - \mu)/r_1 + \mu/r_2$$
$$+ q (1 - \mu)A/2 r_1^3,$$
$$r_1^2 = (x-\mu)^2 + y^2, \quad r_2^2 = (x+1-\mu)^2 + y^2, \quad n^2 = 1+3A/2. \tag{2}$$

The locations of the triangular points, which are solutions of $\Omega_x = \Omega_y = 0$, $y \neq 0$, are given by

$$(2 + 3A) r_1^5 - 2 q r_1^2 - 3A q = 0, \quad r_2^3 = 1/(1 + 3A/2). \tag{3}$$

Restricting upto quadratic terms in A, we obtain, with $h = (q/n^2)^{1/3}$, the series expansion for r_1 in (3) as

$$r_1 = h + A/2h - 3n^2 A^2/4q. \tag{4}$$

3. SOLUTIONS OF THE LINEARIZED EQUATIONS

The characteristic roots of the linearized equations of motion at the triangular points are

$$\Lambda_{1,2} = \lambda^2 = - k \pm n^2 D^{\frac{1}{2}}/2,$$

where

$$D = 9(u^2+4w)\mu^2 - 6 (3 u^2 - u + 6w)\mu + (9 u^2 - 6 u + 1),$$
$$u = 2A/(2 r_1^2 + 3A), \quad v = y^2/r_1^2 r_2^2, \quad w = v (1 + u),$$
$$k = n^2\left[1 - 3 (1 - \mu) u\right]/2.$$

The value of the critical mass, which is obtained from $D = 0$, $k > 0$, is

$$\mu_c = (a - 2 b^{\frac{1}{2}})/c, \tag{5}$$

with

$$a = 3 u^2 - u + 6 w, \quad b = w (9 w + 3 u - 1), \quad c = 3(u^2 + 4 w).$$

Assuming Fp/Fg a small quantity ϵ and restricting only to linear terms in ϵ and A, we obtain from (3), (4) and (5)

$$\mu_c = \mu_0 - (1 + 13/\sqrt{69})A/9 - 2\epsilon/27\sqrt{69} + (4/27 - 320/621\sqrt{69}) \epsilon A,$$

Proceeding as in Sharma - Subba Rao (1979), the existence of retrograde elliptic orbits around the triangular points can be established. The eccentricity of the orbits increases with increase in A and Fp and becomes 1 for $k = 0$. In that case μ_c becomes 0 and $r_1^2 = 3A/2$, $q = (2+3A)(3A/2)^{3/2}/4$. Restricting only to linear terms in A and ϵ, the eccentricity is

$$e = 2^{\frac{1}{2}} (2^{\frac{1}{2}} - 1)^{\frac{1}{2}}\left[1 + \left\{3(2 - 2^{\frac{1}{2}}) + \epsilon\right\}(1 - \mu) A/4\right].$$

4. REFERENCES

Bhatnagar, K.B. and Hallan, P.P. (1979), Celes. Mech., 20, 95-103.
Chernikov, Y.A. (1970), Astron. Z., 47, 217-223.
Sharma, R.K. and Subba Rao, P.V. (1979), Astrophys. and Space Sci., 60, 247-250.
Subba Rao, P.V. and Sharma, R.K. (1975), Astron.Astrophys.,43, 381-383.
Szebehely, V. (1967), 'Theory of Orbits', Academic Press.

NOUVELLES THEORIES DES PLANETES ET DE LA LUNE DANS LES EPHEMERIDES FRANCAISES

P. Bretagnon, J. Chapront, M. Chapront-Touzé
Bureau des Longitudes, Paris, France

ABSTRACT. New theories for the motions of the Moon and the planets, which were developed at the Bureau des Longitudes, are compared with observations and with numerical methods. These new theories will be introduced in 'Connaissance des Temps' from 1984 on.

INTRODUCTION

La Connaissance des Temps publie encore actuellement les théories anciennes de Le Verrier et Gaillot (1855) pour les mouvements du Soleil et des planètes. La théorie du mouvement de la Lune est celle de Brown corrigée par Eckert : Improved Lunar Ephemeris (Eckert et al., 1954). De nouvelles théories des planètes et de la Lune ont été élaborées au Bureau des Longitudes, par les auteurs. Elles ont été comparées et ajustées à l'observation, via des intégrations numériques modernes (Oesterwinter et Cohen, 1972) et (Standish, 1980, Williams, 1980). Elles seront introduites dans la Connaissance des Temps à partir de 1984.

1. THEORIE DU MOUVEMENT DU SOLEIL ET DES PLANETES

Les théories de Le Verrier et Gaillot sont, pour les planètes inférieures et le Soleil, calculées au premier ordre par rapport aux masses et complétées par quelques termes périodiques du deuxième ordre. Pour les grosses planètes, les perturbations sont développées jusqu'au deuxième ordre et les théories des planètes Jupiter et Saturne contiennent une partie des perturbations du troisième ordre.

Pour notre part, nous avons développé nos solutions jusqu'au troisième ordre par rapport aux masses pour l'ensemble des planètes (Bretagnon, 1980, 1982). Nous complétons, de plus, la théorie des grosses planètes par les perturbations d'ordre 4, 5 et 6 déterminées par une méthode itérative (Bretagnon, 1981). Notre solution est complétée par les effets de la Lune sur le barycentre Terre-Lune et par les perturbations relativistes déterminées par Lestrade (1981).

Pour mesurer la qualité de nos solutions, nous avons effectué des comparaisons à des intégrations numériques internes ainsi qu'à des mo-

437

W. Fricke and G. Teleki (eds.), Sun and Planetary System, 437–440.

dèles extérieurs tels que l'intégration numérique d'Oesterwinter et Cohen (1972) et celles du JPL (Standish, 1980).

Pour les grosses planètes Jupiter, Saturne, Uranus et Neptune, ces comparaisons ont été effectuées sur de longs intervalles de temps et dans l'état actuel de notre solution la précision est de l'ordre de quelques 0",1 sur 1000 ans.

Pour ce qui concerne les éphémérides, nous considérons des intervalles de temps plus petits et nous donnons dans le tableau 1 la précision des longitudes vraies héliocentriques de notre solution actuelle sur un intervalle couvrant un siècle de part et d'autre de l'époque J2000.

Tableau 1. Précision des longitudes vraies héliocentriques

Planète	Merc	Vénus	T-L	Mars	Jup	Sat	Ura	Nep
Précision	0",0012	0",0065	0",0061	0",0210	0",0230	0",0350	0",0230	0",0150

En réalité, les nombres qui figurent dans le tableau 1 représentent les plus gros écarts entre notre solution et DE102 pour les longitudes vraies des planètes Mercure, ..., Saturne. Pour Uranus et Neptune nous avons indiqué les plus gros écarts à une intégration interne. Les écarts plus importants, pour ces deux planètes, que nous avons avec DE102 sont, en effet, probablement dus à l'utilisation, dans DE102, d'une masse de Pluton beaucoup trop forte.

Aux précisions du tableau 1 on peut associer les précisions en ascension droite. Nous les donnons dans le tableau 2.

Tableau 2. Précision des ascensions droites

Corps	Soleil	Merc	Vénus	Mars	Jup	Sat	Ura	Nep
Précision	0^s,0004	0^s,0003	0^s,0021	0^s,0054	0^s,0024	0^s,0032	0^s,0018	0^s,0014

Des comparaisons systématiques ont également été effectuées au Bureau des Longitudes entre DE102 d'une part et l'American Ephemeris (AE), la Connaissance des Temps (CDT) d'autre part. Nous donnons dans le tableau 3 les plus gros écarts ainsi obtenus sur un intervalle de 10 ans ainsi que ceux trouvés à partir des nouvelles théories (BDL) pour le Soleil, Vénus et Jupiter.

Tableau 3. Différences avec DE102 sur 10 ans

	α			δ			Dist (10^{-8}UA)		
	AE	CDT	BDL	AE	CDT	BDL	AE	CDT	BDL
Soleil	0^s,0500	0^s,1198	0^s,0004	0",300	0",531	0",004	130	332	3
Vénus	0^s,1700	1^s,2605	0^s,0021	0",610	10",675	0",018	420	2482	5
Jupiter	0^s,0670	0^s,0500	0^s,0024	0",780	1",426	0",023	620	2449	80

On constate sur le tableau 3 les progrès apportés par les nouvelles théories planétaires qui sont 10 à 100 fois plus précises que les théories de Le Verrier et Gaillot publiées jusqu'à présent. Il est toutefois utile de rappeler que la précision concernant les nouvelles théories est une précision interne. Il peut y avoir en effet une incertitude à caractère systématique, comme, par exemple, une rotation d'ensemble due à une meilleure définition du repère, qui ne devrait pas dépasser $0\overset{''}{,}1$.

2. THEORIE DU MOUVEMENT DE LA LUNE

Mis à part les corrections apportées par (Eckert et al., 1966) au seul problème principal du mouvement de la Lune et les modifications des perturbations dues à la forme de la Terre, calculées par Van Flandern (1969), depuis Brown, du point de vue strict de l'éphéméride, aucun travail d'ensemble n'avait été proposé, pour se substituer aux éphémérides conventionnelles. Depuis une dizaine d'années, un certain nombre de recherches ont été entreprises pour améliorer le travail de Brown. Ces études ont été rendues nécessaires, compte tenu des progrès technologiques des mesures (laser). Nous rappelons ci-dessous les contributions les plus récentes à ces recherches.

Problème principal : SALE (Henrard, 1979), ELP (Chapront-Touzé, 1980) et la solution de Schmidt (1980). Récemment Kinoshita (1982) a montré, avec des comparaisons internes à des intégrations numériques, que les erreurs sur 20 ans dans ELP se situaient au niveau de 6 cm. On peut donc considérer la solution du problème principal comme suffisamment bien connue, actuellement.

Formes de la Terre et de la Lune. On dispose présentement de deux solutions précises : celle de Henrard (1980) dans le cadre de SALE et celle de Chapront-Touzé (1982) pour ELP.

Perturbations planétaires. Standaert (1981) a évalué les perturbations planétaires directes de la Lune, à une précision de calcul de l'ordre de 20 cm. Chapront et Chapront-Touzé (1980) ont évalué les perturbations planétaires des cas direct et indirect, utilisant la théorie du Soleil de Bretagnon (1980). Une comparaison avec les séries de Standaert du cas direct, montre des différences sur les inégalités toujours inférieures à 70 cm. Les deux derniers auteurs, rassemblant les éléments de la solution ELP évoqués plus haut, avec l'adjonction des effets relativistes empruntés à Brumberg (1972) et les perturbations dues aux forces de marées dans une modélisation analogue à celle de (Williams et al, 1978), ont calculé une éphéméride de la Lune sur 20 ans et comparé avec une intégration numérique du JPL : LE51 (Williams, 1980). Une analyse des résidus est effectuée dans (Chapront et Chapront-Touzé, 1981). On y propose en particulier de nouvelles valeurs pour les constantes d'intégration, les paramètres lunaires et solaires, en regardant LE51 comme un "modèle d'observations" ; cette intégration est ajustée à 7 années d'observations laser (août 1969 à janvier 1977). Notre ajustement à LE51 nécessite en particulier un choix identique des paramètres physiques (masses, aplatissement, nombre de Love, etc ...).

Le tableau ci-dessous donne les écarts maximum sur 20 ans dans une comparaison brute de LE51 avec les ILE d'une part et ELP d'autre part. Dans ce dernier cas les systèmes de référence de ELP et LE51 ont été

	α	δ	r (distance)
ILE - LE51	$0\overset{s}{.}0645$	$0\overset{''}{.}590$	1360 m
ELP - LE51	$0\overset{s}{.}0008$	$0\overset{''}{.}005$	12 m

amenés en coïncidence, ainsi que les moyens mouvements du nœud et du
périgée.

Il est à noter que ces écarts sont de l'ordre de la précision de
publication de l'éphéméride dans la Connaissance des Temps. Il est clair
que les différences ELP - LE51, données ci-dessus, nécessitent une modélisation semblable, et en particulier, une valeur analogue de l'accéléra-
tion due aux forces de marées.

CONCLUSION

Les précisions internes des théories des planètes et de la Lune,
élaborées au Bureau des Longitudes, sont au moins égales a la précision
de publication. Des différences à l'Observation peuvent exister avec le
choix actuel des conditions initiales du mouvement, des paramètres
physiques ou du système de référence. Celles-ci présentent un caractère
systématique dont la grandeur ne dépasse pas $0\overset{''}{.}1$. Ainsi, les progrès
réalisés justifient l'introduction des nouvelles théories dans la Connaissance des Temps, à partir de 1984.

BIBLIOGRAPHIE

Bretagnon P., 1980 : Astron. & Astrophys., 84, 329
Bretagnon P., 1981 : Astron. & Astrophys. (à paraître).
Bretagnon P., 1982 : Celes. Mech. 26, 161.
Brumberg V.A., 1972 : Relativistic Celestial Mechanics. Nauka. Moscou.
Chapront J., Chapront-Touzé M., 1980 : Astron. & Astrophys., 91, 233
Chapront J., Chapront-Touzé M., 1981 : Astron. & Astrophys. (à paraître)
Chapront-Touzé M., 1980 : Astron. & Astrophys. 83, 86
Chapront-Touzé M., 1982 : Celes. Mech. 26, 53, 63
Eckert W.J., Jones R. and Clark H.K., 1954 : Improved Lunar Ephemeris.
    U.S. Government Printing Office. Washington.
Eckert W.J., Walker M.J. and Eckert D., 1966 : Astron. J., 71, 314
Henrard J., 1979 : Celes. Mech., 19, 337
Henrard J., 1980 : SALE. Facultés Universitaires de Namur. Publ. du
    Dpt. de Math.
Kinoshita H., 1982 : Proceedings :"High Precision Earth Rotation and
    E-M Dynamics". Ed. O. Calame. Reidel, Dordrecht, 245.
Lestrade J.F., 1981 : Astron. & Astrophys., 100, 143
Le Verrier U.J.J., 1855 : Annales de l'Observatoire de Paris.
Oesterwinter C., Cohen C.J., 1972 : Celes. Mech., 5, 317
Schmidt D.S., 1980 : The Moon & the Planets, 23, 135
Standaert D., 1981 : Dissertation Doctorale. Facultés Universitaires Namur
Standish E.M., 1980 : DE-102. Ruban magnétique.
Van Flandern T.C., 1969 : Celes. Mech., 1, 163
Williams J.G., Sinclair W.S. and Yoder C.F., 1978 : Geophysical Research
    Letters, 5, 943
Williams J.G., 1980 : LE-51. Ruban magnétique.

NEW PLANETARY EPHEMERIDES BACK TO 4000 B.C.

Rudolf Dvorak
Astronomisches Institut der Universität Graz, Austria

In this paper new planetary ephemerides for Jupiter and Saturn back to 4000 B.C. are presented as they recently have been developed (H.Hunger and R.Dvorak, 1981).

The Ephemerides for planets are calculated either with numerical integration methods or with analytical or semianalytical perturbation theories. While the first method is used for the positions of the outer planets in the Nautical Almanach (W.J.Eckert et al. 1951) the data in the Connaissance des Temps for the planets Jupiter, Saturn, Uranus and Neptune are determind after the theories of Le Verrier (1855-1861) and Gaillot (1904,1913). Each method has specific advantages, e.g. numerical integration can be corrected rather easily in adapting the position and velocity vector to new observations, while the perturbation theory is more convenient to analyze the planetary motions and give a better understanding of them. Therefore the analytical methods are used for creating tables of positions of the planets as it has been done by Tuckerman (1961,1962) from 1649 A.D. to 600 B.C. and by P.V.Neugebauer (1914) in form of special tables back to 4000 B.C. which covers the historically most interesting time scale concerning astronomical observations.

Whereas the positions of the inner planets are clearly well determined even with older planetary theories, for Jupiter and Saturn the Le Verrier and Gaillot theory (GT) is erroneous as it has been shown in direct comparison with a new theory developed by P.Bretagnon (1978). There are two different main sources of errors in the GT as can be looked up in detail in R.Dvorak (1980):

1. The GT use slightly different polynomial expressions for the mean longitudes due to false integration constants (particularly the masses). The greatest error in the time interval considered is caused by T^2 terms in comparing the GT with the Bretagnon theory (BT); ($-38\overset{..}{}6\ T^2$ for Jupiter and $+69\overset{..}{}87\ T^2$ for Saturn, T expressed in 1000 Julian years).

2. In the GT there are important terms of the great inequality $\{2\lambda(\text{Jupiter}) - 5\lambda(\text{Saturn})\}$ neglected, which lead to important differences in the calculated longitudes and consequently also in the positions of the two planets. In table 1 we present the errors due to both effects in units of $0\overset{.}{.}01$ from 0 to 4000 B.C. in a 500 years intervall.

W. Fricke and G. Teleki (eds.), Sun and Planetary System, 441–442.

Because of the above mentioned errors we decided to calculate for
Jupiter and Saturn the positions from 600 B.C. to 1000 B.C. with the BT
where we can guarantee a precision of 0.01 in the positions (H.Hunger
and R.Dvorak, 1981). In another step we want to determine the position
of Jupiter and Saturn from 600 B.C. on up to the moment until we find
an agreement between BT and GT to 0.01 which should be the time between
100 B.C. and 100 A.D.. In addition a recalculation of the P.V.Neugebauer
tables for Jupiter and Saturn is in preparation to give positions back
to 4000 B.C. with a precision of at least 0.1.

TABLE 1

Errors in the longitudes of Jupiter and Saturn in the Le Verrier –
Gaillot theory due to slightly different integration constants (IC)
and neglected terms in the great inequality (GI) in units of 0.01.

	Jupiter		Saturn	
years B.C.	IC	GI	IC	GI
0	− 4.4	− 1.2	8.1	− 1.1
500	− 7.7	4.3	12.2	1.6
1000	−12.2	− 9.7	17.0	− 0.6
1500	−18.0	17.1	22.5	− 4.3
2000	−25.1	−24.4	28.6	16.0
2500	−33.8	28.1	35.4	−36.0
3000	−44.2	−23.2	42.8	62.6
3500	−56.3	4.6	50.6	−89.2
4000	−70.4	30.9	59.0	103.0

References

Bretagnon,P.: Sur une Solution globale du Mouvement des Planètes, Thèse
    de doctorat d'état, Université Paris 6 (1978).
Dvorak,R.: Genauigkeitsuntersuchungen der Theorien von Jupiter und
    Saturn nach Le Verrier und Gaillot, Sitzungsber.Österreich.Akad.Wiss.
    189.Bd., 129-138
Eckert,W.J., D.Brouwer, and G.M.Clemence: Coordinates of the five outer
    Planets (1653-2020). Astron.Pap. 12 (1951)
Gaillot,A.: Annales de l'Observatoire de Paris 24 (1904)
Gaillot,A.: Annales de l'Observatoire de Paris 31 (1913)
Hunger,H., Dvorak,R.: Ephemeriden von Sonne, Mond und hellen Planeten
    von 1000 bis 600 B.C., (Österreich.Akad.Wiss.; in print)
Le Verrier,U.J.J.: Annales de l'Observatoire de Paris 1-6 (1855-1861)
Neugebauer,P.V.: Tafeln für Sonne, Planeten und Mond, Leipzig 1914
Tuckerman,B.: Planetary,Lunar and Solar Positions 601 B.C. to A.C. 1
    Philadelphia 1962
Tuckerman,B.: Planetary,Lunar and Solar Positions A.D. 2 to A.D. 1649,
    Philadelphia 1964

EPHEMERIDES ET OBSERVATIONS DE MARS A L'ASTROLABE

S. Débarbat
Observatoire de Paris 75014 Paris France
M. Sanchez
Instituto y Observatorio de Marina San Fernando Espana
M. Standish
Jet Propulsion Laboratory Pasadena USA

ABSTRACT. Observations of Mars carried out with astrolabes around the oppositions of 1975 and 1978 have been analysed. Compared were the observations with the ephemerides developed by JPL and with those based on various analytical theories.

Différentes observations de Mars autour des oppositions de 1975 et de 1978, effectuées à des astrolabes, ont fait l'objet d'analyses récentes (Débarbat et al. 1978 et 1979, Débarbat et Sanchez 1981, Standish et al. 1981). Les conclusions confirmaient que l'erreur d'une observation était inférieure à $0\overset{"}{.}3$; malgré le petit nombre des observations (de 2 à 3 dizaines au total), celles-ci permettaient de mettre en évidence des écarts entre observations et éphémérides (de l'ordre d'une demi-seconde de degré), la précision de l'écart se situant, sensiblement, entre $0\overset{"}{.}05$ et $0\overset{"}{.}10$. C'est pourquoi une comparaison des observations effectuées conjointement aux astrolabes de San Fernando et de Paris, au cours des deux campagnes (1975/76 et 1977/78), avec différentes éphémérides a été entreprise, au Jet Propulsion Laboratory (JPL) pour les éphémérides planétaires qui y sont établies, aux Observatoires de Paris et de San Fernando pour des éphémérides basées sur des théories analytiques diverses.

Ephémérides employées :   CDT – Connaissance des Temps ; basée sur les Tables de Leverrier (argument temps uniforme de Leverrier considéré comme voisin du temps des éphémérides).   AE : American Ephemeris and Nautical Almanac ; éphémérides basée sur les Tables de Newcomb avec application des corrections de Ross (équinoxe différant (Duncombe et al. 1974) de celui du FK4 de $+ 0\overset{s}{.}048$ T où T est en siècles juliens depuis 1960.0).   XPEr – Ephéméride analogue à celle de Kaplan, Pulkinen, Emerson (US Naval Circular n° 151) révisée par Seidelmann, basée (Seidelmann 1978) sur des éléments déterminés par Laubscher qui donne pour la correction d'équinoxe (Laubscher 1971) $+ 0\overset{s}{.}089 - 0\overset{s}{.}082$ T (T en siècles juliens depuis 1850.0) ; ce travail est basé sur près de 6000 observations optiques (1751-1969) et près de 800 observations "radar" (1964-1969).   DE 111 et DE 114 - Ephémérides du JPL diffé-

W. Fricke and G. Teleki (eds.), Sun and Planetary System, 443–444.
Copyright © 1982 by D. Reidel Publishing Company.

rant des éphémérides antérieures, DE 96 et DE 108, par l'introduction
de données déduites de tirs "laser" sur la Lune ; l'orientation de DE 96
et de DE 108 était majoritairement tributaire des observations optiques,
DE 111 et DE 114 sont orientées par les mesures lunaires.   BDL -
Ephéméride basée sur la Théorie des planètes élaborée au Bureau des
Longitudes (Bretagnon 1982, Bretagnon et al. 1981) dont l'ensemble est
orienté par calage sur l'éphéméride DE 102 (intermédiaire entre DE 96 et
DE 108) pour la détermination des constantes d'intégration.

           L'étude globale des observations pour chacune des quatre éphé-
mérides du JPL confirme que la correction de phase est sous-estimée,
qu'il existe (comme dans les observations méridiennes) un "effet de phase"
et qu'il faut prendre en compte une correction à l'équinoxe. DE 96 et
DE 108 sont très similaires, de même DE 111 et DE 114. Les premières
sont plus proches des observations (notamment en déclinaison) et deman-
dent une correction du mouvement de l'équinoxe plus grande que les se-
condes ($\Delta\alpha$(E) de l'ordre de 1"50 (0$\overset{s}{.}$100) et 1"05 (0$\overset{s}{.}$070) par siècle ;
valeur déterminée par Fricke : 1"275 (0$\overset{s}{.}$085) par siècle (Fricke 1980)
prise en compte pour les observations étudiées).

           Pour les autres éphémérides, les écarts (O-C) présentent des
variations qui atteignent 0$\overset{s}{.}$8 (CDT) ou 0$\overset{s}{.}$15 (AE) en ascension droite et
plus de 2"(CDT) ou sont quasi nulles (AE) en déclinaison. L'étude par
campagne des différences entre observations et éphémérides (KPEr, DE 111,
DE 114, BDL) s'est appuyée sur un ensemble de courbes de lissage ajustées
selon la méthode développée par Vondrak en 1969, courbes représentant les
observations avec une précision se situant entre 0$\overset{s}{.}$003 et 0$\overset{s}{.}$011 (ascensions droites) et entre 0"03 et 0"09 (déclinaisons). Sont donc signifi-
catifs aussi bien les accords et désaccords entre éphémérides que les
écarts O-C. En ascension droite, on note en particulier des écarts entre
éphémérides atteignant 0$\overset{s}{.}$030 en 1975/76, et se réduisant de moitié en
1977/78. En déclinaison, l'accord entre éphémérides est relativement bon
mais les O-C, faibles en 1975/76, sont nettement négatifs (de l'ordre de
0"10 à 0"20) en 1977/78.

Références

Bretagnon, 1982, Cel. Mech., 26, 161.
Bretagnon, Chapront et Chapront-Touzé, 1981, VIème ERMA.
Débarbat, Pham Van et Sanchez, 1978, Astron. & Astrophys., 64, 281.
Débarbat, Pham Van et Sanchez, 1979, Astron. & Astrophys., 73, 202 et
     77, 370.
Débarbat et Sanchez, 1981, Astron. & Astrophys., 96, 103.
Duncombe, Seidelmann, Van Flandern, 1975, IAU Coll. N° 26, 223.
Fricke, 1981, IAU Coll. N°56, 331.
Laubscher, 1971, Astron. & Astrophys., 13, 426.
Seidelmann, 1978, communication personnelle.
Standish, Débarbat et Sanchez, 1981, Astron. & Astrophys., sous presse.

OBSERVATIONS OF THE SUN AND INNER PLANETS WITH THE LARGE MERIDIAN CIRCLE IN BELGRADE

S. Sadžakov, M. Dačić, D. Šaletić, B. Ševarlić
Astronomical Observatory, Belgrade, Yugoslavia

At the Belgrade Astronomical Observatory since January 1975 they have been regularly observations of the Sun and inner planets. The observations are made according to corresponding instructions (Anonymous, 1963). During the observations of the Sun Sukharev´s filter is used. The object were observed with a hand driven micrometer by the differential method. The choice of reference stars was made according the same criterion as applied in the formation of the differential cataloques (Sadžakov, 1972, 1981). There were no alterations of the observational team; the distribution of observations was on the whole even and personal errors were determined as well. The mean duration of diurnal observations was four hours. During the summer a small number of FK4 stars was observed to a magnitude of 1.2 only, whilst possibilities were greater during the winter period (3.4 apparent magnitude). Because of this more Küstner´s series were observed during the summer period (during the winter period two series corresponding to both positions of the circle were observed).

The ephemeris of the Sun, Mercury and Venus were calculated in Pulkovo observatory according to a programme provided by M.Chubej.

Observed apparent right ascensions and declinations of the Sun and the planets are compared to the ephemeris ones and they are presented in Table 1.

Table 1. Mean values of (0-c) (observed minus calculated) of the period 1975-1981; ε are the mean square error of obtained values; n_1 and n_2 are the number of observations for the determination of right ascension resp. declination.

Object	$(0-C)_\alpha$	ε_α	n_1	$(0-C)\delta$	ε_δ	n_2
Sun	$-0^s.002$	$\pm0^s.006$	155	$-0''.05$	$\pm0''.03$	230
Mercury	$-0^s.005$	$\pm0^s.014$	55	$-0''.04$	$\pm0''.05$	66
Venus	$+0^s.006$	$\pm0^s.006$	171	$+0''.01$	$\pm0''.03$	191

W. Fricke and G. Teleki (eds.), Sun and Planetary System, 445–446.

Since the observations were made everly during a year the formulae proposed by Newcomb (Nemiro, 1963) are used for the calculation of the corrections of the orbital elements of the Sun

$$\Delta\alpha = -\Delta A - \cos\alpha tg\delta\Delta\epsilon - 2\cos\epsilon \ sec^2\delta\cos M \ e\Delta\Pi$$

$$\Delta\delta = -\Delta\delta_0 + \sin\alpha\Delta\epsilon + \sin\epsilon \ \cos\alpha \ (1 + 2e \ \cos M)\Delta L_0$$

and expressions

$$\Delta\lambda = x_1 + y_1\cos(1 - L) + z_1\sin(1 - L)$$

$$\Delta\beta = x_2 + y_2\cos(1 - L) + z_2\sin(1 - L)$$

for the calculation of the corrections of the orbital elements of Mercury and Venus (McClenahan, 1952).

Table 2.   The corrections of the orbital elements of the Sun and the errors (σ) of their determination, based on the observation in the period 1975 - 1981.

	ΔA	$\sigma_{\Delta A}$	$\Delta\epsilon$	$\sigma_{\Delta\epsilon}$	$e\Delta\Pi$	$\sigma_{e\Delta\Pi}$	n_1	$\Delta\delta_0$	$\sigma_{\Delta\delta_0}$	$\Delta\epsilon$	$\sigma_{\Delta\epsilon}$	ΔL_0	$\sigma_{\Delta L_0}$	n_2
	\multicolumn{7}{c}{in $0\overset{s}{.}001$ unis}				in $0\overset{\prime\prime}{.}01$ unis				
Corrections	-1	±7	+53	±42	-8	±5	155	+5	±3	-3	±4	+10	±9	230

Table 3.   The corrections of the orbital elements of Mercury and Venus and the errors (σ) of their determination, based on the observations in the period 1975 - 1981.

	x_1	σ_{x_1}	y_1	σ_{y_1}	z_1	σ_{z_1}	n_1	x_2	σ_{x_2}	y_2	σ_{y_2}	z_2	σ_{z_2}	n_2
	\multicolumn{7}{c}{in $0\overset{s}{.}001$ unis}				in $0\overset{\prime\prime}{.}01$ unis				
Mercury	-20	±13	-10	±19	+42	±15	55	-1	±7	- 2	±11	-2	±9	66
Venus	+ 6	± 6	+ 3	± 9	+ 6	± 8	171	-4	±4	-17	± 6	-5	±5	191

From the data Tables 2 and 3 it is evident that the values of the corrections obtained coincide one with another within the limits of the errors in posision determinations. Determination errors $\Delta\delta_0$ are approximately equal for the Sun and Venus than for the Mercury. The latter fact can be explained by the unequal number of observations and also by the existence of a dependence on accuracy in the determination of zero point corrections to the mean distances to the planets.

REFERENCES:

Anonymos, 1963. Trudy 15 atrometr. Konf. Leningrad, 420-422.
McClenahan,W.S.: 1952, Publ. Dom. Obs. Ottawa, 15, 106.
Nemiro,A.A.: 1963. Trudy 15 astrometr. Konf. Leningrad, 87-89.
Sadžakov, S. Chaletić,D.: 1972. Publ.Obs.astr. Beograd, 17, 1.
Sadžakov, S. Chaletić,D. Dačić,M.: 1981. Publ.Obs.astr. Beograd, 30, 1.

SOME RESULTS OF THE MERCURY TRANSIT OBSERVATIONS IN 1970 AND 1973 AT BELGRADE

V. Benishek-Protitch
Astronomical Observatory, Belgrade, Yugoslavia

1. INTRODUCTION

The transits of Mercury across the solar disk in 1973 have been observed at the Belgrade Observatory too. The photographs in 1970 have been taken in the focal plane of the refractor 650/10550 mm., whereas that in 1973 was achieved with the refractor guidescope of the astrograph 11/128 mm., to which a special solar camera with a teleobjective (magnification 5.25) was attached. Complete material, though partially processed, remains unexploited to the present day.

2. MEASUREMENTS OF PLATES AND RELATIVE COORDINATE DETERMINATIONS

In our enedeavouring to obtain from these observations as much irformations as possible, a selection of photographs was made. Thus, 45 photographs out of 56 in 1970 and no more than 18 of 28 in 1973 were chosen. The plates were measured using the Zeiss measuring engine, which allows a precision of ±0.001 mm.

In order to determine the coordinates of the centre of the Sun's disk we measured the equidistant chords of its image, parallel to the axes of the corresponding rectangular coordinate system. There were 7 chords (the central one inclusive) parallel to the first and as many parallel to second axis. It is assumed that the center of Sun's image is coincident with the intersection of straight lines determined by the middle points of each one of the two sets of chords. Mercury coordinates were obtained by measurements of the east and west, north and south edges of its disk (5 settings in each instance). Proceeding from these data the coordinates of Mercury relative to the centre of the Sun were determined as well as the most probable values of the apparent semi-diameters of the two bodies.

3. SPHERICAL DISTANCES OF THE CENTRES OF MERCURY AND SUN AND ADJUSTMENT OF THEIR VALUES

By a suitable transformation, taking fourth order quantities into account, a very simple analytical relation between relative coordinates Δx and Δy and the angular distance of Mercury from the Sun, Δs, was established. By using this expression we determined, for each of the individual moments of observation, the corresponding quantities Δs and, thereafter, by the least-squares method, their most probable values.

W. Fricke and G. Teleki (eds.), Sun and Planetary System, 447–448.

The adjustment was effected through intermediary of the equations of conditions in the form:

$$\Delta m = \left(\Delta s^2 - \Delta s_0^2 \right)^{1/2} = a_0 + a_1 t + a_2 t^2 + a_3 t^3; \quad (t = T - T_0)$$

where the parameters a should be determined.
The least angular distance Δs and the instant T_0 were found with a satisfactory accuracy, by interpolating the values Δs near the predicted minimum.

4. THE TIMES OF CONTACTS AND RESIDUALS

Even though the revision of the observations of Mercury's transits made at Belgrade has been undertaken principally with the aim of deducing solar parallax in spite of the long prevailing view of the unsuitability of these transits for such refined calculation, we determined also other parameters connected with this phenomenon (Benishek-Protitch, V., 1981). The relation $\Delta m = f(t)$, the correctness of which is proved by the high coeficient of correlation (r=0.999 999), allowed by employing the found mean values of the Sun's and Mercury's radii, the determination of times of all four local contacts. According to our observations the times of these contacts, the time of the least angular distance of centers and the semi-diameters of the Sun and Mercury are as follows:

UT	1970	(O-C)	1973	(O-C)
T_1	$4^h 21^m 01\overset{s}{.}1$	$+50\overset{s}{.}9$	$7^h 47^m 13\overset{s}{.}7$	$-53\overset{s}{.}1$
T_2	4 24 00.0	+49.0	7 48 52.7	-53.4
T_0	8 16 29.0	+23.5	10 32 22.2	-19.8
T_3	12 08 52.9	- 7.5	13 15 20.2	-16.7
T_4	12 11 51.5	- 9.0	13 17 00.5	-16.7
Δs_0	118''.48	+ 0''.41	29''.44	+ 1''.05
R_s	951.72	± 0.19	970.55	± 0.19
r_M	5.99	± 0.11	4.94	± 0.12

The (O-C) values are calculated using the corresponding data from Astronomical Ephemeris. In evaluating these residuals one should keep in mind that our values of the apparent solar radius R_s differs form the ephemeris ones by +1''.3 on the average. This is in agreement with the value of the current and the earlier adopted correction for irradiation.

5. CONCLUSION

From the given data it follows that the duration of the transit in 1970 was shorter from the predicted one by 1^m. This result is quite in accordance with the change in the position of the apparent path of Mercury relative to the center of the Sun.
However, the interval between the begining and the end of the phenomenon in 1973 is longer by $0\overset{m}{.}6$. Such difference is in contradiction with the stated deviation in Δs. Considering this amount, a substantiately shorter duration of the transit is what one would have expected. What is the real cause of this discordance, we are not able to say for the time being.

REFERENCE:

Benishek-Protitch, V., 1981. MA Thesis (unpublished).

CONTRIBUTION OF THE PULKOVO OBSERVATORY TO THE
IMPROVEMENT OF ORIENTATION OF THE FK4 SYSTEM
USING OBSERVATIONS OF SELECTED MINOR PLANETS

L.S. Koroleva   V.I. Orelskaya
Pulkovo Observatory Institute for Theoretical Astronomy
Leningrad, U.S.S.R. Leningrad, U.S.S.R.

The astronomical system of celestial coordinates is in process of
revision. This revision is due to the following facts: the adopted
value of precession in FK4 is not accurate, the equinox has a nonproces-
sional motion, the more distant the epoch of observations the worse is
the accuracy of coordinates of stars. Since 1963 (when the compilation
of the catalogue FK4 was completed) a number of new catalogues have appeared which can be used for an improvement of the system of the fundamental catalogue FK4 and proper motions of fundamental stars; besides,
the catalogue can be extended to the stars up to 9^m, for which purpose
the AGK3R and SRS (compiled in the FK4 system) can be useful.

The 15th General Assembly of the IAU in 1973 recommended the improvement of the FK4 catalogue so that the new FK5 system could represent as closely as possible the true dynamical coordinate system
(see Orelskaya 1980).

For a correction of the FK4 an improvement of the equinox and
equator is needed. So far, most determinations of corrections to the
elements of orientation have been carried out from observations of the
Sun, moon and major planets. But since these bodies of the solar system
don't have point images, systematic observational errors arise. Hence,
the corrections obtained are inaccurate. That is why Dyson (1928) suggested that disc-like bodies of the solar system should be replaced by
point images of minor planets. On the basis of this idéa plans were
made by Numerov (1935) and Brouwer (1935). Numerov was the first to
outline the program; he proposed to use 9 minor planets for correction
not only of the equator and equinox but also of systematic errors of star
positions. The American astronomer Brouwer (1941) put forward a larger
project based on observations of 16 selected minor planets. Observations
on Brouwer's project were taken from 1935 to 1948 and the first results
of the determination of systematic errors of star positions of the Yale
catalogues and the GC were published by Pierce (1971).

The plans by Numerov and Brouwer were adopted at the IAU General
Assembly in Paris in 1935. But it was possible to begin the work on

449

W. Fricke and G. Teleki (eds.), Sun and Planetary System, 449–451.

Numerov's plan only in 1949. The collection of the data and its reduction was begun at the Institute for Theoretical Astronomy in the USSR
in 1965 and by now about 30 000 observations of minor planets have been
collected from 32 observatories in this country and abroad for a period
of 25 years.

Nonprecessional motion of the equinox was discovered from analysis
of proper motions of stars in right ascension. It was caused by   insufficiently accurate determinations of positions of the equinox in the
XIX century due to the fact that the magnitude equation was not taken
into account. It was noticed that moments of registration of the passage of a bright star and of a faint one are essentially different, and
since right ascensions of the other stars are referred to this moment
this causes a motion of the equinox. Thus, different zero points of RA
contributed to the error of the equinox. The motion of the equinox was
obtained from right ascensions in determining the precession from proper
motions of stars. According to Fricke (1967) the FK4 proper motions in
RA require a correction $\Delta e = + 0\overset{s}{.}083$ per century, which means that the
correction ΔA to the FK4 equinox is a function of time.

Fricke (1980) gave the following formula: $\Delta A(T) = \Delta A(T_0) + (T - T_0)$
$d(\Delta A) dT$. With the help of this formula one can calculate the correction
for any epoch T. He collected 35 determinations of the equinox correction from observations of the Sun, the major and minor planets and
occultations of stars by the Moon. On this basis he determined an equinox correction for 1950.0 equal to $0\overset{s}{.}035$ and a centennial variation of
this correction that proved to be equal to $0\overset{s}{.}085$, which is in good
agreement with the determination from proper motions of stars. This
confirms the motion of the equinox.

On the basis of observations of selected minor planets at the Institute for Theoretical Astronomy in 1972-1980 the equinox correction was
determined from 9 minor planets for the epoch 1976.0 by Orelskaya (1980a,
1981). It turned out to be $0\overset{s}{.}056$. This result obtained from observations of minor planets confirms the value determined from Fricke's
formula which is

$$\Delta A(T) = + 0\overset{s}{.}035 \pm 0\overset{s}{.}003 + (0\overset{s}{.}085 \pm 0\overset{s}{.}010) \ (T - 19.50).$$

Due to this correction the right ascensions of the FK5 will define
the zero point as closely as possible to the true intersection of the
ecliptic and equator. The equator of the existing astronomical coordinate system given by the FK4 catalogue is based on declinations of stars
which were determined using absolute catalogues of 1846-1956 and observations of the Sun and major planets. The results obtained from observations of minor planets showed that the equator (FK4) is well fixed
by the declinations of stars and does not need any correction, which is
in agreement with the results obtained by Fricke (1981).

The Pulkovo Observatory contributed greatly to the improvement of
the position of the equinox point and equator. Since 1949 observations

of selected minor planets have been made (1500 observations were analy-
sed). At the Pulkovo Observatory the plates taken at the Cape Observatory in 1956-1963 were processed. At the Nikolaev Branch of the
Pulkovo Observatory 1015 observations of minor planets were made from
1961 to 1975. At Pulkovo about 3000 observations of minor planets were
obtained without taking into account those that were made in expeditions
and have not yet been reduced. Thus, at the Pulkovo Observatory 10 per
cent of all the observations of minor planets were made to which 9
Soviet observatories and 20 observatories in other countries have contributed.

REFERENCES

Brouwer,D.: 1935, A.J. 44, 1022
Brouwer,D.: 1941, Annals New York Acad. Sciences 42, 133
Dyson,V.: 1928, Transactions IAU III, 227
Fricke,W.: 1967, A.J. 72, 1368 (Mitt. Astron. Rechen-Institut,
     Ser. B, No.16)
Fricke,W.: 1980, Celestial Mechanics 22, 133
Fricke,W.: 1981, In "Reference Coordinate Systems for Earth Dynamics".
     p.331. Eds. E.M. Gaposchkin, B. Kolaczek. Reidel-Dordrecht
     (Mitt. Astron. Rechen-Institut, Ser. A, No.135)
Numerov,D.: 1935, Astron. Zh. Acad. Nauk SSR 12, 584
Orelskaya,V.: 1980, Pis'ma Astron. Zh. 6, 318
Orelskaya,V.: 1980a, Pis'ma Astron. Zh. 6, 659
Orelskaya,V.: 1981, Trudy 22, Astron. Conference, Moscow (in press)
Pierce,D.: 1971, A.J. 76, 177

SECTION V

THREE-DIMENSIONAL REFRACTION

PROBLEMS OF THREE-DIMENSIONAL REFRACTION IN ASTROMETRY

G.Teleki
Astronomical Observatory, Belgrade, Yugoslavia,
J.Saastamoinen
National Research Council of Canada, Ottawa, Canada

ABSTRACT. After the discussion of the basic questions, it is demonstrated that it is possible - and indeed necessary - to determine spatial refractional changes appearing in the classical astrometric observations at larger zenith distances. Making use of the local corrections of this kind to high precision astrometric observations is indispensable. Suggestions concerning further researches and application are given. A new global atmospheric model is presented.

1. BASIC PRINCIPLES

In considering the astronomical refraction, the fact that the light ray is passing through an inhomogeneous medium must be taken into account. The light ray traveling through this medium from the point P_1 to P_2 is described by the well-known Fermat-principle:

$$\delta \int_{P_1}^{P_2} n \, ds = 0 \qquad (1)$$

for which an infinitive number of parameters must be known. The strong formulation of this principle is practically impossible in astrometrical practice, therefore we have to consider its weak formulation only (Teleki, 1974). This fact leaves open the question on the strictness of the law of refraction it implies. But at present the Optics does not provide an answer to this dilemma.

The second question concerns the value of refractive index n, appearing in (1), as a function of position. The formulae according to which the refractive index is calculated (Teleki, 1974) are based on results obtained under laboratory conditions, and they give the values n as a function of pressure, temperature, water vapour and carbone dioxic, with the accuracy of the order of 10^{-8}. For the practice this accuracy is sufficient - although the laboratory conditions do not correspond to the dynamical atmospheric conditions (air flow, gradients of meteorological

W. Fricke and G. Teleki (eds.), Sun and Planetary System, 455–462.
Copyright © 1982 by D. Reidel Publishing Company.

elements) to which the formulae are being applied. The main problem is not presented by the formulae, but by their application. If we want to retain the accuracy of 10^{-8}, then the meteorological elements must be known very accurately: the pressure below 0.025 mm Hg, temperature below 0.01, relative humidity below 0.01% (at 20°C), and the CO_2 gas quantity below 0.00006 parts by volume. There are such accuracies which are unattainable in the astronomical practice: the possible accuracy of the determination of n is about 10^{-6}. For this reason in the calculation of the pure (normal) refraction we cannot expect a higher accuracy in accidental respect than $\pm 0''.02$ tg z. There is one more question related to the calculation of n: where to measure the meteorological elements to get the most realistic value of pure refraction.

A third, apparently the most important problem related to (1) is the determination of elements of meteorological field along the optical path from P_1 to P_2. Of particular importance is the knowledge of these fields at the lower layers where the refractional effects are likely to be the greatest and most changeable in time. As temperature fluctuations are known to be the principal "culprit" of the optical inhomogeneity of the atmosphere it is obvious that temperature fields must be the prime object of investigation.

2. ASTROMETRIC PRACTICE AND POSSIBILITIES

In the current astrometric practice, the calculation of the refractional influence is made on the assumption that the path of a light ray is contained in a plane, whereby, besides an atmospheric model, meteorological elements are taken into account at one point only, i.e., at telescope. Moreover, it is assumed that one has to deal with a spherical and symmetrical atmospheric model, which means that the calculated refraction values are independent of the azimuths of the stars.

The treatment of refraction in the way just exposed might be called the two-dimensional approach to this question. However, if astrometry is to pass over into the treatment of refraction as a three-dimensional phenomenon, it must overcome a whole array of difficulties. The astrometry will achieve the required results only gradually; all the solutions of integral (1) impose a necessary condition that the three-dimensional gradients of the refractive index are known. In surface layers it is possible, although very difficult, to measure values of n, while in higher layers only some averaged parameters are what we usually can reckon with. The determination of the position of the surfaces of constant refractivity is yet another problem.

Accordingly, the present state is as follows. In view of what contemporary Optics and Meteorology are capable of offering it is possible to obtain three-dimensional refraction influences using measured data of meteorological fields surrounding the instrument, and by taking into account an atmospheric model (for the free atmosphere separately). This means that, in practice, the most preferable procedure would be to calculate some average influence using the atmospheric model adopted, and then to determine the refraction anomalies according the regional and local measurements. The whole issue, therefore, tends to approx - mations and the practical introduction of local refractional values.

3. POSSIBLE THREE-DIMENSIONAL REFRACTIONAL INFLUENCES IN ASTROMETRY

Three-dimensional refractional influences can be divided into two compo- nents: horizontal component δA (in azimuth), and vertical comporent δz (in zenith distance). In the following, by δz will be understood the anomalous refraction values only, i.e., the differences between the true and the pure refractions. Their possible amounts can be assessed on the basis of measurements already made, and by making relevant calculations.

Harzer (1924), the first author to introduce three-dimensionality into the concept of astronomical refraction, inferred from his calculations that δA, even at the horizon, does not exceed $0\overset{''}{.}4$. His findings are confirmed by geodetic measurements: mean δA values in the measurements of horizontal angles amount to about $0\overset{''}{.}4$ to $0\overset{''}{.}6$, although under exceptional circumstances (in city areas) they may attain 20'' (Yunoshev, 1969). In this, one should bear in mind that δA is variable as a function of cosec z (Moritz, 1967).

The magnidutes of δz can be judged from the calculations of Harzer (1922- 24), Sugawa and Kikuchi (1979), and Saastamoinen (1980). We gather from Harzer's Tables (pp. 26-29), that, for the Kiel meridian, there is even a zenith refraction of $0\overset{''}{.}00170$ and that there is a difference in the refractional values for the same zenith distances to the south and to the north of the zenith. This difference grows with increasing zenith distance. However, up to 85° zenith distance it does not exceed $0\overset{''}{.}06$. At the horizon it varies up to 11''. Sugawa an Kikuchi have found that δz in the zenith zone, in the North Hemisphere, dose not exceed $0\overset{''}{.}003$ in the north-south direction, while this limit in the east-west direction is $0\overset{''}{.}0015$ (it can be assumed that δz is increasing with $\sec^2 z$ up to 45°). Similar amounts of these quantities are found by Saastamoinen - see Table 1.

From the different data referred to above one can notice that the δz corrections, in their absolute values, are close to each other and as far as small zenith distances are concerned, practically negligible. Yet, there appear some differences if one goes into details.

4. CONCLUSIONS

It is obvious that the use of a given type of correction is dependent upon the accuracy required of the astrometric observations. But as long

as we have to deal with the classical astrometry one might say,
considering in the proceeding section, that three-dimensional elements
δA and δz, at smaller zenith distances - even up to 45° - do practically
not impose the necessity of having to be taken into account. But at
larger zenith distances their use is a necessity (for instance in the
determination of the so-called refraction constant correction from the
upper and lower transits of the same star). It should be born in mind
that the analyses, referred to above, relate predominantly to the free
atmosphere and not to the usually disturbed surface layers.

High-accuracy astrometric observations certainly require more rigorous
criteria. For instance, laser ranging to satellites requires that three-
dimensional refraction be applied (Gardner, 1976).

The following conclusions might be drawn:

4.1. A global atmospheric model - like the one described in the following
section - should constitute a foundation to the calculation of the
international refraction tables. These would serve as a general frame in
determining the atmospheric influences (we believe it is needed to
introduce a standard in the calculation of refractional influences, too);

4.2. An accurate (simulated) atmospheric model more fitting to the Earth's
figure should be introduced already now in the high accuracy astrometric
calculations. The mathematical figure of the Earth is an ellipsoid of
revolution whose radius of curvature varies considerably with azimuth A
and latitude ϕ:

$$ r = c \left[V + (V^3 - V) \cos^2 A \right]^{-1} $$

with $V^2 = 1 + e^{-2} \cos^2 \phi$; $c = 6399.593\ 6259$ km; $e^{-2} = 0.006\ 739\ 496\ 775\ 48$
(Geodetic Reference System 1980). In the calculation of pure refraction
it would be more realistic to assume that the surfaces of constant
refractivity, instead of being spherically symmetric, run parallel to the
surface of the reference ellipsoid making the astronomical refraction
slightly smaller in the meridian than in the first vertical (by $0\overset{.}{.}10$ for
$z = 80^\circ$ at the equator and by one-half of that amount in latitude 45°).
In an ellipsoidal atmospheric model, a light ray cannot be considered
strictly as a plan curve.

4.3. No matter which solution is adopted - 4.1. or 4.2. - still closer
approximations are needed, by taking into account regional (see Sugawa-
Kikuchi, 1979, or different standard atmospheres, e.g. Sissenwine, 1969)
along with local influences such as mountains, air surfaces, pavilion
surroundings. Solution 4.1. implies that even the slightest regional
differences are taken into account.

4.4. High-accuracy observations impose the necessity of investigating
the local refraction influences. This is in some instances necessary
with the classical observations also. The question is essentially that
of inquiring of influences of the distrubed surface layers.

About endeavours of introducing the three-dimensional refractional
values in the astrometry, the reader is referred to a brief overview in
the paper of Teleki (1981).

5. GLOBAL ATMOSPHERIC MODEL

Because of refractivity of surface air increases from the thermal equator
toward the colder climates at the poles, the surfaces of constant
refractivity must in the lower atmosphere acquire a general downward
slope toward the equator. Assuming that the atmosphere is in hydrostatic
equilibrium, the slope angle of the surfaces will gradually diminish
with heigh, becomming zero at roughly 8 km above the sea level. At this
height begins a compensating layer, reaching up to the stratosphere, in
which the inclination of the air strata is reversed and the surfaces of
constant refractivity slope down toward the poles.

The fact that astronomical refractions calculated on the assumption of
spherical symmetry of the refracting layers of the free atmosphere have
not been found greatly in error suggests that anomalous refractions
caused by the inclination of air strata in the lower atmosphere are
substantially compensated by the reverse tilt in the upper layers.
Wünschmann (1931) was the first to underline this fact.

Figures 1 and 2 have been taken from a recent quantitative study of this
phenomenon (Saastamoinen, 1980). As source material for the meteorological data global latitudinal average were computed from observations at
543 surface and 102 upper-air stations (U.S.Department of Commerce:
"Monthly Climatic Data for the Wourld", vol.23, 1970).

In view of the scarcity of meteorological data and the simplistic method
of averaging, the obtained meridional tilt distribution may be regarded
only as an estimate of the order of magnitude for systematic departures
in the atmosphere from a spherically symmetric state. Owing to the
stronger latitudinal gradients that occur in the winter distribution of
temperatures (figs. 3 and 4), these departures are generally greater in
winter than in summer.

The anomalous refractions calculated by numerical integration from the
meridional tilt distribution and the rates of diminution of refractive
index with heigh are given in Tabel 1 for different latitudes and zenith
distances. In the Northern Hemisphere, the global anomaly tends to
increase astronomical refraction if a star is observed in the north,
while decreasing it by an equal amount if observed in the south. In the
Southern Hemisphere, the reverse applies.

In each vertical column of the global atmospheric model, the tilting of
air strata in the lower levels is indeed compensated by the reversed
tilt in the upper zones. The compensation, however, changes from one
column to another giving rise to significant anomalous refractions at

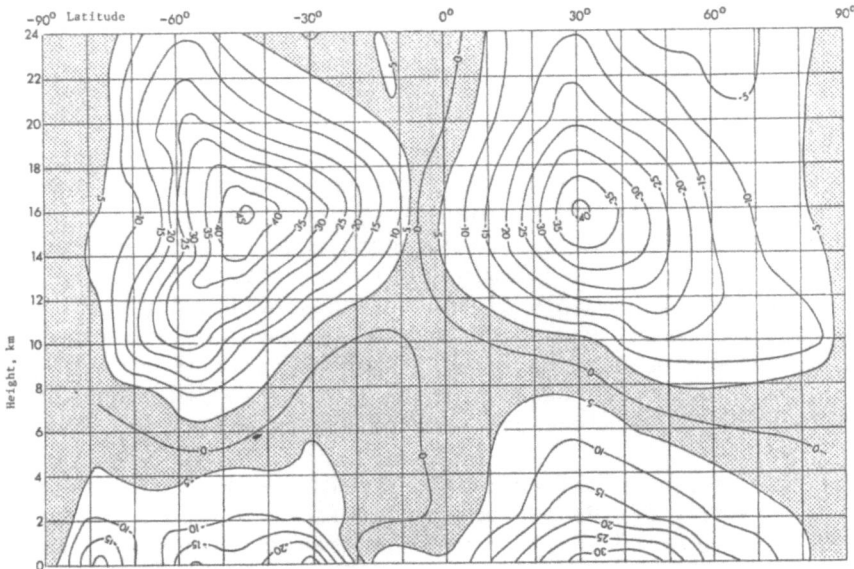

Figure 1.   Meridional tilt of surfaces of constant refractivity, 10^5 ψ, in January over land areas (1 unit $\sim$ 2'').

Figure 2.   Meridional tilt of surfaces of constant refractivity, 10^5 ψ, in July over land areas (1 unit $\sim$ 2'').

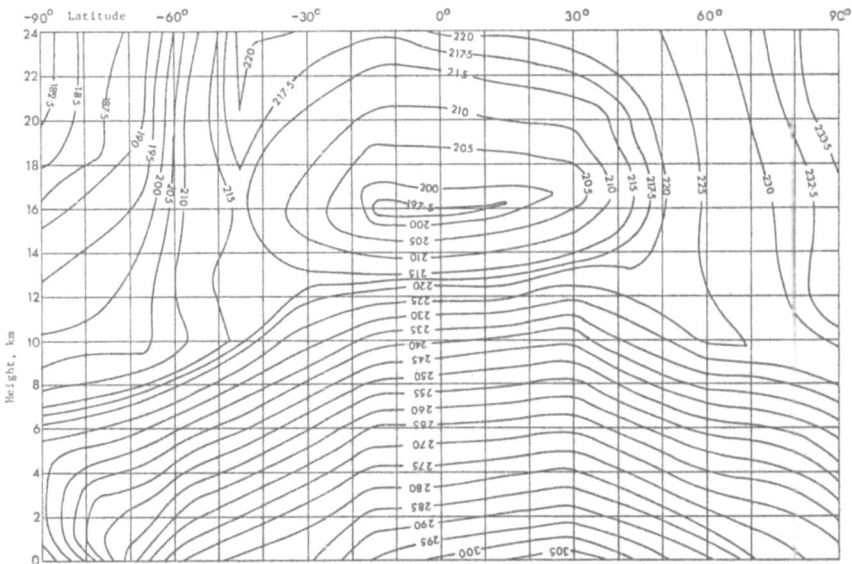

Figure 3.   Distribution of temperature, T, in January over land areas.

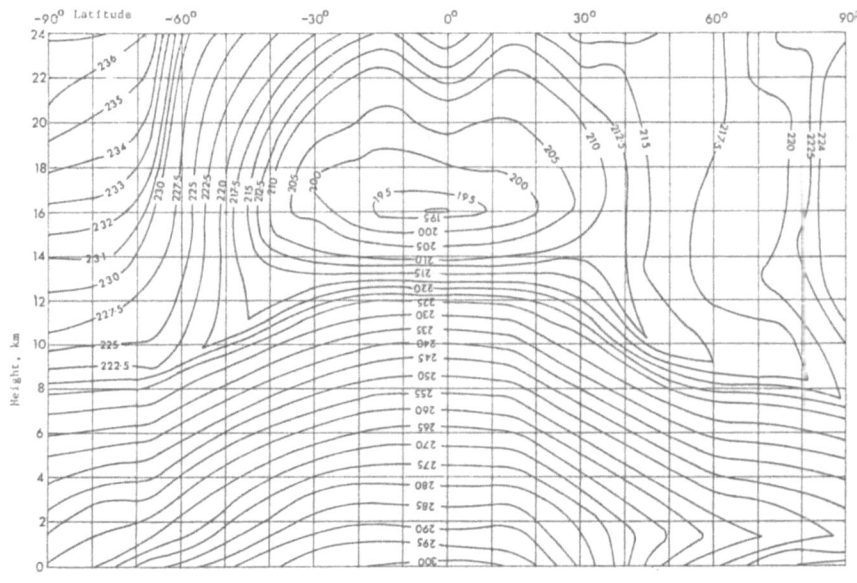

Figure 4.   Distribution of temperature, T, in July over land areas.

greater zenith distances.

TABLE 1. Anomalous Refraction Caused by Meridional Tilt of Atmospheric
          Layers (Minus sign indicates that pure refraction has to be
          decreased).

Latitude	Zenith Distance				
	45°	60°	70°	75°	80°
	January:				
-60°	-0''001	-0''01	-0''03	-0''06	-0''15
-45	-0.001	-0.01	-0.03	-0.06	-0.15
-30	-0.005	-0.02	-0.04	-0.07	-0.18
0	0.001	0.00	0.00	0.01	0.02
30	0.013	0.04	0.09	0.16	0.37
45	0.010	0.03	0.07	0.12	0.28
60	0.005	0.01	0.04	0.07	0.15
	July				
-60°	-0''010	-0''03	-0''08	-0''16	-0''37
-45	-0.001	-0.01	-0.03	-0.05	-0.10
-30	-0.005	-0.02	-0.04	-0.08	-0.20
0	-0.003	-0.01	-0.02	-0.04	-0.09
30	0.000	0.00	0.01	0.02	0.05
45	0.008	0.02	0.06	0.11	0.26
60	0.003	0.01	0.03	0.05	0.13

REFERENCES

Gardner,C.S., 1976: Radio Sci., 11. 1037.
Harzer,P., 1922-24: Publ.Sternw.Kiel, 13, 26.
Harzer,P., 1924: Publ.Sternw.Kiel, 14, 2.
Moritz,H., 1967: Österr.Z.f.Vermessungsw., Sonderheft 25,333.
Saastamoinen,J., 1980: Report of a Study on the Latitudinal Distribution
                       of Meridional Tilts in the Atmosphere, Working
                       Group on Astron.Refr. of IAU Commission 8.
Sissenwine,N., 1969: World Survey of Climatology, Elsevier Publ.Comp.,
                     4, 5.
Sugawa,C. and Kikuchi,N., 1979: Proc.IAU Symp.89, 103.
Teleki,G., 1974: Publ.Obs.Astron. Belgrade, 18, 213.
Teleki,G., 1981: Bull.Obs.Astron. Belgrade, 131, 4.
Wünschmann,F., 1931: Gerlands Beitr.Geophys., 31, 83.
Yunoshev,L.S., 1969: Bokovaya refrakciya sveta pri izmereniyah uglov,
                     Ed.Nedra, Moscow, 7.

THE USE OF LIDAR TO OBTAIN THREE-DIMENSIONAL REFRACTION DATA

J.A. Hughes
U.S. Naval Observatory, Washington, DC

S. DeLateur
SRI International, Menlo Park, CA

Abstract: The need for a real time (or near real time) detailed knowledge of the atmospheric structure necessary for determining astronomical refraction in general is exhibited. Examples involving water
vapor and isopycnic tilts are given.

The possibility of determining the necessary parameters by means of
active atmospheric probing using Light Detection and Ranging (Lidar)
methods is reviewed. A brief discussion of Raman versus Differential
Absorption Lidar (DIAL) techniques is included. Recent advances in
applicable lasing materials such as the Alexandrite crystal offer
interesting possibilities.

At the present time a conceptual design study is underway. This is
intended to culminate in the construction of a system in 1982-83.

INTRODUCTION

The most exasperating characteristic of astronomical refraction
is the fact that while it is a relatively straight-forward matter to
account for nearly 99.9% of the effect, it is currently essentially
impossible to allow for the remaining 0.1% with any assurance of success. We have however, reached the point where that remaining 0.1% is
important, and especially so if any part of it is systematic. This
situation is not caused by a lack of excellent theories of refraction.
We have the Pulkovo Tables (1956) and Garfinkel's (1967) polytropic
theory to name but two of the prominent methods available. The problem
is that it is just too much to expect any theory, even one with many
adjustable parameters, to account for the minute by minute or even
hourly variations in the dynamic planetary boundary layer. On the whole
it is noteworthy how well the theories succeed since they are essentially
static representations of the atmosphere.

WATER VAPOR

Consider the effect of water vapor. It changes both the refractivity

W. Fricke and G. Teleki (eds.), Sun and Planetary System, 463–470.
Copyright © 1982 by D. Reidel Publishing Company.

and the gas constant so as to decrease the refraction. Difficulties
arise when one uses only a ground level value of the humidity without
any knowledge of the actual three dimensional distribution of the water
vapor. Merely using the observed ground level value of the ratio of
the partial pressure of water vapor to the total pressure, P_w/P_t, is
equivalent to assuming that this ratio is a constant throughout the
atmosphere. Since humidity falls rapidly with height this procedure
would generally overestimate the effect of the water vapor.

In order to illustrate this type of problem the changes in refraction (with respect to dry air) caused by various partial pressures
of water were computed using the Pulkovo Tables (IV ED.), and also by
means of numerical integrations of various water vapor profiles. The
differences in the changes using these two methods were then computed
and tabulated. For a total ground level pressure near 760 mm Hg,
with 10 mm being due to water vapor, and for temperatures in the range
of 273° to 293° K, it was found that good agreement occurred when the
water vapor was assumed to decline linearly with height, reaching zero
at seven kilometers. Table I illustrates these differences, Change
(Integration) - Change (Pulkovo), for that profile. Evidently if one
happened to be observing under just the right conditions, a near perfect
allowance for water vapor would be made. However, as Table I shows, the
differences become greater at lower total pressures or higher partial
pressures of water vapor. There is, in addition, no guarantee that the
true profile is that which was assumed here.

TABLE I
Change (Integration) - Change (Pulkovo)

T^oK	273		283		293	
P_t mm	$P_w = 10$	$P_w = 30$	$P_w = 10$	$P_w = 30$	$P_w = 10$	$P_w = 30$
760	" -.001	" -.014	" -.001	" -.012	" -.002	" -.013
640	+.017	+.045	+.018	+.042	+.017	+.042
560	+.032	+.084	+.030	+.081	+.029	+.079

ISOPYCNIC TILT

A second contributor to the types of error considered here is any
systematic tilt of the assumed isopycnics. One of us, Hughes (1979),
has done considerable theoretical work on the generation of such tilts
by urban heat islands. Without going into great detail we present here
two cross sections through the density perturbation field caused by a
realistic heat island. Cross section A lies in the plane of the meridian
of a hypothetical observatory, while section B lies in the plane of the
corresponding prime vertical. Table II shows the values of the tilt
angles in these planes. The observatory is supposed to be located some

6 km southeast of the center of the heat island. The wind is assumed to be from the west. Values are given up to a height of 500 m and for horizontal distances of 400 m and 800 m (negative to the south for section A and to the west for section B).

TABLE II
Isopycnic Tilts

	Section A					Section B				
Meters	-800	-400	0	+400	+800	-800	-400	0	+400	+800
100	+.04°	+.09°	+.15°	+.21°	+.25°	+.43°	+.42°	+.41°	+.39°	+.37°
200	+.11	+.10	+.08	+.05	+.02	+.14	+.11	+.08	+.06	+.03
300	-.02	-.04	-.05	-.06	-.06	-.08	-.09	-.10	-.10	-.09
400	-.05	-.05	-.04	-.03	-.02	-.11	-.09	-.08	-.06	-.05
500	-.01	-.00	+.01	+.01	+.01	-.03	-.01	+.00	+.01	+.01

These results are strictly theoretical, and although the order of magnitude of the tilts is reasonable, there is no guarantee that the true tilt is that which is shown here.

Thus for the two phenomena considered above we are presently forced into what are essentially _ad hoc_ assumptions and procedures. This unsatisfactory situation can only be improved by injecting additional, real time, three dimensional atmospheric data into the various algorithms used for treating observational material.

LIDAR

One method of generating the required data is with a laser detection and ranging (Lidar) system. The approach is similar to conventional radar, except the interactions occur at optical frequencies. With the laser transmitter and receiver systems positioned at a common location, a telescopic optical receiver measures the amount of backscatter from a short transmitted laser pulse. Ranging is performed by calculating the time for the pulse to travel the distance to a region of interest plus the time for the backscattered pulse to return. Thus, by sampling the time-varying received optical intensity after a specific delay, study of a restricted location along the path can be accomplished.

By using models that describe the absorption and scattering properties of the atmosphere at optical frequencies, information about major and trace constituents can be obtained through "inverting" the measurements of backscattered intensity. Depending on the realm of study, the frequency of the transmitter, the frequency response of the receiving system, the optics design, and the data processing can be selected to provide a large area of coverage and rapid measurement capabilities. See Hinkley (1976) for an excellent review of the field.

DIAL AND RAMAN LIDAR SYSTEMS

Two different Lidar techniques will be discussed in this paper: (1)
Differential Absorption Lidar (DIAL) and (2) Raman Lidar. Each of
these are candidates as system approaches to provide vertical profiles
of atmospheric water vapor content and temperature. However, the
scattering and absorption basis for the measurements are distinct in
each system.

The laser transmitter of the DIAL system sends pulses at two wavelengths. One wavelength is selected to coincide with an absorbing line
of the molecular species of interest. The other near-by wavelength is
chosen for negligible absorption. The receiving system is tuned to
measure the elastic Mie and Rayleigh backscatter at the same wavelengths as were transmitted. By forming the difference of the returned
intensities at the absorbed and unabsorbed wavelengths, a time-varying
(and thus, range resolved) concentration of the species is estimated.
This technique has also been called Differential Absorption of Scattered
Energy (DASE) and Differential Absorption and Scattering (DAS).

The transmitter of a Raman Lidar system sends a pulse of a single
wavelength. The receiver system is tuned to the wavelengths that are
uniquely Raman-shifted (from the transmitted wavelength) by the molecular species to be studied. By using the Raman cross-section of the
species, an analysis of the magnitude of the received intensity can be
translated into a concentration profile.

WATER VAPOR MEASUREMENT SYSTEMS

Both Raman and DIAL systems are used successfully to obtain water
vapor measurements. Only those experiments that have supplied vertical
profiles are listed in the following tables; those systems that supply
a column integrated measurement or average along a horizontal path are
not included.

Table III shows the range of Raman systems that have been designed
specifically for water vapor profiles. All of these systems fundamentally
measure the ratio of the number density of H_2O to that of N_2. Thus,
only a water vapor mixing ratio is estimated. This can be translated
into a water vapor profile by assuming a density profile for N_2 with
accompanying error.

Table IV lists recent water vapor profiling DIAL systems. Less
powerful laser transmitters are typically needed than those used in
Raman-based systems because the Rayleigh-Mie backscatter cross-sections
are many orders of magnitude larger than Raman cross-sections.

TABLE III

Raman Lidar Systems for Vertical Water Vapor Profiling

Cooney (1971)

Transmitter: doubled Ruby at 347.15nm. Receiver: 397.5 nm from H_2O and 377.7 nm from N_2. Approach: water vapor profile obtained by normalizing H_2O Raman return with N_2 return, and assuming an N_2 atmospheric content. 13 percent accuracy. Night measurement. Range: 300 m - 2600 m with 100 m intervals.

Melfi (1972)

Transmitter: doubled Ruby at 347.15 nm. Receiver: 397.5 nm from H_2O and 377.7 nm from N_2. Approach: water vapor profile from H_2O return normalized by N_2 return. Night measurement. Range: 200 m - 2500 m with 50 m intervals.

Pourny, Renault, and Orszag (1979)

Transmitter: doubled ruby at 347.15 nm. Receiver: 397.6 nm from H_2O and 377.7 nm from N_2. Approach: water vapor mixing ratio from H_2O return normalized by N_2 return. Night measurement. 15 percent accuracy. Range: 300 m - 1800 m with 30 m intervals.

Renault, Pourny, and Capitini (1980)

Transmitter: quadrupled YAG at 266 nm. Receiver: 277.5 nm from O_2. 283.6 nm from N_2, and 294.6 nm from H_2O. Approach: water vapor mixing ratio from the H_2O return is normalized by the N_2 return. Estimation of ozone concentration is used to compensate for heavy ozone attenuation of the signals in this wavelength region. 10 percent accuracy at 500 m with 30 m intervals. Daytime operation. Range: 150 m - 950 m with 30 m intervals.

TABLE IV
DIAL Systems for Vertical Water Vapor Profiling

Browell, Wilkerson, and McIlrath (1979)
    Transmitter: ruby at 694.3 nm and dye (pumped by ruby) at 724.3
    nm. Receiver: 724.3 nm (water vapor absorbing wavelength) and
    694.3 nm. Approach: difference in returned signals is analyzed
    for H_2O density profile (mol/cm^3). Equipment includes an absorption
    cell that measures the cross-section for H_2O at the dye laser out-
    put for each pulse. 9 percent accuracy. Night operation. Range:
    100 m - 3000 m with 100 - 180 m intervals.

Cahen, Pelon, Flamant, Lefere, Chanin, and Megie (1980); also, Cahen
and Lesne (1981)
    Transmitter: Nd: YAG at 532 nm and dye (pumped by Nd:YAG) at 723
    nm. Receiver: 723 nm (absorbed by water vapor) and 532 nm
    Approach: difference returns to obtain water vapor profiles.
    Wavelength and line-width are servo-controlled. 10 percent accuracy.
    Daytime operation. Range: 8000 m with 30 m intervals.

Werner and Herrmann (1981)
    Transmitter: ruby at 694.28 nm and temperature varied ruby from
    694.2 nm to 694.28 nm. Receiver: 694.215 nm (absorbed by water
    vapor) and 694.237 nm. Approach: difference returns to obtain
    water vapor density profiles. Accuracy of 1 mbar in absolute
    humidity. Daytime operation. Range: 450 m - 1500 m with 100 m
    intervals.

TEMPERATURE PROFILING SYSTEMS

    Table V contains a list of Raman temperature profiling systems.
Common to these experiments is the technique of measuring the temperature
dependence in the magnitude of the Raman spectrum of N_2 and O_2. The
lines in the Raman spectrum have differential changes in magnitude with
respect to temperature. Thus, by comparing the change in two bands, a
temperature value can be deduced.

    There is a lack of DIAL vertical temperature profiling systems.
However, work is presently being performed with system techniques that
show promise. One proposal is to use "Mason's Method" (Mason, 1975:
Schwemmer and Wilkerson, 1979) with the absorption spectra of O_2 in the
region of 760 nm. This wavelength is close to the wavelengths selected
in successful DIAL water vapor measurements (720 nm). Thus a single
system could supply temperature as well as water vapor density.

TABLE V
Raman Lidar Systems for Vertical Temperature Profiling

Cooney and Pina (1976)
    Transmitter: ruby at 694.3 nm. Receiver: 689.0 nm and 691.2 nm
    Approach: measure the difference in the magnitude of the Raman
    spectrum of O_2 and N_2 in two bands which vary with temperature.
    Accuracy is 5 degrees C. Nighttime operation. Range: 600 m –
    1100 m with 300 m intervals.

Gill, Geller, Farina, and Cooney (1979)
    Transmitter: ruby at 694.3 nm. Receiver: 691.6 nm and 689.0 nm.
    Approach: same as above. Accuracy is .85 degrees C. Range:
    1300 m – 2300 m with 75 m intervals.

A COMBINATION SYSTEM

The combined DIAL technique mentioned above is now being researched.
Typically, dye lasers have been used to cover the 650 nm – 790 nm
region as transmitters for such simultaneous temperature and humidity
measurement system studies. Besides covering the important absorption
regions for H_2O and O_2, they are easily tuned to specific wavelengths.
However, solid-state lasers have demonstrated an ease of operation and
reliability that is important for day-to-day meteorological observations.
Recently, a solid-state lasing material, Alexandrite, was developed for
laser output from 700 nm – 815 nm. Thus, the characteristics of this
material are being included within new DIAL system designs.

CONCLUSION

It appears that Lidar offers real possibilities with regard to providing input data for use in refining our current allowances for refraction. Water vapor profiling presents no problem, and temperature measurements certainly are feasible. Since for the small perturbations considered here, the Boussinesq approximation is valid, it follows that isopycnics
follow isotherms. Thus temperature measurements are equivalent to
density measurements, and with, for example, a conically steered laser
beam, the isopycnic tilts can be determined.

If present plans hold, a system to accomplish these ends will be
in operation within two years. The results should be of considerable
interest.

Bibliography

Browell, E.V., T.D. Wilkerson, and T.J. McIlrath, (1979): "Water Vapor
Differential Absorption Lidar Development and Evaluation," Appl. Opt.,
vol. 18, 3474-3483.

Cahen, C., J. Pelon, P. Flamant, J. LeFrere, M.L. Chanin, and G. Megie,
(1980): "French Lidar Facility at the Haute Provence Observatory for

Tropospheric and Stratospheric Measurements," Proc. Tenth Int. Laser
Radar Conf., Silver Spring, Maryland.

Cahen, C., and J.L. Lesne, (1981): "Improvements to Water Vapor Lidar
Using DIAL Techniques," Proc. Conf. Lasers and Electro-Optics, Washington, D.C.

Cooney, J., (1970): "Remote Measurements of Atmospheric Water Vapor
Profiles Using the Raman Component of Laser Backscatter," J. Appl. Met.,
vol. 9, 182-184.

Cooney, J. and M. Pina, (1976): "Laser Radar Measurements of Atmospheric
Temperature Profiles by Use of Raman Rotational Backscatter," Appl. Opt.,
vol. 15, 602-603.

Garfinckel, B. (1967): "Astronomical Refraction in a Polytropic Atmosphere," Astron, J., 72, pp. 235-254.

Gill, R., K. Geller, J. Farina, and J. Cooney, (1979): "Measurement of
Atmospheric Temperature Profiles Using Raman Lidar," J. Appl. Met.,

Hinkley, E.D., (1976): Laser Monitoring of the Atmosphere, Springer-
Verlag, New York.

Hughes, J.A. (1979): "Environmental Systematics and Astronomical
Refraction II," IAU Sym.,,No. 89, Refractional Influences in Astronomy
and Geodesy, Ed. Tengstrom and Teleki, pp. 13, 25.

Mason, J.B., (1975): "Lidar Measurement of Temperature: A New Approach,"
Appl. Opt., vol. 14, 76-78.

Melfi, S.H., (1972): "Remote Measurements of the Atmosphere Using Raman
Scattering," Appl. Opt., vol. 11, 1605-1610.

Pulkovo Refraction Tables (1956), Academy of Sciences Press, IV. Ed.,
Moscow.

Pourny, J.C., D. Renault, and A. Orszag, (1979): "Raman-Lidar Humidity
Sounding of the Atmospheric Boundary-layer," Appl. Opt., vol. 18, 1141-
1148.

Renault, D., J.C. Pourny, and R. Capitin, (1980): "Daytime Raman-Lidar
Measurements of Water Vapor," Opt. Let., vol. 5, 233-235.

Schwemmer, G.C. and T.D. Wilkerson, (1979): "Lidar Temperature Profiling:
Performance Simulations of Mason's Method," Appl. Op., vol. 18, 3539-3541.

Werner, C., and H. Herrmann, (1981): "Lidar Measurements of the Vertical
Absolute Humidity Distribution in the Boundary Layer," J. Appl. Met.,
vol. 20, 476-481.

FINAL REFRACTION PROBLEMS IN TIME AND LATITUDE OBSERVATIONS THROUGH CLASSICAL TECHNIQUES

C. Sugawa and I. Naito
International Latitude Observatory, Mizusawa, Iwate, Japan

Abstract

This paper discusses refraction effects remained so far upon time and latitude observations through the classical techniques such as the VZT, the PZT and the astrolabe from the meteorological points of view. In the planetary boundary layer during the night-time, there exist marked density variations having time-scale of several to several ten minutes which are caused by advections and internal gravity waves. Their effective pressure variations at the ground should be orders of 0.5 mb or more. Despite employing the wellknown Talcott method in the VZT and the method of constant altitude in the astrolabe, the observed values still include refraction errors of order of 0".01 because of almost all the time-scales of the above variations in the planetary boundary layer as the time intervals of the star observations.

1. Introduction

The classical techniques in the positional astronomy accompany the two kinds of fatal problem concerning meteorological effects upon their observations. The one is the atmospheric refraction corrections and the other is the meteorological environment effects upon the instruments. In the latter problem little successful improvements have done so far to remove effects upon the instruments, in spite of the recent rapid development of temperature and strain measurements. Because everyone has known that to answer these problems it is easier to build new instruments based on modern techniques than to improve the present instruments. However, everyone believes that these classical techniques will still have to play main roles at least for more than several years from now. It must therefore be desirable to make efforts for a while to grasp non-instrumental and meteorological errors in the observed values. In the meteorology, atmospheres to which the refraction studies have so far treated have often been called the normals including the seasonal and diurnal variations. Since the time intervals of observing at least a pair of stars seem to be in around several to several ten minutes, the

W. Fricke and G. Teleki (eds.), Sun and Planetary System, 471–474.

refraction effects of such normals would completely be removed. It consequently turns out that the refraction problems would remain only in the atmospheric phenomena having time-scales of variation around the time intervals of the star observations. The shorter time-scale variations than these are called the turbulences which induce the so-called scintillations in the astrometry.

This paper goes to one of the roots of the actual atmospheric refraction problems in time and latitude observations through the classical techniques such as the VZT, the PZT and the astrolabe, in the bases of their observing systems to remove refractions. These instruments account for about thirty percent of all instruments employed now for the IPMS, respectively.

2. Refraction problems in VZT, PZT and astrolabe observations

In the case of the VZT, the observing system called the Talcott method is a high accuracy system which can remove almost all the refraction effects by observing a pair of northern and southern stars with nearly the same zenith distances during four to ten minutes, if the atmospheric field were stationary with no pressure gradient during the period. While the actual atmosphere fluctuates with time-scales near four to ten minutes too, so even the Talcott method cannot remove the refraction effects of such atmospheric fluctuations completely. Therefore, it turns out that the atmospheric time-scales which generate refractions in the VZT are four to ten minutes. Two hours needed for each star group are not so long enough to statistically reduce such refraction effects to negligible amounts.

In the case of the PZT, it seems that no refraction effect remains because of observing zenith stars. However, since four images are photographed to each star during about seventy seconds, a small refraction problem remains due to the reduction process if a rectangle formed with four star images is distorted. The atmospheric time-scale generating a small distorsion to the rectangle through refractions would be in around several ten seconds. Such a time-scale apparently prevails in the planetary boundary layer. The other effects are all the scintillations.

In the case of the astrolabe, the observing system employs the method of constant altitude. For time and latitude observations, the Danjon astrolabe has mostly been driving now, in which about thirty stars of thirty degrees zenith distance are employed for a star group during about two hours. The mean time interval of star observation thus becomes three to four minutes. If each star group consisted of large number of pair of star and if each pair of star has inverse azimuths each other, the circumstance in the refraction problem would be regarded the same as the VZT case. Since, however, the actual star prorgams for observations are not so complete, atmospheric variations having several to several ten minutes time-scales would generate refraction effects. The star program should be easily improved.

The above mentioned problems are summarized in Table 1.

Table 1

	VZT	PZT	Astrolabe
Observing system	Talcott method		Constant altitude
Zenith distance	~5°	0°	30°
Time interval of star observation	4 - 10 min.	20 - 30 sec.	3 - 4 min.
Refraction effect	0".005 - 0".01	not small	$\lesssim$ 0".03

3. Actual atmospheric structure and its effects upon refraction

It is obious after the modern meteorology that the time-scale of several to several ten minutes of atmospheric variation mentioned in the preceeding section appears in the lower troposphere. The lowest layer of the troposphere is called the planetary boundary layer, and its lowest several ten meters near the ground is called the surface boundary layer. These two boundary layers hold stationary when the surface roughnesses are homogeneously overspread in all directions. However, the actual surface roughnesses are not homogeneous due to existences of city, forest, crop field and various other environments.

When the ground surface is covered by more than two kinds of roughness height along the mean wind direction, internal boundary layers appear above the surface boundary layer. Structures and thickness of such internal boundary layers vary with a change in the mean wind direction, so discontinuities in temperature and wind velocity occur at the boundary of these layers. Such discontinuities usually accompany small density gaps. When these layers vary or move with the changes in wind direction, small pressure gradients should be observed at the ground amounting to about 0.2 mb.

In the night-time a cold air mass is mostly formed near the ground and grows up. When this cold air mass moves as an advection, small pressure variations happen to be observed amounting to about 0.2 mb or more. Such an advection occurres once several minutes in the actual planetary boundary layer in the night-time. At the Mizusawa observatory, this kind of advection is frequently observed because of the observatory location near the bottom of a valley.

When the stability increases and approaches the critical Richardson number, the turbulences are not maintained and instead internal gravity waves appear. The fundamental frequency of the internal gravity waves is given by the Brunt-Vaisala frequency which is known the free oscillation frequency of air in a stably stratified atmosphere. Since the Brunt-Vaisala frequency is equall to root of the static atability, the fundamental period of the Brunt-Vaisala frequency decreases with increase in stability. A standard value of such a fundamental period is about nine minutes, but the value tends to become longer in the actual planetary boundary layer. The observed pressure variation at the ground due to such an internal gravity wave frequently amounts to 0.5 mb or more. A typical example of the internal gravity wave is a lee wave. At the Mizusawa observatory, such lee waves or their broken waves are often observed.

In any way, the prevailing time-scales of all the above phenomena are several to several ten minutes that are confirmed in the boundary layer meteorology. The longer time-scales than such variations up to the semidiurnal variation are called the spectral gap of atmospheric motion. The above prevailing time and horizontal scales are summarized in Table 2.

Table 2

	Time-scale	Horizontal-scale	Intensity
Turbulence	10 - 100 sec.	10 - 100 m	$\lesssim 0.5°C$
Change in internal boundary layer	10 - 100 min.	1 - 10 km	$\lesssim 0.2$ mb
Advection of cold air mass	10 - 100 min.	1 - 10 km	$\lesssim 0.5$ mb
Internal gravity wave	1 - 30 min.	1 - 30 km	$\lesssim 1.0$ mb

Finally, we will estimate the refraction effects of the above various phenomena not removed by the observing systems. All the phenomena in Table 2 are effective to the refractions, because their prevailing time-scales are very close to twice the time intervals of star observation in Table 1. The estimations are easily done by a simple refraction formula from the effective pressure variations at the ground in Table 2, provided that the atmosphere above the planetary boundary layer is stationary during several ten minutes. The typical results of such refraction effects are shown in Table 1. In the case of the PZT the time interval of star observation nearly corresponds to a turbulence scale, so the effects cannot be easily estimated but are not so small to be neglected.

4. Concluding Remarks

The present paper has shown that the refraction effects now remain in the order of 0".01 in the VZT and the astrolabe observations. In order to make such errors small as far as possible, it needs not only to keep the meteorological environments such as the roughness height homogeneous in all directions around the observatory but also to improve the observing system of stars. At the end of this paper, we would like to comment that even the new techniques such as the VLBI and the Laser Ranging have an errors not small originating from the meteorological environment near the observational sites, in which case the method of meteorological observation becomes more important to remove the refractions completely by using their reduction formula.

ASTRONOMICAL REFRACTION CALCULATED FROM AEROLOGICAL DATA IN JAPAN

R. Fukaya and H. Yasuda
Tokyo Astronomical Observatory, Mitaka, Tokyo, Japan.

NUMERICAL COMPUTATION.

Using actual aerological data in Japan, we calculated numerically the astronomical refraction. We study the refraction which optical path suffers in passing from the i-th layer ($i = 0, 1, \cdots, k$) to the (i-1) the layer. Here, we assume the earth as a sphere and set successively the concentric air strata. For the purpose, we put

n_i ; the refractive index for the i-th layer,

Z_i^i ; the zenith distance at the point O where optical path passes the boundary between the i-th and (i-1)th layers,

and Θ_i ; the angle between the radius of the sphere at O and the vertical direction, parallel to the radius at the observation point at the ground surface.

Here, $\Theta_0 = 0$ at the ground surface. By the sine law in a triangle and Smell's law, we have

$$\Theta_i = Z_{i-1} - \sin^{-1}\{\frac{r_{i-1}}{r_i} \sin (Z_{i-1} - \Theta_{i-1})\} \qquad (1)$$

and

$$Z_i = \Theta_i - \sin^{-1}\{\frac{n_{i-1}}{n_i} \sin (Z_{i-1} - \Theta_i)\} \qquad (2)$$

where r_i is the radius of the sphere at O. The zenith distance at the last atmospheric layer, Z_k , is corrected for the refraction ;

$$\Delta Z = Z_k - Z_0 . \qquad (3)$$

By the equations (1),(2) and (3), we can calculate the astronomical refraction, when we know the coefficient c in Dale-Gladston equation and the air density at each atmospheric layers. We calculate the monthly mean values of the atmospheric density for 1975 by using pressure and temperature data for 21 standard pressure levels from the earth surface to the height of 30 mb at Tateno Aerological Observatory near Tokyo.

W. Fricke and G. Teleki (eds.), Sun and Planetary System, 475–476.

Up to a height of 91 km above a 30 mb pressure level, the pressure and
temperature data in " U.S. Standard Atmosphere 1976 " were used for
four different heights. Using Edlén's equation and the equation of
Barrel and Sears and adopting the density of a standard dry air (t=0°C,
p=760 mm Hg and vapor pressure = 0 mm Hg) at a standard gravity to be
0.001293, we obtained the coefficient c in Dale-Gladston equation for
the wavelength λ=5753 Å to be 0.22633. We similarly obtained c for a
standard humid air (vapor pressure = 5.5 mm Hg) to be 0.22610. It
nearly coincides with the value of c (= 0.22607) in Pulkovo Refraction
Table. Thus, we used the value of 0.22607 in our computations for humid
air. Table lists the results of our numerical calculation for dry and
humid air in the 2nd and 5th columns respectively.

COMPARISON WITH PULKOVO REFRACTION TABLE.

Our results were compared with astronomical refraction calculated by
PULKOVO REFRACTION TABLE (4th Edition). The 4th and 7th columns in
Table give the differences between the two calculations. The differences
for humid air show a distinct seasonal variation, smaller in winter and
larger in summer. It is perhaps due to the difference of distribution
of water vapor in atmospheric layers at Tateno and Pulkovo. The dry air
does not show any seasonal variation, though the average of the
differences is larger than that for humid air.

The air density and the refractive index for actual airstrata at each
of 21 standard pressure levels were calculated by using both the
aerological data at Tateno and the equation of Barrel and Sears. Then
we could obtain the value of c at each height. Though the values of c
in the layers below the height of 10 km are larger than 0.22630, they
remain between 0.22630 and 0.22620 in the layers above the height of 10
km. Their average coincides with c for dry air (0.22633). However,
the astronomical refraction calculated by Pulkovo Table is smaller by
0.1 arc second at zenith distance 60° than that calculated numerically
by the aerological data and c = 0.22633.

Table.   Comparison Between Numerical Calculation
         and Pulkovo Table (at Zenith Distance = 45°)

Month	Dry Air			Humid Air		
	N.C.	Pulkovo	Diff.	N.C.	Pulkovo	Diff.
Feb.	59''804	59''730	+ 0''074	59''702	59''686	+ 0''016
Apr.	57.698	57.624	+ 0.074	57.551	57.525	+ 0.026
Jun.	55.684	55.611	+ 0.073	55.468	55.433	+ 0.035
Aug.	55.067	54.993	+ 0.074	54.814	54.772	+ 0.042
Oct.	57.139	57.066	+ 0.073	56.964	56.925	+ 0.039
Dec.	59.765	59.691	+ 0.074	59.656	59.638	+ 0.018
	c = 0.22633			c = 0.22607		

EXPERIMENTAL MODEL FOR DIURNAL ASTRONOMICAL REFRACTICN

A.Poma*, E.Proverbio*
and S.Mancuso**
* Cagliari Astronomical Observatory
** Napoli-Capodimonte Astronomical Observatory

ABSTRACT

Refraction changes in the troposphere during day-time impose
severe limitations in the accuracy of optical and radio
astronomical observations. A technique for the determination
of empirical model of diurnal refraction based on the measu
re of the effects of differential refraction of the solar
disk is presented. Preliminary results are compared with
those calculated from the model of standard atmosphere.

1. THE DIURNAL ASTRONOMICAL REFRACTION

The astronomical refraction plays, as it is well-known, a
fundamental role in the optical and radio astronomical obser
vations; besides, at the lower elevation angles, the horizon
tal refractivity gradients can introduce centimeter level
errors in laser ranging data (Gardner, 1977).
The theoretical mean refraction corresponding to an ideal
gas atmosphere in hydrostatic equilibrium in a newtonian gra
vitational field and with spherical simmetric distribution
of density is based on the formulae of the pure refractive
index and on the mean values of aerological data correspon
ding to mean atmospheric condition.
Actually, none of the conditions known for defining the mean
refraction seem to be satisfactory. This is true especially
when the normal refraction formulae are used in order to re
duce the observations carried out under conditions in which
the anomalous refraction components become more sensitive,
as in the meridian or positional observations of the Sun,
the Moon, stars and planets carried out during the day. The

W. Fricke and G. Teleki (eds.), Sun and Planetary System, 477–482.
Copyright © 1982 by D. Reidel Publishing Company.

difference in temperature and density between the dark side
of the earth, tracked during nocturnal observations, and the
sunlight hemisphere tracked during the diurnal observations,
cannot explain the systematic effects depending on the diur
nal refraction observation. Probably, the existing formulae
for the calculation of the refractive index and the simplified model of the atmosphere are not suitable to the calcu
lation of the mean refraction during solar time. The solu
tion of such a problem becomes more complicated because of
the existence of anomalous and accidental fluctuations in
the diurnal mean refraction and because of the presence of
personal and instrumental errors (Shamaev, 1977).
The idea of determining the deviation of the refraction
observed during the day from that computed from the conven
tional tables of the refraction dates back to the observa
tion carried out by Argelander and Kovalskii in the past
century. The analysis of these observation pointed out by
Radau (1882 and 1889) emphasized that the values of the
astronomical refraction obtained from the observation of the
Sun result considerally smaller because the air density va
ries gradually in height more during the day than night.
Studies of diurnal refraction anomalies and corrections to
the refraction constant have been attempted in the last
years by several astronomers (Teleki 1967, Nefedeva 1974,
Takagi & Goto 1975).

2. THE EFFECTS OF DIFFERENTIAL REFRACTION ON SOLAR DISK

The aim of this study is to offer a model for the calcula
tion of the diurnal refraction starting from the measurement
of the differential refraction affecting the solar disk at
various zenith distances. Taking as valid the formula for
the calculation of the refraction

$$\varrho = A \, tg \, z + B \, tg \, z \, \sec^2 z$$

where

$$A = a + bP + c \, PT(273 + T)^{-1}$$
$$B = a' + b'P + c'PT(273 + T)^{-1}$$

being P and T the pressure (in mm) and the temperature (in
Celsius degrees) of the air at the ground, the differential
refraction in height is given at less than 0".01 (up to

zenith distance of 75°) by the relation

$$\Delta \varrho = A \sec^2 z + B (\sec^4 z + 2 \, \text{tag}^2 z \, \sec^2 z) \, \triangle z$$
$$+ A \sin z \sec^3 z \, \triangle z^2$$

From this last one, if $\triangle z$ is the vertical diameter of the
sun and $\triangle \varrho$ the value observed of the differential refra
ction to the mean zenith distance z(center of the solar
disk), it is possible to obtain by the method of the least
squares, the mean values of the constant a,b,c,a', b' and c'.
The rough values of $\triangle \varrho$ given in table 1, show that the dif
ferential refraction on the solar disk can be observed even
with telescopes of modest dimensions.

Table 1. Rough values of the differential refraction on
solar disk

z	$\triangle \varrho$	z	$\triangle \varrho$
0°	0".6	60°	2".4
20	0".7	70	5".1
40	1".0	80	19".9

In order to use the apparent image to the solar disk for the
measurement of the differential refraction, it is interesting
to estimate the effects that, on each point of the solar
limb, determine:

(i) the parallactic correction $\triangle$ p
(ii) the differential aberration $\triangle$ s

Considering the well known formulae of the spherical astro
nomy, the displacements in right ascension and declination
of each point of the solar disk for geocentric parallax and
annual aberration are given by

$$d(\triangle \alpha)_1 = -8".8 \sec \delta \cos \varphi \, \triangle H \quad \text{(for meridian obs.)}$$
$$d(\triangle \delta)_1 = -8".8 \cos z \, \triangle \delta \quad \text{(for meridian obs.)}$$
$$d(\triangle \alpha)_2 = 2".3 \sin 2\lambda (1 + 0.32 \sin^2 \lambda) \, \triangle \lambda$$
$$d(\triangle \delta)_2 = 0".5 \sin 2\lambda (1 + 0.24 \sin^2 \lambda) \, \triangle \lambda$$

where H and λ are respectively for the hour angle and the
longitude of the Sun.
Putting in the last $\triangle\delta \approx \triangle H \approx \triangle \lambda \approx 30' = 0^R.009$ it can be
seen that the effects of the parallactic correction and the
differential aberration upon the coordinates of the solar
limb given by the foregoing two pairs of equations are very
small and in any case always smaller than 0".1.
This allows us to consider the refraction as the only appre
ciable cause of the variation of the shape of the solar disk.

3. A MODEL FOR THE REFRACTED SOLAR DISK.

Let us assume that the Sun is a circular disk of radius D.
Let us take the center of the solar disk as the pole and
vertical semidiameter as a polar axis of a system of polar
coordinates; as a consequence of the refraction any point P
(D,θ) of the solar disk is displaced to an apparent position
P'(D',θ')and the disk undergoes a complex transformation in
its shape. Assuming ϱ=Atg z, we can write (after correcting
the algebric mistake of the formula on page 210 of the book
by Woolard and Clemence, 1966)

$$D' = D (1 - A) (1 - f \cos^2 \theta')$$
$$\theta' = \theta$$

the equation of an ellipse with semimajor axis a = D(1 -A)
and flattening f = (A/1 - A) tg^2 z.
As a test for this model, some preliminary photographic ima
ges of the Sun has been obtained with the Cooke equatorial
refractor (40 cm diameter, 5.915m focal lenght) at Teramo
Observatory, Italy.
The apparent diameter of the image of the Sun on the plate
results about 55 mm (i.e. 1" = 30 /um).
The plate reduction has been carried out with the microdensi
tometer PDS at Napoli Observatory, Italy.
Relative coordinates of the points of the apparent solar
disk have been derived by measuring the optical density of
the plate; but it is difficult, in general, to measure it
with sufficient accuracy and therefore absolute determina
tions of the solar diameters on the plate have probably a
somewhat lower precision because of various systematic er
rors; to avoid it, at least in part, we prefer derive values
of the flattening f of the refracted solar disk by using the

Table 2 Observed values of the flattening f of the refracted solar disk (the internal error in f is less than 0.5×10^{-3})

z	f	$(A/1-A) tg^2 z$
40°–50°	$1.7.10^{-3}$	$0.2 \ 10^{-3}$
50°–55°	1.8	0.4
55°–65°	1.3	0.8
65°–70°	1.7	1.5
70°–75°	2.9	2.9
76°.5	4.7	4.5
78°.8	6.0	6.6
86.8	14.5	83.5

following method based on the geometric properties of the co
nics: each plate has been scanned in 11 parallel lines; assu
ming the curve of the optical density of each line to be sym
metrical, the coordinates (x, y) of the midpoints are not af
fected by systematic errors; besides, from the properties of
the chords of the ellipse it follows:

$$x = ky + q$$

where the angular coefficient k is related to the flattening
f by

$$k = f \sin 2p$$

and where p is, in this case, the parallactic angle.
We used the above relations to derive k and f from (x, y) by
the least squares method.
The mean values of the flattening thus obtained are compared
in table 2 with those resulting from the model of the stan
dard refraction. The consistency of the results is in part
satisfactory.
For values of z between 65° and 77° the observed and computed
values agree fairly well one to another; of course, the are
not agreement for z > 80° and, it is less explanable, also for
40° < z < 65°.

4. CONCLUSIONS

The results presented here are preliminary and a number of problems still remain to be solved before the method can be considered reliable. The largest potential source of error is probably depending upon the reduction method of the photo graphic plate. A better fitting procedure of the data obtai ned by microdensitometer would enable errors to be reduced by a factor 5 or more. In addition the reduction of other sources of errors in the observations and a second-order mo del of the refraction will provide much more accurate data from which we can derive a better knowledge of the parame ters of the refraction.

The authors wish to express their thanks to Mrs. E.Acampa, Mr. S.Pilloni and Dr. S.Uras for computational help. Help by Mr. A.Dipaoloantonio in making the observations is greatly acknowledged.

REFERENCES

Gardner, C.S., 1977, Applied Optics, Vol.16, N.9, <u>2427</u>
Nefedeva, A.I., 1974, Sov. Astron., Vol.13, N.3, <u>379</u>
Radau, M.R., 1882, Ann. Obs. Paris, Mem. 16, Part B
Radau, M.R.,1889, Ann. Obs. Paris, Mem. 16, Part G
Shamaev, V.G., 1977, Sov. Astron., Vol. 21, N.1, <u>129</u>
Takagi, S., Goto, Y., 1975, Publ. of the ILOM, Vol.X, N.1,41
Teleki, G., 1967, Publ. Obs. Astron. Beograd., N. 13
Woolard, W.E., Clemence M.G., <u>Spherical Astronomy</u>, Academic
        Press, 1966

ASTROMETRIC SITE SELECTION

G. Teleki
Astronomical Observatory, Belgrade, Yugoslavia

ABSTRACT: The requirements for the most convenient astrometric observing sites are discussed. Among properties of a good astrometric observing site, besides those demands set forth for the astrophysical sites, there must be also: the maximum stability of the ground, of the plumb line and of the atmospheric layers of constant refractivity; as homogeneous as possible temperature field around the instruments and pavilions; clear days during the year distributed as evenly as possible. The author aralyses the global influences of several atmospheric, geophysical and geolcgical factors on astrometric observations and on the basis of these considerations he comes forward with the suggestion bearing on the most favourable locations for the best astrometric observations.

1. INTRODUCTION

The results of the astrometric observations are influenced, among others, by atmospheric, geophysical and geological factors of the most different kinds. These influences might be divided into local ones - whose sources and action are in the immediate surroundings of the observing stations, within a territory of about 10 x 10 km^2 -, the regional (1000 x 1000 km^2) and the global ones, the latter extending over the whole globe. In the present analysis we shall endeavour to provide a presentation of possible global influences in order to find out the basic criterion for the selection and assessment of the location of a particular observatory. One may on the whole be pretty sure that the amount of these global influences must be inferior to the regional ones and certainly inferior to the very irtricate local influences. Nevertheless, some of them may be very pronounced: the cloudiness may serve as an example. The knowledge of only global influences is not sufficient for the assessment of the quality of a particular location, but it does provide starting information on the potential ones.

The impact of a given influence is dependent not solely on its amount and the frequency of its occurence, but also on the accuracy with wich the quantity, affected by it, is determined. The following might there-

W. Fricke and G. Teleki (eds.), Sun and Planetary System, 483–491.

fore be stated: the current accuracy of astrometric results is of the order of 0."01, that is, approximately 0.1 m on the Earth's surface. It is expected that by the introduction of new technics in the astrometric observations from the Earth's surface (VLBI, etc.) their accuracy will be increased by an order of magnitude at least (Tucker, Teleki, 1978). It is accordingly necessary, particulary so if the future is kept in mind, to analyse all the atmospheric, geophysical and geological influences apt to affect astrometric results by more than about 1 cm.

2. ASTROMETRIC SITE SELECTION CRITERIA

No unique selection criteria of the astronomical observing stations are given. This is explanable not only by the very delicate character of the selection in general, but also by the different requirements regarding the nature, programme and the time of observation. Particularly poorly treated are the questions concerning selection of the astrometric stations, or, to put it more correctly: there has been little concern in the past over the location where the instrument was to be installed and in the majority of cases the same is true of the present-day instrument mounting. Higher accuracy required of the terrestrial astrometric observations makes inavoidable a more profound inquiry concerning these questions and their application to the practice.

In considering locations for the astrometric stations one should proceed from the general astronomical - in fact astrophysical - criteria, as most of the experience has been gained just through the latter. Here again a differentiation must be stated. From the previous general criteria of astronomical site selection - see, for instance, Stock (1964) - a transition has occured to the "special" ones, such as those of the dark sky (Walker, 1975) or those of the solar observatories (Brandt, 1975). Most deserving in our view is the combination of criteria as used in selecting location of the North Hemisphere Observatory of U.K. (Anonymous, 1976). This observatory will comprise an astrometric instrument also.

But all of these criteria, the last on inclusive, are not sufficient to meet astrometric demands in their totality. What they fail to take into account is the stability of the observing site (the ground, the pillar) and of the direction of the vertical, as well as the inclination of the atmospheric strata of equal density and their variability in the course of time. What is also lacking is the demand of an even distribution of clear days (nights) during the year (of particular importance in the observations where continuity is required) and the demand of the most homogenous possible temperature field around the instrument and the pavilion.

The following criteria in selecting locations for astrometric stations should therefore, in our view, be taken into account:

2.1. The maximum possible percentage of clear days and nights;
2.2. Even distribution of clear days and nights during the year;

2.3. Dark sky in the night and a suitable sky by the day;
2.4. Even distribution of dark nights and suitable days during the year;
2.5. High transparency;
2.6. Minimum optical turbulence;
2.7. Good seeing;
2.8. As homogeneous temperature field around the instrument and the pavilion as only possible;
2.9. Steady inclination of the atmospheric strata of equal density and their minimum variability with time;
2.10. The surrounding ground of the instrument should be grassy (with low bush), with no buildings, bare of woods and roads, etc.;
2.11. No strong (blow) winds;
2.12. The location should be dry;
2.13. Maximum possible stability of the observing site (of the ground in general and of the pillar in particular) and of the plumbe line;
2.14. Such locations should be chosen where geophysical, geological and urban conditions are not likely to change for a sufficiently long time in the future;
2.15. Do not plane observations at zenith distances over 60°;
2.16. Good living conditions should be secured.

It is certainly hard to meet simultaneously demands for favourable observing conditions by night with favourable conditions by day - but such demands do occur in astrometry. But if night observations only are intended, the demands set forth may be less rigorous.

The above set of criteria comprehends, compared with the one valid for astrophysical observations, the following additional demands: 2.2, 2.4, 2.8, 2.9, 2.10, 2.12 and 2.13. Clearly, it is hardly possible to meet these demands in their totality, thus a reasonable compromise must be accepted.

3. THE CONSIDERED INFLUENCES

In connection with criteria 2.1, 2.9, 2.11, 2.13 and 2.14 importance must be attached to the discussion of the following global influences:

3.1. Seismic;
3.2. Atmospheric (refraction, cloudiness),
3.3. Strain,
3.4. Plate tectonic,
3.5. Gravimetric,
3.6. Tidal.

Some of these influences on the astrometric data can be accounted for by corresponding formulae - for instance the pure (normal) refraction, gravimetric and tidal influences on the vertical. In these influences we should analyse only their variability relative to the calculated values.

If we are, therefore, to proceed from this reasoning, then the gravimetric and tidal effects can be neglected on account of their variability causing the changing in the direction of the plumb line of less than $0\rlap{.}''001$ (Pellinen, 1978). In contrast, the amount of the effects of other influences can exceed this value and this is the reason of our submiting them to closer inquiry.

The question arises of how to rank the influences from 3.1 do 3.4 as it is apparent that their effects are not of the same amount. We therefore decided to take Kaula's (1978) space-time spectrum of geodynamic processes as a basis in ranking the influences 3.1, 3.3 and 3.4. We calculated the average rates of the process developing in km/year units, on the premise that the larger these values are, the more dominant effects on the astrometric data must be forthcoming from the corresponding processes.

In order to include also atmospheric influences 3.2 in this ranking we proceeded in the following fashion:

- the anomalous refraction is taken as $0\rlap{.}''2$ per day,
- an order of magnitude higher influence of the cloudiness than the one stemming from anomalous refraction.

In this way we achieved the (relative) rating list of any-one influence's importance: seismic 7, cloudiness 6, anomalous refraction 5, strain 3, and plate tectonic 1.

In this article a special attention will be paid to the three first influences: seismic ones, those connected with cloudiness and the anomalous refraction. Of the influences originating from the strain, in connection with the seismic ones, only elastic strain of a global character will receive treatment.

Because of limited space we are not going to speak here about the influences of the plate tectonic. Yet, the need should be stressed of the astrometric stations to be located at a distance not below 100 km from the active plate boundaries and of the known deformation zones in it. These conditions were set by Mao and Mohr (1976) in their discussing the site selection for laser satellite - tracking stations.
According to them, the stations should be distant enough of the glaciation areas and the coastal regions. In their view, the most convenient locations of the stations are those on the shield areas which are free of the regional crust deformation, apart from gentle epeirogenic mouvements excepted, since about 2.5 to 2 billion years (we refer to these shield areas in the concluding part of this article). We deem it quite warrantable to set the same demands on the astrometric stations, for then tectonic risks would be avoided.

4. SEISMICITY

The strains in the Earth's crust are potential inciters of the ground shiftings which can exert substantial influence on the astrometric results. The magnitude of the strain and its field can, on the whole, bear regional and local characteristics. Elastic strain are, on the contrary, global in character (Vvedenskaya, 1969). By the elastic strain field of the Earth, associated with the earthquakes, we understand the totality of the seismic areas. This field is not homogeneous, exhibiting different strain characteristics at various places.

Of great importance is the fact that both compression, which is prevalent, and expansion, are horizontally oriented in all the parts of the Earth. Consequently, horizontal drifting of the stations, relative to the Earth as a whole, is what one rightly expects to be occuring(naturally, in local ranges the vertical drifts can be also expected). For this reason these areas, in particular those most active seismically, as for instance the Pacific and Mediterranean-Himalayan seismic belts, are not suitable for astrometric observatories.

The information on seismic areas can be found in several books (for detailed information see, for instance, "Seismicity of the World", U.S. Department of Commerce).

5. ATMOSPHERIC INFLUENCES

Atmospheric factors exercise influence on the path of the light ray, alter instrumental characteristics and create various meteorological fields around and inside pavilions. These atmospheric influences must be allowed for by astrometrists (Teleki, 1978).

Of all these influences reference will be made here only to refraction disturbances and to the cloudiness as the dominant limiting factors in optical observations. General characteristics will be given but we also indicate, where possible, the seasonal and some regional features. The remaining factors should, in our view, be taken into account in treating regional, and more so in treating local aspects of this question.

5.1. Refractional disturbances

A complete presentation of the refraction experienced by the light ray in the Earth's atmosphere - in various regions of the Earth, in the course of a year and day, is still lacking. But what failed to be performed in the optical wave range, was performed in the radio wave range (Bean et al., 1966). The refractional effects in both being similar and comparable, in spite of the fact that the influences caused by humidity in the radio wave range are more pronounced (Bean, Teleki, 1974), there is every reason to make use of Bean's et al. atlas in this analysis.

In making comparison of astrometric results, obtained within a chain of stations - those of the International Latitude Service are one examplean essential role is certainly played by the systematic differences of refractional influences at equal zenith distances. Moreover, in assessing individual stations, not these differences, but their variability in time and the occurence of extreme conditions will be taken into account.

In order to find out possible locations where such disturbances do accur
we borrowed from the above cited atlas the following data, i.e. results
of the corresponding analysis:

5.1.1. Correlation between $N_s=(n_s-1)10^6$, where n_s is the refractive index
in the surface layer, and ΔN(the difference N_s-N at the height of 1 km)
taken as a mean for a month, and
5.1.2. Extreme values of the gradients N in the surface layer up to 100m
height.

On inspecting global characteristics of these data, we note their diversity depending on the position on the Earth's surface as well as their
variability during the year. Whereas the predominant role must be ascribed
to the locations in 5.1.1., in the 5.1.2. data a stronger influence depending in the season is evident.

The data obtained are summarily illustrated in Fig. 1. We find there
those areas marked which are not suitable, in terms of refraction, for
astrometric observations. This selection is not exact, but it shows
roughly the areas which are important for us. We see also that delimi-
nation territories with pronounced temperature gradients in the surface
layer (ocean-land, areas surroundings deserts, etc.) - in the tropical
regions in particular - are ill suited to the astrometric observations.

How far one can go in relying on the atlas of Bean et al.? There certainly
is some foundation to. Here is an example. Teleki (1976) inferred from
this atlas that lower refraction values were to be expected in the Kitab
region, but higher ones in the Ukiah region. In this he was vindicated
by the results of latitude observations at the two stations. The same
conclusion follows from the results of Sugawa and Kikuchi (1979), obtained
by a completely different method of treating atmospheric data.
5.2. Cloudiness
The Earth as a whole is covered by clouds by 55% on the average. But
this particular information by itself cannot reproduce the true state
and diversity found in different areas on the Earth. So, for instance,
the average cloudiness over the oceans is 58% and that over the dry land
is 48%; the cloudiness over the south hemisphere is heigher than that
over the north hemisphere; a cloudy or clear weather is more frequent
over the continental regions at moderate latitudes than is partially
cloudy weather; there is, in addition, considerable variability from
season to season and in the course of a day, ect. (Haurwitz, Austin,1944).

Three belts of higher cloudiness, whose position vary during the year,
are distinguishable. Two belts of strongest cloudiness are stated at
about 60° north and south latitude in the areas of high frequency of
cyclons, where stratified clouds extend over large areas. There, the
frequency of the completely cloudy sky is above 50%. Roughly estimating,
these cloudy belts extend from between 35° to 40° latitude to the poles.

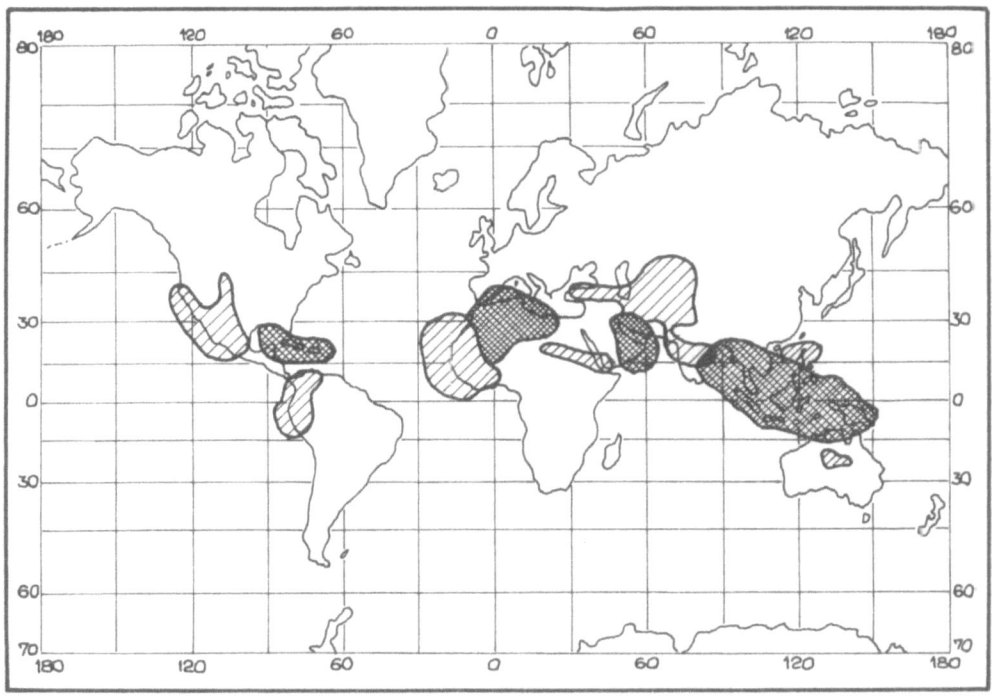

Figure 1. Areas (inclined lines) of doubtful appliciability of using
N_s to predict ΔN (see the influence 5.1.1) and the areas
(cross-lined) of disturbed refractivity variations (see 5.1.2)

The third belt is around equator, in the intertropical zone of convergence, where the clouds are predominantly convective. Characteristics of
this belt is that the clouds inside it do not cover the entire sky. It
is estimated that the frequency of semiclouded sky is above 50%. Thus,
this equator belt - extending approximately from 10° north to 5° south
latitudes - is not as pronounced as the belts at 60° latitude in the
north and south hemisphere respectively.

More suitable astrometrically are the zones between the belts of stronger
cloudiness. In order to define them better we chosed the zones, according
latitude, where the cloudiness is below 55%. In this we used data
presented to by Figures 5 through 8 in the paper of Winston (1969). The
two zones thus established are: the one between 8° and 40° north latitude
and the second between 6° and 27° south latitude. As a matter of course,
the limits of these zones should be considered as only approximate ones,
variable during the year and day, and even inside these zones no complete
homogeneity can be expected. Using the data of Winston (1969) and Kravcov
(1972) it is possible to find also areas where the number of clear days
during the year is 50% on average or above.

6. CONCLUSION

The Figure 2 shows regions which are potentially suitable for astrometric observatories. For this selection we took into account the following global influences: seismicity, refraction disturbances, cloudiness and active plate boundaries (Vvedenskaya, 1969). In this figure we find indicated also the shield areas, i.e. the areas of great stability. For this, the data of Belousov (1975) were used.

As evident, there are few locations that are suitable, in this or another way, for astrometric observations. More complete information on individual potential location should be provided by through analysing regional and, in particular, local conditions.

As for ourselves, we are going to continue researches into this matter.

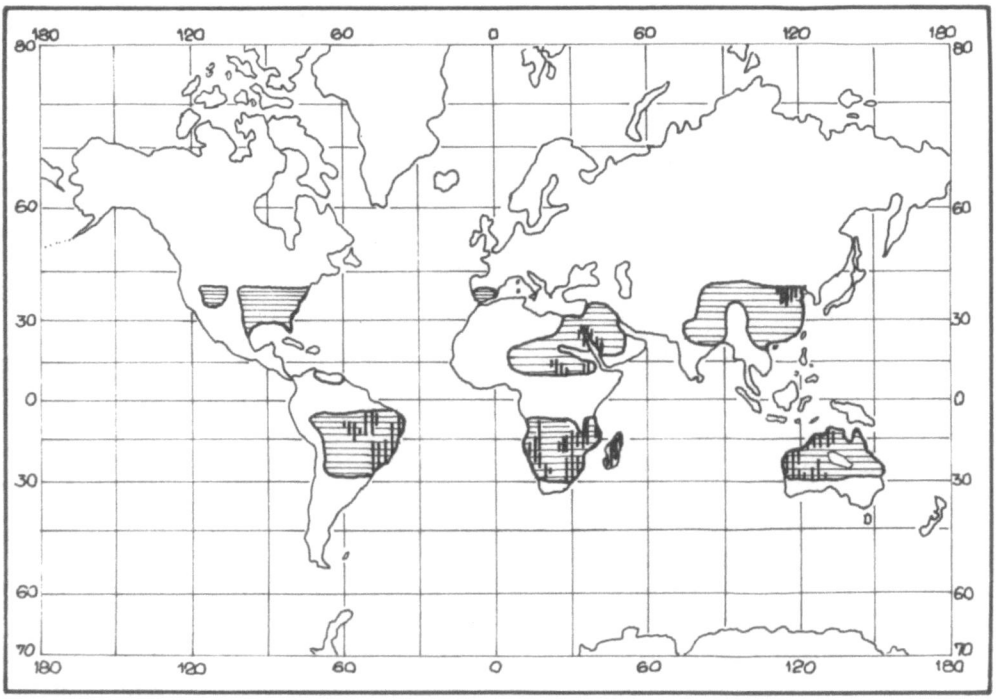

Figure 2. Regions which are potentially suitable for astrometric
          observatories (horizontal lines). The shield areas is
          indicated (vertical lines) in these regions.

REFERENCES

Anonymous, 1976: A Proposal for an Observatory in the Northern Hemi-
          sphere, Sc.Research Council, London, 35.

Bean, B.R., Cahoon, B.A., Samson, C.A. and Thayer, G.D., 1966: A World
        Atlas of Atmospheric Radio Refractivity, ESSA Monograph 1,1.
Bean, B.R. and Teleki, G., 1974: Publ.Obs.Astron.Belgrade, 18, 21.
Belousov, V.V., 1975: Osnovy geotektoniki, Ed.Nedra, Moscow.
Brandt, P.N., 1975: JOSO Annual Report 1975, 55.
Haurwitz, B. and Austin, J.M., 1944: Climatology, Ed.McGraw-Hill Book
        Comp., 92.
Kaula, W.M., 1978: Reports Dept.Geod.Sci., Ohio St.Univ., 280, 349.
Kravcov, L.M. (Red.), 1972: Oblachnij pokrov Zemli, Ed.Gidrometeoizdat,
        Moscow, 1.
Mao, N.H. and Mohr, P.A., 1976: Smithsonian Astrophys. Obs., Special
        Report, 371, 1.
Pellinen, L.P., 1978: Vysshaya geodeziya, Ed.Nedra, Moscow, 230 and 234.
Stock, J., 1964: Bull.Astron., 24, 119.
Sugawa, C. and Kikuchi, N., 1979: Proc. IAU Symp. 89, 114.
Teleki, G., 1976: Mitt.Lohrmann Obs., Dresden, 33, 916.
Teleki, G., 1978: Bull.Obs.Astron.Belgrade, 129, 1.
Tucker, R.H. and Teleki, G., 1978: Proc. IAU Coll. 48, Vienna, 548.
Vvedenskaya, A.V. 1969: Issledovanie napryazhenij i razryvov v ochagah
        zemletryasenij pri pomoshchi teorii dislckacii,
        Ed.Nauka, Moscow, 127.
Walker, M.F., 1975: IAU Commission 50 Circular Letter, 11.
Winston, J.S, 1969: Climate of the Free Atm., Elsevier Publ.Comp., 4,247.

THE INFLUENCE OF IONOSPHERIC REFRACTION ON RADIO ASTRONOMY INTERFEROMETRY

T.A.Th. Spoelstra
Netherlands Foundation for Radio Astronomy
Dwingeloo
The Netherlands

ABSTRACT

Observations with radio astronomy interferometers (VLBI and local interferometers) may suffer severe phase errors due to ionospheric refraction and its variations. Techniques to correct these phase errors are discussed. Results of the application of these correction procedures to observations done with the Westerbork Synthesis Radio Telescope are presented.

1. OBSERVABLES AND CALIBRATION

The prime 'observables' generated in radio astronomy interferometry are the interferometric amplitude and phase from each pair of antennas as a function of time during the observation. The calibration of the data involves the determination of corrections to be applied to the amplitudes and phases to account for various systematic errors, as well as for variable effects due to the troposphere and ionosphere. Systematic errors are a function of baseline, delay, pointing, and gain for each antenna. The necessary corrections can be determined from observations of specific calibration sources and these corrections are usually applied to the observations. After calibration the data need to be accurate to better than a few percent in amplitude and a few degrees in phase to be useful for radio astronomy imaging. This is typically achieved by local interferometers, except under unusual circumstances; however, for VLBI the phase uncertainties due to ionosphere and troposphere are much larger, necessitating special techniques to be employed.

2. EFFECTS OF THE IONOSPHERE

The effects of the ionosphere on radio astronomical observations are mainly Faraday rotation, phase and group delays and refraction (Hagfors, 1976). Faraday rotation is important in the study of polarization of radio sources in which case a correction has to be implied.

W. Fricke and G. Teleki (eds.), Sun and Planetary System, 493–496.

The other effects imply only phase correction. Phase irregularities are produced in an incident wavefront by variations in the refractive index which cause changes in the phase velocity of the signal from its free space value. The variations in the refractive index are caused by changes in the electron content of the ionosphere. Estimates of the electron content as a function of time and of position must be obtained by probing the ionosphere. These effects increase with wavelength squared.

Phase and group delays may be a cause of concern in VLBI. They are less important for local interferometry.

3. IONOSPHERIC REFRACTION

During recent decades a number of authors have estimated ionospheric refraction assuming that the ionosphere is a spherical shell concentric with the Earth (e.g.: Belyaev 1955; Link 1957 and Chvojkova 1958a, 1958b). Smith (1952) showed that appreciable displacements of radio sources are due to departures from spherical symmetry. Komesaroff (1960) derived expressions for the apparent displacement of radio sources taking into account the effects of both a spherically symmetric ionosphere and horizontal electron density gradients.

Komesaroff (1960) showed that the total change in declination is the sum of refraction due to the spherically symmetric component of the ionosphere and due to north-south electron density gradients. The change in right ascension depends to first order only on the east-west gradients in the electron density. His results indicate that a very marked improvement in declination can be obtained, although systematic errors in right ascension remain. One reason for this was the poor determination of east-west gradients in the electron density.

Komesaroff's approach has been applied to observations with the Westerbork Synthesis Radio Telescope (WSRT). This instrument consists of an array of 14 equatorially mounted 25 m parabolic antennas on a 3 km east-west line (for a full description see Bos et al. (1981)). Because of its east-west orientation it was assumed that to first order the east-west gradients are the primary influence on the observations. Furthermore, the availability of f_0F_2 and y_pF_2-values only once every integer hour from one station (the Dutch Meteorological Institute, K.N.M.I., at De Bilt), makes it necessary to interpret time variations of the observed parameters only in terms of east-west gradients in the electron density. Preliminary results were presented by Spoelstra and Schilizzi (1981). However, careful analysis showed that good results could only be expected between a few hours after sunset and a few hours before sunrise when the ionosphere behaves rather regularly. An additional correction was obtained by taking into account the spherically symmetric component of the ionosphere. Also, the F_2-layer was divided into several zones to account for the variation of electron density with height. This approach is used to correct observations made with the WSRT. A full description of the whole procedure is given by Spoelstra (1981,1982)

4. RESULTS

Observations done with the WSRT can be corrected for ionospheric refraction. Figure 1 shows results of an observations of 3C48 at 50 cm wavelength. The radio source 3C48 is essentially a point source. Therefore, the phase should be constant and $0°$ during the observation (when the instrument is steered properly and the right corrections are applied). However, the phase errors are clearly very serious (the instrumental phase accuracy is $\leq 0°5$). Calibration with corrections for ionospheric refraction resulted in a substantial improvement of the observation. However, since ionospheric parameters are only available for each integer

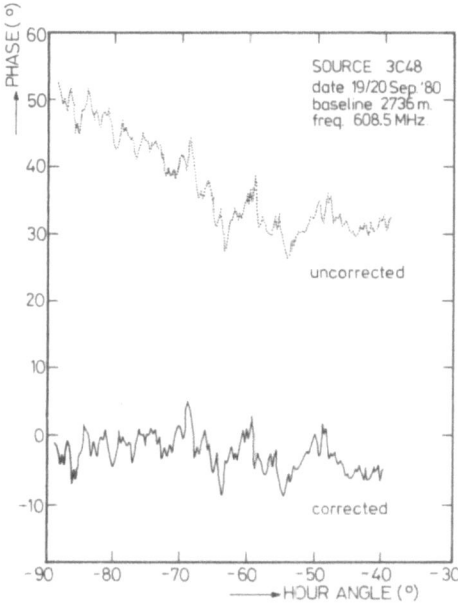

Fig. 1: Phase variations as a function of hour angle for 3C48 as observed and as corrected.

hour, it is not possible to correct for variations on a time scale of less than a hour (e.g. due to acoustic-gravity waves). It is expected that in the near future the use of satellites (instead of the present ionosonde of the K.N.M.I. at De Bilt) for ionospheric probing on shorter time scales will alleviate this problem. The present correction technique for ionospheric refraction clearly fails during several hours around sunrise. This is due to vertical drifts of the plasma during sunrise (Hargreaves, 1979, 66f). We hope that the use of satellite observations will help alleviate this problem.

REFERENCES

Belyaev, N.A., 1955, Astr. J. Moscow 32, 357.
Bos, A., Raimond, E., Van Someren Gréve, H.W., 1981, Astron. Astrophys. 98, 251.
Chvojkova, E., 1958a, Bull. Astr. Inst. Czech. 9, 1.
Chvojkova, E., 1958b, Bull. Astr. Inst. Czech. 9, 133.
Hagfors, T., 1976, in: "Methods of Experimental Physics", vol. 12B, ed. M.L. Meeks, New York, p. 119.
Komesaroff, M.M., 1960, Australian J. Phys. 13, 153.
Link, F., Bull. Astr. Inst. Czech. 8, 112.
Smith, F.G., 1952, J. Atmospheric Terrest. Phys. 2, 350.
Spoelstra, T.A.Th., 1981, Netherlands Foundation for Radio Astronomy, Internal Technical Report 162.
Spoelstra, T.A.Th., Schilizzi, R.T., 1981, in: Proceedings of the COSPAR/URSI symposium on "Scientific and Engineering Uses of Satellite Radio Beacons", Warsaw, Poland, May 19-23, 1980, ed.: A.W. Wernik, p. 315.
Spoelstra, T.A.Th., 1982, Radio Science (in preparation).

DISREGARD OF SOME RAY TRACING PRINCIPLES
IN OPTICAL AND RADIO SPECTRAL BANDS

E.Woyk (Chvojková)
Astronomical Institute, Czechoslovak Academy of Sciences,
12023 Praha 2., Czechoslovakia

The figures of many papers show that the authors ignore some basic ray tracing principles which substantially simplify the study of the propagation of electromagnetic waves and which would also simplify the study of the three-dimensional problem.

In the investigated regions, or their parts described below, levels of equal index of refraction n are assumed to be concentric spheres - not necessarily concentric with the Earth.(If not concentric see Woyk 1969.) The radius r is measured from the common centre of curvature and the angle of incidence,i, is given by the Snell's law, $rn \sin i = const = r_t$, valid for each point r on the ray path. Some deductions follow on Fig.1:

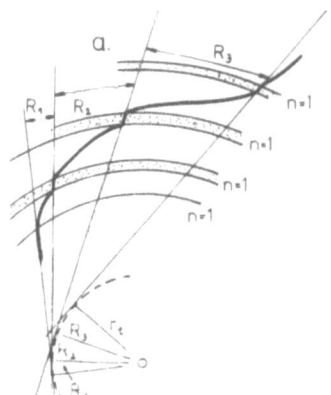

Figure 1. If a ray crosses a fictive (concentric) segment at which the real n has been substituted by $n=1$ (by $N=0$ in the radio band) then, in each of these segments, the path is always tangential to the same sphere of radius r_t to which it was tangential before having entered (ionised) layers and after having left them (supposing perfect sphericity, and $n=1$ outside the layer). Each ray propagates through such a layer as though it were turned round the common centre of curvature by the angle of refraction R which is rather proportional to the number of all deviating elements (electrons) over unit area.

There exist two circular levels r_{∞} along which a horizontal ray theoretically propagates around the Earth, as long as the layer remained perfectly spherical. They occur when $d(nr)/dr=0$, i.e. when $dn/dr=-n/r$. See Woyk 1954 and Fig.2. One of these circular paths, r_{∞}^{II}, lies below the minimum of n (.i.e. below the maximum of electron density N) and it is dispersive (rays reflected above or below r_{∞}^{II} always deviate from it.) The lower one, r_{∞}^{I}, occurs above the maximum of n (minimum of N),

W. Fricke and G. Teleki (eds.), Sun and Planetary System, 497–498.
Copyright © 1982 by D. Reidel Publishing Company.

usually closely above the lower boundary of the ionised layers (the curvature of the ray is so small that it coincides with the earth curvature.)
The propagation along r_{∞}^{I} is very stable: rays reflected above or below
it periodically deviate back towards r^{I}. Note: A ray passing through the
the whole layer has i=min at r_{∞}^{I} and i=max at r_{∞}^{II}. Some deductions:

A ray from the outer space can penetrate below the maximum of elec-
tron density and be reflected from a level $r_r^{\vee}$ which is above r_{∞}^{II} but below the maximum of the layer.

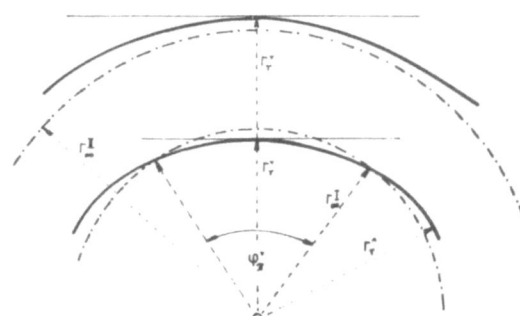

Figure 2. Dot-and-dashed lines are
the two circular paths, the full
lines are two paths. Rays reflec-
ted above r_{∞}^{II} or below r_{∞}^{I} often
lie below the tangents at the points
of reflection even though they rise
to higher altitudes r.
    Rays oscillating along r_{∞}^{I} are
reflected from two levels $r_r^{\wedge}$ and $r_r^{\vee}$
and a beam horizontal at r_{∞}^{I} is
periodically focused after the distance φ_{*}.

A ray oscillating along the lower r_{∞}^{I} need not leave the layer at
all. - Rays from below cannot be reflected but between r_{∞}^{I} and r_{∞}^{II} which,
however,, approach each other if the ray frequency rises. Thus when, theoretically, $r_{\infty}^{I} > r_{\infty}^{II}$ the layer becomes totally transparent from below.

A theoretical electron density profile N(r) has been adopted applicable to any ionised layer. It makes it possible to gain analytical
path formulae,Woyk 1965,1969. It appears from these formulae that no
unique path formula can be constructed for trajectories through the whole
layer - since, if the ray propagated from the I-zone (containing n_{max},
N_{min}) to the II-zone (with n_{min},N_{max}) then cos $k\varphi$ turns into sinh $k\varphi$
(for rays passing the whole layer) and into cosh $k\varphi$ (reflected rays),
or vice versa.

The described formulae and principles have also been applied to the
visual light (Woyk 1978) and can be used for tropospherical propagation
as well.How to treat the propagation in not perfectly spherical layers,
see Woyk 1969.

Woyk,E.: 1954, Bull.Astronom.Inst. Czechoslovakia, 5, pp.104-11.
Woyk,E.: 1965, Jour.Res.NBS, 69D, pp.453-7.
Woyk,E.: 1969, Radio Science, 4, pp.23-33.
Woyk,E.: 1978, Applied Optics, 17, pp.2108-13.

REFRACTION EFFECTS ON GEODETIC MEASUREMENTS
IN THREE-DIMENSIONAL TERRESTRIAL NETS

Ludvík Hradilek
Charles University
128 43 Praha 2, Albertov 6
Czechoslovakia

Abstract: The biasing influence of terrestrial refraction
on geodetic three-dimensional nets can be substantially reduced by an appropriate design of the observational procedures complemented by relevant mathematical models expressing refraction and statistical tests eliminating major bias.

1. STEPS TAKEN TO IMPROVE ACCURACY

At the IAG Symposia in Stockholm in 1974 and in Wageningen
1977, and IAU Symposium in Uppsala 1978, a method was presented for evaluation of refraction by the adjustment of
three-dimensional nets or trigonometric leveling traverses
and an application of such nets was illustrated by determining crustal movements in the High Tatra Mts. The uplifts
of the mountain peaks surrounding the Žiar Valley reached
6.5-8.5 mm/year and were estimated with a standard deviation
of 2 mm/year in the Height Reference System of the Slovak
Socialist Republik. This accuracy corresponds to the first
measurement of the three-dimensional net in 1961/62. Recent
efforts for eliminating bias in geodetic leveling and a substantial increase in precision of range-finders yield a
justified hope for reducing the above standard deviation
below 1 mm/year.
Therefore we have also tried to refine our methods determining refraction by an adjustment of three-dimensional
nets. In the last three years we have 1/ generalized mathematical models expressing refraction at every station or for
every line of sight, 2/ completed stochastic models defining
a priori variances of zenith distances, and 3/ investigated
systematically the functional models for adjustment of the
observed and transformed data.

W. Fricke and G. Teleki (eds.), Sun and Planetary System, 499–502.
Copyright © 1982 by D. Reidel Publishing Company.

2. MATHEMATICAL MODELS EXPRESSING REFRACTION

When adjusting a three-dimensional net, all the data measured contribute to the evaluation of refraction. Spatial distances - as much inclined as possible - and spirit leveling height differences - provided the latter are available between some points of the network - diminish especially systematic refractional bias over larger areas, and the zenith distances supply the bulk of information determining refraction. However, the zenith distances should be measured three times at least in time intervals larger than two hours and carried out as one observational unit at every station.
The coefficient of refraction k_{ij} of a line of sight P_iP_j is given in a general form

$$k_{ij} = f(P_i) + g(P_iP_j). \qquad\qquad /1/$$

The function $f(P_i)$ expresses the main part of the refraction at the observation station P_i and the function $g(P_iP_j)$ represents a refractive correction for the particular line of sight. The special forms of the functions $f(P_i)$, $g(P_iP_j)$ may be obtained from previous investigations on refraction or deduced from the actual changes of zenith distances measured; the unknown coefficients included in these functions may be evaluated as parameters by the adjustment of the network. When adjusting zenith distances measured under normal observational conditions, i. e. not during periods of sudden refractional changes, a zero hypothesis in the form

$$f(P_i) = k_i \quad , \quad g(P_iP_j) = 0 , \qquad\qquad /2/$$

(i. e. one coefficient of refraction for one station) proved to be satisfactory for 96%-99% of lines of sight. The anomalous 4%-1% lines of sight were mostly constituted by grazing lines. We check the zero hypothesis by two statistical tests: 1. The test of actual changes of refraction as given by repeated measurements of zenith distances at the station (if the refraction is to be equal for all lines of sight at a station, then the actual changes of refraction should be equal too). 2/ The test of outlying values of misclosures of condition equations. Note that it is not necessary to compile the condition equations, their misclosures are given by an algebraic sum of absolute terms of observation equations. For expressing the alternative hypothesis the actual changes of refraction are of crucial significance; we may also employ elevations of the stations, inclinations of the lines of sight, equivalent heights of the lines of sight etc. (Hradilek, 1980).

3. STOCHASTIC MODEL FOR THE ADJUSTMENT OF ZENITH DISTANCES

The a priori variance of the average of measured zenith distances may be expressed in the form

$$m_z^2 = \frac{m_a^2}{n} + m_s^2 + (C \gamma\, m_k)^2 \,, \qquad\qquad /3/$$

where n is the number of zenith distances in the average and m_a represents the accidental errors, m_s^2 expresses instrumental systematical errors and $(C \gamma\, m)_k^2$ residual refraction errors. C is a coefficient evaluated within the limits ⟨0.5 , 1.5⟩ according to the number of measured zenith distances and their changes, γ = arctan (t/2R) where t is the horizontal distance (i. e. chord) ,   R denotes the mean radius of the Earth, and m_k is the standard deviation of a coefficient of refraction as determined by previous adjustments and given in Fig. 1. In case of consideration of only one coefficient of refraction for the whole network, we increase the values m_k about three times in order to fit them (for distances 6 - 7 km) to the usually assumed value $m_k = 0.03$

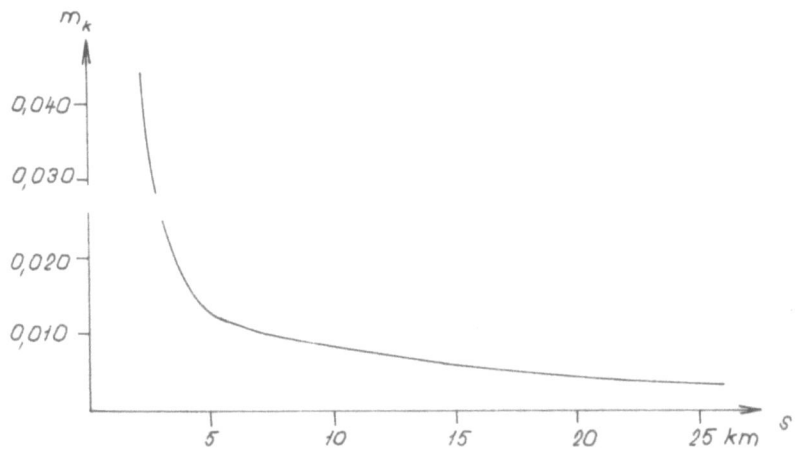

Figure 1. Standard deviation m_k of the coefficients of refraction estimated by the   adjustment of trigonometric leveling nets and three-dimensional nets. s is an average length of lines of sight in the network (Hradilek, 1980).

4. FUNCTIONAL MODELS FOR THE ADJUSTMENT OF THREE DIMENSIONAL NETS

The influence of refraction on the accuracy of three-dimensional nets also depends on the choice of functional model

for the adjustment. The functional models expressed in form
of observation equations or condition equations for directly
observed data (without data transformation) yielded satisfactory results. Difficulties arose when transforming measured zenith distances and horizontal angles into facial
angles. It must be admited that the simulated data (without
the influence of refraction) gave very good results and the
precision of adjusted elevations changed surprisingly very
little with decreasing inclinations of slope distances; however, the adjustment of facial angles deduces from actual
observations (influenced by refraction) gave significantly
larger and less reliable values of coefficient of refraction
than the adjustment of original and not transformed measurements.
A promising change for substantial increase in precision of
both horizontal and elevation coordinates is given by three-
dimensional nets defined only by ranging of top precision.
When adjusting spatial distances simulated with an accuracy
of 5.10^{-7} s in three network models, namely 1/ the mountain
side model, 2/ the mountain ridge model and 3/ the double
ridge model, corresponding to the topography of the High
Tatra Mts, the double ridge model yielded the same accuracy
in elevations as did the adjustment of a much larger number
of the measured zenith distances, horizontal angles and spatial distances (being observed with an accuracy
$5 \text{ mm} + 1.10^{-6}$ s) .

5. TRIGONOMETRIC LEVELING TRAVERSES VERSUS SPIRIT LEVELING
   LINES

The ease of trigonometric leveling encourages our prognosis
concerning the replacement of the spirit leveling by trigonometric leveling where the latter answers the accuracy requirements. For the evaluation/elimination of refraction in
trigonometric leveling traverses with distances up to 2 km,
we point reciprocally to the special targets fitted on the
telescopes of KERN DKM 3A theodolites and obtain differences
of a few mm against the geodetic leveling height differences.

REFERENCES

Hradilek,L.:1974, Proceedings IAG Symp.Stockholm, Vol.5.
Hradilek,L.:1978, in P. Richardus (ed.), Proc. IAG Symp. Wageningen, pp. 185-190, Rijkscommissie voor Geodesie, Delft.
Hradilek,L.:1979, in E. Tengström and G. Teleki (eds), Refractional Influences in Astrometry and Geodesy, pp. 195-
201, D. Reidel.
Hradilek,L.:1980, Mitt.Inst.Theor.Geodäsie Univ.Bonn, Nr.
61, pp. 1-33, Bonn.

EXPERIMENTAL INVESTIGATION OF REFRACTION ABOVE WATER CROSSINGS

V.S. Milovanović,
Institut za geodeziju, Beograd, Bulevar revolucije 73

INTRODUCTION. The investigation of tectonic movements in the area of the artifical lake of the hydroelectric power station "Djerdap" (length 130km width 1 km) requires periodical levelling in about ten places. The accuracy requirements of this project call for the application of high precision levelling. For the water crossings level instruments and theodolits are employed. The asymmetrical refraction error is the limiting factor in the achievement of the required accuracy. Therefore it is obviously necessary to investigate the sources of the asymmetrical refraction and to find possible means to eliminate its influence.

BASIC CONSIDERATIONS. The simultaneous reciprocal observations with sidechanging of the observing set (observer and instrument) will eliminate the main part of the refractional influence as well as the error due to the observing set. The remaining error, the asymmetrical error, has to be eliminated by one of the following procedures:
1. The instrumental solution via the dispersion effect using lasers of different wavelengths,
2. The meteorological solution via observing the vertical temperature gradient and computing the refraction correction,
3. Observing during the time intervals for which the sum of the asymmetrical refraction influence is zero,
4. Observing during the time periods of the minimum of the influence of the asymmetrical refraction.

Without the availability of the equipment for the direct determination of refraction or the determination of the temperature gradient the third or the fourth procedure has to be used.

For the use of the third procedure the essential question is to determine the observation period yielding an elimination of the asymmetrical refraction effect. According to the experimental investigations of Henneberg (1974) this period is 24 hours. It means that in this period the influence of the asymmetrical refraction caused by periodical variation of the temperature gradient in the part of the optical sight with asym-

W. Fricke and G. Teleki (eds.), Sun and Planetary System, 503–504.

metrical situation - in general at the ends of the optical sight-tend to
zero.

For the use of the fourth procedure the knowledge of the diurnal variation of the refraction is essential. The Figures 1 and 2 show the general picture of diurnal variation of refraction above land and water
areas (Kazanskij, 1966; Mozžuhin, 1978).

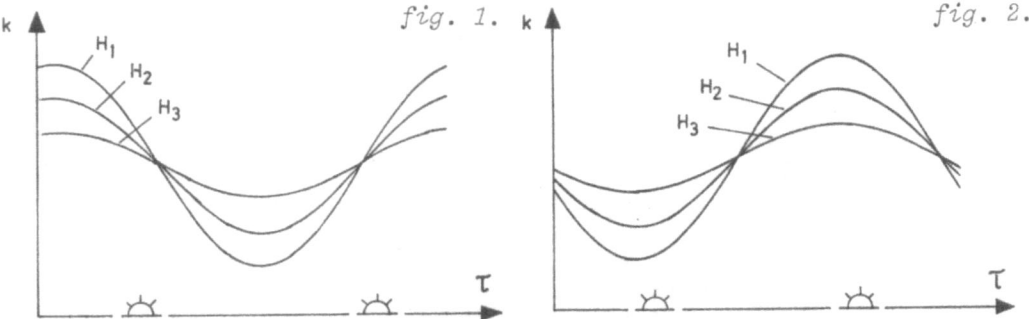

Diurnal variation of k above land   Diurnal variation of k above water
areas for different heights $H_1 < H_2 < H_3$ areas for different heights $H_1 < H_2 < H_3$

IDENTIFICATION OF OPTIMUM METEOROLOGICAL CONDITIONS FOR WATER CROSSING.
The main source of the variation as well as the asymmetrical refraction
is the variation of the temperature gradient. Therefore it is resonable
to assume that the influence of the asymmetrical refraction is reduced
by performing observations in the periods when the temperature gradient
is zero. These periods are rather short, as they persist generally for
a few tens of minutes only, for clear to light cloud conditions. Preferable conditions of longer periods are given when cloud cover is very
thick and there is a moderate or high wind. This procedure is adequate
if it is possible to perform a great number of observations during these
short periods. The availibility of electronic theodolites such as ET1
Kern and TC1 Wild make the fourth procedure feasible.

THE TEST-AREA. The test-area was chosen for the determination of the
periods when the vertical temperature gradient is close to zero. The
test-area is the part of the artificial lake in the vicinity of the town
Ram. The left bank of the Danube is flat and the right bank is steep.
The determination of the periods of the minimal temperature gradients
is made by observing verticale angles of the targets at different heights
(right bank). The processing of the observations is underway and the
results will be reported later.

REFERENCES

Kazanskij,V.K., 1966 Zemnaja refrakcija, Gidrometeoizdat,Leningrad,p.76.
Henneberg,H.G., 1974 Proc.Symp. of Terrestrial Electromag...,Stockholm.
Mozžuhin,A.O., 1978 Vermessungstechnik, 6, p. 195.
Fraser, C.S., 1979 The Canadian Surveyor, 33, p. 27.

ATMOSPHERIC TURBULENCE AND ITS EFFECTS ON DIRECTION MEASUREMENTS

F.K. Brunner
Dept. of Geodetic Science
University of Stuttgart
D-7000 Stuttgart 1, FRG.

ABSTRACT.   Optical direction measurements to terrestrial and extra-terrestrial targets are affected by atmospheric turbulence. The known formulae for the variance and the spectrum of the angle-of-arrival fluctuations and experimental results are summarised. The ultimate precision of direction measurements is estimated as a function of instrumental design parameters, the strength of atmospheric turbulence, and the length of the averaging period.

INTRODUCTION

In this review paper the image degradation effects on telescope observations are addressed which are caused by atmospheric turbulence. Terrestrial geodetic and astronomical *direction observations* using telescopes will be the main concern. The refraction effects caused by the mean refractive index field and the known methods to evaluate these effects are not considered here.

The atmosphere is rarely static but rather turbulent in its motion. The turbulent fluctuations of the various atmospheric parameters will give rise to random fluctuations of the refractive index as functions of both space and time. An electromagnetic wave propagating through such a turbulent medium will show fluctuations in phase, amplitude and angle-of-arrival.

Geodesists are well aquainted with the image fluctuations as seen through the telescope of a theodolite or level, and with the sight length limitations caused by strong turbulence. In astronomy the quivering of star images is a well known phenomenon. Image blurring is caused by the corrugation of the wave front and exists even in an instantaneous observation. Whilst image quivering, which will result also into blurring in a long term photographic exposure, is caused by the rapid tilting of the wave front at the aperture of the receiving telescope, i.e. angle-of-arrival fluctuations.

W. Fricke and G. Teleki (eds.), Sun and Planetary System, 505–510.

The domineering theme of the paper is "pointing" (direction observation) to a single object rather than "resolution" (two point resolution) of a telescope. Geodetic and astrometric applications are covered.

FORMULAE

In general, the refractive index of air, n, shows large scale spatial variations as well as random fluctuations caused by turbulent motion (wind speed and evolution of eddies). Its behaviour is most conveniently described in terms of the refractive index structure function which is a statistical second moment of n, and is defined by

$$D_n(\rho) \equiv < \left[n(\underline{v}_2) - n(\underline{v}_1) \right]^2 > \tag{1}$$

where $n(\underline{v}_i)$ is the value of n at the position $\underline{v}_i$. The spatial difference between the positions v_i is ρ. For separations ρ within the inertial subrange scales of *locally isotropic* atmospheric turbulence, $D_n(\rho)$ has the form (Tatarskii, 1971; Clifford, 1978):

$$D_n(\rho) = C_n^2 \ \rho^{2/3} \tag{2}$$

where C_n^2 is the refractive index structure parameter. C_n^2 is a measure of the magnitude of the fluctuations of the refractive index, and is given in units of the -2/3 power of metres.

The inertial subrange is bounded by two scales, which are termed the inner (ℓ_0) and outer (L_0) scale of turbulence. ℓ_0 is the size of the smallest atmospheric inhomogeneities (eddies), and typical values for ℓ_0 are between one and three millimetres. L_0 is the size of the largest atmospheric inhomogeneities for which atmospheric turbulence may still be considered isotropic. As an aid, the turbulent atmosphere is often visualised as consisting of lenses (actually the eddies) being of different scale sizes between ℓ_0 and L_0, and moving randomly through space.

In general, the values of C_n^2 range from 10^{-17} or less when atmospheric turbulence is very weak, to 10^{-12} for strong turbulence, e.g. close to the ground on a sunny day. Excellent reviews of the distribution of C_n^2 with height and time were given by Clifford (1978), Fante (1980) and Lawrence et al. (1977).

Let the incident light have a plane wave front normal to the propagation direction x. The wave propagation problem is to solve the scalar wave equation for the propagation of $U = U(\underline{v}, t)$, which can be derived from the Maxwell equations:

$$\Delta U - (n/c)^2 \ \partial^2 U/\partial t^2 = 0 \tag{3}$$

where Δ is the three-dimensional laplacian operator, c is the speed of light in vacuum, and $n = n(\underline{v}, t)$ is the local refractive index. The

scalar wave equation cannot be solved directly because of the random
character of n. The method of smooth perturbations is usually applied
to find approximate solutions. (Tatarskii, 1971; Strohbehn, 1978;
Ishimaru, 1978; Fante, 1975, 1980; Lawrence and Strohbehn, 1970).

Image dancing or quivering (angle-of-arrival fluctuation) is caused by
fluctuations in the tilt of the mean phase front at the receiving aperture of a telescope. Let the electromagnetic wave be monochromatic with
the wavenumber $\kappa = 2\pi/\lambda$ were λ is the wavelength. For a propagation
distance L, the radius of the first Fresnel zone f is given as
$f = (\lambda L)^{1/2}$. The variance of the angle-of-arrival fluctuations, σ_α^2, can
be calculated for a plane wave propagating through weak turbulence with
sufficient accuracy as (Tatarskii, 1971):

$$\sigma_\alpha^2 \cong \begin{Bmatrix} 1.46 \\ 2.92 \end{Bmatrix} b^{-1/3} C_n^+ \quad , \qquad \begin{matrix} \ell_o \ll b \ll f \\ L_o > b \gg f \end{matrix} \tag{4}$$

where b is the aperture diameter of the receiving telescope. σ_α^2 refers
here to the angle-of-arrival fluctuations in one coordinate direction,
such as the vertical one. C_n^+ is the symbol for the appropriate expression
of $C_n^2(x)$.

C_n^+ may express the following three cases:
(i) $C_n^2(x)$ being constant along the wave path

$$C_n^+ = L \; C_n^2 \tag{5}$$

(ii) $C_n^2(x)$ varying smoothly along the wave path

$$C_n^+ = \int_0^L C_n^2(x)\,dx \tag{6}$$

(iii) celestial light propagating through the whole atmosphere along a
slant path at a zenith angle ζ (the effect due to the curvature of the
earth can be neglected)

$$C_n^+ = \sec \zeta \int_{z_A}^\infty C_n^2(z)\,dz \tag{7}$$

where z_A is the height above the ground of the receiving telescope.

The temporal power spectrum of the angle-of-arrival fluctuations, $W_\alpha(\omega)$,
of a plane wave propagating through a locally homogeneous and isotropic
turbulent atmosphere can be derived as (Tatarskii, 1971; Lawrence and
Strohbehn, 1970):

$$W_\alpha(\omega) = \begin{Bmatrix} 0.066 \\ 0.132 \end{Bmatrix} C_n^2 \; L \; v_T^{5/3} \sin^2 \left(\frac{\pi b \omega}{v_T}\right) \omega^{-8/3}, \qquad \begin{matrix} \rho \ll f \\ \rho \gg f \end{matrix} \tag{8}$$

where v_T is the transverse component of the horizontal wind speed. The
derivation of Eq. (8) is based on the validity of the frozen turbulence

hypothesis which postulates that the turbulent eddies are swept across the line-of-sight with a wind speed v_T.

DISCUSSION

The experimental investigations of σ_α^2 were briefly summarized by Lawrence and Strohbehn (1970). Tatarskii (1971) gives an extensive review of the determinations of σ_α^2 and $W_\alpha(\omega)$ over various terrestrial path lengths, and of image quivering from stellar and solar limb observations. In the geodetic context, Brückner (1979) has investigated angle-of-arrival fluctuations over a 1.5 km and a 22 km long path using a telescope of 6.8 m focal length. All known experimental results seem to corroborate the theoretical derivations.

Eq. (4) expresses the dependence of σ_α^2 on the aperture diameter, b, and via C_n^+ on the path length, L, and on the average value of the strength of atmospheric turbulence, C_n^2. A reduction of the aperture diameter b results in a sharper image (reduced blurring effect) but at the same time in larger image fluctuations. The selection of time periods which are associated with small C_n^2 values is a well known method to reduce image fluctuations.

From Eq. (8) it is apparent that $W_\alpha(\omega)$ depends on the transverse wind speed v_T in addition to the above mentioned parameters. Tatarskii (1971) showed that the peak of the normalised power spectrum of the angle-of-arrival fluctuations should occur at about the frequency $0.22\ v_T/b$. After this peak the spectrum decays sharply with a slope of $-8/3$ to higher frequencies. Very little energy is contained in this part of the spectrum. The slope of the spectrum in the lower frequencies before the peak should be around $+2$.

In the previous sections the effects of atmospheric turbulence on direction measurements are investigated. The stochastic behaviour of atmospheric turbulence inhibits correcting the measurements for these effects. Therefore the fluctuating part of the refraction effect represents the fundamental limit in the ultimate precision of direction measurements. Recently the concept of the ultimate precision of geodetic measurements was discussed by the author (Brunner, 1979).

The precision of direction measurements can be estimated as a function of instrumental design parameters and the strength of atmospheric turbulence, expressed by C_n^2. It is important to realise that σ_α is the standard deviation of a single observation. σ_α is typically 1 to 5 µrad for stellar observations (through the whole atmosphere) at ζ of 60° to 80°. During the day, when atmospheric turbulence is generally larger, values for σ_α of 10 to 40 µrad are commonly observed. Assuming an aperture diameter of 0.04 m for a theodolite telescope, σ_α may be observed to be as high as 95 µrad along a horizontal path of 1 km length during strong turbulence ($C_n^2 \cong 1 \cdot 10^{-12} \text{m}^{-2/3}$), and 65 µrad for a 15 km horizontal path during medium turbulence ($C_n^2 \cong 5 \cdot 10^{-14} \text{m}^{-2/3}$).

For geodetic and astrometric direction measurements using either electronic recording or visual observation systems a certain averaging process has to be considered. The variance of the mean of a random variable σ_m^2, was derived by Bendat and Piersol (1971) for a continuous record length T. σ_m^2 can be expressed with sufficient accuracy as

$$\sigma_m^2 = (2\ BT)^{-1} \tag{9}$$

where B is the bandwidth of the noise. Hence the variance of the mean value decreases linearly with increasing record length. Of course, stationarity has to be assured during T. The case of equally spaced visual observations with each observation being of a certain duration was treated by Hog (1968) with an application to star image motion. See also in this context Ivanov (1979).

The above brief discussion of the precision of direction measurements should only highlight the importance of selecting sufficiently long signal averaging periods in order to achieve the required precision. The power spectrum $W_\alpha(\omega)$ will provide important information for this task. The upper limit of the atmospherically induced noise can be estimated from equations for σ_α^2 where the magnitude of the refractive index structure parameter C_n^2 will play a key role. An assessment of the attainable precision given as the degradation of the system performance by atmospherically induced noise must be considered a fundamental question in geodesy and astrometry (Brunner, 1979).

ACKNOWLEDGEMENTS

The paper was written whilst the author was on leave from the School of Surveying, University of New South Wales, Australia, at the Department of Geodetic Science, University of Stuttgart, FRG, with Prof. Erik W. Grafarend as his host. This leave was made possible through a fellowship from the Alexander von Humboldt-Foundation.

REFERENCES

Bendat, J.A., and Piersol, A.G.: 1971, *Random Data: Analysis and Measurement Procedures*. Wiley – Interscience, New York.
Brückner, R.: 1979, *Zur Szintillation bei terrestrischen geodätischen Messungen*, Wiss. Arb. Fachrichtung Vermessungswesen Univ. Hannover, Nr. 39.
Brunner, F.K.: 1979, *Aust. J. Geod. Photo. Surv.*, 31, pp. 51–64.
Clifford, S.F.: 1978, in Strohbehn, J.W. (ed.), *Laser Beam Propagation in the Atmosphere*, Springer-Verlag, pp. 9–41.
Fante, R.L.: 1975, *Proc. IEEE*, 63, pp. 1669–1692.
Fante, R.L.: 1980, *Proc. IEEE*, 68, pp. 1424–1443.
Hog, E.: 1968, *Z. Astrophysik*, 69, pp. 313–325.
Ishimaru, A.: 1978, *Wave Propagation and Scattering in Random Media*, Vol. 2, Academic Press, New York.

Ivanov, V.I.: 1979, *Proc. IAU Symp. No. 89*, Uppsala, pp. 67-72.

Lawrence, R.A., and Strohbehn, J.W.: 1970, *Proc. IEEE*, 58, pp. 1523-1545.

Lawrence, R.S., and Clifford, S.F., and Ochs, G.R.: 1977, *Proc. URSI Commission F*, La Baule, pp. 1-6.

Strohbehn, J.W.: 1978, (ed.) *Laser Beam Propagation in the Atmosphere*, Springer Verlag, New York.

Tatarskii, V.I.: 1971, *The Effects of the Turbulent Atmosphere on Wave Propagation*, (Translated from the Russian), NTIS, Springfield, Va.

LOCAL GEODYNAMICS WITH TWO-COLOUR INSTRUMENTS

Erik Tengström
Institute of Geophysics
Dept. of Geodesy
Hällby
S-755 90 Uppsala, Sweden

ABSTRACT.   For realistic studies of spatial movements in local or regional crustal domains, repeated accurate determinations of relative three-dimensional positions for selected points, are essential. Such points, which should be carefully chosen and marked, might be located along the Earth's surface – and that is the common case – but occasionally also in subsurface cavities (e.g. in mines). 3D measurements in surface networks is the subject of this review. Such measurements comprise both distance- and direction observations. As the observations are made in the actual atmosphere, they must be corrected for refractional effects, due to this atmosphere at the 'moment' of measurement. Some pieces of advise for future improvements of existing methods and instruments, to be used for obtaining geometrical directions, are given.

INTRODUCTION

To achieve relative 3D positions at any time, one must think of utilizing instruments, which observe optical pathlengths and apparent directions with the same accuracy as the one they can obtain for determining simultaneously the refractive corrections.

In short time-interval studies of the dynamics of certain domains it seems necessary to guarantee an accuracy of the order of 10^{-7} (± 0.1 mm/km) both in distance- and direction measurements and in the respective atmospheric corrections.

DISTANCE MEASUREMENTS

As regards distance measurements, I think it has been proved now, that the terrameter, developed by Huggett and Slater (1977), which uses two optical and one radiofrequency modulated signal in observing pathlengths, has already reached this goal.

The principle of the terrameter approach might be understood from the

W. Fricke and G. Teleki (eds.), Sun and Planetary System, 511–518.
Copyright © 1982 by D. Reidel Publishing Company.

following (Huggett and Slater 1977): The refractivity of humid air

$$N = n - 1 \; ,$$

where n is the total refractive index, is proportional to the dry air- and water vapor densities in different ways, so that we may write

$$N = \alpha \rho_S + \beta \rho_W \tag{1}$$

ρ_S being the density of dry air, ρ_W the one of the water vapor.

By laboratory experiments, it has been shown, that, for optical signals, α and β are only depending on the wavelength used. This is the case for the α-values of radio signals, too, but their β-values are additionally temperature dependent. However, an error in mean temperature of the order of $10^\circ C$ does not affect the computed geometrical distance D by more than $D \times 10^{-7}$.

Now, by the total pathlengths $2R_1$, $2R_2$ and $2R_3$, observed with red, blue, and microwaves, respectively, travelling forth and back to the modulator at A via the cat's eye retroreflector at B, the R_1, R_2 and R_3 should be regarded as observation quantities, and taking the pathlength differences between the blue light and the red one, and the corresponding difference between the radio signal result and the red one, we get the equations:

$$\Delta R_{2-1}/D = (\alpha_2 - \alpha_1)\overline{\rho}_S + (\beta_2 - \beta_1)\overline{\rho}_W \; , \text{ and}$$
$$\Delta R_{3-1}/D = (\alpha_3 - \alpha_1)\overline{\rho}_S + (\beta_3 - \beta_1)\overline{\rho}_W \tag{2}$$

The geometrical distance D, using red light, will be

$$D = R_1 - D(\alpha_1\overline{\rho}_S + \beta_1\overline{\rho}_W) \; , \tag{3}$$

where, of course, $\overline{\rho}_S$ and $\overline{\rho}_W$ are mean densities along the path AB. This is justified, because

$$R = \int_A^B n \; ds \simeq \overline{n}_{AB} \; D \quad , \text{ and} \tag{4}$$

$$\overline{n}_{AB} = 1 + \alpha\overline{\rho}_S + \beta\overline{\rho}_W$$

for the path.

The α- and β-values are being calculated from existing standard formulas to within 10^{-7}, or better.

The observed quantities ΔR_{2-1} and ΔR_{3-1}, and also the left members of (2) (using approximate values of D) can be obtained with the same accuracy.

So, inserting $\overline{\rho}_S$ and $\overline{\rho}_W$ from (2) into (3), we can solve for D and obtain

$$D = R_1 - A(1 + A't)\Delta R_{2-1} - B(1 + B't)\Delta R_{3-1} \qquad (5)$$

where the constants A, B, A', B' can be calculated from known expressions of α and β.

The terrameter, which sends at A red (He-Ne 6328Å) and blue (He-Cd 4416Å) optical signals plus a microwave signal to the retroreflecting target at B, and receives them back at A, thus gets information about the values of $2R_1$, $2R_2$ and $2R_3$.

With calculated A, A', B and B' coefficients and the 'observed' ΔR_{2-1} and ΔR_{3-1}, the geometrical distance D at the 'moment' of observation is given through (5).

Huggett claims, that one can rely on D-values accurate to 10^{-7} from each observation with the instrument, at least in local areas.

Repeated measurements 1975-76 (Huggett and Slater 1977) show results over distances of at least 5 km, month for month, which seem to indicate changes in distances between points in a network around the central station of the town Hollister in California, where fault-slidings have been experienced earlier. Also new information about deepseated slidings beyond the ends of known fault profiles is provided by the terrameter results.

It is clear, that the terrameter approach is essential for the 3D-studies of the geodynamical behaviour of certain geodynamical areas. And I think it has already proved to give us a 10^{-7} D accuracy.

DIRECTION MEASUREMENTS

As regards the possibility of obtaining refraction-free directions from angular observations, using two signals of widely separated optical wave-lengths, it has been demonstrated by experiments in the late sixties, and after 1970, that this two-colour (or dispersion) method, under favourable atmospherical conditions, might yield results of refraction corrections at the 'moment' of observation with an accuracy, corresponding to a mean square error of the order of $\pm 0''.5$. This implies, that we should be able to determine refraction-free directions with the same accuracy, relying on the power of our theodolite, and this is not a problem in practice. Vertical and horizontal displacements could then be detected to within 3 mm/km. See works by Owens (1967), Brein (1968), Prilepin (1974), Tengström (1974, 1975, 1977), Glissmann (1976, 1977), Williams (1974, 1977). (It should be emphasized here, that Williams' work on his dispersometer is a continuation of the original efforts by Dyson at NPL in the early sixties).

But, even under non-turbulent (or mainly macro-turbulent) atmospheric

conditions, and with sufficient knowledge of the average water vapor
gradient along the line of sight, we are far from the accuracies, obtained
by the terrameter. In fact, 0".5 means 2.5 10^{-6}.

To achieve 10^{-7} we must increase the accuracy in determining the angle
between the red and blue ray (the dispersion) by a factor of 25.

The Uppsala team has, during long time, been aware of the possibility of
using Michelson's stellar interferometer principle for this purpose. We
have therefore constructed a prismatic system, with a magnifying power
of 24, to be placed in front of the Cassegrain receiving telescope with
its 6 slit grating (Tengström 1977). Experiments along an 1 km long base,
have shown, photographically recorded, wonderful pictures of interference
patterns, the lines of which can be easily measured in a good comparator.

It is therefore of importance to continue these experiments over distances,
which are relevant in geodynamical test nets.

The principle of using the measured dispersion-angle δ for determining
the refraction angle r can be understood from Figure 1 (Tengström 1967),
and the following simple reasoning:

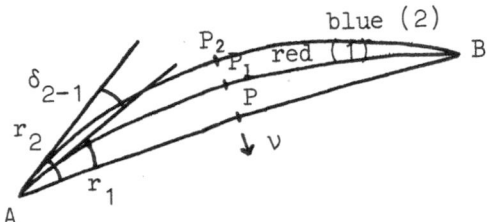

Figure 1. Refraction and Dispersion.

$$\delta_{2-1} = r_2 - r_1 = \int_A^{P_2} \frac{1}{n_2} \frac{\partial n_2}{\partial \nu_2} ds_2 - \int_A^{P_1} \frac{1}{n_1} \frac{\partial n_1}{\partial \nu_1} ds_1 \qquad (6)$$

Now $n = 1 + \alpha\rho_S + \beta\rho_W$,

where, as before, α and β only depend on the wavelength, and can be
computed to within 10^{-7}, at least, using very accurate values of the
red and blue light frequencies.

Any refraction angle r_λ is derived from the expression

$$r_\lambda = \int_A^{P_\lambda} \frac{1}{n_\lambda} \frac{\partial n_\lambda}{\partial \nu_\lambda} ds_\lambda =$$

$$\approx \int_A^{P_\lambda} [1 - (\alpha_\lambda \rho_S + \beta_\lambda \rho_W)] \left(\alpha_\lambda \frac{\partial \rho_S}{\partial \nu_\lambda} + \beta_\lambda \frac{\partial \rho_W}{\partial \nu_\lambda}\right) ds_\lambda \qquad (7)$$

Here P_λ is the point on the λ-path, where its tangent is parallel to the chord AB.

Because of r being small, and assuming, that the projections of P_1 and P_2 on $\overline{AB}$ yield the same point P, we have

$$ds_1 = ds_2 = ds_\lambda = ds \quad , \text{ and } \quad \nu_1 = \nu_2 = \nu_\lambda = \nu \text{ , and we get:}$$

$$\delta_{2-1} = \int_A^P [(1 - (\alpha_2 \rho_S + \beta_2 \rho_W)) [\alpha_1 \frac{\partial \rho_S}{\partial \nu} + \beta_2 \frac{\partial \rho_W}{\partial \nu}] -$$

$$- [1 - (\alpha_1 \rho_S + \beta_1 \rho_W)] [\alpha_1 \frac{\partial \rho_S}{\partial \nu} + \beta_1 \frac{\partial \rho_W}{\partial \nu}]] ds$$

$$= [\alpha_2 \frac{\partial \overline{\rho}_S}{\partial \nu} + \beta_2 \frac{\partial \overline{\rho}_W}{\partial \nu} - \alpha_1 \frac{\partial \overline{\rho}_S}{\partial \nu} - \beta_1 \frac{\partial \overline{\rho}_W}{\partial \nu}] \ \overline{AP} \ .$$

Neglecting squares and other products of α- and β-terms, we have

$$\delta_{2-1} = [\frac{\partial \overline{\rho}_S}{\partial \nu} (\alpha_2 - \alpha_1) + \frac{\partial \overline{\rho}_W}{\partial \nu} (\beta_2 - \beta_1)] \ \overline{AP} \ . \qquad (8)$$

As $\beta_2 - \beta_1$ is very small in the optical case, we might neglect the second term, and then simply obtain

$$\delta = (\alpha_2 - \alpha_1) \overline{AP} \ \frac{\partial \overline{\rho}_S}{\partial \nu} \quad , \text{ from which}$$

$$\overline{AP} \ \frac{\partial \overline{\rho}_S}{\partial \nu} \approx \frac{\delta_{2-1} \quad (meas)}{(\alpha_2 - \alpha_1) \ (comp)} \qquad (9)$$

can be derived with an accuracy, which is certainly high enough for satisfying all demands of the order of 10^{-7}.

As $r_\lambda \simeq \alpha_\lambda \overline{AP} \dfrac{\partial \overline{\rho}_S}{\partial \nu} + \beta_\lambda \overline{AP} \dfrac{\partial \overline{\rho}_W}{\partial \nu}$ (simplified (7)), we have

$$r_\lambda \simeq \frac{\alpha_\lambda}{\alpha_2 - \alpha_1}\, \delta_{2-1} + Q_\lambda \equiv \kappa_\lambda \delta_{2-1} + Q_\lambda \tag{10}$$

Using accurate laboratory formulas, we can compute κ_λ, but must evaluate Q_λ by means of available β-values, approximate values of $\overline{AP}$ and humidity gradient information. If we use λ_2 = 4416Å (blue He-Cd laser light), and λ_1 = 6328Å (red He-Ne light), the value of κ is for λ = 5900Å (yellow light of Na, close to the effective wavelength in all field observations), equal to 63.4 . The same wavelengths are used by the terrameter.

In Uppsala, we have, as lightsources, preferred to use λ_1 as above, and λ_2 = 3250Å (He-Cd ultraviolet-line, which can only be photographically recorded). For this λ-combination, κ = 23.4, which means an essential increase of the accuracy of the r-determination.

For the accurate calculation of the dispersion, our optical interferometric recording (fringe-pattern) is also important.

It is with our observation techniques and utilized frequencies, that we have obtained - on our 20 km testbase - , under favourable atmospherical conditions, a direct - and absolutely tested - refraction determination, which is accurate to ±0".7 of arc. In these observations the Michelson magnification was not used.

It is hoped, that, for shorter distances, at least, our Michelson prismatic system will solve the problem of obtaining 25 times as great resolving power in determining dispersion, necessary in certain local geodynamical research.

Also another method of increasing the accuracy of δ has been suggested in Uppsala (Mårtensson 1979). He uses in the focal plane of the Cassegrain receiver a very accurate position-sensitive photodetector, and he expects - with this device - to increase the resolving power of the instrument, or its accuracy in measuring dispersion, so that he might be able to compete with the Michelson magnification approach.

More important, however, is it, that Mårtensson does not observe δ itself, but through changes of the distance between the lightsources tries to bring the received signals into coincidence in the focal plane. This is of great value, because the microturbulent effects on the two beams are never the same when they are arriving apart, but might be equal or almost equal if they reach the instrument in coinciding wave fronts.

Additionally, it has been foreseen, for the future experiments at Uppsala, that suitable time- and spatial averagings should be tried to achieve refraction values, which are accurate enough for various purposes, especially at short and medium distances (local geodynamics).

The laser sources were originally chosen to enable observations at
greater distances because of their high intensity, but conventional
lightsources are better suited for defining a mean wave front without
local disturbances in narrow beams.

We now intend - for such distances - to use low energy lasers, and perform a spatial integration, obtained by a concave mirror, which collects
most of the light from the incoming wave front into the focus of this
mirror before it - after a second mirror-reflection - enters the small
aperture of the instrument in the direction of the mean wave front. The
interference patterns will probably be less distinct, but the differential effect of turbulence on the two monochromatic fronts should at least
be decreased for any integration time.

INSTRUMENTS FOR REFRACTION-FREE 3D OBSERVATIONS

Future developments of field instruments, capable of directly determining
refraction-free directions (observed directions instantaneously corrected
for refraction) are indicated by Glissmann (1978) and Williams (1978).
Glissmann relies on reflecting, Williams on lens systems. The theoretical
ideas by both are very convincing, but there are many technical problems
to solve, before these instruments can show their abilities in the geodynamical networks, I was talking about.

Let us give free space to a competition between various suggested theoretical solutions, and let their pleaders show the practical power of
their instruments in the field.

Our dream for the future is that we should be able to construct an instrument, which gives, at any moment, refraction-free distances and directions. May we hope to witness the success of all work being done today
to reach this goal.

There are still many obstacles on our way. The direction problem is
especially a severe one, but I believe in its solution by means of the
Michelson magnification approach and by realisticly using the experience
and knowledge by our colleagues in micrometeorology. I hereby like to
stress the importance of the research, done by people, interested in
Meteorological Geodesy or Geodetic Meteorology, terms, due to Brooks,
for defining a new field of interdisciplinary research, e.g. persons
like Fritz Brunner, who have made great contributions to our work, what
concerns the evaluation of turbulence effects. E.g. (Brunner 1981).

Again, and again, I like to plead for the use of Michelson interferometry
for direction determinations in local geodynamics. My experience with a
big stellar interferometer (0.5 - 1.5 m) - already in the sixties - showed
clearly, that small angles could be determined to within 0".003, which
means a very accurate method of studying the creeping of material in
rocky domains. And also (Tengström 1978), refraction-free directions will
be able to realize (with sparse gravimetric information) the determination
of very accurate geoidal pictures over local areas.

REFERENCES

Brein,R., 1968: "Die Bestimmung der atmosphärischen Refraktion aus der
Dispersion des Lichtes". DGK, Reihe B, Nr.165.

Brunner,F.K., 1981: "Determination of line averages of sensible heat
flux, using an optical method". Boundary Layer Meteorology. In print.

Glissmann,T., 1976: "Zur Bestimmung des Refraktionswinkels über die Dispersion des Lichtes mittels positionsempfindlicher Photodioden".
Wiss.Arb.Lehrst.Geod., Phot.u.Kart., TU Hannover, Nr.62.

Glissmann,T., 1978: " Ein Koinzidenzverfahren zur Messung von refrak-
tionsfreien Richtungen". ZfV, Heft 5.

Huggett,G.R., Slater,L.E., 1977: "Recent advances in multiwavelenght
measurement". in Proc.Int.Symp. 'EDM and Influence of Atmospheric
Refraction', ed. P. Richardus. Publ. of Neth.Geod.Comm., Wageningen.

Mårtensson,S.-G., 1979: "Refraction. Proposal for a new measuring technique". Comm. from Dept. of Geodesy, University of Uppsala. (in Swedish).

Owens,J.C., 1967: "Optical refraction index of air: Dependence on pressure, temperature and composition". Applied Optics, Vol. 6, No.1.

Prilepin,M.T., 1974: "Elimination of angular refraction by means of
multiwavelength methods". in Proc.Int.Symp. on 'Terr.Electromagn.Dist.
Meas. and Atm.Eff. on Angular Meas.', Vol. 5, Stockholm.

Tengström,E., 1967: " Elimination of refraction at vertical angle
measurements, using lasers of different wavelengths". in Proc.Int.
Symp. 'Figure of the Earth and Refraction', Vienna.

Tengström,E., 1974: " Report on the results of IDM experiments at Uppsala
1970, at the Finnish base of Niinisalo 1971, and of further experiments
with He-Ne and He-Cd lasers". in Proc.Int.Symp. on 'Terr.Electromagn.
Dist.Meas. and Atm.Eff. on Angular Meas.', Vol. 5. Stockholm.

Tengström,E., 1975: "Report of IAG Special Study Group No.1.23 through
the period 1971-1975 on 'Studies of Atmospheric Refraction'". IUGG,
Gen.Ass., Grenoble.

Tengström,E., 1977: "Some absolute tests of the results of IDM measurements in the field, with a description of formulas, used in the tests".
in Proc.Int.Symp. 'EDM and Influence of Atmospheric Refraction', ed.
P. Richardus. Publ. of Neth. Geod.Comm., Wageningen.

Williams,D.C., 1974: "A dispersometer for the measurement of angular refraction". in Proc.Int.Symp. on 'Terr.Electromagn. Dist.Meas. and
Atm.Eff. on Angular Meas.', Vol.5, Stockholm.

Williams,D.C., 1977: "First field tests of an angular dual wavelenght
instrument". in Proc.Int.Symp. 'EDM and Influence of Atmospheric Refraction', ed. P. Richardus. Publ. of Neth.Geod.Comm., Wageningen.

Williams,D.C., 1978: "A coincidence procedure for the measurement of
refraction-free directions". ZfV, Heft 5.

TERRESTRIAL REFRACTION AND VERTICAL TEMPERATURE GRADIENT

L.N. Mavridis
Department of Geodetic Astronomy, University of Thessaloniki,
Thessaloniki, Greece.

ABSTRACT

Current work on the determination of the coefficient of terrestrial
refraction and its diurnal and seasonal variation is reviewed. For the
computation of the refractive effects on the basis of meteorological
measurements the vertical temperature gradient is needed. Recent empirical determinations of this gradient and corresponding theoretical atmospheric models are discussed.

TERRESTRIAL REFRACTION

The determination of the coefficient of terrestrial refraction can
be made with the help of measurements of vertical angles (or zenith distances) between two geodetic stations, using either the method of one-
way observations (station refraction) or the method of reciprocal observations (line refraction). On the basis of these measurements the diurnal
and the seasonal variation of K can investigated. Gounaris et al. (1981)
used this methodology for a study of terrestrial refraction in the area
of Thessaloniki. To this purpose 1924 hourly values of K covering 128
days of the years 1969-74 have been determined with the help of both one-
way and reciprocal vertical angles observations, carried out between two
geodetic stations situated at a distance of about 15 km and absolute
heights equal to 177 m and 361 m. On the basis of this material a thorough study of the diurnal variation of K and its seasonal changes has
been made.

Another methodology is to determine K in the framework of the adjustment of trigonometric leveling traverses or three-dimensional networks.
Blažek and Hradilek (1979) have carried out a determination of station
refraction for each of the stations of three double trigonometric traverses i.e. traverses with additional vertical angles observations. The
improvement in elevation accuracy thus obtained, was substantial in traverses with longer distances (about 4 km) between stations. Hradilek (1979)
investigated the accuracy achieved in the determination of K with this
methodology. His conclusion was that the line refraction method is super-

519

W. Fricke and G. Teleki (eds.), Sun and Planetary System, 519–522.

ior to the station refraction approach, when elaborating trigonometric leveling traverses, whereas the station refraction method is more convenient for trigonometric and three-dimensional networks of a larger extent, especially those designed in high mountain regions. Ramsayer (1979) measured reciprocal zenith distances and astronomical longitude and latitude in a test network with lines of 4 km to 23 km. The measurements of the reciprocal zenith distances were repeated every hour 12-60 times for the various lines, and the network was adjusted rigorously three dimensional in an ellipsoidal reference system. The values of the mean refraction coefficient thus obtained vary between K = 0.10 at day and K = 0.34 at night. The values of K and the corresponding refraction angle can be determined with higher accuracy, if the two reciprocal angles are measured simultaneously, and with day light rather, than at night.Furthermore, for observations with day light the deviation of the true effective refraction coefficient in the observation station from the mean refraction coefficient of the observed line is inverse proportional to the distance.

A third possibility is to use the dispersion effect of light waves propagating through the atmosphere for the determination of the vertical refraction angle. Several technical solutions have been proposed (Prilepin, 1974) and few have resulted in actual prototypes (Tengström, 1978; Glissmann, 1976; Williams, 1978). It is not unrealistic to predict that these instruments will yield an accuracy of $0".5$ for the vertical refraction angle in the near future. However, test measurements have shown that atmospheric turbulence causes considerable problems in measuring the small dispersion angle, and precise measurements are only possible during favourable observation times.

Brunner (1979) has developed the theory of another method for evaluating the vertical refraction angle from the variance of the angle of arrival fluctuations, assuming a horizontally homogeneous turbulent atmospheric surface layer. An advantage of this method, which has not yet been tested in field experiments, is that the effects of the turbulent medium on wave propagation which have been found adverse to other techniques are utilized here to advantage.

Finally, we have the possibility of the meteorological solution, which is based on the well-known relation:

$$K = 504 \frac{P}{T^2} (0.0342 + \frac{dT}{dh}), \tag{1}$$

where T,P are the air temperature and atmospheric pressure and $\tau = dT/dh$ is the vertical temperature gradient. For the determination of the value of K on the basis of this relation the values of P,T and especially the value of τ, should be known.

VERTICAL TEMPERATURE GRADIENT

The most direct method for the determination of the vertical temperature gradient $\tau = dT/dh$ would be to measure the air temperature in various altitudes. In lower altitudes this can be done with the help of meteorological towers, while for higher altitudes radiosonde observations should be used. Both methods, however, can be used only in very limited number of meteorological stations. For geodetic purposes we are obliged to use atmospheric models, which are constructed either empirically on the basis of air temperature measurements in various altitudes or theoretically on the basis of the laws of Atmospheric Physics. It is obvious, however, that even the theoretical models must be calibrated with the help of observational data. Some atmospheric models are briefly discussed in the following:

1) Kukkamäki (1938) proposed the model:

$$\frac{dT}{dh} = ah^b,$$ (2)

where b was determined from existing air temperature observations.

2) Brocks (1948) adopted also relation (2) for the layer up to about 30m. For the altitude range between 30 m and a few hundred meters Brocks adopted the adiabatic gradient $\tau = -1^\circ C/100\,m$. For the values of the parameters a and b different solutions were made for the various altitudes, times of the day, and seasons of the year, with the help of existing air temperature measurements.

3) Levalois and Masson d'Autume (1953) proposed the formula

$$T(h,t) = T(0,m) + \lambda h + e^{-h/h'} \cdot f(t-h/v) + \varphi(t),$$ (3)

where t is the time of the day, λ is the vertical gradient of the diurnal mean $T(0,m)$, h' is the height at which the daily temperature range is reduced to 1/2.7 of its surface value, v is the velocity with which the diurnal maximum temperature is transmitted upwards, and $\varphi(t)$ is the estimated effect of radiation, constant for all heights.

4) Angus-Leppan (1971, 1979) and Angus-Leppan and Webb (1971) adopted the Heat-Balance Model. Unlike the earlier, basically empirical models, this model is based on the theory of turbulent transfer in the surface layer. According to this model a distrinction has to be drawn between stable, neutral and unstable thermal stratification in the atmosphere. For unstable (daytime) conditions, there are three strata, defined in terms of a stability parameter L (Obukhov length). L varies with the cube of wind speed and inversely with the upward heat flow. On a typical breezy summer day the lowest stratum might average up to 1m, and the middle stratum 1 to 30 m. In the upper stratum τ is equal to $-1^\circ C/100\,m$, as in Brocks' model. Parameter b is -1 in all cases except the middle stratum, where it takes the value -1.33. Parameter a is different in each stratum, being a function of meteorological parameters, including those which determine turbulent heat flow (sun and sky radiation, cloudiness, heat properties of the surface, moisture etc.), wind velocity, surface roughness and air temperature. This approach was developed further into the Turbulent Transfer Model (TTM) for application to EDM reductions by Brunner and Fraser (1977). They have compared the values from the TTM against test data with very favourable results.

In connection with the determination and the study of the diurnal and seasonal variation of τ, the work by Gounaris et al. (1981) should be mentioned. The authors inverted the meteorological solution and used relation (1) for the determination of the values of τ from the observed values of K. Then, used their rich observational material concerning K for the determination of 1924 hourly values of τ covering 128 days of the years 1969-74. On the basis of these data a study of the diurnal and seasonal variation of τ in the area of Thessaloniki has been made.

CONCLUSIONS

From the above discussion the following conclusions can be drawn:
1) In spite of the great progress made during recent years, more accurate atmospheric models for temperature are badly needed.
2) A closer co-operation of astronomers, geodesists, and meteorologists would be very benefical for the further advancement of atmospheric re-fraction studies.
3) Geodesists should try to make full use of the fact that geodetic ob-servations could be very valuable in determining spatially integrated values of meteorological parameters.

REFERENCES

Angus-Leppan, P.V.: 1971, in Conference of Commonwealth Survey Officers, Cambridge.
Angus-Leppan, P.V.: 1979, IAU Symp. No. 89, 165.
Angus-Leppan, P.V. and Webb, E.K.: 1971, IUGG General Assembly, Moscow, Travaux, IAG, Section I.
Blažek, R. and Hradilek, L.: 1979, IAU Symp. No. 89, 195.
Brocks, K.: 1948, Ber. des Deutschen Wetterdienstes in der U.S.- Zone Nr. 5, Bad Kissingen.
Brunner, F.K.: 1979, IAU Symp. No. 89, 227.
Brunner, F.K. and Fraser, C.S.: 1977, Unisurv. G, 27, Univ. of New South Wales, Sydney.
Glissmann, T.: 1976, Wiss, Arbeiten Geod., Photogramm., Kartogr. Univ. Hannover 62.
Gounaris, A.I., Mavridis, L.N., and Papadimitriou, A.L.: 1981, Proc. Acad. Athens 56, 91.
Hradilek, L.: 1979, IAU Symp. No. 89, 191.
Kukkamäki, T.J.: 1938, Publ. Finnish Geod. Inst. No. 25.
Levalois, J.J. and Masson d'Autume, G.: 1953, Inst. Geogr. Natl., Paris.
Prilepin, M.T.: 1974, Int. Symp. on Terrestrial EDM and Atmospheric Effects on Angular Measurements, Stockholm.
Ramsayer, K.: 1979, IAU Symp. No. 89, 203,
Tengström, E.: 1978, in P. Richardus (ed.), Proc. Int. Symp. on EDM and Influence of Atmospheric Refraction, Wageningen, 101.
Williams, D.C.: 1978, in P. Richardus (ed.), Proc. Int. Symp. on EDM and Influence of Atmospheric Refraction, Wegeningen, 163.

TERRESTRIAL REFRACTION AND VERTICAL TEMPERATURE GRADIENT IN THE AREA
OF THESSALONIKI

A.I. Gounaris
Chair of Geodesy, Univ. of Thrace, Xanthi, Greece
L.N. Mavridis
Dept. of Geodetic Astronomy, Univ. of Thessaloniki, Greece
A.L. Papadimitriou
Dept. of Geodesy and Geod. Applications, Univ. of Patras, Greece

INTRODUCTION

The coefficient of terrestrial refraction and numerous relevant meteorological parameters were determined in the area of Thessaloniki,
Greece, during 128 days of the years 1969-74. The determination of the
coefficient of terrestrial refraction was made with the help of simultaneous reciprocal zenith distance measurements carried out between two
geodetic stations (Faner Toumba and Panorama) situated in the area of
the measurements (Fig. 1). The distance of the two stations is equal to
15122.55 m and their absolute heights are respectively equal to H≈177 m
and H≈361 m. The exact value of the height difference between the two
stations was found with the help of precision spirit leveling equal to
183.68 m. In the middle of the distance of the two geodetic stations
lies the Thessaloniki Airport, which is indicated in Fig. 1 by TA.

The results of the simultaneous reciprocal zenith distance measurements were analyzed by two methods, i.e.: a) the method of one-way observations, which gives the values K_1, K_2 of station refraction at each of
the geodetic stations Faner Toumba and Panorama, and b) the method of
simultaneous reciprocal observations, which gives the value K of line refraction along the line connecting the two stations. The measurements
were carried out in five observing periods A-E. Table 1 gives for each
observing period the time interval covered by the measurements, the total
number of days during which measurements were performed, the frequency
of the measurements, and the observing hours of each day. From this table
we see that during observing periods B and C the measurements were performed every 30^m (at the hour and at the half hour), while, during periods
A, D, and E, the measurements were carried out every 60^m (at the hour). In
the following, only the measurements carried out at the hour will be considered for all observing periods A-E. In this way a total of 1924 determinations of the coefficient of terrestrial refraction covering 128 days
of the years 1969-74 are available. As mentioned already, each determination gives the following data: a) the values K_1, K_2 of station refraction
at each of the geodetic stations Faner Toumba and Panorama and b) the
value K of line refraction along the line connecting the two stations.

523

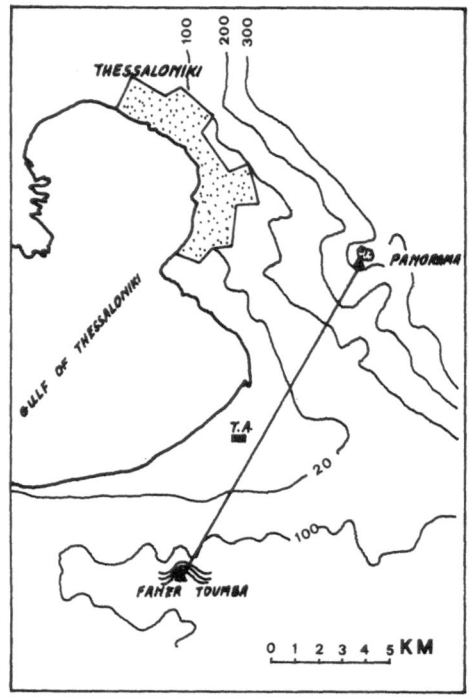

Fig. 1. The area of the measurements

    The meteorological data available for the area of the measurements
come from two independent sources, i.e.;
a) The values t_1, t_2 of air temperature and P_1, P_2 of atmospheric pressure
corresponding to the two geodetic stations Faner Toumba and Panorama were

Table 1

Observing Periods

Period	Time Interval	Days	Frequency	Observing Hours
A	Summer 1969	28	Every 60^m	7^h to 18^h
B	Spring 1971	26	Every 30^m	8^h to 17^h
C	Summer 1971	32	Every 30^m	8^h to 17^h
D	Summer 1972	18	Every 60^m	0^h to 23^h
E	Summer 1974	24	Every 60^m	0^h to 23^h

determined with the help of meteorological observations carried out in
these stations during the simultaneous reciprocal zenith distance measurements used for the determination of the coefficient of terrestrial refraction. With the help of these data the mean values $t = (t_1 + t_2) : 2$ and
$P = (P_1 + P_2) : 2$ of air temperature and atmospheric pressure corresponding
to the geodetic stations were computed.

b) A meteorological station of the National Meteorological Service of Greece is in operation at Thessaloniki Airport. At this station numerous meteorological parameters (air temperature, atmospheric pressure, relative humidity, wind speed and direction, nebulosity etc.) are measured every 3^h. Moreover, radiosonde observations are carried out at the same station twice a day (at 0^h and 12^h UT). As the station lies in the center of the area of the measurements, the results of the meteorological observations carried out in it, which have been kindly put at the disposal of the authors by the National Meteorological Service of Greece, can be considered as representing the average meteorological conditions prevailing in the area under consideration.

From the above discussion we see that rich observational material, including determinations of station and line refraction and relevant meteorological data, is available for the area of the measurements. Part of this material was used already for a study of terrestrial refraction in this area (Mavridis and Papadimitriou, 1971, 1973). Another discussion of terrestrial refraction and the vertical temperature gradient in the same area, using again only part of the observational material described above, was published recently (Gounaris et al., 1981) and will be summarized in the present paper. A more thorough study of terrestrial refraction and the vertical temperature gradient in this area, using the entire observational material available, will be given in a forthcoming paper.

TERRESTRIAL REFRACTION

We first start with the study of terrestrial refraction. Fig. 2 gives the mean diurnal variation of the coefficient of line refraction K for each of the observing periods A-E, together with the corresponding variation of air temperature t. A general pattern of the diurnal variation of K, common to all observing periods A-E, is clearly seen from this figure: the value of K assumes an almost constant maximum value during night; after sunrise the value of K decreases, assumes a nearly constant minimum value between 10^h-17^h, then increases until sunset, and assumes again its maximum value during night. From the same figure we see that the extreme values of K during each day vary from season to season. For a more thorough study of this phenomenon the average K^M of the values of K corresponding to the time interval 10^h-17^h, which gives a measure of the minimum value of K during the corresponding day, has been plotted in Fig. 3 as a function of the corresponding average t^M of air temperature for each of the 128 days of all the observing periods A-E taken together. From Fig. 3 we see that the value of K^M decreases with increasing value of t^M until a minimum value $K^M \approx 0.12$, corresponding to the value $t^M=27^o$-28° C, and then increases again with increasing value of t^M.

VERTICAL TEMPERATURE GRADIENT

Besides the coefficient of terrestrial refraction, the vertical temperature gradient τ is also of considerable importance for the study

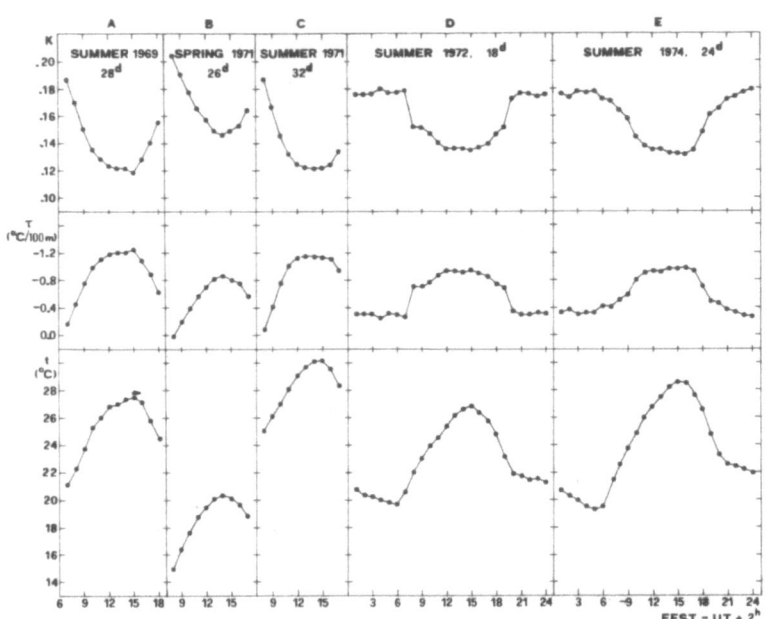

Fig. 2. Mean diurnal variation of the coefficient of terrestrial refraction (K) and the vertical temperature gradient (τ) for each of the observing periods A-E, together with the corresponding variation of air
temperature (t). The time is expressed in East European Standard Time
(EEST = UT + 2^h).

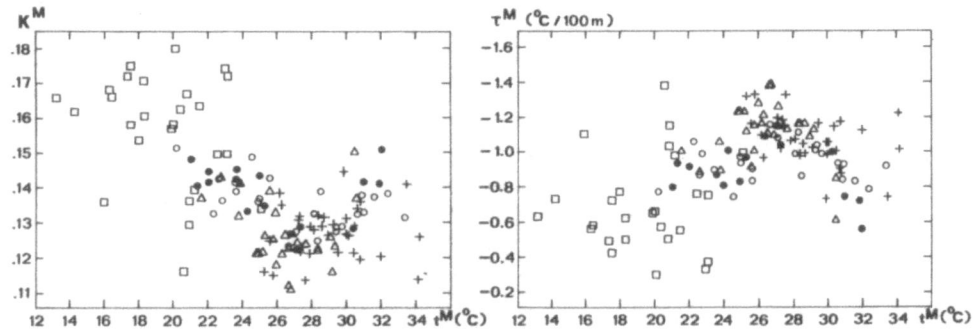

Fig. 3. The averages K^M and τ^M of the values of terrestrial refraction
and vertical temperature gradient corresponding to the time interval
10^h-17^h plotted as a function of the corresponding average t^M of air
temperature for each of the 128 days of all the observing periods A-E
taken together.

of the atmospheric influences on the geodetic measurements. The most
direct method for the determination of τ would be to measure the air
temperature in various altitudes. In lower altitudes this can be done

with the help of meteorological towers, while for higher altitudes radiosonde observations can be used. Both methods, however, can be used only in a very limited number of meteorological stations. For many applications it would be of importance to develop methods for the determination of τ on the basis of simpler measurements. In a previous paper (Gounaris et al., 1981) a method for the determination of τ on the basis of very simple measurements was used (Brocks, 1940, 1954). Indeed, if we consider the well-known formula;

$$K = 504 \frac{P}{T^2} (0.0342 + \frac{dT}{dh}),$$ \hfill (1)

where P is the atmospheric pressure in millibars and $T = 273^{O} + t$ the air temperature in ^{O}K, we can use the value of line refraction K along the line connecting the two geodetic stations, which was found with the help of simultaneous reciprocal zenith distance measurements between these stations, in order to determine the value of the vertical temperature gradient τ = dT/dh in the atmospheric layer, contained between the two stations. This method presents among others the following advantages: a) the value of τ is determined with the help of very simple measurements, and, therefore, the method can be applied easily during geodetic field work, and b) the value of τ is determined primarily on the basis of the value of K and is, therefore, a spatially integrated value.

The above method was used for a study of the vertical temperature gradient τ in the area of Thessaloniki. To this purpose 1924 hourly values of τ in the atmospheric layer contained between the geodetic stations Faner Toumba (H≈177 m) and Panorama (H≈361 m), covering 128 days of the years 1969-74 were determined with the help of relation (1), using the corresponding values of K,P, and T. This material was then used for a study of the diurnal variation of τ in the area of Thessaloniki and its seasonal changes. Fig. 2 gives the mean diurnal variation of the vertical temperature gradient τ for each of the observing periods A-E, together with the corresponding variation of air temperature t. A general pattern of the diurnal variation of τ, common to all observing periods A-E, is clearly seen from this figure: the value of τ assumes an almost constant maximum value during night; after sunrise the value of τ decreases, assumes a nearly constant minimum value between 10^h-17^h, then increases until sunset and assumes again its maximum value during night. From the same figure we see that the extreme values of τ during each day vary from season to season. For a more thorough study of this phenomenon, the average τ^M of the values of τ corresponding to the time interval 10^h-17^h, which gives a measure of the minimum value of τ during the corresponding day, has been plotted in Fig. 3 as a function of the corresponding average t^M of air temperature for each of the 128 days of all the observing periods A-E taken together. From Fig. 3 we see that the value of τ^M decreases with increasing value of t^M until a minimum value $\tau^M \approx -1.2 \, ^{O}C/100m$, corresponding to the value $t^M = 27^O - 28^O, C$ and then increases again with increasing value of t^M.

REFERENCES

Brocks, K.: 1940, Meteorol. Z. 59, 19.
Brocks, K.: 1954, Arch. Meteor. Geophys. Bioklim. A 6, 370.
Gounaris, A.I., Mavridis, L.N., and Papadimitriou, A.L.: 1981, Proc.
    Acad. Athens 56, 91.
Mavridis, L.N. and Papadimitriou, A.L.: 1971, IUGG General Assembly,
    Moscow, Travaux, IAG, Section I.
Mavridis, L.N. and Papadimitriou, A.L.: 1973, J. Geophys. Res. 78, 2679.

A STUDY OF THE REFRACTIVITY N OF THE AIR IN THE AREA OF ATHENS

P. Savaidis
Department of Geodetic Astronomy, University of Thessaloniki,
Thessaloniki, Greece

In a previous paper (Badellas et al.,1980) a study was made of the variation of the refractivity N of the air in the microwave region with altitude in the area of Thessaloniki, using radiosonde observations carried out at the Mikra Airport of Thessaloniki during the period 1971-74. In the present paper a similar study has been undertaken for the area of Athens using the radiosonde observations carried out at 12^h UT in the Hellinikon Airport of Athens ($\varphi=37^{\circ}54'N$, $\lambda=23^{\circ}44'E$) during the years 1970-74. The total number of observations used is equal to 1113, corresponding to 42 months of the above period. From the record of each observation the values of air temperature, relative humidity, and altitude corresponding to the levels of 1000 mb, 980 mb, 960 mb, ..., 700 mb, as well as to the ground level have been obtained. With the help of these data and the formula of Essen and Froome, the mean monthly values of the refractivity N of the air corresponding to the altitudes of 200 m, 300 m, ..., 3000 m have been computed.

For the mathematical representation of the variation of N with altitude following models have been used:
1. the linear model
$$N(h) = N(0) + Lh, \tag{1}$$
2. the parabolic model
$$N(h) = N(0) + Bh + Ch^2, \tag{2}$$
3. the exponential model
$$N(h) = N(0) \cdot e^{-h/H}, \tag{3}$$
where L,B,C, and H are parameters, the values of which have been computed for each of the months considered through a least squares solution. N(h) and N(0) are respectively the values of N at the altitudes of h m and 0 m. The values of L,B, and H thus obtained are systematically higher during the colder months of the year, than during the warmer ones. For a more thorough study of this phenomenon, following solutions were considered:
1. Solution I in which the values of L,B,C, and H were taken equal to the averages of the values of these parameters corresponding to all the 42 months under consideration.
2. Solution II in which different values of L,B,C, and H were used for the warmer (May through October) and the colder (November through April)

W. Fricke and G. Teleki (eds.), Sun and Planetary System, 529–530.
Copyright © 1982 by D. Reidel Publishing Company.

months of the year, which were taken respectively equal to the averages
of the values of these parameters for the corresponding months.
3. Solution III in which the values of L,B,C, and H were assumed as
functions of the corresponding mean monthly air temperature at ground
level t(0). The parameters of these functions were determined through
least squares fittings.

In order to decide which of the three models and solutions represents better the observed variations of N with altitude, values of N were
computed with the help of the formulae (1), (2), and (3) using the values
of the parameters L,B,C, and H corresponding to each of the solutions
I,II and III. The values of N thus obtained were compared with the values
given by the radiosonde observations. In this way following conclusions
could be drawn:
1. The approximation given by all models and solutions is better for the
colder months of the year, than for the warmer ones.
2. Out of the three models, i.e. the linear, parabolic, and exponential
model, the exponential one gives the best approximation for all solutions
I,II, and III.
3) The highest accuracy obtained by the exponential model corresponds
to solution III, i.e. to the case in which parameter H is considered as
a function of the air temperature at ground level t(0), of the form

$$H = 8233 - 58.15t(0). \tag{4}$$

The corresponding dispersion of the discrepancies $N^{obs} - N^{co}$ is equal
to ±4.71.

It should be noted that conclusions Nos. 1 and 2 are also valid for
the variation of N with altitude in the area of Thessaloniki. In this
case, however, the highest accuracy obtained by the exponential model
corresponds to solution II rather, than to solution III, although the
difference in the accuracy given by these two solutions is fairly small.

A more thorough discussion of the variation of N with altitude in
Greece will be given in a forthcoming paper (Savaidis, 1981).

REFERENCES

Badellas, A., Mavridis, L.N., Savaidis, P., and Tsioumis, A.: 1980, A
    Study of the Refractivity N of the Air in the Area of Thessaloniki,
    Meteorologika, No. 69, 189.
Savaidis, P. 1981, A Study of the Refractivity N of the Air in the Area
    of Greece, PhD Thesis, University of Thessaloniki (in preparation).

LIST OF PARTICIPANTS

AUSTRALIA
Brunner,F.K.

AUSTRIA
Dvorak,R.
Gruber,M.
Haupt,H.
Pfleiderer,J.
Schroll,A.
Schober,H.J.
Tscharnuter,W.M.

BELGIUM
Bosman Crespin,D.
Bertiau,F.
Demoulin,P.
Ledoux,P.
Pauwels,T.

BULGARIA
Bonov,A.
Ivanova,V.G.
Shkodrov,V.G.

CANADA
Saastamoinen,J.

CZECHOSLOVAKIA
Chvojková,E.
Hajduk,A.
Hradilek,L.
Kresák,L.
Rušin,V.
Sykora,J.

EGYPT
Asaad,A.A.

FINLAND
Oterma,L.
Tuominen,J.

FRANCE
Bel,N.
Capitaine,N.

Celnikier,L.M.
Chollet,F.
Connes,J.
Connes,P.
Débarbat,S.
Edelman,C.
Feissel,M.
Kovalevsky,J.
Lafon,J.-P.J.
Morando,B.
Nouel,F.
Pecker,J.-C.
Ribes,E.
Robley,R.
Stellmacher,I.

GERMAN DEMOCRATIC REPUBLIC
Pflug,K.

GERMAN FEDERAL REPUBLIC
Brosche,P.
Enslin,H.
Fahr,H.J.
Fernández,J.A.
Fricke,W.
Höppner,W.
Rahe,J.
Roth,M.L.
Schmadel,L.D.
Scholl,H.
Stix,M.
Tüg,H.
Völk,H.J.

GREECE
Antonopolou,E.
Arabelos,D.
Asteriadis,G.
Deliyannis,J.
Mavridis,L.N.
Petropoulos,B.
Savaidis,P.
Theodossiou,E.
Tsioumis,A.G.
Tsoga,M.

Tziavos,I.

HUNGARY
Balázs,B.A.
Barta,G.
Illés,E.
Marik,M.
Márki-Zay,L.
Marx,G.
Pap,J.
Varga,M.

IRELAND
Elliott,I.
Kiang,G.

ITALY
Bertola,F.
Blanco,C.
Burchi,R.
Carusi,A.
Cassacchia,R.
Chiuderi,C.
Chiuderi Drago,F.
Einaudi,G.
Farinella,P.
Ferrini,F.
Forti,G.
Manara,A.
Mancuso,S.
Nobili,A.M.
Poma,A.
Proverbio,E.
Scalatriti,F.
Uras,S.
Valsecchi,G.B.
Zappalà,V.
Zuccarello,F.

JAPAN
Sugawa,C.

MEXICO
Maupome,L.

NETHERLANDS
Atanasijević,I.
Hovenier,J.W.
Spoelstra,T.A.Th.
Van Herk,G.
Van Herk Kluyver,H.A.

NORWAY
Elgaröj,O.
Hauge,O.
Jensen,E.

PEOPLES REPUBLIC OF CHINA
Xiao,N.

POLAND
Brancewicz,H.
Dulinski,G.
Gaska,S.
Gorgolewski,S.
Grudzinska,S.
Kolaczek,B.
Sikorski,J.

SPAIN
De Buitrago,J.G.

SWEDEN
Hahn,G.
Lagerkvist,C.-I.
Rickman,H.
Tengström,E.

USSR
Babadzhanov, P.B.
Batrakov,Yu.V.
Brumberg,V.A.
Chernishev,V.I.
Cyrkulenko,S.G.
Demidov,M.L.
Gavrilov,I.V.
Gigolashvili,M.Sh.
Grigoryev,V.M.
Gurtovenko,E.A.
Khetsuriani,Ts.S.
Kocharov,G.E.
Korobova,Z.B.
Koroleva,L.S.
Kosin,G.S.
Kosovichev,A.G.
Kulidzanishvili,V.I.
Levkovskiy,V.I.
Mateshvili,G.G.
Merkulenko,V.E.
Morozhenko,A.V.
Nazarchuk,G.K.
Ojringel,I.M.
Oskanyan,V.A.

Peshcherov,V.S.
Pljusnina,I.A.
Polozova,N.
Protasov,Yu.
Sattarov, I.
Schamaev,V.G.
Shakht,N.A.
Shchukina,N.G.
Shevchenko,V.V.
Signa,L.A.
Sochilina,A.S.
Tyagun,N.F.
Viik,T.L.-F.
Zajtsev,A.L.
Zhugzhda,J.D.
Yatskiv,Ya.S.

UNITED KINGDOM
Kopal,Z.
Message,P.J.
Morrison,L.V.
Napier,W.M.
Taylor,G.E.
Wilkins,G.A.

USA
Arrhenius,G.
Gehrels,T.
Helin,E.F.
Hughes,J.A.
Ness,N.F.
Smith,C.A.

YUGOSLAVIA
Arsenijević,J.
Atanacković,O.
Benišek-Protić,V.
Čabrić,N.
Čadež,A.
Čadež,V.
Čelebonović,V.
Dačić,M.
Dimitrijević,M.
Djokić,M.
Djurašević,G.
Djurović,D.
Grujić,R.
Jovanović,Bož.
Jovanović,Bor.
Jovanović,M.
Jugin,M.

Karabin,M.
Kapetanović,N.
Kontić,R.
Knežević,Z.
Krga,R.
Kubičela,A.
Kuzmanoski,M.
Martić,M.
Milošević,V.
Milovanović,V.
Olević,D.
Pakvor,I.
Popović,G.
Randić,L.
Ruždjak,V.
Sadžakov,S.
Sekulović,V.
Stančić,Z.
Teleki,G.
Tomić,A.
Trajkovska,V.
Vince,I.
Vršnak,B.
Vujnović,V.

IAU
West,R.M.

Aerosol stratification ... 213

Asteroidal belt ... 403

Asteroids
- Aten types ... 271
- binary systems ... 277, 293
- collision ... 295
- color variations ... 285
- cometary origin ... 275
- earth-crossers ... 269
- evolution ... 265
- magnitudes ... 299
- phase-magnitude relation ... 267
- pole determination ... 303
- rotation ... 285, 289, 291
- search with CCD ... 279
- surface properties ... 265

Astrolabe observations
- Sun ... 35
- Mars ... 443

Astrometric site selection ... 483

Astrometry
- refraction effects ... 457

Atmospheric refraction ... 511

Atmospheric turbulence
- effect on directions ... 505

Comets
- dynamical evolution ... 361, 371
- encounters with Jupiter ... 379, 389, 395, 397
- encounters with outer planets ... 385
- evolution ... 364
- Halley ... 307, 335
- life times ... 368
- long-period ... 321
- numerical integration of orbits ... 415
- origin ... 362, 371, 375
- outbursts ... 337
- relationship with asteroids ... 369
- resonance motions ... 391, 397
- search with CCD ... 279
- space missions ... 315
- ultraviolet spectroscopy ... 323

Earth
  -    atmospheric model   459, 521
  -    atmosphere: vertical
          temperature gradient   520, 523
  -    crust motions   192
  -    magnetic field variation   205
  -    non-rigidity   185
  -    rotation   149, 165, 173, 179
Earth-moon system   20
Earth-moon system dynamics   430
Eclipses of Sun   173
Ephemerides
  -    planetary   437, 441
  -    Mars   443

Gravimetric geoid determination   203

Halley   307, 335
  -    meteor streams   335
Hipparcos satellite   197

Interplanetary magnetic field   215
Interplanetary matter   253
  -    dust   255
  -    grains   339
  -    plasma-dust   331

Jupiter
  -    magnetosphere   243
  -    satellites   354

Kirkwood gaps   405, 411

Latitude observations   195
Latitude variations   191
LIDAR   463
  -    DIAL and RAMAN systems   466
  -    temperature profiling   468
  -    water vapor systems   466
Lunar occultations   173
Lunar photometric constant   263

Mars
  -    astrolabe observations   443
Mercury
  -    meridian observations   447
MERIT project   163
Meteor showers   401
Meteorites   227
Minor planets
  -    51 Nemausa   305

Minor planets
- for equator point   305, 449
- for equinox   305, 449
- mass determination   405
- numerical integration
      of orbits   415
- occultations   287
- perturbation theory   413
- resonances   405, 411

Moon
- tidal acceleration   173

Nutation   151

Oceanic tides   179

Planetary system
- formation   233
Planets
- atmospheres   259
- chemical composition   18
- hypsometry   261
- internal structure   18
- perturbation theory   421, 431, 437
- potential   209
- rings   353
Polar motion   151
Precession   151
Project MERIT   163

Ray tracing principles   497
Refraction
- aerological data   475
- altitude variation   529
- data by LIDAR   463
- diurnal effects   477
- diurnal and seasonal variations   519
- effects on geodetic measurements   499
- effects on local geodynamics   511
- effects on radio interferometry   493
- effects on solar disk   478
- effects on time and
      latitude observations   471
- mathematical models   500
- three-dimensional   455

Satellites
- dynamics   353, 429
- geosynchronous   199, 201
- Navstar type   217

Saturn
- E-ring grains   251
- magnetosphere   243
- rings   356
- satellites   355
Savić-Kašanin theory   249
Solar active regions   115
Solar activity   189
Solar atmosphere   103, 105
Solar chromosphere   113
Solar corona   145, 257
Solar coronal loops   141
Solar cycle   77
Solar flares   48, 115
Solar gamma rays   49
Solar granulation   82
Solar limb   41
Solar magnetic fields   81, 84, 119, 143
Solar models   59
Solar nebula
- excitation   221
- gas accretion   235
- grain coaculation   234
- grain condensation   234
- grain evaporation   234
Solar neutrinos   47, 53, 59
Solar photosphere   99
Solar prominences   131, 133, 135, 137
Solar radiation emission   139
Solar radio emission   97, 139
Solar radio granulation   109
Solar spectrum   29, 97, 101, 103
Solar system
- chemical composition   15
- evolution   13
- isotopic composition   227
- origin   222
- primordial matter   221
Solar wind   331
Sun
- as a star   25
- astrolabe observations   35
- chemical composition   31
- color indices   45
- convection   63, 82
- dynamo theory   63
- luminosity   29
- magnetism   68
- mass   28
- meridian observations   445
- noisy   89

Sun
- oblateness   27
- oscillations   27, 81, 93
- photometry   46
- positions   35
- radius   26
- rotation   63, 73, 77, 79
- semi-diameter   39
- tracers   71
- velocity fields   79
- U.S.S.R. solar observatories   123
Sunspot groups   77, 79
Sunspot spectrum   107
Sunspots
- fine structure   129
- internal motions   125
- proper motions   127

Terrameter   511
Terrestrial planets   17, 249
Terrestrial refraction coefficient   519, 523

Universal time variations   189
Uranus rings   357
UT 1   157

Zodiacal light   331, 343, 345